Lecture Notes in Computer Science 16290

Founding Editors

Gerhard Goos
Juris Hartmanis

Editorial Board Members

Elisa Bertino, *Purdue University, West Lafayette, IN, USA*
Wen Gao, *Peking University, Beijing, China*
Bernhard Steffen, *TU Dortmund University, Dortmund, Germany*
Moti Yung, *Columbia University, New York, NY, USA*

The series Lecture Notes in Computer Science (LNCS), including its subseries Lecture Notes in Artificial Intelligence (LNAI) and Lecture Notes in Bioinformatics (LNBI), has established itself as a medium for the publication of new developments in computer science and information technology research, teaching, and education.

LNCS enjoys close cooperation with the computer science R & D community, the series counts many renowned academics among its volume editors and paper authors, and collaborates with prestigious societies. Its mission is to serve this international community by providing an invaluable service, mainly focused on the publication of conference and workshop proceedings and postproceedings. LNCS commenced publication in 1973.

Khuong An Nguyen · Zhiyuan Luo

Editors

The Importance of Being Learnable

Essays Dedicated to Alexander Gammerman

 Springer

Editors
Khuong An Nguyen
Royal Holloway University of London
Egham, UK

Zhiyuan Luo
Royal Holloway University of London
Egham, UK

ISSN 0302-9743 ISSN 1611-3349 (electronic)
Lecture Notes in Computer Science
ISBN 978-3-032-15119-3 ISBN 978-3-032-15120-9 (eBook)
https://doi.org/10.1007/978-3-032-15120-9

This Springer imprint is published by the registered company Springer Nature Switzerland AG
The registered company address is: Gewerbestrasse 11, 6330 Cham, Switzerland

If disposing of this product, please recycle the paper.

Preface

On 2 November 2024, friends, colleagues, and former students gathered at Royal Holloway, University of London, to celebrate Professor Alexander Gammerman's 80th birthday. In recognition of his contributions across disciplines, a Festschrift was published to reflect the breadth and depth of his intellectual influence. This volume is dedicated to the impact of his scientific legacy. Professor Gammerman's career exemplifies the transformative power of interdisciplinary research. From pioneering mathematical models of plant photoreceptors to advancing the formal treatment of uncertainty in artificial intelligence, his work embodies a rare confluence of analytical precision, conceptual depth, and visionary application.

The title, *The Importance of Being Learnable* has both a technical meaning and a philosophical dimension. In machine learning, "learnability" is a formal property - a model's capacity to generalise from data. But in life, to be learnable is to be open to revision, to contradiction, and to insight. The title also echoes Oscar Wilde's satire: machine learning hints at being apolitical and universal but is often shaped by funding priorities and ideological biases.

Professor Alexander Gammerman's early research on **phytochrome** - the light-sensitive protein that governs plant development - offered predictive models of photomorphogenesis. These mathematical frameworks, grounded in differential equations, revealed how plant development can be controlled by changing the spectrum of irradiation. Gammerman's models laid the foundations for quantitative analyses of the phytochrome dynamics.

In the 1990s, Professor Gammerman turned to **Bayesian inference and belief networks**, helping to formalise probabilistic reasoning in expert systems. His work emphasised the importance of dealing with uncertainty and updating beliefs. Together with his colleagues, Professor Gammerman was involved in several large projects that applied Bayesian techniques. One of these projects was in the medical field, and another was a project with the Home Office on statistical profiles of offenders.

His most enduring contribution came with the co-creation of **conformal prediction**, a framework that provides valid confidence measures for machine learning models. Conformal predictors offer distribution-free guarantees, quantifying uncertainty in a way that is transparent, adaptive, and mathematically robust. This innovation has transformed how we think about trust in AI, especially in safety-critical domains such as medicine, pharmaceuticals, and anomaly detection systems.

Beyond his research, Professor Gammerman has been a devoted educator and mentor. He taught across all levels of Computer Science - from first-year undergraduate to Master's level - instilling in students a mathematical culture rooted in rigour and clarity. His lectures were known for their precision, accessibility, and occasional dry wit. Over the course of his career, he supervised over three dozen PhD students, many of whom have gone on to lead research in machine learning and AI around the world.

His administrative contributions are equally notable. He was the Head of the Computer Science Department at Royal Holloway from 1995 to 2005. He was also the founding director of the Centre for Reliable Machine Learning, a hub for interdisciplinary research focused on trustworthy AI. And he was also involved in establishing the Kolmogorov Lecture and Medal, honouring excellence in algorithmic learning and statistical reasoning. These initiatives reflect his commitment to building institutions that endure - and to celebrating the intellectual traditions that shaped his own work.

This volume is organised into five parts with a total of 18 chapters:

- **Part I: Personal Recollection (3 chapters)**

    This part recollects the memories in the early 1980s that shaped Prof Gammerman's research agenda.

    - "Life in Computer Science at Heriot-Watt University 1983–1993" (Greg Michaelson).
    - "Intelligent Data Analysis" (Xiaohui Liu).
    - "Statistical Modelling in Specific Case Analysis with Bayesian Belief Networks" (Colin Aitken).

- **Part II: Theoretical Developments (4 chapters)**

    This part lays down rigorous foundations for modern AI uncertainty quantification.

    - "Randomness, Exchangeability, and Conformal Prediction" (Vladimir Vovk).
    - "An Investigation of Conformal Test Martingales" (Henrik Boström).
    - "Bootstrapped Chains for Learning Using Privileged Information" (Evgueni Smirnov and Filip Schlembach).
    - "VC-theoretical Explanation of Double Descent" (Vladimir Cherkassky and Eng Hock Lee).

- **Part III: Reviews of Conformal Prediction Applications (4 chapters)**

    This part reviews the recent progress in trustworthy AI and the developments and applications of Conformal Prediction worldwide.

    - "Conformal Prediction and Trustworthy AI" (Tony Bellotti and Xindi Zhao).
    - "A Review of Theoretical Advances and Practical Applications of Conformal Prediction in China" (Shuo Zhao, Zhirui Zhang, Yifan Liu, Khuong An Nguyen, You Wang, Zhiyuan Luo and Guang Li).
    - "Application of Confidence and Probabilistic Models to Practical Problems" (Lars Carlsson, Johan Hallberg Szabadváry, Ernst Ahlberg and James Gammerman).
    - "Conformal Prediction in the Age of Multimodal Foundation Models: A Survey" (Christian Stavan, Kunal Tilaganji, Sai Mathura Krishnan, Sai Srinivas Kancheti and Vineeth N. Balasubramanian).

- **Part IV: Predictions with Guarantees (5 chapters)**

    This part provides detailed industrial use cases of Conformal Prediction and its impact.

    - "Conformal Prediction for Offensive Security" (Giovanni Cherubin).

- "Temporal Distribution of Clusters of Investors and Their Application in Prediction with Expert Advice" (Wojciech Wisniewski, Yuri Kalnishkan, David Lindsay and Siân Lindsay).
- "Known Unknowns: Trading Scheduled Surprises with Conformal Prediction" (David Lindsay and Siân Lindsay).
- "Reliable Train Delay Forecasting with Conformal Prediction" (Xu Feng, Khuong An Nguyen and Zhiyuan Luo).
- "Conformal Mining for Multiple Error Correction in the Data" (Ilia Nouretdinov).

- **Part V: Python Packages for Conformal Prediction (2 chapters)**
  This part releases ready-to-use tools for practitioners and researchers to further the research area.

  - "Protected Probabilistic Classification Library" (Ivan Petej).
  - "GPyConform: Conformal Prediction with Gaussian Process Regression in Python" (Harris Papadopoulos).

Each section includes technical papers, retrospective essays, and reflections from colleagues and students. Together, they illuminate a career defined by intellectual rigour, philosophical depth, and a touch of humor. Each contribution was peer reviewed by a team of experts.

The Festschrift includes an interview "A Conversation with Alexander Gammerman", and his Inaugural Lecture "Machine Learning: Progress and Prospects". The Conversation offers a glimpse into Professor Gammerman's life and remarkable scientific journey that spans two countries (the Soviet Union and the United Kingdom) and two centuries: the second part of the 20th century and the first quarter of the 21st century. His Inaugural Lecture was delivered at Royal Holloway University of London in 1996 (almost 30 years ago), but it gives an interesting perspective on the development of machine learning over that period.

The editors and contributors present these papers with admiration and gratitude.

Acknowledgements. We are grateful to Alex Gammerman and Volodya Vovk for their help and advice.

November 2025

Khuong An Nguyen
Zhiyuan Luo

Contents

A Conversation with Alexander Gammerman 1
  Zhiyuan Luo

Inaugural Lecture Machine Learning: Progress and Prospects 28
  Alexander Gammerman

Life in Computer Science at Heriot-Watt University 1983-1993 46
  Greg Michaelson

Intelligent Data Analysis 56
  Xiaohui Liu

Statistical Modelling in Specific Case Analysis with Bayesian Belief
Networks 75
  Colin G. G. Aitken

Randomness, Exchangeability, and Conformal Prediction 87
  Vladimir Vovk

An Investigation of Conformal Test Martingales 118
  Henrik Boström

Bootstrapped Chains for Learning Using Privileged Information 144
  Evgueni Smirnov and Filip Schlembach

VC-Theoretical Explanation of Double Descent 157
  Vladimir Cherkassky and Eng Hock Lee

Conformal Prediction and Trustworthy AI 177
  Anthony Bellotti and Xindi Zhao

A Review of Theoretical Advances and Practical Applications
of Conformal Prediction in China 198
  *Shuo Zhao, Zhirui Zhang, Yifan Liu, Khuong An Nguyen, You Wang,
  Zhiyuan Luo, and Guang Li*

Application of Confidence and Probabilistic Models to Practical Problems 227
  *Lars Carlsson, Johan Hallberg Szabadváry, Ernst Ahlberg,
  and James Gammerman*

Conformal Prediction in the Age of Multimodal Foundation Models:
A Survey 269
    Christian Stavan, Kunal Tilaganji, Sai Mathura Krishnan,
    Sai Srinivas Kancheti, and Vineeth N. Balasubramanian

Conformal Prediction for Offensive Security 306
    Giovanni Cherubin

Temporal Distribution of Clusters of Investors and Their Application
in Prediction with Expert Advice 325
    Wojciech Wisniewski, Yuri Kalnishkan, David Lindsay, and Siân Lindsay

Known Unknowns: Trading Scheduled Surprises with Conformal
Prediction 345
    David Lindsay and Siân Lindsay

Reliable Train Delay Forecasting with Conformal Prediction 387
    Xu Feng, Khuong An Nguyen, and Zhiyuan Luo

Conformal Mining for Multiple Error Correction in the Data 411
    Ilia Nouretdinov

Protected Probabilistic Classification Library 432
    Ivan Petej

GPyConform: Conformal Prediction with Gaussian Process Regression
in Python 449
    Harris Papadopoulos

Author Index 467

A Conversation with Alexander Gammerman

Zhiyuan Luo[✉]

Department of Computer Science, Royal Holloway University of London,
TW20 0EX London, UK
Zhiyuan.Luo@rhul.ac.uk
https://cml.rhul.ac.uk/people/zhiyuan/

Abstract. Alexander Gammerman is a British computer scientist and Professor at Royal Holloway University of London. He is the co-inventor of conformal prediction. Alex is the founding director of the Centre for Reliable Machine Learning at Royal Holloway, and a Fellow of the Royal Statistical Society. He has published 9 books and about 200 research papers. He ranks amongst the top 1% researchers in Artificial Intelligence and Machine Learning.

Alex's work spans two countries (the Soviet Union and the United Kingdom) and two centuries: the second part of the 20th century and the first quarter of the 21st century. It has been an extremely interesting time with the rapid development of Artificial Intelligence, Machine Learning, the Internet and the Large Language Models. How are these discoveries, inventions, and developments influencing the life of an academic?

I have known Alex personally for many years. I first came as a PhD student to Heriot-Watt University, Edinburgh, in 1988. I then worked in various places before arriving at Royal Holloway University of London as a Lecturer in 2002. Alex has been more than a mentor and colleague to me—he has been a guiding figure and a close friend, with his pioneering work continuing to inspire both me and my PhD students.

The following conversation, which took place both face-to-face and remotely in July–August 2025, offers a glimpse into Alex's remarkable scientific journey, his reflections on decades of research, and his perspective on the ever-evolving world of Artificial Intelligence.

1 Family

ZL: Alex, thank you for agreeing to have this conversation. I am sure you have many interesting stories to tell. To begin with, could you tell us a little bit about your family background?

AG: I was born in Alma-Ata (USSR, Kazakhstan) in 1944, where my parents were evacuated in 1941, when the Germans approached Moscow in the 2nd World War. After the war, the family returned to Moscow and soon after, my

K. An Nguyen and Z. Luo (Eds.): Alexander Gammerman Festschrift, LNCS 16290, pp. 1–27, 2026.
https://doi.org/10.1007/978-3-032-15120-9_1

dad, an expert in the textile industry, was invited to Tallinn to help with the development of the light industry in Estonia; so the family moved to Tallinn. That's where I grew up and finished school.

My parents were born in Ukraine and met in Odessa at Odessa National Economics University in 1924. Dad graduated from the University, but my mother was expelled since she was the daughter of a cantor and therefore, by the "politically correct" communist ideology, did not have the right to study at any university.

My dad - Yakov Moiseevich Gammerman (1895–1978) - was working all his life in the textile industry eventually became the director of a large textile factory Marat in Tallinn. The factory was famous in the USSR for its high-quality men's and women's underwear. There is a website describing my dad's life (in Ukrainian)[1].

My mum - Susannah Grigoryevna Gammerman, (née Elisheva Darer, 1906–1996) - stayed at home while my brother and I were growing up. After we left home for education and work, she worked as a Librarian at the Library of the Academy of Sciences of Estonia (Fig. 1).

Fig. 1. Dad and Mum with my brothers, Mikhail and George, 1939.

My parents had 4 children - 4 boys - but two of them died during the 2nd World War, and only my eldest brother Misha and me, the youngest, survived.

---

[1] http://berdychiv.in.ua/гаммерман-яків-мойсейович/

My older brother Misha - Mikhail Yakovlevich Gammerman (1931–2006) - was the Chief Designer of flowmeters, the devices to measure the flow of heat and liquid. He received many awards and medals for his inventions. The flowmeters were of very high accuracy and used all over the Soviet Union. You can find more details in a Wikipedia page devoted to him (in Russian)[2] (Fig. 2).

Fig. 2. My brother Misha, 1975.

Apart from that, my brother was a great chess player and participated in various tournaments in the USSR. He had a great influence on my upbringing and education.

I met my wife, Suzy, in 1979 in St. Petersburg, while she was a visiting student from Cambridge University. We married in 1982 and emigrated to the UK in 1983. Our 3 children were born and educated in Britain: James (b.1990), graduated from Imperial College London, specialist in Data Science; Anya (b.1992), graduated from King's College London in neurosciences; Sonia (b.1996), graduated from University College Maastricht, specialist in communications and marketing, but her real passion is electro swing jazz music and singing (Fig. 3).

---

[2] https://ru.wikipedia.org/wiki/Гаммерман,_Михаил_Яковлевич

Fig. 3. L-R: Anya, Suzy, me, James, Sonia in Trafalgar Square, 2022.

2 School, University and Postgraduate Education, 1963-1972, the Soviet Union

ZL: Your parents lived through the Russian Revolution and the Second World War. That must have shaped their outlook and yours. Their resilience and drive to succeed clearly influenced your academic path. What was school like for you? Any favourite subjects?

AG: School was fine, nothing remarkable. Most lessons were dull, to be honest - except for Chemistry and Literature. We had two genuinely passionate teachers, and that made all the difference. Outside of class, I kept busy with basketball, volleyball, and athletics.

ZL: After school, you chose to study physics at university. What drew you to it? And what was life like at St. Petersburg University back then?

AG: In my family, the cult of education was very high: my father was educated as an economist and then continued learning various aspects of the cotton industry all his life. He is the author of two books devoted to textiles and rose from a foreman to the director of a big factory. My brother Misha graduated from school with a gold medal, graduated from the University (he studied at Moscow Aviation Institute and then at Tallinn University of Technology), created his own laboratory and was an outstanding innovator. My grandparents from my mum's side (and several previous generations) were also well-educated, religious people. So yes, you could say my path was laid out early.

I got hooked on physics in high school after reading Feynman's Lectures. Quantum physics fascinated me - and that curiosity stuck.

Back then, physics was everywhere - in films, books, public discourse. It had a kind of glamour. After the war and the rise of nuclear science, physics felt more vital than art. Studying it was like being part of the future. Like doing AI or machine learning today. I participated in various Olympiads in physics and was even among the winners at some competitions. I also read a very interesting book written by Daniil Danin, called *The Inevitability of a Strange World*. This book is about how the basic ideas of quantum mechanics were developed and who were the people behind it: Bohr, Schrödinger, Heisenberg and others. That sealed the deal.

ZL: And once you got to university - what did you actually learn?

AG: I got a solid grounding in all kinds of physics - mechanics, quantum electro-dynamics, relativity. Gell-Mann's quark theory was emerging, but it wasn't part of our curriculum yet. Oddly enough, the subjects I'd need later - probability theory, programming, information theory - were barely covered. I had to pick those up as a graduate student.

And yes, I learned a few extracurricular things too: to drink vodka, play cards in all-night sessions, chase girls ... the usual student fare. [Laughs].

But mostly, I spent time in the library, reading math and physics books. The lectures were hit-or-miss. None of the professors really sparked my imagination. Eventually, I grew disillusioned with physics.

ZL: Still, were there advantages to studying it?

AG: Absolutely. Physics gave me a broad education and a strong foundation in maths and science. That opened doors to all sorts of fields - mathematical biology, computational science, machine learning. I don't think any other discipline would've offered such range. It taught me how to model problems, make generalisations, and trust physical intuition—even when the maths didn't quite behave. Eventually, I started carving out my own path and chose to pursue mathematical biology in postgraduate studies.

2.1   Mathematical Models in Biology; Agrophysical Research Institute, St. Petersburg, 1969-1972

ZL: Switching from physics to mathematical biology sounds like quite a leap. What prompted that change?

AG: Well, theoretical physics jobs were scarce. I had to choose - either stick with experimental physics or find something new. Laser tech was popular, and many of my classmates jumped on that bandwagon. But I've never liked following the crowd, and pure experimentation didn't appeal to me. For my final-year, I found an interesting project in biophysics—modelling protein structures. It involved using computers for calculations, which nudged me toward programming. That was a turning point.

ZL: So that's when your academic life really began?

AG: Yes, properly so. I enrolled in postgraduate studies at the Agrophysical Research Institute (AFI) in St. Petersburg. There were entrance exams—Physics, English, Philosophy—and in April 1969, I got in as a PhD student. The scholarship was 69 rubles a month, about 17 dollars. Not exactly enough to live on, so I started tutoring physics and maths on the side.

ZL: What was the Institute like?

AG: It turned out to be a great place to study. Founded by Academician Abram Joffe (he also set up the Institute of Physics and Technology in St. Petersburg) it became a hub for Mathematical Biology, and it attracted some brilliant minds. The scientific life was buzzing - seminars, conferences, lively discussions. We had visits from top researchers like Alexey A. Lyapunov (a mathematician), Nicolay V. Timofeev-Resovsky (a biologist), Raisa L. Berg (a geneticist) and others. It was intellectually rich.

Social life was vibrant too - skits, wall newspapers, concerts. But of course, everything was under Party control. Eventually, a new Party boss arrived and started purging the "unreliable" ones - people who were too independent, too free-thinking. It was a shame. Many talented researchers were forced out.

ZL: It sounds like a stimulating place against the political trends in the country. Who supervised your PhD?

AG: Dr. Leonid Fukshansky. He was just a few years older than me and had done impressive work on modelling the biological clock in plants. I read his thesis and was thrilled when he agreed to supervise me. He proposed a project on modelling phytochrome - a plant receptor involved in photomorphogenesis. I had no idea what morphogenesis even meant! But I went to the library and, by sheer luck, stumbled upon a little-known Alan Turing's 1952 paper *The Chemical Basis of Morphogenesis* [24]. That was a good enough motivation and sparked my interest.

ZL: That's quite a discovery. Did Leonid remember things the same way?

AG: Not exactly! [Laughs]. According to him, a colleague asked if he'd take on another PhD student. He was hesitant - already had several. But when he heard the student was a USSR schoolboy athletics champion, he changed his mind.

ZL: You were a champion?

AG: Yes, back in school I did sprints, hurdles, long jump. I was part of Estonia's 4×100 relay team in the 1962 Youth Spartakiad. We won the final. After school, I left athletics behind, but apparently Leonid was still impressed.

ZL: And as a supervisor?

AG: He was excellent. He pointed me toward interesting open problems and useful references. He also taught me how not to be the arrogant mathematician barging into biology with grand theories but little understanding. And equally, how not to be the biologist who ignores mathematical rigour. That balance was crucial (Fig. 4).

ZL: Could you explain the *phytochrome* system? How does it work, and why is it important in plant biology?

AG: Phytochrome was discovered, mainly by the work of Harry A. Borthwick and Sterling B. Hendricks in 1959 in Beltswill (USA). The presence of phytochrome allows us to understand how plants control their germination, stem growth, leaf growth, and flowering [2].

Hendricks proposed that the phytochrome exists in two inter-convertible forms, one absorbing in the red (P_r) spectrum at a maximum of 660 nm, and one absorbing in the far red (P_{fr}) spectrum at approximately 720 nm, and concluded that the red and far-red light causes transformation between the two forms - a photoreversible reaction[3].

ZL: Fascinating. Let's talk about your PhD. What were the main questions you tackled, and what do you see as the key contributions of your research?

AG: My work focused on two main problems. First, I wanted to understand the dynamics of phytochrome—a light-sensitive pigment in plants. Second, I aimed to find a link between those dynamics and actual physiological responses, like how plants grow.

---

[3] https://www.degruyter.com/document/doi/10.1515/ci-2016-0506/html? lang=en.

Fig. 4. Leonid Fukshansky, c.2020.

To model the dynamics, I had to build a system of first-order differential equations. The tricky part was that the coefficients weren't just numbers—they were functionals that depended on the light spectrum and the reactions happening in the phytochrome system [8,13,14].

Later, I extended the model to handle more complex situations, like when the light spectrum changes over time or varies with depth in the plant tissue. That turned the equations into non-autonomous differential equations, which don't have neat, closed-form solutions. So I turned to numerical simulations and started using computers to solve them.

Most of my time was spent running simulations - changing initial conditions, tweaking light properties, adjusting parameters - to see how the system behaved. That's how we could predict the behaviour of phytochrome under different lighting conditions.

ZL: And the second part - connecting that to plant responses?

AG: Yes, once we had a handle on the dynamics, we looked for correlations with actual plant behaviour. One clear example was stem elongation. We found a measurable relationship between the concentration of the active form of phytochrome and how much the plant grew. That was a big step. It showed that the mathematical model wasn't just abstract—it could explain real biological effects. We published those results in a Soviet journal, and they were later translated into English [3].

ZL: That must have drawn some attention.

AG: It did. Our work was noticed by researchers abroad, in Europe and the USA - people like Professor Hans Mohr in Germany and Professor Harry Smith in the UK. When I eventually came to England, Harry Smith even tried to get me a position in his lab in Leicester. But by then, I was already a lecturer in Edinburgh.

ZL: And your PhD?

AG: I defended my thesis and received my PhD in physics and mathematics (кандидат физико-математических наук). Looking back, I think the real contribution was laying the groundwork for quantitative modelling of phytochrome. Today, of course, we know much more - phytochrome has even been found in microorganisms. But those early models helped open the door.

2.2 Pattern Recognition; Regional Research Computer Centre, St. Petersburg, 1972-1983

ZL: What came next after your PhD?

AG: My first job was actually just down the corridor - from the same Agrophysical Research Institute, but in a different department: the Information Systems Lab. Later, around 1975–76, it evolved into the Regional Computing Centre of the Agricultural Academy of the USSR.

The Centre was in a beautiful old building right on St. Isaac's Square in the heart of St. Petersburg. It belonged to the All-Union Institute of Plants, or VIR, which housed the World Collection of Plants. Our job was to process and analyse that vast collection.

ZL: That sounds like quite a shift.

AG: It was. The head of the Centre, Dr. Mark Lanin, was a remarkable man - brilliant researcher, charismatic, and, incidentally, a fantastic chess player.

He'd developed two major systems. One was TABIA, an accounting system that came out in the mid-70s - just before the first spreadsheet appeared in the West. The other was SL-740, a search algorithm that was probably the fastest in the USSR at the time.

ZL: And what was your role?

AG: I was tasked with developing statistical and pattern recognition algorithms. We applied them to medical diagnostics and plant classification. That's when I really started diving into statistics and what we now call machine learning.

Back then, AI wasn't mainstream in the USSR. The buzzword was "Cybernetics" - which had just recovered from being labelled a "bourgeois pseudoscience" in the 1950s. By the 70s, it was all the rage, along with control theory.

ZL: So that's when the shift toward machine learning began?

AG: Exactly. I didn't know it at the time, but a book by Vapnik and Chervonenkis on pattern recognition had just come out [31]. Years later, I'd meet both

Fig. 5. Mark Lanin, c.2000.

of them and even work alongside them. But back then, I was just starting to explore the field - without knowing where it would eventually lead (Fig. 5).

ZL: Could you walk us through the kinds of projects you worked on at the Centre? And what sort of computing tools did you have at your disposal?

AG: Sure. One of our first major projects was in pediatric oncology. We collaborated with Dr. Genrich Fedoreev at the St. Petersburg Oncological Institute—he headed the children's department. It was quite pioneering at the time: using statistical methods and computers to help diagnose and treat cancer.

We focused on hemangiomas, which were usually treated surgically right after birth. Sadly, that often led to permanent cosmetic or functional damage. But by applying pattern recognition techniques, we showed that most of these cases didn't need surgery at all. They tended to regress naturally by the age of five or six [32]. That was a big moment-research saying, "Wait, maybe less is more."

We later extended the same statistical approach to study Hodgkin's and non-Hodgkin lymphoma in children - looking at diagnosis, treatment strategies, and prognosis.

ZL: That sounds like a lot of data to work with.

AG: It was. And most of it was qualitative - categorical features, not numbers. So we had to figure out which features mattered most, and how they were related. We developed a multidimensional contingency coefficient based on the Kullback-Leibler information measure [17]. Quite elegant, actually.

As for computing - well, we had a MINSK-32. It was the Soviet version of the IBM 360. We programmed in Assembler and Fortran, and everything was stored on punched cards. You had to be very careful - drop a box of cards and your entire week's work could scatter across the floor.

ZL: And this was all happening at the Centre?

AG: Yes, the Computer Centre was buzzing with activity. We didn't just work on medical data - we also applied similar techniques to the World Plant Collection housed at VIR. That led to several publications [16, 17, 33].

The atmosphere at the Centre was wonderful. Lively discussions, visitors from all over the USSR and abroad. It felt like a real intellectual community.

But outside those walls, things were changing. The Afghan war had started. Antisemitism was creeping back. Dissidents were being jailed. The mood was darkening. We were refining predictive algorithms while living in a country where, as we joked, "the past was more unpredictable than the future." [Laughs].

ZL: Was the Centre eventually shut down?

AG: Yes. The Party decided we were too free-spirited, not ideologically reliable enough. Institutions like ours were quietly dismantled. It was a sobering moment. Many of us realised the country no longer had a future we could believe in. In the end, it was clear: in the battle between "Bayesianism and Bolshevism"[4], science didn't stand a chance.

3   Machine Learning, Heriot-Watt University, Edinburgh, Great Britain, 1983-1993

ZL: You arrived in the UK in 1983. What was that experience like - and how did your first year unfold?

AG: We arrived in early May - Suzy and I, fresh from St. Petersburg. It was a train journey across Europe, and the contrast between East and West was striking. I still remember the border crossing in East Berlin: dim lights, grim-faced soldiers. Then suddenly - West Berlin! Bright, bustling, full of music and life. It felt like stepping into another world.

Once in England, I started applying for jobs. To my surprise, I discovered that some of my papers from Soviet journals had already been translated into English by American publishers. I listed those in my CV, and that helped me land my first job: Lecturer in Computer Science at Heriot-Watt University in Edinburgh, Scotland starting October 1983.

ZL: Why Heriot-Watt?

AG: Funny story. At the interview, they asked why I'd applied. I said, "Well, your university is famous for AI and machine learning - people like Donald

---

[4] Bolshevism - Political theory and practice of the Bolshevik Party which, under Lenin, came to power during the Russian Revolution of October 1917. The Bolshevik (meaning 'majority') radical communist faction within the Russian Social Democratic Labour party (*Oxford Reference*).

Michie work here." The panel looked puzzled. "That's Edinburgh University," they said. I blinked. "But this is Edinburgh University, isn't it?" "No, no - we're Heriot-Watt. Different place."

In the Soviet Union, each big city had just one university. The rest were polytechnic institutes of higher education. So my confusion was understandable - and we all had a good laugh. Somehow, despite that mix-up, they offered me the job.

ZL: What was the department like?

AG: Heriot-Watt was keen to build up its research in AI. The so-called "AI winter" was ending, and Japan's push for fifth-generation computing had reignited interest. Professor Howard Williams, who headed the department, secured funding to launch a new MSc course in Knowledge-Based Systems. Several of us were hired to teach and develop research in that area.

ZL: And you stayed for ten years. What do you remember most?

AG: The department was in Grassmarket, right in the heart of Edinburgh. It was a magical place - medieval gates, lively pubs, the avant-garde Traverse Theatre just around the corner, and even a tiny dog cemetery hidden behind a wall near Heriot School. The 1st of May bookshop was next door, where my friend and colleague Greg Michaelson volunteered. It was a wonderful mix of history, culture, and eccentric charm (Fig. 6).

Fig. 6. North Berwick, Scotland, c.1985

Academically, I kept pursuing machine learning. But at the time, it wasn't fashionable. Everyone was into logical programming and Prolog. I was pretty much a one-man band in machine learning at HWU.

I even remember a memo from the AI department at Edinburgh University advising PhD students not to choose machine learning - it was considered too ambitious for a three-year thesis. That made me smile. Sometimes, being out of step with fashion is exactly where you need to be.

ZL: Aside from your research, what courses did you teach at HWU?

AG: The very first course I was asked to teach at Heriot-Watt was in Artificial Intelligence. That sounded exciting - until I tried to find a textbook. This was 1983, and to my surprise, there were simply none to be found. I scoured bookshops in London - Dillon's and several others - and then tried my luck in Edinburgh. Nothing.

So I had to build the course from scratch. I started by writing notes in Russian, then painstakingly translated them into English. After that, I transferred everything onto transparency slides - those old acetate sheets we used with overhead projectors. It was quite a task, but also strangely satisfying. You learn a lot when you have to teach from your own notes.

Most of my teaching revolved around the MSc in Knowledge-Based Systems. I supervised a good number of final dissertations, and in the first year, I also taught a course in Computer Vision.

At the time, Edinburgh was buzzing with AI research. Donald Michie's team at Edinburgh University was doing fascinating work - developing algorithms and testing them on chess-playing programs. The idea was simple: if a computer could beat a human at chess, we'd achieved AI. Looking back, it was a bit naïve, but it captured the imagination.

My brother, as I've mentioned, was a serious chess player. He had an enormous collection of chess literature, and when he sent me some of it, I passed it along to Michie's team. They were thrilled. I became a regular visitor to their lab, and those exchanges - between science, games, and curiosity - were some of the most enjoyable moments of that time.

3.1 Bayesian Inference Without Assuming Independence

ZL: You've done some fascinating work on Bayesian inference at Heriot-Watt. Could you tell us more about those projects?

AG: Once I'd settled in - got my lecture notes in order, learned my way around UNIX and the hardware - I was keen to dive back into research. My focus was still pattern recognition and machine learning.

One thing that struck me early on was the divide in British statistics between the Bayesian and frequentist camps. In Russia, that split wasn't nearly as pronounced. What surprised me even more were papers claiming that probability theory wasn't suitable for AI or reasoning under uncertainty. Some of those arguments were, frankly, quite muddled-confident, but conceptually off.

ZL: So you leaned into Bayesian methods?

AG: Yes, almost inevitably. At Heriot-Watt, I started a project to challenge a peculiar claim floating around—that Bayesian models couldn't be used in knowledge-based systems because they required independence between features. We set out to show that wasn't true.

We used a substantial medical dataset, thanks to Dr. Tony Gunn and Dr. Steve Nixon from Western General Hospital in Edinburgh. Their clinical insights were crucial. We compared two models: Proper Bayes, which made no independence assumptions (affectionately dubbed the G&T system, though not for Gin and Tonic), and Simple Bayes, which did assume independence.

ZL: And you worked with your father-in-law on this?

AG: Yes, Roger Thatcher. He was an extraordinary statistician of remarkable breadth, whose intellectual curiosity ranged from demography to cosmology, machine learning to archaeology, and who, despite never holding an academic post, published widely with great clarity and rigour across disciplines.

One of his most elegant contributions, the 1964 paper "Relationships between Bayesian and confidence limits for predictions" [22], extended the notion of confidence limits to apply to predictions about future observations and not just to unknown parameters - a subtle but profound shift. In doing so, he built a conceptual bridge between Bayesian and frequentist paradigms, showing that under certain conditions, frequentist procedures could be interpreted as Bayesian ones with specific priors.

It was a beautiful insight, especially in prediction problems. At a time when the two camps were seen as fundamentally opposed, Roger's work quietly demonstrated that they weren't so different after all (Fig. 7).

Fig. 7. Roger Thatcher, 1980.

The project, indeed, demonstrated that the assumption of independence is not required. The results were summarised in our joint paper "Bayesian diagnostic probabilities without assuming independence of symptoms" [9].

ZL: When I joined Heriot-Watt as a PhD student in 1988, you were working on Bayesian Belief Networks. Could you tell us a bit about that line of research?

AG: Yes, that was an exciting time. Bayesian Belief Networks - or BBNs - were a kind of middle ground between the Simple Bayes and Proper Bayes models. The idea was that not all features had to be independent - just some of them.

If you imagine a graph where the nodes represent variables and the edges represent dependencies, then any two nodes not connected are assumed to be conditionally independent. Add a set of conditional probabilities to that graph, and voilà - you have a Bayesian Belief Network. It's a directed acyclic graph, meaning no loops, and it gives you a structured way to reason under uncertainty.

Judea Pearl had just developed his message-passing algorithm in 1982 [18], which made inference on these networks computationally feasible. That really opened the door for practical applications.

ZL: And that led to the Offender Profiling project?

AG: Exactly. It was called the *CATCHEM* project - a collaboration with Professor Colin Aitken from Edinburgh University's Statistics department, along with the Home Office and Derbyshire Police. Colin was interested in applying statistical methods to forensic science, and subsequently founded the journal *Law, Probability and Risk*, serving as its inaugural chief editor.

The goal was to build statistical profiles of offenders using both data and the detectives' domain knowledge. It was a fascinating mix of hard data and human intuition. The results were published in a Home Office report titled "Predicting an Offender's Characteristics using Statistical Modelling", and Colin and I co-authored a paper on it as well [1].

ZL: I remember struggling with recovering the causal structure in my own PhD work. Pearl suggested the Chow-Liu algorithm, but our experiments with medical data were only partly successful. Still, it was a rich area to explore. What are your memories of those years in Edinburgh?

AG: Oh, I have very fond memories. There were several other projects - EPSRC, EU-funded - and the intellectual atmosphere was lively. But it wasn't just the research. The social life was rich too. I made many friends and colleagues in Scotland, and some of them have contributed to this Festschrift, which is lovely.

After ten years at Heriot-Watt, I moved on to Royal Holloway in 1993. Then, in 1996, I gave my Inaugural Lecture there. It was a chance to reflect on everything I'd done since arriving in the UK. That lecture is actually included as a chapter in this book.

4 Reliable Machine Learning, Royal Holloway University of London, 1993 - Present

ZL: You were appointed to the established chair in Computer Science at Royal Holloway in 1993 and also served as the Head of Computer Science department from 1995 to 2005. In 1998, the Centre for Reliable Machine Learning was established. What was that period like?

AG: It was a big transition. In Autumn of 1992, I was offered a job at Royal Holloway, University of London as a Professor of Computer Science at the University of London[5]. I started the new job on the 1st of September 1993 - it took a year to move the family from Edinburgh to London. The UK was in recession, and selling our house in Edinburgh wasn't easy. So the move was slow and a bit stressful.

Once I settled in, I dove into teaching and managing research grants. I also began organising a series of seminars on Bayesian inference with Unicom Seminars. Among the speakers were Vladimir Vapnik from AT&T, Judea Pearl from UCLA, Phil Dawid from UCL, David Spiegelhalter from Cambridge, and many others. Those seminars ran from 1993 to 1995 and eventually led to three edited volumes published by Wiley and Springer [4,5,7].

ZL: That's when you met Vapnik?

AG: Yes, during those seminars. I invited him to join our department, and in 1995, he accepted a part-time professorship while continuing his work in the USA. Around that time, he published "The Nature of Statistical Learning Theory" and later "Statistical Learning Theory" [25,26]. These were landmark texts. They introduced Support Vector Machines—SVMs—which built on the Generalised Portrait method he'd developed with Alexey Chervonenkis[6], originally presented in their joint monograph [31].

The key innovation was the use of kernel functions within Hilbert-Schmidt theory. That allowed for powerful non-linear classification. Our department was one of the first to implement SVMs. We built a demo system for binary classification, and it really took off - downloaded by over 300 institutions worldwide.

4.1 Prediction with Confidence; Conformal Predictors

ZL: Apparently, your and Volodya Vovk's interests in SVM and its limitations led to your work in the development of Conformal Predictors. Could you say a little bit about this link? And where did you first meet Volodya Vovk?

AG: I met Volodya at the EuroCOLT conference in Barcelona in 1995. Talking to him, you can immediately see that this unassuming and quietly spoken man is ingenious. He always finds an original point of view and converts his ideas into a

---

[5] At the time, all professorial appointments were at the University of London (UL) rather than for individual Colleges of UL.

[6] A historical account of the Generalised Portrait can be found in [28].

rigorous and powerful theory. His mathematical instinct and brilliant technique led him to reconsider the foundations of classical probability theory, proposed by his University supervisor Andrey Kolmogorov, and to establish a new calculus and interpretation of probability [20, 27].

Another area of research in which Volodya has been very active is machine learning, where he has also been pushing the boundaries in developing the Conformal Predictors and related methods. He also laid the game-theoretic foundations of the Prediction with Expert Advice method. I won't go into all his achievements here - there's not enough space - but I'll say this: he's a close friend, and working with him over the past 30 years has been a privilege (Fig. 8).

Fig. 8. L-R: Vladimir Vovk, Alexander Gammerman, Vladimir Vapnik, 1996.

AG: The idea of developing confidence measures really took shape as we explored both the strengths and the limitations of Support Vector Machines. That's what led us to Conformal Predictors - or CP for short. It's a statistically rigorous framework for quantifying predictive uncertainty, and it offers something quite rare in machine learning: provable validity.

Back in 1995–96, Volodya Vovk and I had a series of conversations with Vladimir Vapnik. We were discussing SVMs and Vapnik's broader concept of *transduction*- the idea of making predictions directly for new data points, rather than building a general model first. It was a subtle but powerful shift in thinking.

Then further discussions between Volodya and me happened while I was preparing my 1996 Inaugural Lecture [6] - it also includes a section on transduction. Subsequently, a paper was published [12] reflecting these discussions.

From there, the ideas evolved. Over the years, we formalised the theory and published it across a number of papers, talks, and books [10, 19, 29, 30]. Conformal Prediction represents, in many ways, a paradigm shift in how we think about uncertainty in machine learning. It doesn't just give you a point prediction - it tells you how confident you can be in that prediction, with statistical guarantees.

That's something most machine learning methods still struggle to offer. And it's one of the reasons I believe CP has such lasting value (Fig. 9).

Fig. 9. Volodya Vovk, 2018.

ZL: For the benefit of our readers, could you summarise the basic features of Conformal Predictors?

AG: First of all, CP provides provably valid measures of confidence. It also does not make any additional assumptions about the data beyond the i.i.d. assumption - or even a weaker assumption of exchangeability.

What is also important that any known machine learning algorithm can be used to develop a CP. It will produce the well-calibrated prediction regions.

Several other techniques, we developed, that keep the validity property; they include *Venn-Abers reliable probabilistic* prediction; *Conformal Predictive Distribution*; *Conformal Testing Martingales* for testing the data to satisfy the i.i.d. assumption (and for detecting changes in distribution) and others (Fig. 10).

ZL: What role do think CP could play in the new and exiting research in Large Language Models?

AG: Recent development of *Deep Learning* and *Large Language Models* (LLMs) has already a considerable impact on the conformal predictors. CPs have evolved to address the needs of these powerful but often lacking calibration techniques [15]. I think that by enhancing calibration, interpretability, and trust, Conformal Predictors will play an increasingly critical role in making machine learning more reliable and statistically valid.

One persistent limitation of LLMs is their lack of a grounded notion of truth—they generate text, but they are not inherently reliable for factual accuracy and

Fig. 10. Algorithmic Learning in a Random World: 2005 (first edition) and 2022 (second edition).

require strict verification. We often face this problem. We hope, however, that Conformal Prediction, with its rigorous treatment of uncertainty, might help instill in these models something resembling a statistical conscience.

4.2   Conformal Predictors Applications

ZL: Conformal Predictors have been successfully applied in many fields. Could you mention some of the most interesting ones from your point of view?

AG: Yes, indeed, we have been involved in many applications and grants from the UK, EU, China, Cyprus, etc. One of the latest projects was under the Amazon Research Awards programme entitled "Conformal Martingales for Change-Point Detection". There are applications in: Pharmaceutical industry (drug discovery); Medicine (diagnostics and treatment, including ovarian cancer, heart problems, gastroenterology, depression, etc.); and many others. Some of them can be found on our website: https://cml.rhul.ac.uk/projects.html.

For example, in a collaborative project with AstraZeneca (AZ), Conformal Predictors were employed to help prioritise which chemical compounds should be tested - to speed up the search for promising candidates in drug development. According to AZ report, this approach allowed to make a reduction of approximately one-third in both the time and cost associated with the discovery process [11, 23].

While the pharmaceutical companies are understandably discreet about the financial implications of replacing traditional assays with predictive models. However, we estimated that such efficiencies could translate into savings ranging from $250 million to $1 billion per drug.

ZL: The results for AstraZeneca are impressive. Indeed, it is already clear that Conformal Predictors have a wide circle of applications and can have a significant societal impact.

5 Teaching and Supervision

ZL: Let's talk briefly about teaching and supervision. What was your experience like of teaching in both countries?

AG: When I first arrived in Britain, I hadn't done much teaching. In the USSR, teaching and research were kept separate—universities taught, research institutes did the research. The British model, where both happen under one roof, struck me as very sensible.

It's healthier, really. You're not constantly haunted by the pressure to publish. And from a teaching perspective, it's great—students get exposed to current research, not just dusty textbook material.

ZL: What courses did you teach? Did you use any particular teaching techniques?

AG: I taught across the board - from first-year undergraduates to Master's students. Courses ranged from theoretical topics like computational theory to hands-on programming in C++. There's always debate about what kind of mathematics computer science students should learn - predicate logic, analysis, discrete maths. Personally, I think the specific content matters less than the mindset. What's important is teaching students how to frame problems precisely, solve them logically, and interpret results clearly. That's the real goal.

ZL: Many students say you're great at explaining complex ideas - and that you often tell jokes or stories. Is that something you learned, or does it come naturally?

AG: I don't know. All I am trying to do is to pass on my enthusiasm for the subject. Teaching demands a skill set beyond research—it's a performance art. One must learn to engage, to play the role of lecturer-as-actor before a student audience. Obviously, a joke or anecdote can work: it relaxes the room, and makes learning a little bit easier.

ZL: You also taught in different countries while on your sabbaticals or as a visiting professor. Any interesting comparison with the UK teaching?

AG: Yes, I gave lectures - or full courses - in the USA, Spain, France and China between 1995 and 2011. In some places, students were very strong in mathematics but less confident in programming. That contrast made me think: perhaps we should be teaching more maths to our own students. I even wrote a report to the Royal Society about it, who sponsored some of my travels. But I'm not sure anything changed as a result.

ZL: And your main contribution to teaching?

AG: Definitely supervision. I've supervised many PhD students - sometimes solo, sometimes with colleagues - and it's been one of the most rewarding parts of my academic life. Watching students grow into independent researchers is a joy.

6   Administration

ZL: We've covered your research and teaching. Let's turn to your time as Head of Department at Royal Holloway. What was that experience like?

AG: In 1995, the Principal of Royal Holloway, Professor Norman Gowar, invited me to take on the role of Head of Department. At the time, Professor Chris Mitchell, who headed the department, had completed his five-year term and was ready to pass the baton. I was less than eager - my plate was already full - but Professor Gowar had a remarkable talent for persuasion. After several conversations, I agreed, on the condition that the College support the expansion of the Department and the creation of new research groups: one in Machine Learning, the other in Bioinformatics.

At that time, the department's reputation was largely built on its work in Information Security. Important work, of course, but I felt we needed to broaden our research base (Fig. 11).

Fig. 11. With the staff members of Computer Science Department, Royal Holloway, 1998.

ZL: And did the College support that vision?

AG: They did. With their backing, we gradually expanded. We built up a strong Machine Learning group and began assembling a promising Bioinformatics team. We recruited some excellent people - lecturers and professors with real energy and vision. Sadly, after 2005, the Bioinformatics direction was put on hold. Priorities shifted, and the momentum we'd built in Bioinformatics was lost. That was disappointing, but such is the nature of institutional life - winds change.

ZL: Beyond the day-to-day admin, you also launched some initiatives to raise the department's profile - like the Computer Learning Research Centre (renamed later to the Centre for Reliable Machine Learning - CRML) and the Kolmogorov Lecture and Medal.

AG: What the Department truly needed was greater visibility - both nationally and internationally. Our first major target was the 1996 Research Assessment Exercise (RAE), a periodic audit in which every university department in the UK is graded on its research output, with scores ranging from 1 to the coveted 5*.

I'd only just taken over, so we had very little time to prepare. But we worked hard—really hard—and managed to secure a grade 4. That was a solid result, given the circumstances.

Then came the 2001 RAE. This time, we were ready. We earned a grade 5, and the EducationGuardian ranked us 9th in the country out of more than 100 Computer Science departments. That was a proud moment for all of us.

ZL: Yes, I found the RAE 2001 Panel Feedback Report. It was very complementary about our ML research. Here is what the Report said: *"The Panel were impressed by the strength within the Computer Learning Group and the profound impact that their work has had on theory and applications"*.

AG: It was good to get appreciation for our work in ML.

Looking back, I do think our efforts helped raise the Department's profile—no question about that. But I've always had mixed feelings about the RAE, and later the REF. For all the time, energy, and strategic manoeuvring they demand, I remain unconvinced that they've genuinely improved the quality of research.

What they've certainly done is create a framework - a kind of bureaucratic ritual - for allocating funding and ranking departments. It gives administrators and funding bodies a sense of control, a veneer of objectivity. But whether it fosters real intellectual progress? I'm not so sure.

In some ways, it's like judging a symphony by the number of notes played. You can measure output, citations, impact factors - but the deeper qualities of research, the originality, the risk-taking, the long gestation of ideas - that's much harder to quantify. And often, it's precisely those things that get squeezed out when everyone's chasing metrics.

Still, we played the game. And we played it well. But I've always felt that the real work - the kind that matters - happens in the quiet corners, in conversations, in long walks with a notebook, and in the minds of people who aren't thinking about scores.

ZL: In 1998, the College established a new Machine Learning Centre. You were clearly instrumental in this - how did it come about, and what were your arguments?

AG: By 1997, it had become clear that our machine learning activities needed a more coherent structure. Research was branching into several promising directions - Statistical Learning Theory, Prediction with Expert Advice, Reinforcement Learning, Conformal Prediction, and others. The momentum was there, but without a dedicated hub, we risked fragmentation. Establishing a centre to consolidate and focus these efforts seemed both timely and necessary.

So I began conversations within the department and with the College leadership. We needed a dedicated hub - a centre that could bring everything together, give us focus, and signal to the wider world that we were serious about machine learning.

We drafted a proposal for the RHUL Academic Board and in February 1998, the Board approved the creation of the Computer Learning Research Centre (CLRC), which was later renamed the Centre for Reliable Machine Learning[7]. The Centre was staffed by a team of full-time academics (Volodya Vovk, Chris Watkins, John Shawe-Taylor and myself); there were some part-time academics, including Vladimir Vapnik (also at AT&T, USA), Alexey Chervonenkis (also at Institute of Control Sciences, Russia), and Glenn Shafer (also at Rutgers University, USA) and the visiting professors Chris Wallace (Monash University, Australia), Ray Solomonoff (Oxbridge Research, USA), Jorma Rissanen (IBM Almaden Research Centre, San Jose, USA), Leonid Levin (College of Arts and Sciences, Boston University, USA) and others.

If I'm not mistaken, back in 1998, there were only a handful of machine learning centres worldwide. Ours at RHUL, and Carnegie Mellon's, were among the very few.

ZL: The Centre has trained a number of PhD students over the years. How did it start?

AG: Over the years, the College supported us with a number of PhD studentships. This allowed us which allowed to train an excellent set of researchers. Among them were Ilia Nouretdinov, Jason Weston, Craig Saunders, and others. They completed their doctorates and have since gone on to lead machine learning research in major tech companies and universities. Today, you'll find them at Google, Meta, Microsoft, Amazon, and many other institutions around the world. It's been gratifying to see their work shape the field.

ZL: In the same year, 1998, you organised a colloquium with several world-leading academics and researchers. Could you describe the event and the invited speakers?

AG: Yes, in 1998 we marked the 30th anniversary of the Computer Science Department - founded in 1968 - with a series of colloquia. One of them was

---

[7] More information about the Centre for Reliable Machine Learning is available at https://cml.rhul.ac.uk/.

devoted to Machine Learning and Inference with the title *The Importance of Being Learnable.* The event attracted nearly 100 participants from across the UK, Europe, the USA, Israel, and Australia, from academia and industry.

There were five invited speakers: Ray Solomonoff, Vladimir Vapnik, Alexei Chervonenkis, Jorma Rissanen, and Chris Wallace - towering figures in the fields of inductive inference and machine learning. Their work, dating back to the 1960s, helped ignite the intellectual revolution that shaped the modern theory of learning (Fig. 12).

Fig. 12. Colloquium 1998: The Importance of Being Learnable. L-R: J. Rissanen, V. Vapnik, A. Gammerman, A. Chervonenkis, C. Wallace, R. Solomonoff.

The originators of the VC theory, Vladimir Vapnik and Alexei Chervonenkis spoke on developments in statistical learning theory over the past 30 years. Jorma Rissanen outlined the theory of the Minimum Description Length principle and stochastic complexity. Chris Wallace, raised the fascinating open question of the relationship between different inductive approaches. Ray Solomonoff, discussed the application of many effective ML techniques to a wide variety of problem areas.

ZL: You were one of the founding organisers of the Kolmogorov Lecture and Medal. How did this initiative begin?

AG: It started in 2003, when we realised that Kolmogorov's centenary was approaching. It felt like the right moment to honour his legacy - not just as a towering figure in probability theory, but as someone whose ideas continue to shape modern mathematics and computer science.

So we launched the University of London Kolmogorov Lecture and Medal. The very first recipient was Ray Solomonoff, one of the pioneers of artificial intelligence[8] and co-creator of Kolmogorov-Solomonoff complexity [21]. It was a fitting tribute.

Over the years, the event grew into something quite special. We welcomed an extraordinary lineup of speakers - Per Martin-Löf, Leonid Levin, Jorma Rissanen, Yakov Sinai, Robert Merton, Vladimir Vapnik ... each one brought their own perspective, and each had, in some way, built on the intellectual foundations that Kolmogorov laid[9].

We certainly enjoyed those meetings. The conversations were rich, sometimes surprising, and always deeply rooted in the kind of thinking Kolmogorov inspired—rigorous, curious, and beautifully abstract (Fig. 13).

ZL: You were also involved in organising a conference called COPA - Conformal Prediction with Applications. How did it come about?

AG: In 2011, during a conversation with our former PhD student Harris Papadopoulos, I floated the idea of launching a conference dedicated to Con-

Fig. 13. Kolmogorov Lecture 2005. L-R: A. Chervonenkis, V. Vovk, R. Solomonoff, Per Martin-Lof, A. Gammerman, Z. Luo.

---

[8] Ray Solomonoff was among the pioneering minds at the 1956 Dartmouth Conference where the term "Artificial Intelligence" was first coined. He joined RHUL as a Visiting Professor in 1997.

[9] Further details are available at: http://kolmogorov.cml.rhul.ac.uk/.

formal Predictors. The concept quickly gained interest, and by 2012 we had organised the inaugural symposium, Conformal and Probabilistic Prediction with Applications (COPA), held in Greece, as part of the AIAI conference. And so, COPA was born.

Since then, it has grown into an independent annual conference in the machine learning calendar. This year marks the 14th COPA, which will be held in London. Accepted papers have been published in the Lecture Notes in Artificial Intelligence, and since 2017, in the Proceedings of Machine Learning Research (PMLR), reflecting the symposium's continued academic impact and evolution.

ZL: Thank you again for taking the time to talk with me and share your experiences and insights, Alex. There is a summary of your work on your personal webpage www.gammerman.com and Wikipedia https://en.wikipedia.org/wiki/Alexander_Gammerman.

It has been a wonderful opportunity to learn about your fascinating research journey from the corridors of Soviet research institutes to British academia, spanning from 20th to 21st centuries! From physics to mathematical biology, to pattern recognition, and ultimately to conformal prediction - your journey has been fascinating. Please carry on!

AG: Thank you! Indeed, research is a wonderful thing. But as Albert Einstein said once: *If we knew what it is that we are doing, it wouldn't be called research.* [Laughs].

References

1. Aitken, C.G.G., et al.: Bayesian belief networks with an application in specific case analysis. In: Computational Learning and Probabilistic Reasoning, pp. 169–184. John Wiley & Sons (1996)
2. Borthwick, H.: The high-energy light action controlling plant responses and development. Proc. Natl. Acad. Sci. U. S. A. **64**(2), 479–486 (1969)
3. Gammerman, A.: Correlation between processes in phytochrome and photomorphogenic reactions of plants. In: Doklady Biophysics, Proceedings of the Academy of Sciences of the USSR, vol. 226, pp. 13–15 (1976)
4. Gammerman, A.: Probabilistic Reasoning and Bayesian Belief Networks. Alfred Waller, Henley-on-Thames (1995)
5. Gammerman, A.: Computational Learning and Probabilistic Reasoning. John Wiley & Sons, Chichester (1996)
6. Gammerman, A.: Machine Learning: Progress and Prospects. Royal Holloway University of London, Inaugural Lecture Series (1996)
7. Gammerman, A.: Causal models and intelligent data management. Springer-Verlag (1999)
8. Gammerman, A., Fukshansky, L.: A mathematical model of phytochrome - the receptor of photomorphogenic processes in plants. Soviet J. Dev. Biol. **5**, 122–129 (1974)
9. Gammerman, A., Thatcher, R.: Bayesian diagnostic probabilities without assuming independence of symptoms. Methods Inf. Med. **30**(1), 15–22 (1991)
10. Gammerman, A., Vovk, V.: Hedging prediction in machine learning (with discussion). Comput. J. **50**, 151–170 (2007)

11. Gammerman, A., Vovk, V., Luo, Z.: Novel machine learning methods for improving drug discovery efficiency. Royal Holloway University of London REF 2021 Impact Report (2021)
12. Gammerman, A., Vovk, V., Vapnik, V.: Learning by transduction. In: UAI'98: Proceedings of the Fourteenth Conference on Uncertainty in Artificial Intelligence, vol. 14, pp. 148–155 (1998)
13. Gammerman, A.Y., Fukshansky, L.: Theoretical analysis of processes in the phytochrome pigment system. Stud. Biophys. **52**(1), 65–72 (1975)
14. Gammerman, A.Y., Fukshansky, L.: Theory and calculations of dynamics of phytochrome transformations in the green leaf. Soviet Plant Physiol., 18 (1976)
15. Giovannotti, P., Gammerman, A.: Calibrated large language models for binary question answering. Proc. Mach. Learn. Res. **230**, 218–235 (2024)
16. Lanin, M.I., Gammerman, A., Sheetova, I.P.: Development of a set of programs using information-theoretic methods for solving systematic problems. In: Abstracts of reports of the 12th International Botanical Congress, vol. 1, p. 27 (1975)
17. Lanin, M.I., Gammerman, A., Sheetova, I.P., Bykov, O.D.: Application of multidimensional information-theoretic analysis in the study of the contingency of qualitative features. vol. 61, pp. 99–104. VIR (1978)
18. Pearl, J.: Probabilistic Reasoning in Intelligent Systems. Morgan-Kaufmann (1988)
19. Saunders, C., Vovk, V., Gammerman, A.: Transduction with confidence and credibility. In: Proceedings of the 16th International Joint Conference on Artificial Intelligence, Stockholm, Sweden, 1999 (1999)
20. Shafer, G., Vovk, V.: Game-theoretic foundations for probability and finance. Wiley, Hoboken (2019)
21. Solomonoff, R.: Universal coding, information, prediction, and estimation. IEEE Trans. Inf. Theory **30**(4), 629–636 (1984)
22. Thatcher, A.R.: Relationships between Bayesian and confidence limits for predictions. J. Roy. Stat. Soc. B **26**(2), 176–192 (1964)
23. Toccaceli, P., Gammerman, A.: Combination of inductive mondrian conformal predictors. Mach. Learn. **108**, 489–510 (2019)
24. Turing, A.: The chemical basis of morphogenesis. Philos. Trans. R. Soc. B **237**(641), 37–72 (1952)
25. Vapnik, V.: Nature of Statistical Learning Theory. Springer (1995)
26. Vapnik, V.: Statistical Learning Theory. Wiley, Hoboken (1998)
27. Vovk, V.: A logic of probability with application to the foundations of statistics (with discussion). J. Royal Stat. Soc. B **55**, 317–351 (1993)
28. Vovk, V., Gammerman, A., Papadoupolus, H.: Measures of Complexity. Festchrift in honor of Alexey Chervonenkis. Springer (2015)
29. Vovk, V., Gammerman, A., Saunders, C.: Machine learning applications of algorithmic randomness. In: Sixteenth International Conference on Machine Learning, pp. 444–453 (1999)
30. Vovk, V., Gammerman, A., Shafer, G.: Algorithmic learning in a random world; 1st edn. (2005); 2nd edn. (2022). Springer
31. Вапник, В., Червоненкис, А.: Теория распознования образов. Москва (1974)
32. Малинин, А.П., Сафонова, С.А., Пунанов, Ю.А.: Развитие децкой онкологии в Ленинграде и Санкт-Петербурге. Санкт-Петербург (2015)
33. Пименов, М.Г., Малкина, Р.М., Гаммерман, А.Я.: Анализ Внутривидовой Химической и Морфологической Изменчивости (adenostyles rhombiofolia (willd.) m.pimen. Методами Многомерного Информационного Анализа и Автомотической Группировки. Растительные Ресурсы, **17**(1), 23–36 (1981)

Inaugural Lecture Machine Learning: Progress and Prospects

Alexander Gammerman$^{(\boxtimes)}$

Department of Computer Science, Royal Holloway University of London,
TW20 0EX London, UK
A.Gammerman@rhul.ac.uk
https://cml.rhul.ac.uk/people/alex/

Abstract. This Inaugural Lecture was delivered at Royal Holloway University of London in 1996 - almost 30 years ago. The Lecture here is presented in its original format to give readers a perspective on the development of machine learning over that period. A few remarks have been added to reflect recent developments, and the list of references has been updated to enhance the convenience and accuracy for readers.

1   Introduction

My subject in this lecture is machine learning: how to write **algorithms** and **programs** that learn. In the spirit of our time, just before I started to write this lecture, I searched the Internet to see whether there were any entries under "Machine Learning". I expected to find several hundred papers and other documents written on the subject, but found the astonishing total of 400,000 documents written on Machine Learning and accessible on the Internet in November 1996. Surely, no-one can read that many documents, and indeed probably 99% of them are some trivial programs, or re-inventing the wheel. Perhaps it would be worth writing a machine learning program to analyse this set of documents.

So I may be trying to do an impossible task: to review the progress in this field. I did not read one tenth, or one hundredth or even one thousandth of the available papers, and inevitably this talk will be a very personal view on the subject.

When did machine learning start? Maybe a good starting point is 1949 when Claude Shannon suggested a learning algorithm for chess playing programs. Or maybe we should go back to the 1930s when Ronald Fisher developed discriminant analysis – a type of learning where the problem is to construct a decision rule that separates two types of vector. Or could it be the 18th century when David Hume discussed the idea of induction? Or the 14th century when William of Ockham formulated the principle of "simplicity" known as "Ockham's razor"? (Ockham, by the way, is a small village not far from Royal Holloway). Or it may be that, like almost everything else in western civilisation and culture, the origin of these ideas lies in the Mediterranean? After all, it was Aristotle who said that "we **learn** some things only by doing things".

K. An Nguyen and Z. Luo (Eds.): Alexander Gammerman Festschrift, LNCS 16290, pp. 28–45, 2026.
https://doi.org/10.1007/978-3-032-15120-9_2

I would like, however, to start from the middle of this century – the computers have just arrived and, perhaps, this topic (ML) is as old as computer science. In 1950, Alan Turing had just published one of his best papers (apart from his mathematical papers) entitled "Computing Machinery and Intelligence" [12]. In order to make the machine intelligent (in some sense) he suggested that the machine might be programmed to simulate a child's brain, then equipped with a **learning program** and taught like a child. This brings out the point that **thinking (intelligence)** is closely connected with **learning**.

The field of machine learning has been greatly influenced by other disciplines and the subject is in itself not a very homogeneous discipline but includes separate, overlapping subfields. There are many parallel lines of research in ML: inductive learning, neural networks (NN), clustering, learning by analogy, genetic algorithms (GA), and theories of learning - they are all part of the more general field of machine learning - see Fig. 1.

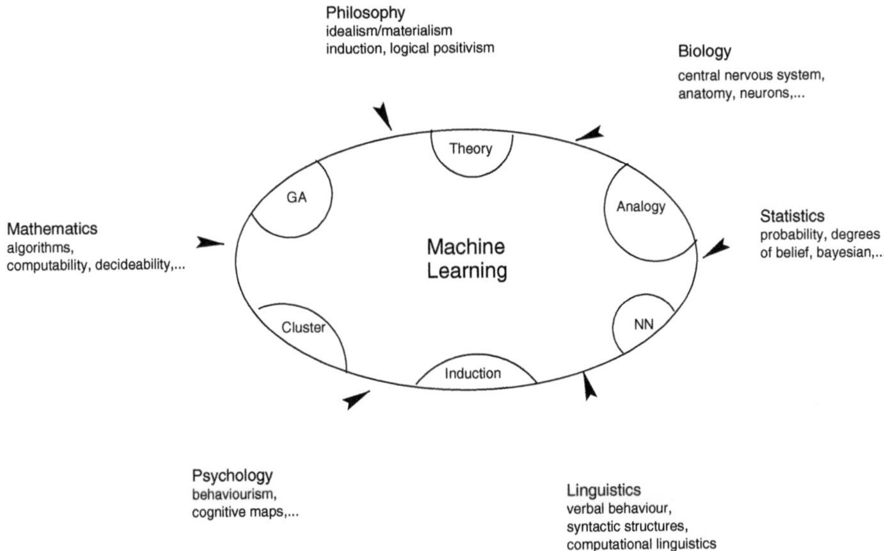

Fig. 1. Machine Learning.

Let me start by showing you several examples of achievements in machine learning.

The first example is a system called ALVINN [9] which learns to steer a vehicle along a motorway by observing the performance of a human driver. The results of the training are very impressive – ALVINN has driven at speeds up to 70 mph for a distance of up to 90 miles on a motorway[1].

---

[1] After 30 years of development in the field, by 2025, the self-driving cars reached a commercial pilot stage.

Another example is the chess playing program. Let us compare the ratings of human and machine chess champions. Back in 1960 the World Champion was Botvinnik and he had a rating of 2,616, while the best machine in 1965 had a rating of 1,400. By the year 1992 Kasparov, the current World Champion, had a rating of 2,805 while the best computer in 1994, Deep Thought 2, had a rating of 2,600 approx. For comparison, a county player has a rating in the region of 2,100, an international master of 2,400, and an international grandmaster of 2,500. During the last 30 years or so, the performance of these programs has improved enormously, and perhaps one can extrapolate this trend further[2].

There are many other successful applications of machine learning. In particular, there are the machine learning programs working in the discrimination of credit card applications; in recognising handwritten zipcodes; in counting small volcanoes in images of Venus; and in speech recognition using hidden Markov models. There are many medical applications, for example in the diagnosis of abdominal pain which I am going to talk about today. There are several automated programs for establishing human genome sequences. There are also machine learning programs used for predicting seismic events like earthquakes, in classification of tissue samples for breast cancer screening, in prediction of financial indices such as exchange rates, text categorisation and many others.

In this lecture I will discuss several important developments in the field of machine learning and consider some prospects: what is the future of the subject? I will start by explaining **what learning is**, then I will consider several different inductive learning models using the **Bayesian approach**. After that I will move to our current interest in developing a new type of universal learning machine called **Support Vector Machine**. Then I would like to spend some time talking about the prospects offered by ML. In particular, I am going to draw your attention to a new and very promising development called **transductive learning**, which may allow us to deal with previously unformalised concepts such as insight, intuition etc.

What I would also like to emphasise in this lecture is that machine learning is not just an experimental science, nor is it just a theory of learning. Only by engaging with both the theory and the experiments can one really make progress in this subject. What I aim to show is that the theory allows us to design good systems and the experiments allow us to validate the theory.

---

[2] In 1997, just the following year after this Lecture was delivered, IBM's *Deep Blue* beat the champion Garry Kasparov. Between the 2000s - 2010s, new programs like *Fritz and Stockfish* were developed and they became accessible on consumer hardware. The introduction of deep learning has further advanced the development of chess-playing programs, such as *AlphaZero*. These engines learn and imitate playing styles that resemble human intuition. The current World Champion, Magnus Carlsen has a rating around 2882. By contrast *Stockfish and AlphaZero* are estimated to have ratings above 3500, depending on hardware and testing conditions - that is, they are simply too strong even for the best players.

2   What Is Learning?

So what is learning? According to Webster's dictionary "to learn" means "to gain knowledge, or understanding, or skill, by study, or instruction, or experience".

I can try to represent this idea in a simple graphical form where in the figure below there is a box called **experience** or **data** or **set of examples** and a box called **knowledge**. The arrow between these two boxes is called **learning**, the process whereby we learn something out of experience – out of data, and we gain some new knowledge. Knowledge gained through learning partly obtained from a description of what we have observed, and partly obtained by making inferences from (past) data in order to predict (future) examples. Obviously if data have no regularities, any law incorporated into them, we won't be able to find any new knowledge. In other words, in random data, there is no knowledge to be found (Fig. 2).

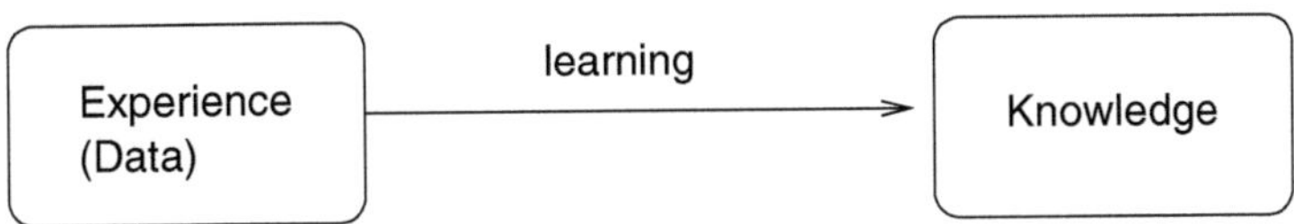

Fig. 2. Learning from experience

In answer to the question, what do we do?: we are trying to find some regularity, some knowledge in data.

The next question we may ask is, why do we do this? and the answer is, of course, we would like to make some predictions for future examples or to make some decisions or just simply to understand. If we would like to answer the question how, how do we do learning? the simplest answer is, we do it by searching – we are physically searching for "good" models in our data and then we use these models for future predictions. There are also different types of machine learning such as supervised/unsupervised learning or online/offline learning but they can all be described by the same scheme.

Let me now give you an example of a real application of supervised learning [2]. This is an application in the medical domain, in the diagnosis of abdominal pain, and the basic set of data was collected at a hospital in Scotland. There were about 6,000 patient records, of patients who suffered from abdominal pain. Each patient had 135 binary symptoms, such as are represented in the figure below, and there were 9 possible diagnostic classes to which each of these patients was assigned: classes like appendicitis (App), dyspepsia (Dys), perforated peptic ulcer (Ppu) etc. The task was to extract out of this information the set of relevant symptoms for each diagnostic class – for each disease – so that the program would be able to predict with certain probability the disease of a new patient. So the input information – the input matrix – looked like Table 1.

There is an input set of symptoms – attributes or feature vectors, or random variables, $X \in \{x_1, x_2, ..., x_n\}$, and each variable has a set of values. The classes

Table 1. Medical Database: Scotland; 6,387 patients; 135 symptoms; 9 diseases

ID	Age	Sex	Pain-on-site	Nausea	....	Diagnosis
84136	18	M	Y	N	......	App
65140	34	F	N	N	......	Dys
71853	61	M	Y	Y	......	Ppu
......	...	...	...	...	......	...

C_1, C_2, ..., C – or we can call them groups or diseases, or diagnostic groups – are all mutually exclusive and exhaustive. So basically we are trying to find a mapping between symptoms and classes. Our task is to find a combination of symptoms which will indicate a certain class with a certain probability. An algorithm called G&T was developed and the learning part of this algorithm is shown below.

Algorithm (G&T)

Matching-with-Selection Learning (examples, attributes, classes)

    input: *examples*

                 attributes

                 classes

    if   *examples* is empty then *terminate*

                 else if all examples have the same

                 class then **return** *class*

    else

                 1. for each class c calculate χ^2 values for each attribute and choose the attribute with the highest χ^2

                 2. partition the current set of data into two subsets including and excluding the selected attribute

                 3. repeat for each subset until the termination is met

                 4. calculate probability p and confidence limits

    end

    output: combination of attributes with probabilities of classes and confidence limits

The first step in this algorithm is to find a set of the most important symptoms for each diagnostic class. This is done by using a set of statistical tests, in our case the χ^2 test. The second step is to partition the current set of our data, our training examples, into two subsets of *including* and *excluding* the selected attribute. And the third step is to repeat the algorithm for each subset until the termination condition is met. Once this is done we can calculate how many patients who had a certain combination of symptoms also developed one of the diseases, and we can also calculate how many patients had the same combination of symptoms but didn't develop the disease. Once we take a ratio between the two we shall have the best estimate or a probability.

This is repeated for each diagnostic class so at the end of the day we shall have all the combinations of symptoms important for each diagnostic class with the corresponding probabilities.

After this we should be able to make a prediction for a new patient and classify this new patient according to his or her symptoms. Once we have classified all the new patients – it is called the testing set – we should be able to estimate how good the results are, by comparing the results of the algorithm with the performance of the doctors (in percentage of correct diagnoses), and in Table 2 you can see the result of this investigation. In the same table you can see the results of performances of several other algorithms re-implemented using the G&T model. These include the "Simple Bayes" model, and the CART model - for details see [2]. The consultants are still doing better but the performance of the programs is not that far off.

Table 2. Performances of doctors and programs

Consultants	76%
Registrars	65%
Junior Doctors	61%
G & T	65%
G & T (Simple Bayes)	74%
G & T (CART)	64%

I will use this example now in order to describe the most important characteristics of a learning algorithm. The first is the **representation of the data** (a matrix in our example). Another important characteristic of a learning algorithm is the **search procedure**. And the last one is the **performance** of the algorithm – we need some sort of criteria for estimating the performance of the chosen algorithm.

In addition to these three important issues there are also some important points to decide, in particular about the **learning principle** which lies at the heart of the design of a learning algorithm. When I say learning principle, I mean we can use inductive learning or a neural network model or we can use, for example, Minimum Description Length principle [10].

But among all these important issues perhaps the most difficult problem lies with the search for the "good" models. For example, in the chess game programs there could be up to 10^{40} legal positions, or if we consider that each player can make about 50 moves in a game, and the branching factor of the game is about 35, so the tree will have 35^{100} possible moves, an unimaginable number. In our medical example if each of the symptoms could have only 2 values, yes and no, then with about 135 possible symptoms we would have 2^{135} possible combinations. So the question arises about the computational efficiency of our algorithm: is it possible to build an efficient search procedure?

This is not a new question. In the 1930s and 1940s, work by Gödel, Kleene, Church and Turing showed that there are truths that are not deduceable, and functions that are not computable. It started from recursion theory and now it is a whole branch of computer science called complexity theory. Basically, it can be shown that certain problems are harder to compute than others. Many problems can be computed in polynomial time (time which is a polynomial function of the size of the problem), other problems can take exponential time or even worse. This problem of computation is called "combinatorial explosion", or "exponential growth", or the problem of dimensionality. Richard Bellman back in the 1960s called this the "curse of dimensionality".

So what I would like to present now are some possible solutions to this problem. In particular I am going to talk about two possible solutions. As you can see in Fig. 3, we have the data, the information processing act called learning and then the box called knowledge, but in addition to this I can always try to use, somehow, some prior knowledge – prior information, and therefore to reduce the amount of calculation.

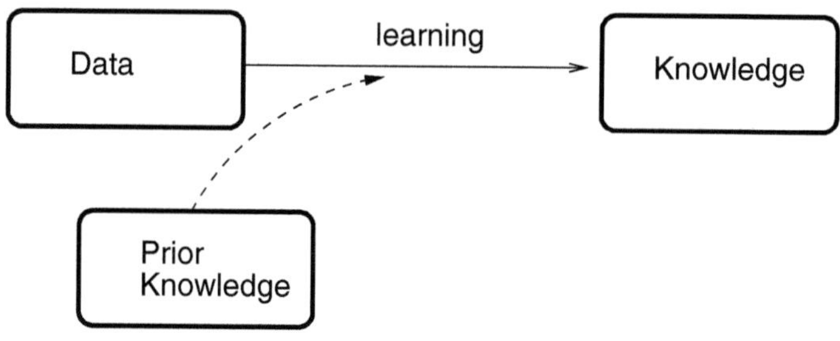

Fig. 3. Solution 1.

Another way of tackling the problem, of reducing the amount of computation, is just to reduce the original set of data - our experience - to a very essential set of examples, called support vectors, and then use just that subset of data in order to learn and gain new knowledge - see Fig. 4.

In the first approach the prior information could be a set of additional assumptions such as the assumption of conditional independence or the assumption that all our examples are distributed identically and independently. And in the second approach, when we use only essential information, for example, support vectors, it means that somehow we find a mechanism to compress the data up to the only "important" set of vectors in order to solve our problem.

3   Bayesian Approach

Let me now show you how the first approach works, when we use prior information - prior knowledge. The first model I would like to show you is the so-called

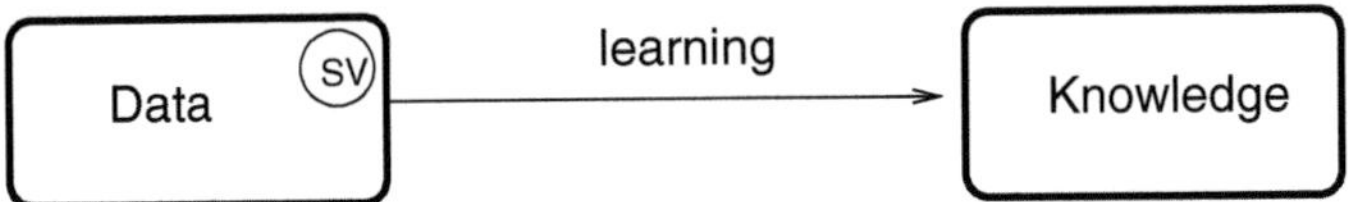

Fig. 4. Solution 2.

"simple" Bayes model when we can make an assumption of independence, that is, all our symptoms X_1, X_2, up to X_n are independent given the disease, and this can be represented in graphical form where on the top level is our node with all possible diagnostic groups or possible diseases and on the next level down are all possible symptoms, so we assume that each of the symptoms X_1, X_2 is independent given the disease. This is represented graphically in Fig. 5. The performance of "simple" Bayes model in our medical example is shown in Table 2.

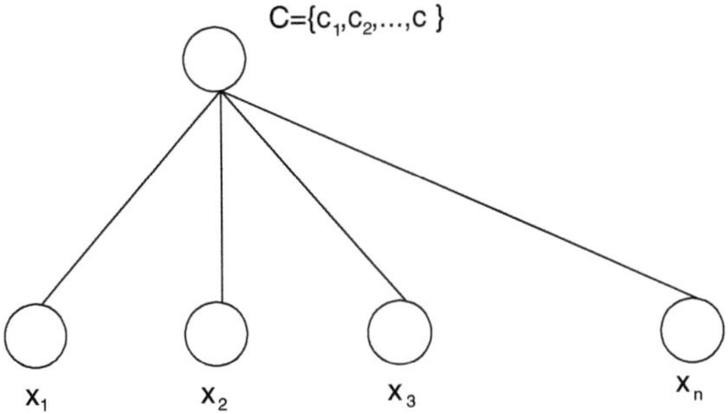

Fig. 5. Simple Bayes model.

We can expand this idea further and make a graphical representation where not all symptoms are independent but only some of them (the nodes that are not connected are assumed to be conditionally independent). If we represent this idea using a graph, and if we also supply a set of conditional probabilities to this graph, then it is called a Bayesian Belief network (BBN). By definition, a Bayesian belief network is a directed acyclic (no loops) graph, and a set of conditional probabilities on that graph.

In 1982 Pearl developed an algorithm [8] for a tree structure where he considered only parents and children for each node on the tree. If you look in Fig. 6 you can see that node X has one parent U and two children Y and Z. If an observation is made then we would have to revise our original probabilities, and calculate posterior probabilities for each node. What Pearl suggested was, let's consider information coming from children as *lamda* messages or likelihoods, and information coming from a parent as a *pi* message or prior probability, and fuse this information in node X. These are local computations because we consider

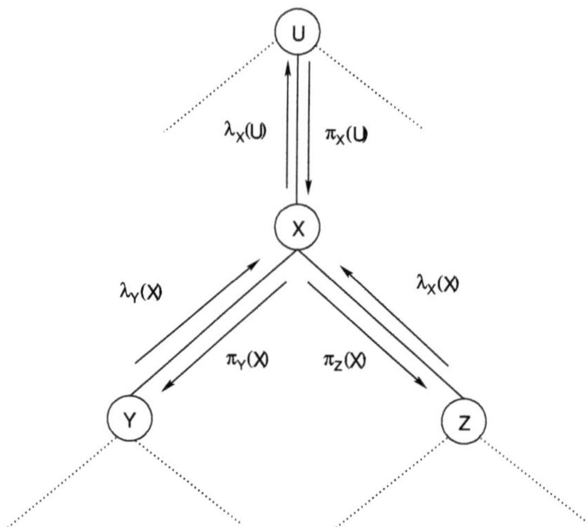

Fig. 6. Local computations on a tree structure.

only a node with its parents and its children and we are not considering nodes which are widely separated on the tree. And if we do this procedure step by step down the tree, we can build up and calculate the overall structure

The same idea has been generalised for a general graphical structure, as shown below, and by decomposing the graph into small groups called cliques and making a clique tree; this allows us to calculate the posterior belief or posterior probabilities as soon as we make an observation. Behind this method lies the method called Gibbs potentials [6] and it guarantees that we always calculate our results correctly (Fig. 7).

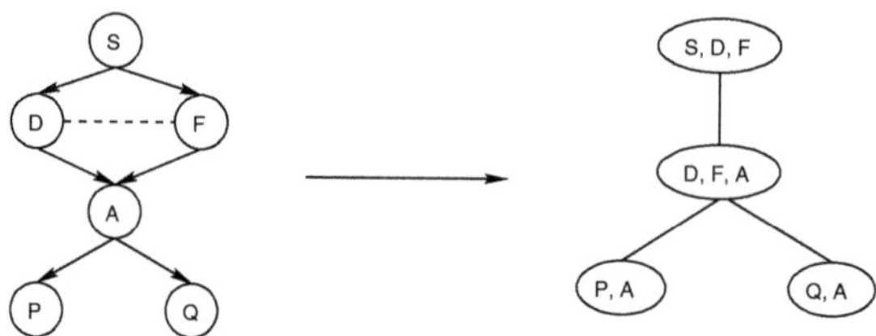

Fig. 7. Local computations on a graphical structure.

A number of computational systems have been developed using these ideas. In 1987 we developed a Causal Probabilistic Reasoning System which was later (1992) redeveloped into a shell system called PRESS [7]. PRESS has a number of important and useful options that allow a user to deal with a mixture of discrete and continuous variables, to construct a BBN out of data (structural learning), to determine the most important variables, etc. The systems have been

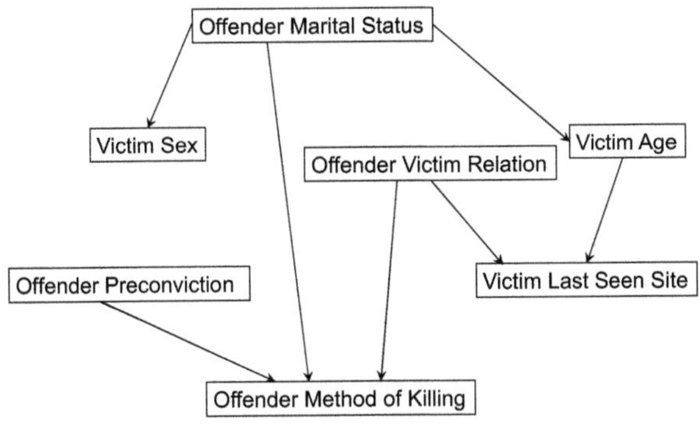

Fig. 8. BBN for offender profiling.

tried in a number of applications and one of the most recent, shown in Fig. 8, is Offender Profiling – a collaborative project with the Home Office [1]. The aim of the project was to develop statistical profiles of offenders using some data and knowledge held by the detectives. The data set was initiated by Derbyshire Constabulary in 1986, and there were 320 cases of offenders described by 5 attributes such as offender preconviction, offender age, etc., and 8 attributes related to the victims and scenes of crimes. Several Bayesian Belief networks have been created by the detectives and one of them is shown here, with some of the corresponding probabilities in the table below. Then as soon as information arrives (an observation is made) about a particular node – victim age etc. – this information, the probabilities can be revised and we can make some predictions - see Table 3 - about the offender's preconviction, marital status, etc.

Table 3. Conditional probabilities for offender profiling

Probabilities for offender characteristics for a
female victim, aged 0–7 years, found strangled
outside her own home

Characteristic	Outcome	Initial	Revised
Living with partner	Yes	0.24	0.36
	No	0.76	0.64
Relationship	Known	0.57	0.11
	Unknown	0.43	0.89
Preconviction	Yes	0.73	0.70
	No	0.27	0.30

I can summarise this approach by saying that we use additional prior informa-tion in the form of either making some assumptions of independence in the simple Bayes model or using Bayesian belief networks to simplify the computational procedure and make it much more efficient. Obviously, the gain in computational efficiency is just one of several benefits in using additional (prior) information. In general, the use of prior knowledge allows us to develop very powerful algorithms. For example, here is a general Bayesian algorithm developed by V. Vovk [14]. Suppose that *a priori* we have a set of possible hypotheses: H_1, H_2, ...; more generally we can have some family of hypotheses $\{H_\theta/\theta\epsilon\Theta\}$ – it can be finite, or countable, or continuous, and instead of finding the best hypothesis, we compute the "weights" for all the hypotheses (H_θ). These weights would reflect how much we trust (or like) each of our hypotheses. We start by assuming the prior weights $P(d\theta)$, and when a new example arrives, every H_θ gives a prediction P_θ. Then the Aggregating Algorithm merges all P_θ with accordance of their weights $P(d\theta)$. When the actual hypothesis is disclosed we compute the loss l_θ supported by hypothesis H_θ, and then recompute the weights:

$$P(d\theta) := \frac{e^{-\eta l_\theta} P(d\theta)}{\int e^{-\eta l_\theta} P(d\theta)}$$

where η is a learning rate. As one can see if a hypothesis is not "correct" (after disclosure of the real class) - this is reflected in a big value of loss function l_θ - then the weights are slashed.

Special cases of this algorithm are: Bayesian merging and weight updating, the weighted majority algorithm in pattern recognition, Cover's universal portfolios algorithm and many others.

4   Support Vector Machine

Another way, as I have already mentioned, is just to compress the data itself and make use of only an "essential" set of examples - support vectors. This approach was originally developed by V. Vapnik [13], who is now leading this research in our Department.

Let me start again from the familiar picture of the learning process (Fig. 4). We shall follow our scheme again: from data to new knowledge through inductive learning.

The **data** are represented with a **set of examples** $(x_1, c_1), ..., (x_l, c_l)$; x_i is a vector of attributes and c_i is a class which takes value either positive (+1) or negative (-1). The task is to find a decision rule that separates examples into positive (+1) and negative (-1) classes. The **rule** would then represent the **knowledge**. This is a typical pattern recognition problem.

The main idea is to map our original set of vectors into high dimensional feature space, and then to construct in this space a linear decision rule (an "optimal hyperplane"). Then we can find the maximal margin between the vectors of the two classes. The vectors that determine the margin are called "**support vectors**", and they can replace the whole training set of examples.

Let me use a physical analogy to explain the main ideas. I would represent a decision rule with a magnetic rod. This rod can move freely in the space between positive and negative examples, and it generates a field around itself with sharp boundaries. But as soon as the field reaches several nearest negative and positive examples, it stops propagating, and the rod will be fixed at this point. Those negative and positive examples that are on the boundary of the field are called the "support vectors". We can now make predictions by determining on which side of the margins the new examples come. The idea is illustrated in Fig. 9 where black dots represent positive (+1) examples, and white dots represent negative (-1) examples.

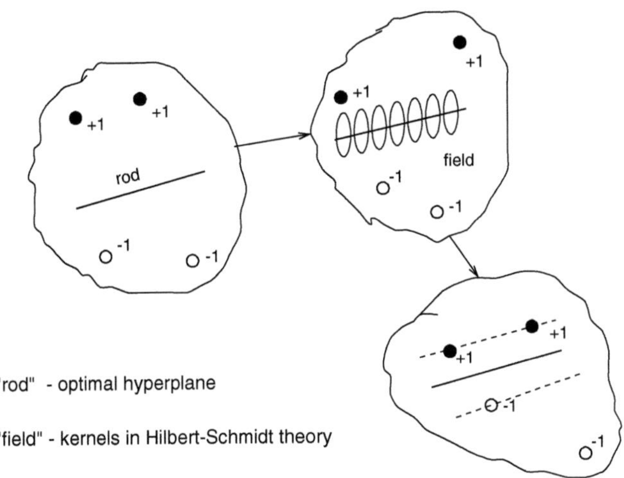

Fig. 9. Support Vector Machine for linear separable case.

This is an exact description of the linear and separable case. In general we could present our "field" as the **kernel functions** $K(x, x')$ of the Hilbert-Schmidt theory.

The training procedure in this algorithm, to find the optimal hyperplane, amounts to solving a constrained quadratic optimisation problem. The solution of this problem is a unique global minimum.

One can also assess the performance of the algorithm, and it has been shown that the probability of error is bounded by the ratio of the expected number of support vectors to the size of the training set:

$$Pr(err) \leq \frac{E(\#SV)}{\#training\_examples}$$

This is a constructive implementation of the main idea, and it boils down to the compression of the original set of data to "only" the support vectors set (usually 3%—5% of the original set).

The next question we may ask is, why are the SV machines so successful in compressing the data, where does their power of generalisation lie? The answer to this question turns out to be a feature of the problem called **VC-dimension** (Vapnik-Chervonenkis). One of the main results of learning theory has been the discovery of the following inequality that holds with probability at least $1 - \eta$ and gives the bound for probability of errors in the testing set:

$$Pr(err\_testset) \leq Pr(err\_trainset) + \Phi(\frac{l}{h}, \frac{-\ln \eta}{l})$$

where h is the VC-dimension of the set of functions, and l is the number of examples in the training set.

Recall our example when the field becomes more powerful, and the band becomes wider and wider, and the rod gets more and more restricted. It turns out that the VC-dimension of the set of possible rod positions becomes smaller and smaller. And when at the limit we reach a point where we cannot move the rod any more, the VC-dimension becomes minimum, and thus, the SV-machine implements the principle of minimizing the VC-dimension. V. Vapnik obtained some accurate estimates of the VC-dimension of this set depending on the width of the band.

The idea of learning using SV-machines is currently being tried in several application areas: speech recognition, faces recognition, tomography, medical diagnosis and it is already clear that this new approach has very strong generalisation abilities and predictive power.

5   Transduction

The last point I would like to touch on in this lecture is a different type of learning. So far we have been discussing mainly inductive learning.

Let me now reformulate the problem and ask why do we need the "knowledge" at all? One possible answer to this is that we want to say something about (to predict) future examples. So, we can say that we use "knowledge" to predict "new examples" (Fig. 10):

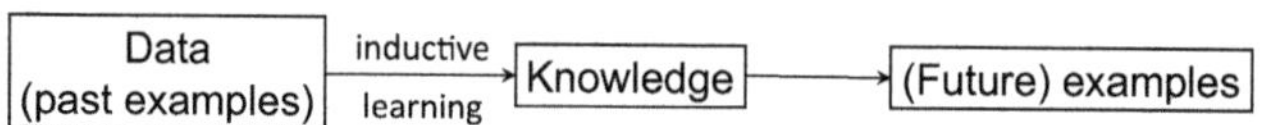

Fig. 10. Prediction in inductive learning.

The "knowledge" in our last example was a decision rule or optimal hyperplane. Obviously we spent a great deal of (computational) effort in order to construct the rule and to find the support vectors. When the rule is constructed it of course allows us to predict the classes (positive or negative) of new examples. In fact, we can then predict any new examples, even an infinite number of new examples can be classified. In reality, however, we are interested only in predicting a finite number of new examples. If we require to predict *only some* new examples, but have already *knowledge* to predict *any* new examples, then, perhaps, we have "overkilled" the problem.

Then, the interesting question arises: do we always need to go from the particular (examples – the data) to the general (knowledge) – this is usually called induction, and then back to the particular (future) examples – this is usually called deduction?

Can we make a short cut, and go from particular directly to particular? We shall call this short cut a **transduction** [13]. It looks as if this way we shall make some "savings" – instead of having "general knowledge" it is more efficient to get "specific knowledge" about particular instances. So, the transduction is going from particular (past) examples to particular (future) examples without any attempt to "generalise" our experience and get "general knowledge" - see Fig. 11 (taken from [13]).

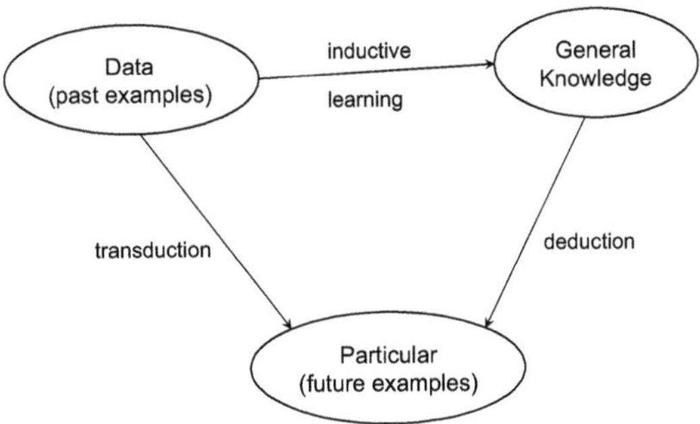

Fig. 11. Prediction in transductive learning.

We hope that transduction will (1) allow us to make the learning process computationally more efficient; (2) facilitate some theoretical advance, since it will be easier to prove some properties of a learning algorithm (inferring from particular to particular).

Let me illustrate this last point with an example. Recall our picture of positive and negative examples (black and white dots) separated by the optimal hyperplane (Fig. 9). Perhaps one of the disadvantages of this induction learning is that this is a "black and white" picture, with no "grey" area.

When we found our separating surface (decision rule) we obtained some "black and white" picture; we have black points, and white points, and we classify all points of one side of the margin as black, and on the other side of the margin as white. What we would like to have is some tints of "grey" colour, expressing our *confidence* - see Fig. 12.

For example, if a new point is near to the black points area, we expect it to be a black point, with significant confidence. If it is near to the white points, we are confident that it should be a white point. And if it is near the separation surface, we are not sure, this is a grey region.

Here is another example: a clear digit 2 (could correspond to the "black" area), clear digit 7 – "not 2" – corresponds to the "white" area, and something that is not clear – "grey" area:

Fig. 12. Uncertainty quantification by *confidence* in digit recognition problem.

With induction, no efficient procedure for finding this "grey" picture is known. Let me now explain how this idea of confidence can be expressed using transduction [4]. Assume that we now have our training set of data and also a new example which has not been classified yet. We can represent it with the following picture - see Fig. 13.

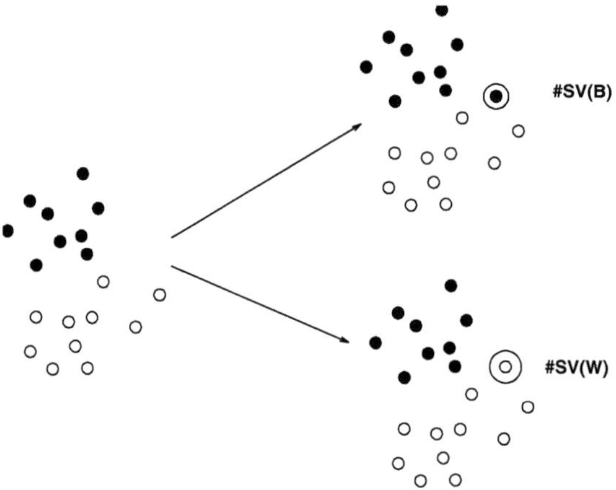

Fig. 13. Prediction with confidence.

In the picture $\#SV(B)$ is the number of support vectors in the black picture, and $\#SV(W)$ is the number in white. We now have two pictures: in one the new example is classified as "black", and in the other, it is classified as "white". Recall that there is no general separating curve any longer, because there is no "general knowledge" in this transduction process. All we want is to predict whether the new example is white, or whether it is black, i.e. to make a prediction for that particular point. All we can say is that this new example is a support vector in at least one of the two pictures. Then, we can formulate a simple rule that classifies our new example with a certain confidence:

Prediction rule (l is the number of examples and SV(B) and SV(W) are the support vectors in the black and white pictures respectively):

(1) WHITE if

$$new\_example \in SV(B) \& new\_example \notin SV(W)$$

or

$$new\_example \in SV(B) \cap SV(W) \& [\#SV(B) < \#SV(W)]$$

CONFIDENCE : $1 - \frac{\#SV(B)}{l}$

(2) BLACK if

$$new\_example \in SV(W) \& new\_example \notin SV(B)$$

or

$$new\_example \in SV(B) \cap SV(W) \& [\#SV(B) > \#SV(W)]$$

CONFIDENCE : $1 - \frac{\#SV(W)}{l}$

(3) NO PREDICTION

$$new\_example \in SV(W) \cap SV(B) \& [\#SV(B) = \#SV(W)]$$

NO CONFIDENCE (any prediction with no confidence)

This prediction rule works well, and full confidence is expressed with 1, and no confidence with 0. For example, if a *new example* is a support vector in the "black" picture, and it is not in the "white" picture, we classify it as white with the confidence $1 - \frac{\#SV(B)}{l}$; if the fraction of support vectors in the "black" picture is small, then the confidence is high (close to 1).

Remark 1. These ideas were discussed extensively, and this led us to the development of *conformal predictors* [3,4,11,16]. The method is ideally suitable as a framework for uncertainty quantification; the most definitive exposition of Conformal Predictors theory is given in [15].

Remark 2. Recent development of *Deep Learning* and *Large Language Models* (LLMs) has a considerable impact on the conformal predictors. CPs have evolved to address the needs of these powerful but often overconfident and lack calibrated uncertainty estimates [5]. By enhancing calibration, interpretability, and trust, Conformal Predictors play an increasingly critical role in making modern machine learning systems more reliable and statistically valid.

One of the weaknesses of LLMs is that they do not have a notion of the ground truth; there is a hope that CPs could bring this notion to the large language models.

6 Conclusion

Let me now return to our original picture of inductive learning and claim that following the success of 20th century science (e.g. quantum mechanics), true knowledge is inaccessible and the best we can hope for is to substitute for it the best hypothesis. But even the best hypothesis is very often difficult to find, and this could lead to the position of agnosticism. However, I will abandon this line of argument for now since it will take us into a new and deep philosophical discussion about the nature of knowledge.

There is one last comment I would like to make in conclusion. I started this lecture by saying that learning is intimately connected with intelligence. For centuries people have argued about what lies at the basis of human intelligence. Is it our logical abilities and generalisation skills that lead us to gain new information and discover laws of nature? Or is it human intuition and insight that play the most important role in different discoveries as well as in the huge area of human intelligence called arts and humanities? It is clear that logical ways of thinking and generalisation abilities can be connected with deduction and induction, while transduction can be a way to formalise intuition and insight. This opens a fascinating prospect for future research but perhaps this is a subject for future lectures.

References

1. Aitken, C.G.G., et al.: Bayesian belief networks with an application in specific case analysis. In: Gammerman, A. (ed.) Computational Learning and Probabilistic Reasoning, pp. 169–184. John Wiley & Sons, Chichester (1996)
2. Gammerman, A., Thatcher, R.: Bayesian diagnostic probabilities without assuming independence of symptoms. Methods Inf. Med. **30**(1), 15–22 (1991)
3. Gammerman, A., Vovk, V.: Hedging predictions in machine learning (with discussion). Comput. J. **50**(2), 151–163 (2007)
4. Gammerman, A., Vovk, V., Vapnik, V.: Learning by transduction. In: Proceedings of the Fourteenth Conference on Uncertainty in Artificial Intelligence (UAI98), University of Wisconsin Business School, pp. 148–155, Madison, Wisconsin, USA, July 24-26 (1998). (also Technical Report CSD-TR-96-21, CS RHUL, 1996)
5. Giovannotti, P., Gammerman, A.: Calibrated large language models for binary question answering. Proc. Mach. Learn. Res. **230**, 218–235 (2024)
6. Lauritzen, S.L., Spiegelhalter, D.J.: Local computations with probabilities on graphical structures and their application to expert systems. J. Royal Stat. Soc. Seri. B **50**(2), 157–224 (1988)
7. Luo, Z., Gammerman, A., Aitken, C.G.G., Brewer, M.: Exact and approximate algorithms and their implementations in mixed graphical models. In: Gammerman, A. (ed.) Probabilistic Reasoning and Bayesian Belief Networks, pp. 33–53. Alfred Waller, Henley-on-Thames (1995)
8. Pearl, J.: Probabilistic Reasoning in Intelligent Systems, Morgan-Kaufmann (1988)
9. Pomerleau, D.A.: ALVINN: an autonomous land vehicle in a neural network. Proc. Neural Inf. Process. Syst. (NeurIPS) **1**, 305–313 (1989)
10. Rissanen, J.: Universal coding, information, prediction, and estimation. IEEE Trans. Inf. Theory **30**(4), 629–636 (1984)

11. Saunders, C., Vovk, V., Gammerman, A.: Transduction with confidence and credibility. In: Proceedings of the International Joint Conference on Artificial Intelligence, vol. 2, pp. 722–726, Stockholm, Sweden (1999)
12. Turing, A.: Computing machinery and intelligence. MIND **59**, 433–460 (1950)
13. Vapnik, V.: Nature of Statistical Learning Theory. Springer (1995)
14. Vovk, V.: Aggregating strategies. In: Proceedings of the 3rd Annual Workshop on Computational Learning Theory, pp. 371–383 (1990)
15. Vovk, V., Gammerman, A., Shafer, G.: Algorithmic Learning in a Random World. Springer, 1st edn. (2005). 2nd edn. (2022)
16. Vovk, V., Saunders, C., Gammerman, A.: Machine learning applications of algorithmic randomness. In: Proceedings of the Sixteen International Conference on Machine Learning (ICML'99), pp. 444–453 (1999)

Life in Computer Science at Heriot-Watt University 1983-1993

Greg Michaelson[(⊠)]

Emeritus Professor of Computer Science, Heriot-Watt University, Edinburgh, UK
`G.Michaelson@hw.ac.uk`

Abstract. This chapter provides a brief reminiscence of Alex Gammerman's time at Heriot-Watt University in Edinburgh, between 1983 and 1993.

1   Heriot-Watt University

In Edinburgh, when people refer to "The University", they are widely understood to mean the University of Edinburgh, founded after the Reformation in 1583, the youngest of the six original UK universities [3]. In fact, there are four universities in Edinburgh: the other three are Heriot-Watt University, Edinburgh Napier University and Queen Margaret University.

Heriot-Watt University, the second oldest in Edinburgh, was established in 1966, after a long, and at times problematic, relationship with its elder sibling. The University's roots go back to the foundation of the Edinburgh School of Arts in 1821. The first Mechanics' Institute in the world, the School offered unrivaled technical higher education to working class men. This was driven both by the need to provide a skilled workforce as industrialisation accelerated, and to channel potentially disruptive energies at the end of the Napoleonic Wars [8].

The School of Arts was renamed the Watt Institution and School of Arts in 1852, after the Watt Subscription Fund donated money to the School, as Edinburgh's monument to James Watt, who developed the steam engine regulator. Women were first admitted in 1869. In 1885 it became the Heriot-Watt College, after it was merged with the Heriot's Trust, endowed from the money amassed by George Heriot, the late 16th/early 17th century goldsmith and money lender [8].

The College offered an outstanding technical education, and, while it did not have degree awarding powers, its graduates received a degree-equivalent qualification: the Associate of Heriot-Watt College (AH-WC). Furthermore, the College was the de facto engineering faculty of the University of Edinburgh, whose students could take College courses towards their degrees [8].

In 1963, the Robbins Report proposed a major expansion of UK higher education, involving initiating new universsities, and granting advanced institutions university status. The Royal College of Science and Technology in Glasgow, with similar roots to Heriot-Watt, and a longstanding rival, had already become Strathclyde University in 1962. Heriot-Watt College had been offered affiliation

K. An Nguyen and Z. Luo (Eds.): Alexander Gammerman Festschrift, LNCS 16290, pp. 46–55, 2026.
https://doi.org/10.1007/978-3-032-15120-9_3

with Edinburgh, but preferred to go its own way, and became a University in 1966. There was, however, widespread skepticism about whether Heriot-Watt and Strathclyde were truly Universities, as they lacked the full range of traditional faculties [8].

The College, and then the University, was centered on the sprawling Chambers Street complex in central Edinburgh, with the main building first opened in 1873. The University's modernity may be seen in the busts of the founders of phrenology around the bay window on the west of the building, which now houses the Edinburgh Sheriff Court and Justice of the Peace Court. However, there was limited space on the site, and the College, and then the University, expanded into markedly less salubrious buildings in the Grassmarket, directly under the Castle. This was where, eventually, the Computer Science Department and the Computer Centre were based [8].

Heriot-Watt's first computer, a Ferranti Sirius, was installed in 1962, under the joint control of the Departments of Mathematics and Electrical Engineering, perhaps the origin of damaging rivalry between them. The BSc Computer Science programme, one of the first full undergraduate provisions in the UK, was started by Mathematics in 1966. The Computer Science Department, which had been the Computer Unit, became independent in 1967 under Alex Balfour. As well as the BSc, the Department offered an MSc in Computer Science, from 1970, and contributed to a BSc in Accountancy and Computing, which began in 1973 [10].

2 Computer Science and Artificial Intelligence in Edinburgh

In the UK, Computer Science and Artificial Intelligence have tangled, and often antagonistic, histories, no more so than at the University of Edinburgh.

The University of Edinburgh's Computer Unit was founded in 1963 by Sidney Michaelson, a computer pioneer from Imperial College in London, and my dad. A postgraduate Diploma in Computer Science was introduced in 1964, and a first year undergraduate course was started in 1965, eventually leading to a full undergraduate programme in the early 1970s. In 1966, the Computer Unit was split into the Computer Science Department, led by Michaelson, and the Edinburgh Regional Computer Centre (ERCC). ERCC was supposed to service all the east of Scotland Universities, but, in practice, it mostly supported Edinburgh [2].

In 1971 the Edinburgh Department and the ERCC both moved to the new James Clark Maxwell Building, on the King's Buildings science and engineering campus, to the south of the City. Edinburgh Computer Science grew and became world leading, rapidly overshadowing Heriot-Watt, which remained small until the 1980s. As Heriot-Watt Computer Science lingered in the City centre, regular informal contact between the Departments was curtailed, though both employed each other's graduates and staff [2].

Edinburgh was one of the first UK centres of Artificial Intelligence (AI), and quickly gained a leading international reputation. The history is rather complex, marked by a proliferation of smaller units, which slowly melded together.

Donald Michie, who had worked with Alan Turing at Bletchley Park, came to Edinburgh as a research immunologist in 1961, and developed a strong interest in AI. He set up a small computer research group in 1963, which became the Experimental Programming Unit in 1965. In 1967, he was made Professor of Machine Intelligence. In 1968, the University established the Department of Machine Intelligence and Perception, with three research groups, Michie's Experimental Programming Unit, the Bionics Research Laboratory under Richard Gregory, and the Theoretical Section under Christopher Longuet-Higgins. The University also established the separate Metamathematics Unit, concerned with automated theorem proving, under Bernard Meltzer, who had translated Gödel's work into English. In 1971, the School of Artificial Intelligence was formed from the components of the Department of Machine Intelligence and Perception, along with the Metamathematics Unit, now renamed the Department of Computational Logic. Furthermore, in 1969, the cross disciplinary School of Epistemics was created, renamed the Centre for Cognitive Science in 1982 [2].

It is important to note that Edinburgh AI was focused on symbolic computing and robotic engineering, with relatively little activity in machine learning (ML). This may be because Minsky and Papert's robust criticism of neural network research [6], an important focus for AI ML activity, furthered a split in the AI community between "computationalists" and "connectionists", with Edinburgh primarily in the "computationalist" camp.

Alas, UK AI, like AI world-wide, failed to deliver the over-hyped promises of its protagonists. In 1972, the Science Research Council's highly critical Lighthill Report severely hindered AI's progress in the UK, by curtailing research funding [7]. This was particularly damaging at Edinburgh, where provision was heavily research based. In a major reorganisation, Machine Intelligence under Michie became an independent Research Unit, and the Artificial Intelligence Department became part of a new School of Computer Science and Artificial Intelligence. However both Departments remained wholly separate. In what Birse terms "the downward spiral in the [AI] Department's fortunes" (p177), many senior AI staff left, some to Computer Science [2]. The whole process was complicated by long-standing personal animosity between Michie and Michaelson.

AI's component groups were housed in a variety of buildings in central Edinburgh, quite close to Heriot-Watt Computer Science. When Edinburgh Computer Science moved to the King's Buildings, AI stayed put. Thus, there was ongoing informal contact between Heriot-Watt Computer Science and Edinburgh AI staff, particularly over lunch.

In 1983, the fortunes of UK AI were revived by the government's Alvey Programme, a response to the Japanese 5th Generation Programme [7]. However, Michie had left Edinburgh for Glasgow, where he established the Turing Institute under the aegis of the University of Strathclyde [11].

3   79 Grassmarket

When the University was formed, it was planned for it to be rehoused on a green field site at Riccarton, to the west of the City. However, ongoing national financial problems through the 1970s, as the UK's steady economic decline was accelerated by free-market Thatcherism, meant that the move was piecemeal, as funds became available. Thus, when Alex started at Heriot-Watt University, in 1983, CS was still housed in a curious collection of buildings at 79 Grassmarket.

The Grassmarket runs east-west along the valley at the base of Edinburgh Castle. In CS's time, it had a reputation as one of the roughest areas in the Old Town, lined with pubs on the north side, and with hostels for homeless people on the south side. During the day, it was bustling and busy: at night, dark and forbidding to the unwary, though perfectly safe. Today, the entirely inoffensive homeless folk have been decanted into what passes for care in the community, and the Grassmarket, now thoroughly up market, is well lit, almost traffic free, and thronged with tourists year round.

Towards the west end of the Grassmarket was the 1969 Mountbatten Building [4], built on the site of the old Corn Exchange [5]. A plaque still marks where the Corn Exchange had been hit by a bomb from a Zeppelin in April 1916. Before the move to Riccarton, Mountbatten housed the Department of Electrical and Electronic Engineering (EEE), as well as Accountancy and Finance, Business Organisation, Economics, and Languages, as well as a library and the TV Centre. Curiously, EEE had installed the second Unix system in the UK, in 1974 [9]. Today the building is a classy hotel with outstanding views of the Castle [8].

The Computer Service was next door at 37 Grassmarket, the former Heriot Brewery [4]. This housed the Burrough 5700 mainframe computer, replacing the ICL 4130 that had replaced the Ferranti Sirius, and serving the whole University with punched card input: latterly VDUs [10].

79 Grassmarket had been the site of the Castle Brewery [4], and of a Female Shelter (Magdaline Asylum) [5], for the incarceration of unmarried mothers. It was taken over by the College in 1928 and redeveloped for the teaching of Mining [8].

The buildings were arranged around a courtyard, abutting onto the Greyfriars' Church graveyard and George Heriot's School on the south side. Accommodation for Pharmacy was added in 1964 [8]. Entry from the Grassmarket was up Hunter's Close and through an archway, which was closed off at night by two large padlocked iron gates, see Fig. 1.

The old lecture theatre on the north side of the courtyard had an octagonal glass turret in the ceiling, with pulleys to lower a metal plate to shut out the light during use of the brass epidiascope. We used chalk and overhead projectors. Below this lecture theatre were mocked up mine galleries which had been used for mine rescue training, long kept closed.

In front of the building stood the "Old Ben" steam engine shown in Fig. 2, built by Walkinshaw in the 1790s for colliery use, taken out of service in 1956, and currently on loan to the National Mining Museum in Newtongrange [1]. In the later 1980s, the Dean of Science and Engineering instigated the production of a video to promote the Faculty's cognate Departments. I wrote the script. In the final cut, the words "...with modern computer facilities..." were read over a shot of Old Ben. The video was never released.

Fig. 1. 79 Grassmarket entrance [GJM]

The Department was based in Building 6 on the western side of the courtyard. Maintenance staff and store rooms were on the ground floor. The first floor was mostly computer laboratories and staff accommodation, with the computer room at the far end. The second floor had most of the staff accommodation, the Departmental library and office, the common room, and a raked lecture theatre. Many of the rooms had corridor walls with translucent glass in brown wooden frames. The whole building was scrupulously clean and well appointed, but otherwise unloved. Substantial improvements were deemed pointless, even as the move to Riccarton was endlessly deferred.

Computer provision was very good, if eclectic. All Departmental staff had VDUs on their desks, which could be connected to an arbitrary minicomputer running Unix. All VDUs and machines were linked to a plugboard in the machine room. To change from one machine to another, the VDU link had to be physically changed, provided there was a free connection. Machines were connected via uucp, but each had separate storage, and, with the mix of CPUs, it was necessary to recompile code, so copies of sources proliferated. In 1983, there were a PDP 11/45 and two Z8000 based machines. Computer Science student access was via labs of VDUs.

Fig. 2. Old Ben.

In Building 6, there was also a lab full of BBC Micros used to teach non-specialist students. These were on fixed wooden lab benches which still had gas taps and sinks.

The other buildings in 79 Grassmarket were occupied by the Department of Pharmacy, and Computer Applications Services (CAS), a CS based technology transfer service.

In 1983, there were ten academic CS staff, with Professor Howard Williams as Head of Department, appointed in 1980. There were also two secretaries and three technical support staff.

However, the Department began to expand. In the wake of the Japanese 5th Generation Programme, but before Alvey, Howard had led the introduction of an MSc in Knowledge Based Systems, which had gained funded places from the Science Research Council (SRC).

Further, the University Grants Commission (UGC), the body that funded all UK universities, had short-sightedly decided to close the Department of Pharmacy, and the funds were to be redistributed to programmes in optoelectronics in Physics, and in IT between EEE and CS, a source of considerable contention. Nonetheless, these all enabled staff increases and the occupation of the Pharmacy, and then the CAS, buildings on the north side of the courtyard.

In the next two years, Howard was fortunate to gain three Alvey programme grants, concerned with logic databases and hardware, with support from the UK's leading computer company International Computers Limited. Mike Norman, who'd joined the Department in 1982, gained two further Alvey grants, to set up the Scottish Man-Machine Interface Centre. Funding was jointly with Jim Alty at the University of Strathclyde, and the Centre had sites in Edinburgh and Glasgow. Shortly thereafter, an MSc in Human Computer Interaction was introduced, again with funded places. The grants brought postdocs, and the MSc programmes proved a steady source of strong PhD students.

Around that time, Professor Fred Heath transferred from EEE. Fred, a computer pioneer, ran a startup company called Memex, which made search engine hardware that was sold to libraries and newspapers. He also had a collection of player pianos.

4   Alex at Heriot-Watt

Alex arrived from the USSR in the spring of 1983. That autumn, I was appointed from the University of Glasgow alongside Chris Miller from the University of Leeds[1], and Alex, some of whose publications had been translated into English. My interest was in functional programming, Chris' in logic programming, and Alex's in pattern recognition, so we all complemented well the KBS MSc.

Alex recalls coming for interview in mid-September of 1983. It was a very warm September day and the interview was in the afternoon. So, he went to a Grassmarket's pub and had a pint to cool himself off. Then he went to the Meadows where he dozed off. When he woke up there were only a few minutes before the interview, so he ran to the 79 Grassmarket, and arrived sweaty and smelling of beer. Nonetheless, despite confusion over which University in Edinburgh was interviewing him, Alex got the job.

Alex remembers Heriot-Watt Computer Science as "a brilliant place". There was strong camaraderie amongst staff, which Alex enjoyed. The longstanding weekly highpoint was the Friday "drinkers" excursion to a nearby hostelry, originally The Fiddlers' at the west end of the Grassmarket, and latterly The Bow Bar on Victoria Row, also frequented by lawyers.

79 Grassmarket had a very limited car park, so when Alex arrived in the morning, it wasn't easy to find a parking space. Courtyard parking was managed by Harry, a Scottish maintenance man who was not infrequently drunk. Observing Alex's problems, Harry would observe: "Pressure of work, Mr Gammerman? No time to find parking?" So Alex, in typical Russian style, bribed him with a box of very expensive Cuban cigars[2]. After this, Alex had no more problems, as Harry acted as his personal parking valet.

The new Mountbatten Building, to house both EEE and CS at Riccarton, was finally completed in 1992, see Figs. 3 and 4.

Alas, shortly after CS left the Grassmarket, it was announced that the Department was to be rationalised into a new Department of Computing and Electrical Engineering. The key outcome was that CS lost all autonomy, despite, or perhaps because of, sustained opposition from staff to the merger. So, Alex and Sue decided that this was a good time for their burgeoning family to move south, and Alex became Professor of Computer Science at Royal Holloway, University of London.

5   Alex and I

Alex and I got on well from the start. As well as a quizzical view of the world, we shared a love of arguing to establish a position, rather than to score points.

---

[1] One of Chris and my first achievements was successfully lobbying for a modem to link CS to the burgeoning Internet, via Leeds.

[2] The cigars were originally given to Alex by his father-in-law, who got them from a Burmese man who was a member of an official delegation from Burma. Later Alex found out that the man was accused by his government of corruption and executed by firing squad.

Fig. 3. Goodbye Grassmarket: Alex and colleagues, 5th May 1992 [GJM].

Fig. 4. Goodbye Grassmarket: Howard and Alex, 5th May 1992 [GJM].

Alex seemed to take an anthropological view of the British in general, and the Scots in particular. Much of what passed for a way of life was irritating, and much was puzzling. When he didn't get some nuance, he would comment ruefully that he was just a "bloody foreigner". I felt some sympathy, as the bearer of an English accent in Scotland, in a period when nationalism was ascendant, and "English go home!" was not uncommon graffiti.

While I had never known anyone from the Soviet Union, I had a rather more sympathetic view of it than many Brits, having been brought up by parents who had been communists before I was born. So, for me, it was engaging to discover more about what ordinary life was like there. And I think Alex enjoyed talking with someone who was neither rabidly anti-Soviet, right or left, or a useful idiot, apologising from the safety of an armchair.

Alex and I also shared Jewish heritage, mine through my dad. Further, my grandfather was from Melitopl in Ukraine. But I don't think these were significant for our friendship, other than adding to some sense of dislocation in a Christian, nay Presbyterian, world, as we had both had thoroughly secular upbringings.

In 1984, my partner Nancy and I visited the Soviet Union on an Intourist package tour, a holiday we'd planned before we met Alex. At Moscow airport, I was taken to one side, and questioned by a bored looking official. When he asked me if I had any contacts there, I pulled out the piece of paper on which Alex had written "Dom Knigi, Leningrad" - the house of books, where I might find English language texts. This seemed to satisfy them.

The Soviet Union felt thoroughly ordinary. Memorably, the food was awful, and a young couple, on what seemed like an eccentric choice of honeymoon venue, were hospitalised with food poisoning. Still, I was impressed by the uniformity of standard of living in Moscow, Leningrad, and Trans Caucasian cities, quite unlike the stark British class contrasts the Grassmarket exemplified. I was also impressed by how safe it felt. We were largely left to our own devices, and each time we got lost, local folk were unfailingly friendly and helpful.

My biggest disappointment was how hard it was to see the vibrant revolutionary period art and design I only knew from books. All the relevant galleries seemed to be closed, and we were urged to go and look at Tsarist period and Western European paintings. It was almost as if they were ashamed of their past, or worried that its spirit might be rekindled.

On our return, during one discussion with Alex, I recounted how, on every floor of our Leningrad hotel, there was a VDU next to the lift, with a screen but no keyboard. Alex was most amused, and said I now understood the problems of central planning.

Fig. 5. Alex and Nancy, 1992 [GJM].

And, Alex, like me, was something of an AI skeptic. Memorably, he once remarked to me that Perceptrons, a prominent predecessor of contemporary machine learning models, were just bad statistics.

While we never worked directly with each other, nonetheless, we enjoyed lots of social time together, with our partners, eating, drinking, and putting the world to rights (see Fig. 5). We were sad when Alex and Sue left; I'm honoured to have been asked to assemble this account of my good friend's life at Heriot-Watt.

Acknowledgments. I am grateful to Alex for permission to use details of his time at Heriot-Watt University. Without his anecdotes, this would be a dull account.

Our friends, and former colleagues, Peter King, Nick Taylor and Howard Williams have kindly provided many details of Heriot-Watt Computer Science.

References

1. Allen, C.: NT2573 : Steam engine "Old Ben". geograph (2008). https://www.geograph.org.uk/photo/663873
2. Birse, R.M.: Science at the University of Edinburgh 1583-1993. University of Edinburgh (1994)
3. Curtis, S.J.: History of Education in Great Britain. University Tutorial Press (1967)
4. Gifford, J., McWilliam, C., Walker, D.: The Buildings of Scotland. Penguin Books, Edinburgh (1984)
5. Godfrey, A.: Old Ordnance Survey Town Plans. Edinburgh Castle 1877. A. Godfrey (1984)
6. Minsky, M., Papert, S.: Perceptrons. MIT Press (1969)
7. Montealegre, A.A.: AI in the UK: ready, willing and able? Appendix 4: Historic Government policy on artificial intelligence in the United Kingdom. Artif. Intell. Committee (2017). https://publications.parliament.uk/pa/ld201719/ldselect/ldai/100/10018.htm
8. O'Farrell, P.N.: Heriot-Watt University. An Illustrated History, Pearson (2004)
9. Salus, P.H.: A Quarter Century of UNIX. Addison-Wesley (1994)
10. Taylor, N.: Department of Computer Science, Heriot-Watt University. School of Mathematical and Computer Sciences, Heriot Watt University (2015). https://www.macs.hw.ac.uk/cs/History/History.html
11. Wikipedia. Turing Institute. Wikipedia (2025). https://en.wikipedia.org/wiki/Turing_Institute

Intelligent Data Analysis

Xiaohui Liu$^{(\boxtimes)}$

Department of Computer Science, Brunel University of London, Uxbridge, UK
`xiaohui.liu@brunel.ac.uk`

Abstract. Artificial intelligence, data science, and optimization are three distinct scientific disciplines, each shaped by its own history, methodologies, and guiding principles. Over the past few decades, however, these fields have increasingly converged, much like the three musketeers joining forces to confront the challenges posed by big data in the modern world. Together, they have laid the foundation for Intelligent Data Analysis (IDA), an interdisciplinary field dedicated to the effective and trustworthy analysis of data. In this chapter, I offer a personal reflection on the evolution of IDA, highlight key milestones in its development, and examine the opportunities ahead.

Keywords: Artificial intelligence · data science · optimization · intelligent data analysis · trustworthy decision making

1 Introduction

In an era of unprecedented uncertainty, the ability to make sense of complex, large-scale data for trustworthy decision-making has become both critically important and increasingly challenging. Whether in science, policy, or everyday life, we are constantly faced with decisions that depend not only on data, but also on our capacity to interpret that data meaningfully and responsibly.

Over the past few decades, data analysts have drawn on a rich array of methods from artificial intelligence (AI), data science, and optimization to meet this challenge. Each of these scientific disciplines has its own history, methodologies, and guiding principles, yet they have increasingly converged, joining forces to confront the demands of big data in the modern world. This convergence has not only enabled innovative, real-world solutions but has also enriched each field through the cross-fertilisation of ideas and techniques. Together, they have laid the foundation for *Intelligent Data Analysis* (IDA), an interdisciplinary paradigm dedicated to the effective, reliable and ethically grounded data interpretation.

In this chapter, I reflect on IDA's evolution, highlight its key milestones, and explore the opportunities that lie ahead. It is especially fitting to do so in honour of Alex Gammerman's 80th birthday, a pioneering scholar whose work has shaped the intellectual and methodological foundations of trustworthy data analysis. Across a distinguished career spanning several decades, Alex has not only advanced the frontiers of IDA but also inspired generations of researchers – myself included – to pursue innovation and rigour in equal measure.

K. An Nguyen and Z. Luo (Eds.): Alexander Gammerman Festschrift, LNCS 16290, pp. 56–74, 2026.
https://doi.org/10.1007/978-3-032-15120-9_4

2 Background

In this section, I present a brief history of data science, optimization and artificial intelligence, three key pillars in the evolution of intelligent data analysis.

2.1 Data Science

Humans have been analysing data for as long as data has existed, but the practice began to take on a more formal and systematic character in the 17th century with the emergence of probability theory and the early foundations of what we now recognize as statistics. The pioneering work of Pierre de Fermat and Blaise Pascal marked the birth of probability theory as a mathematical discipline, while Thomas Bayes and Richard Price later introduced a complementary, Bayesian interpretation of probability.

Around the same time, John Graunt published his seminal study on mortality patterns [1], widely regarded as the first statistical analysis of public health data. His work marked the beginning of modern demography and helped define statistics as a distinct area of inquiry.

The 19th century and the first half of the 20th century are often seen as a golden era in the development of statistics as a scientific discipline. Pioneering contributions by British mathematicians Francis Galton, Karl Pearson, and Ronald Fisher [2] established many of the field's key mathematical foundations. Pearson, notably, founded the world's first department of statistics at University College London, institutionalizing the field within academia.

The advent of electronic computing was a transformative moment for data analysis. It enabled the development and application of a new class of statistical techniques that had previously been infeasible, including advanced time series modelling [3] and Markov Chain Monte Carlo methods. The same computational advances also paved the way for breakthroughs in machine learning, evolutionary optimization, Bayesian networks, and knowledge discovery in databases.

Among the key figures bridging traditional statistics and modern data science was John Tukey. He championed the idea of Exploratory Data Analysis [4] as a complement to Confirmatory Data Analysis, reshaping analytical practices and emphasizing the importance of understanding and exploring data before applying formal models. Tukey's vision laid important conceptual foundations for what we now call data science, and his influence can be seen in subsequent developments in data mining, cloud analytics, and the broader big data movement.

2.2 Optimization

Like data science, the origins of optimization can be traced back centuries. One of the foundational developments came in the 17th century with the invention of calculus by Isaac Newton and Gottfried Leibniz, which provided the mathematical framework necessary for continuous optimization. Another important but often overlooked contributor is Charles Babbage. Though best known for his work on the Analytical Engine, recent studies have highlighted his significant influence on early optimization practices, particularly in industrial settings.

The two World Wars provided fertile ground for the development and practical application of optimization techniques. During both World War I and World War II, optimization methods proved invaluable in logistics, military planning, and operational efficiency. These efforts laid the groundwork for a more formal study of the field. A major milestone came in the 1940s with the independent development of linear programming by George Dantzig in the United States and Leonid Kantorovich in the Soviet Union. Their work marked the beginning of modern optimization theory and had a profound impact on economics, engineering, and operational research.

Following this, the field expanded to include a broader set of techniques and applications, evolving into what became known as mathematical programming. This broader domain included nonlinear programming, integer programming, and other specialized optimization approaches. It also formed the mathematical core of operational research, a discipline that also integrated simulation, decision analysis, and heuristic methods for real-world problem solving. One of the key figures in the establishment of operational research as a field was Patrick Blackett, whose wartime work demonstrated the power of analytical decision-making [5].

Beginning in the 1960s and gaining momentum in the following decades, researchers began to draw inspiration from nature to develop a class of techniques now known as metaheuristics. These include genetic and evolutionary algorithms [6], inspired by biological evolution; simulated annealing [7], modelled after the physical process of metal cooling, and particle swarm optimization [8], based on the collective behaviour of bird flocking or fish schooling.

These approaches offered powerful tools for tackling complex, non-convex, or poorly defined problems that resisted classical mathematical optimization techniques. Today, metaheuristics play a key role in solving a wide variety of industrial, engineering, and artificial intelligence problems.

2.3 Artificial Intelligence

The concept of creating machines emulating human intelligence traces back to ancient times, appearing in myths, automata, and philosophical thought. However, the formal foundations of AI emerged in the 20th century, with Alan Turing as one of its pivotal figures. In his seminal 1950 paper, Turing asked, *"Can machines think?"* and introduced the Imitation Game (later called the Turing Test), proposing that a machine's ability to exhibit human-like conversation could serve as a benchmark for intelligence [9]. This framework remains influential in AI research today.

The field of artificial intelligence was established as a formal discipline in 1956 during the historic Dartmouth Conference, where key figures such as John McCarthy, Marvin Minsky, Allen Newell, Claude Shannon, Herbert Simon and Ray Solomonoff[1] laid its intellectual foundations. Buoyed by early successes and optimism, researchers believed that human-level AI might soon be within reach. Symbolic approaches, notably

---

[1] Renowned for his foundational work in Algorithmic Information Theory and Kolmogorov-Solomonoff Complexity, Ray Solomonoff was the first recipient of the Kolmogorov Award from Royal Holloway, University of London. Invited by Alex Gammerman as a Visiting Professor, he delivered the inaugural Kolmogorov Lecture in 2003.

Newell and Simon's *Logic Theorist* [10], showcased the ability of machines to perform logical reasoning in well-defined environments. However, these systems proved brittle when confronted with the complexity of real-world problems, gradually leading to a loss of confidence and in part to the onset of the so-called "AI winters" during the 1970s.

A resurgence came in the 1980s with expert systems (e.g., MYCIN for medical diagnosis) and the rise of connectionism. The refinement of backpropagation [11] enabled efficient training of multi-layer neural networks, reviving the 1940s vision of the mcculloch-pitts neuron and laying groundwork for modern machine learning. Moreover, the 1990s and early 2000s saw the development of powerful statistical and machine learning methods, including support vector machines, random forests, Bayesian networks, and reinforcement learning. These approaches expanded the capabilities of AI systems and broadened their application across industries.

The deep learning revolution of the 2010s, powered by big data and GPU computing, transformed AI once again. Landmark achievements, such as AlexNet's success in the ImageNet competition and DeepMind's AlphaGo defeating a world Go champion, demonstrated the unprecedented performance of deep neural networks on complex perceptual and strategic tasks.

Today, AI spans machine learning, natural language processing, robotics, computer vision, generative models, and innovative applications [12], while also confronting ethical and societal challenges—many foreshadowed by Turing's early inquiries. As the field advances, it continues to grapple with questions that bridge technology, philosophy, and human values.

3 Foundational Years (1984–1993)

In this section, I reflect on my personal journey as an early-career researcher and explore the path toward intelligent data analysis.

3.1 PhD Training

I began my PhD in AI at Heriot-Watt University's Department of Computer Science in 1984, under the supervision of Alex Gammerman. Nestled in the heart of the city's historic Grassmarket, with the majestic Edinburgh Castle looming above, the setting was both intellectually stimulating and visually inspiring. Our proximity to the University of Edinburgh's AI Department provided access to world-class seminars and research discussions, as well as a steady flow of technical reports from major AI centres around the globe.

The early 1980s were an interesting era in AI. Symbolic approaches such as expert systems and logic programming still dominated, but connectionist models were beginning to gain serious momentum. Seminal contributions like Hopfield Networks [13] and Self-Organizing Maps [14] signalled a renewed interest in neural networks, foreshadowing the revival triggered by the backpropagation algorithm in 1986 [11].

Initially, I was fascinated by the concept of artificial neural networks, particularly Hopfield Networks. This enthusiasm led me to make a special trip to the British Library in London to photocopy Hopfield's original paper, which inspired me to carry out a series of experiments. While these networks could store and retrieve patterns like a content-addressable memory, scaling them to more complex tasks proved difficult, especially given the constraints of the PDP-11 and mini-VAX machines I was using at the time. Progress was slow, and with mounting funding pressures, I ultimately abandoned this line of investigation. Still, this marked my first hands-on encounter with neural networks, laying a foundation that would later influence my life-long interest in the topic and research on deep learning.

At a critical juncture, Alex suggested that I explore the management of uncertainty in knowledge-based systems, a field where he was already an established leader. That advice was transformative. Under his guidance, I transitioned into this area and became particularly interested in representing uncertainty using probability intervals rather than single-point estimates, a direction that offered both conceptual richness and practical utility for AI inference systems. This early work planted the seeds for my subsequent interest in managing data quality and learning from imperfect data.

Looking back, one of the most formative aspects of my PhD experience was Alex's open-door policy. I could walk into his office at any time to discuss ideas, without appointments or formality, a privilege I only came to truly value when I became a supervisor myself. His consistent encouragement and intellectual generosity supported me at every stage of the research process: from idea generation and planning to evaluation and writing. In every sense, Alex's mentorship was invaluable and central to my development as a researcher.

3.2 Statistics no Longer Sufficient for Data Analysis

After completing my PhD, I undertook postdoctoral research in two engineering departments, first at Heriot-Watt University in Roy Leitch's lab, and later at Durham University under the supervision of Michael Sterling. Both Roy and Michael provided not only generous mentorship and support but also the freedom to pursue my own research interests. Most importantly, they offered invaluable opportunities to work directly with real-world data. During this period, my research focused on analysing industrial process data for the diagnosis and control of manufacturing systems, as well as time series data for forecasting power and water demands. These projects, conducted in close collaboration with leading manufacturing and energy companies across the EU and UK, firmly rooted my work in practical, data-driven problem solving.

Later, during my early academic career at Birkbeck, University of London, I led several applied data analysis projects, e.g., in the context of optic nerve disease screening in close collaboration with Barrie Jones's team at Moorfields Eye Hospital and the identification of variant haemoglobins with Sally Davies's group at Central Middlesex Hospital, both in London. Across all these projects, a consistent pattern emerged: there had been a heavy reliance on traditional statistical techniques and legacy software environments. While these methods had long been the backbone of data analysis, they were increasingly strained by the demands of modern applications and advanced computational techniques were being investigated as a result.

Two key developments were driving this shift. First, real-world datasets were rapidly growing in scale and complexity, challenging the assumptions and scalability of classical statistical methods. Second, a new generation of computational tools and techniques was gaining ground: data warehousing, high-performance computing, machine learning, evolutionary optimisation, and advanced visualisation technologies, among others. These approaches offered unprecedented analytical power and adaptability.

Taken together, these developments marked a turning point. What emerged was a broader realisation: traditional statistics, while foundational, was no longer sufficient on its own. The future of data analysis would lie in combining statistical thinking with advances from AI, optimisation, and computing to meet the growing challenges of the data-rich world.

4 Origins of Intelligent Data Analysis (1994–2003)

In this section, I explore the origins of IDA, describe its early development, and highlight its interdisciplinary ethos, particularly the synergies between computer science, statistics, and applied domains.

4.1 AAAI-94: A Pivotal Moment

A pivotal moment in the conception of IDA came when I presented a paper in AAAI-94, the 12th US National Conference on Artificial Intelligence, held at Seattle in 1994. The paper focused on "outlier analysis", but it was artificially packaged in the more traditional AI language of "uncertainty management" in the title to better align with the conference's themes [15]. Wherever possible in the methodology, I opted for *neural networks* over more conventional *statistical* methods. At the time, I wasn't proud of these small academic contortions. But that experience sparked an idea: what if we could create a conference that welcomed interdisciplinary approaches to data analysis—where AI, statistics, and beyond could gather on equal footing, without the need for such disguises?

At that same conference, I had the pleasure of meeting Vladimir Vapnik. Over a long and engaging conversation, he generously shared the challenges he faced in gaining recognition for his early work on statistical learning theory [16] and the subsequent development of Support Vector Machines, both in the US and internationally. His curiosity about European and UK academic culture led me to suggest he connect with a fellow countryman who was then Chair Professor of the Computer Science Department at Royal Holloway, University of London.

Returning from AAAI-94, I acted quickly on three fronts. First, I began preparations for what would become the inaugural IDA symposium (IDA-95). Second, I set up a research group on IDA at Birkbeck. Third, I shared my ideas and experiences with Alex, who offered his wholehearted support. He remarked that Vladimir was among the top five statisticians in the Soviet Union. Within a year, Vladimir was invited as a keynote speaker at ADT-95, an event Alex co-organised under UNICOM Seminars at the Brunel Science Park and was soon appointed Professor of Computer Science and Statistics at Royal Holloway in 1995.

4.2 IDA-95 and Beyond

Having decided to launch an interdisciplinary conference that would welcome researchers from multiple domains relevant to data analysis, two immediate decisions needed to be made: what to call the event, and where to hold it.

When considering the name, it was clear that *data analysis* had to be at its core, but something more was needed to capture the evolving nature of the field, particularly the rising influence of artificial intelligence at the time. That was when the phrase "Intelligent Data Analysis" came to my mind. The term "intelligent" was especially appealing: it could signify the integration of AI and other relevant disciplines alongside statistics in performing data analysis. It could also suggest intelligent *ways* of conducting analysis, or the use of intelligent *systems* for analysing data. And, quite simply, "IDA" had a memorable and distinctive ring to it, made even more charming when someone noted its connection to *Princess Ida*, the title of a Gilbert and Sullivan opera from the 19th century. And so, IDA it was.

As for the venue, an opportunity arose shortly after AAAI-94. I was attending the International Conference on Systems Research, Informatics and Cybernetics in Baden-Baden, organised by George Lasker from the University of Windsor, Canada. During the event, there was an open invitation to run independent symposiums or workshops alongside the main conference. The arrangement was highly attractive: symposium organisers had full control over the scientific programme, while the host conference managed logistics and finances. It felt like an ideal starting point, and I seized the opportunity.

With excellent support from Alex Gammerman, Vladimir Vapnik, and other members of the international programme committee that also included Paul Cohen and Doug Fisher (USA), Henri Prade (France), Colin Shearer (UK), and Hans-Jürgen Zimmermann (Germany), IDA-95 took place in Baden-Baden from 17 to 19 August 1995. It attracted over 60 participants from more than 20 countries, representing both academia and industry. Disciplines ranged from computer science and statistics to engineering and biomedical research. The papers presented spanned topics from novel data preprocessing techniques to machine learning methods, and from model selection and evaluation to early explorations of human–machine collaboration. The symposium was marked by stimulating discussions and a palpable sense of shared purpose (see [17] for a summary).

A post-event survey strongly endorsed the idea of making IDA a recurring conference. And so it became. Time has flown—we celebrated the 30th ANNIVERSARY of IDA at IDA-2025 in Konstanz, Germany, a testament to the conference's enduring contributions to the global research community. Since its inception, the IDA series has been hosted in over ten countries across Europe, the UK, and the USA, fostering international collaboration and interdisciplinary exchange. Its proceedings have been consistently and proudly published in Springer's *Lecture Notes in Computer Science* series, from Volume 1280 in 1997 to Volume 15669 in 2025. This long-standing publication record reflects not only the growth and continuity of IDA, but also its lasting impact on the evolving field of data analysis.

4.3 Interdisciplinary Ethos of IDA

As the IDA symposium series gained momentum, so too did the broader development of Intelligent Data Analysis as an interdisciplinary field grounded in collaboration across computing, statistics, and domain sciences.

In 1995, Alex and I embarked on a three-week research visit to China, supported by the Chinese Association of Science and Technology and The Royal Society in the UK. Our itinerary included the Institute of Computing Technology at the Chinese Academy of Sciences, Tsinghua University, and Xi'an Jiao Tong University (Fig. 1). IDA was a central theme throughout the visit, sparking engaging scientific dialogue and helping to shape future research directions for both sides.

Fig. 1. Alex visiting the Great Wall while in Beijing (1995).

One unforgettable highlight was our meeting with Xia Peisu[2] (夏培肃), widely recognised as the "Mother of Computer Science in China." A pioneering figure, she co-founded China's first computer research team in 1952 and led the development of its first general-purpose digital computer in 1960. I had first met Professor Xia in 1985, when she was awarded an honorary doctorate by Heriot-Watt University in Edinburgh. Over the years, her encouragement, often in beautifully written correspondence, was a constant source of inspiration. In 1995, she hosted us at the Institute of Computing Technology in Beijing and graciously invited us to her home for dinner. That evening remains one of the most cherished memories for both Alex and myself.

Back in the UK, the dialogue between statistics and computing continued to deepen. In 1998, the Royal Statistical Society (RSS) hosted its first discussion meeting on "Data Mining: Threat or Opportunity?". David J. Hand, twice President of the RSS and one of IDA founders [18], shared his views from a statistical perspective, while I offered insights from computer science. In the discussions that followed, concerns were voiced about computing encroaching into statistics' traditional domain. Four years later, those

---

[2] https://en.wikipedia.org/wiki/xia_peisu.

concerns were revisited when I was invited to deliver a keynote address at the RSS International Conference 2002 in Plymouth, addressing the pointed question: "Has statistics been left behind by computing?".

From these and other events such as the Dagstuhl discussion meeting on IDA [19] in Germany, August 2000 and the European Spring School on IDA [20] in Italy, March 2001 it had become increasingly clear that statistics and computing are the dual parents of modern data analysis and each on its own is incomplete: computing without statistical rigor invites error; statistics without computational power is limited in scope. Rather than ask which contributes more, we saw the importance of fostering integrated approaches across disciplines, precisely what IDA set out to do. A key reflection on the progress made by the end of that decade can be found in the excellent volume edited by Michael Berthold and David J. Hand [21].

A major application domain that embodied the interdisciplinary ethos of IDA in the late 1990s and early 2000s was bioinformatics, computational and systems biology, catalysed by the Human Genome Project. From that period onwards, for instance, Alex became extensively involved in interdisciplinary projects aimed at analysing a variety of biological and medical data to understand complex diseases such as depression [22] and breast cancer [23], in collaboration with biomedical institutions. At the core of these initiatives was the IDA approach, bringing together biologists, clinicians, computer scientists, lab-based researchers, and statisticians to address some of the most difficult questions in biology and medicine.

5   Progress in Intelligent Data Analysis (2004-)

Over the past two decades, foundational technologies underpinning intelligent data analysis have advanced dramatically, propelled by increasingly sophisticated algorithms, expanded computational resources, and the explosive growth of data. The field has progressed from simply extracting insights to engineering autonomous, adaptive systems that learn from data and experience to drive trustworthy, intelligent decision-making. Key breakthroughs include:

- the ability to handle the volume, velocity, and variety of modern data streams through big data technologies such as Hadoop, Spark, and cloud platforms like AWS, Azure, and Google Cloud;
- the maturation and widespread adoption of machine learning, followed by rapid progress in deep learning;
- increasingly automated analysis pipelines that significantly reduce the need for manual intervention;
- a growing emphasis on interpretability, robustness, and trustworthiness, aimed at mitigating bias and supporting more equitable, ethical outcomes; and
- groundbreaking AI applications in diverse areas covering science, technology, engineering, medicine, and more, e.g., protein folding with AlphaFold.

In the remainder of this section, I will highlight three major advances: (1) data analysis automation, (2) trustworthy decision-making, and (3) deep learning from imperfect data.

5.1 Automating Data Analysis

An interesting phenomenon in data analysis is that, although its foundational principles and technologies stem from statistics and computing, most real-world analysts are not statisticians or computer scientists. Instead, they come from applied domains such as science, engineering, business, and medicine. As a result, many application areas have developed their own specialized branches of statistics and AI. Examples include biostatistics, medical statistics, and econometrics on the statistical side, and bioinformatics, medical informatics, and business analytics on the computing side. These subfields often support their own dedicated conferences and journals.

This fragmentation has both advantages and drawbacks. While specialization fosters domain-specific advances, it may also hinder the recognition, adoption, and cross-pollination of methods across disciplines. For instance, the Receiver Operating Characteristic (ROC) analysis was developed in the 1940s in the context of radar signal detection but only gained traction in psychophysics during the 1960s. It wasn't until the 1980s and 1990s that ROC curves became standard in biostatistics and medical diagnostics for evaluating binary classifiers, and only in the 2000s did they receive more rigorous and widespread statistical treatment in fields such as machine learning and econometrics.

One response to this fragmentation has been a concerted effort to automate data analysis, making it more accessible to non-technical users across disciplines. Both the statistics and computer science communities have worked to automate routine components of data analysis, such as data preparation, structure discovery, and insight generation, so analysts can focus on higher-level concerns: formulating the right questions, interpreting results, and applying findings meaningfully.

Analysts often bring exogenous domain knowledge into play: deciding how to approach an analysis, interpreting intermediate results, evaluating data sufficiency, and assessing the informativeness of methods. They consider which aspects of a model are most uncertain, determine where to focus their attention, and sometimes decide to do further data collection. Automating these strategic aspects of analysis remains an active area of research, though significant progress has been made in automating tasks like data cleaning, feature engineering, model selection, and workflow design.

A major development in this space is Automated Machine Learning (AutoML). Early systems such as Auto-WEKA [24] and Auto-sklearn [25] used Bayesian optimization to automate model selection and hyperparameter tuning. These systems introduced concepts like pipeline synthesis, which are now foundational in tools like TPOT [26] and commercial platforms. For deep learning, Neural Architecture Search (NAS) has emerged as a powerful approach to automatically design neural network architectures. Examples include NASNet [27] and DARTS [28], which leverage reinforcement learning and gradient-based search to identify high-performing architectures with minimal human input.

Today, AutoML is widely supported by major technology platforms including Google Cloud AutoML, H2O AutoML, KNIME, Microsoft Azure AutoML, and Amazon SageMaker Autopilot. These tools cover an increasingly broad range of tasks: data quality assessment, preprocessing, feature selection, model training, hyperparameter optimization, evaluation, reporting, and even workflow orchestration (See Table 1 for

Table 1. Automation capabilities across the Python and R ecosystems.

Task	Python Ecosystem	R Ecosystem
Data Quality Checking / EDA	sweetviz, ydata-profiling, dataprep.eda	DataExplorer, skimr, summarytools, validate
Data Preprocessing	scikit-learn Pipelines, feature-engine, pyjanitor, dataprep.clean	recipes (tidymodels), caret::preProcess, dplyr, data.table
Feature Selection	sklearn.feature_selection, Boruta, TPOT, SelectKBest, MLxtend	Boruta, caret::rfe, mlr3filters, FSelector
Model Building	Auto-sklearn, TPOT, H2O AutoML, PyCaret, FLAML, MLJAR, XGBoost, LightGBM	H2O, mlr3, caret, parsnip (tidymodels), autoML
Hyperparameter Optimisation	Optuna, Ray Tune, Hyperopt, GridSearchCV, RandomizedSearchCV	mlr3tuning, caret::train, tune (tidymodels), ParBayesianOptimization
Model Evaluation & Reporting	scikit-learn.metrics, Yellowbrick, MLflow, PyCaret, SHAP, LIME	yardstick, mlr3measures, caret::confusionMatrix, easystats, vip, DALEX
Workflow Automation	PyCaret, MLflow, Metaflow, Kedro, Airflow, Prefect	drake, targets, workflowsets, mlr3pipelines

their manifestations in Python and R ecosystems), many of which previously required pages of code but can now be executed with a single function call.

While automation has advanced rapidly, not all aspects of data analysis are easy to automate. Tasks such as understanding, cleaning, and transforming messy, context-rich data remain among the most difficult challenges. These steps require not just algorithms but also domain expertise, intent interpretation, and human judgment - areas where current AI is still limited. Even more elusive is the automation of problem framing: defining what questions are worth asking and why. This stage is fundamentally creative and requires responsibility, foresight, and values—traits not easily replicated by machines.

In summary, the automation of data analysis has matured from rule-based tools to robust, modular frameworks grounded in statistical theory and AI-driven optimization. While AI accelerates scalability and model discovery, statistical methods remain critical for transparency, inference, and trust. The future of automated data science lies in integrating these strengths to build systems that are not only powerful, but also interpretable, responsible, and adaptive - a theme explored in the next section.

5.2 Trustworthy Decision Making

The automation of data analysis for non-expert users brings a new and pressing challenge: analysis pipelines often become sequences of opaque black boxes. Without insight into how these components work, users may struggle to interpret results, assess reliability, or make informed decisions. This opacity, combined with the challenges of bias and uncertainty in data, complicates the path toward trustworthy decision making.

In today's world, where decisions increasingly rely on data-driven systems, ensuring trustworthiness has become a central concern across both AI and statistics. While these two fields have traditionally prioritized different goals—automation and scalability in AI, versus inference and uncertainty quantification in statistics—they are now converging around shared concerns: uncertainty, interpretability, causality, fairness, robustness, and human oversight in intelligent data analysis.

A foundational component of trustworthy decision making is uncertainty quantification. In this regard, conformal prediction (CP) offers a powerful, practical, and theoretically grounded solution. Alex Gammerman, in collaboration with Vladimir Vovk and others [29–31], was instrumental in laying the foundations for CP that has recently gained renewed interests due to its scalable variants such as split conformal and cross-conformal. CP provides formal, distribution-free guarantees on marginal prediction coverage under the assumption of exchangeable data. It generates prediction intervals (or sets) that contain the true outcome with a user-specified coverage level (e.g., 95%), regardless of the underlying model's architecture, accuracy, or training procedure. Unlike Bayesian approaches or deep ensembles, CP makes no assumptions about model correctness and instead relies solely on the exchangeability of data. This makes CP particularly well-suited for safety-critical applications such as autonomous driving, healthcare, and energy, where reliable uncertainty quantification is vital.

Beyond uncertainty, other pillars of trustworthy decision making have seen notable progress:

- Interpretability and Explainability: Tools such as LIME [32], SHAP [33], and counterfactual explanations provide local interpretability for complex models like deep neural networks.
- Causal Inference: Libraries like DoWhy [34] and EconML (Microsoft Research) allow machine learning models to estimate treatment effects while retaining causal meaning.
- Fairness and Bias Mitigation: The emerging field of Fairness, Accountability, and Transparency in Machine Learning (FATML) [35] has introduced fairness metrics (e.g., equalized odds, demographic parity), bias-mitigation algorithms, and tools such as Fairlearn and Aequitas.
- Robustness and Reliability: Following the discovery of adversarial examples in deep learning [36], methods for enhancing robustness have proliferated.
- Human-in-the-loop methods: active learning [37], interactive machine learning, and reinforcement learning from human preferences [38].

Importantly, there is rarely a single optimal solution in data analysis. Instead, the best solution depends on multiple, sometimes conflicting, objectives: the research question, practical constraints, modelling assumptions, and the trade-offs acceptable to stakeholders. In modern data analysis, this calls for multi-objective optimization, a formal framework to balance multiple goals such as accuracy, fairness, interpretability, privacy, robustness, and speed (see Table 2).

For example, in the European project HarmonicAI (2024–2028), which aims to develop ethical, open, and trustworthy AI for digital health, we focus on optimizing three

Table 2. Typical tensions between analysis objectives.

Objective	Potential Conflicts With
Accuracy	Interpretability (e.g., deep models vs. simple models); Speed; Fairness; Privacy
Fairness	Accuracy (enforcing parity may reduce performance); Speed; Interpretability
Interpretability	Accuracy (simple models may underperform); Robustness; fairness
Privacy	Accuracy (differential privacy introduces noise); Speed (encryption adds overhead)
Robustness	Speed (robust training is computationally expensive); Interpretability
Speed	Accuracy, Privacy, and Robustness (may be compromised for efficiency)

objectives: explainability, fairness, and privacy. These objectives are deeply intertwined and often conflict with each other. Explainability may require extensive data and feature engineering, which conflicts with privacy-preserving goals. Fairness can benefit from broad population data, again in tension with privacy. Fairness mechanisms themselves (e.g., adversarial debiasing) may reduce model transparency.

Multi-objective optimisation (MOO) offers a principled way to resolve such trade-offs. Rooted in operational research [39] and evolutionary computation [40], MOO algorithms aim to approximate the Pareto front—a set of solutions representing different trade-offs among objectives. Evolutionary MOO approaches have proven especially effective due to their flexibility, scalability, and ability to explore complex search spaces.

Still, many-objective optimization (MaOO) [41], with four or more objectives, introduces unique challenges:

1. Pareto dominance becomes less effective, as most solutions become non-dominated.
2. Balancing convergence and diversity become increasingly more difficult as the number of objectives rises.
3. Increasing time and space requirements as well as parameter sensitivity and performance evaluation difficulties become more pronounced.

Although successful applications of MaOO remain relatively limited compared to those of traditional multi-objective optimization, progress is rapidly accelerating. Emerging methods such as decomposition-based, indicator-based, and meta-objective algorithms [42] are increasingly enabling MaOO across diverse domains, ranging from software testing [43] to migration logistics [44].

In summary, the last two decades have seen growing synergy between AI, statistics and optimisation in the pursuit of trustworthy, transparent, and human-aligned decision systems. Advances in uncertainty quantification, interpretability, fairness, robustness, and causal reasoning are redefining best practices. As these innovations continue to converge, they hold the promise of shaping the next generation of intelligent data analysis: systems that are not only powerful and scalable but also principled, interpretable, and accountable.

5.3 Deep Learning from Imperfect Data

Deep learning, a powerful subset of machine learning, has undergone remarkable advances over the past two decades, transforming fields such as computer vision, natural language processing, drug discovery, and autonomous systems. This revolution has been fuelled by breakthroughs in neural architectures, training methodologies, and optimization techniques, along with the availability of massive datasets and increasing computational power.

Key deep learning paradigms include Restricted Boltzmann Machines (RBMs), Deep Belief Networks (DBNs), Autoencoders (AEs), Convolutional Neural Networks (CNNs), Recurrent Neural Networks (RNNs), Long Short-Term Memory (LSTMs), Transformers, Graph Neural Networks (GNNs), Generative Adversarial Networks (GANs), and Deep Reinforcement Learning (DeepRL). For brevity, this section highlights three of the most influential developments: CNNs, Transformers, and GANs. For a more comprehensive review of all major methods, see [45, 46].

Convolutional Neural Networks is a class of deep neural networks especially designed for processing structured grid data like images, leveraging *convolutional layers* to automatically and hierarchically learn spatial features (e.g., edges, textures, objects) through local receptive fields, weight sharing, and pooling operations. These layers are typically followed by pooling layers to reduce dimensionality and fully connected layers for classification or regression. Introduced in the seminal paper on LeNet-5 by LeCun et al. [47] and popularized by AlexNet [48], CNNs have become the backbone of modern computer vision tasks, excelling in applications like image classification, object detection, medical image analysis, and more.

Transformers, designed to handle sequential data without relying on recurrence or convolution, marked a paradigm shift in natural language processing (NLP). Introduced by Vaswani et al. [49], the Transformer architecture became the dominant approach in sequential modelling and NLP, often outperforming previous dynamic neural networks such as LSTMs [50], RNNs [51], and Gated Recurrent Unit (GRU) [52]. Using self-attention mechanisms, Transformers enabled parallel processing of sequences and more effective modelling of long-range dependencies. This innovation led to powerful pre-trained models such as BERT and the GPT series, which demonstrated state-of-the-art performance across a wide range of tasks.

Introduced by Ian Goodfellow et al. in 2014 [53], Generative Adversarial Networks (GANs) are a class of deep learning models designed to generate realistic data by setting up a game between two neural networks: a *generator* and a *discriminator*. The generator attempts to produce data (such as images or text) that mimic real examples, while the discriminator evaluates whether a given input is real (from the training set) or fake (produced by the generator). Through this adversarial process, the generator learns to create increasingly convincing outputs, while the discriminator improves at spotting fakes. Over time, the two networks ideally reach a point where the generated data are indistinguishable from real data, making GANs particularly powerful for applications like image synthesis, data augmentation, and creative content generation.

Despite these advances, deep learning models face substantial challenges when applied to real-world data, which are often imperfect in both quality and quantity. Imperfections may include missing, noisy, inconsistent, biased, or untimely data [54,

55]. Furthermore, deep learning models are notoriously data-hungry, and many applications struggle to provide sufficient labelled data for effective training and evaluation [56, 57]. This section highlights two common and pressing challenges: noisy labels and insufficient data.

Deep Learning with Noisy Labels. High-performance deep learning typically requires large volumes of accurately labelled data, but manual annotation is often expensive, time-consuming, and error prone. Labels obtained from crowdsourcing or automated tools are susceptible to errors too, resulting in noisy training data that can misguide learning and degrade model performance. This "noisy label problem" is particularly detrimental during supervised learning, where the model relies heavily on the correctness of training labels.

Two main strategies have emerged to address this challenge. The first focuses on improving data quality, by either correcting mislabelled samples or filtering out noisy examples [58]. The second strategy seeks to make models more robust to label noise by modifying architectures, loss functions, or training procedures in the spirit of robust statistics. Recent methods include noise-tolerant loss functions, curriculum learning, and confidence-based sample selection. For an in-depth review, see [59].

Deep Learning with Limited Data. Large datasets often yield better generalisation and reduce overfitting, yet many domains such as medical imaging struggle with data scarcity due to privacy concerns, rarity of conditions, and expert annotation requirements. In such cases, *data augmentation* is a vital technique for synthetically expanding training data and enhancing deep learning performance, particularly for image-based models such as CNNs. Common augmentation strategies include geometric transformations, colour-space adjustments, kernel filtering, and GAN-based sample generation [56, 60].

However, data augmentation is both domain-dependent and task-dependent. For instance, time series data exhibit temporal dependencies that invalidate many image-based augmentation techniques. Moreover, augmentations effective for classification may be inappropriate for anomaly detection or forecasting tasks. Advanced augmentation methods for time series include window warping, amplitude and phase perturbations, and time series synthesis using GANs [57]. While augmentation has led to significant gains, most notably in ImageNet classification with AlexNet, automating the generation of high-quality synthetic data remains a challenge across many domains.

6 Conclusion

The challenges of analysing large and complex data have changed profoundly over the past few decades. While today's data analysts benefit from a rich and expanding toolkit, including statistical methods, machine learning, and AI-driven optimisation, the core demands of effective, trustworthy, and efficient data analysis remain unchanged. As data volumes soar and automated decision-making becomes more pervasive, the need to move beyond traditional paradigms has never been more pressing.

Deep learning, large language models (LLMs), and large reasoning models (LRMs) have transformed many application domains. However, traditional statistical modelling remains indispensable—particularly for hypothesis-driven inquiry, transparent inference, and causal reasoning. At the same time, the growing focus on explainable AI and trustworthy data analysis has gained urgency, especially considering recent findings that expose limitations in the behaviour and reliability of state-of-the-art LLMs [61] and LRMs [62].

Within this evolving landscape, the IDA community continues to play a critical role: advancing foundational and applied methods (e.g. for agentic, multimodal, and sustainable AI), fostering interdisciplinary collaboration, and ensuring that data analysis remains both principled and pragmatic, striking right balances between multiple competing objectives.

Looking ahead, the field must tackle enduring and emerging challenges alike: improving efficiency, mitigating bias, deepening automation without sacrificing oversight, and enhancing interpretability. The future lies in building integrated, ethical, and human-centric AI systems that significantly extend human capabilities while ensuring that automated insights align with real-world goals and societal values.

References

1. Graunt, J.: Natural and Political Observations Made upon the Bills of Mortality. Royal Society, London (1662)
2. Fisher, R.A.: Statistical Methods for Research Workers. Springer (1925)
3. Box, G.E.P., Jenkins, G.M.: Time Series Analysis: Forecasting and Control. Holden-Day, San Francisco (1970)
4. Tukey, J.W.: Exploratory Data Analysis. Addison-Wesley (1977)
5. Blackett, P.M.S.: Studies of War: Operational Research. Oliver & Boyd (1962)
6. Bäck, T., Fogel, D.B., Michalewicz, Z.: Handbook of Evolutionary Computation. Oxford University Press and IOP Publishing, New York (1997)
7. Kirkpatrick, S., Gelatt, C.D., Vecchi, M.P.: Optimization by simulated annealing. Science **220**(4598), 671–680 (1983)
8. Kennedy, J., Eberhart, R.: Particle swarm optimization. In: Proceedings of IEEE International Conference on Neural Networks, vol. 4, pp. 1942–1948 (1995)
9. Turing, A.M.: Computing machinery and intelligence. Mind **59**(236), 433–460 (1950)
10. Newell, A., Simon, H.A.: The logic theory machine: a complex information processing system. IRE Trans. Inf. Theory **2**(3), 61–79 (1956)
11. Rumelhart, D.E., Hinton, G.E., Williams, R.J.: Learning representations by back-propagating errors. Nature **323**(6088), 533–536 (1986)
12. Russell, S.J., Norvig, P.: Artificial Intelligence: A Modern Approach, 4th edn. Pearson (2020)
13. Hopfield, J.J.: Neural networks and physical systems with emergent collective computational abilities. Proc. Nat. Acad. Sci. (PNAS) **79**(8), 2554–2558 (1982)
14. Kohonen, T.: Self-organized formation of topologically correct feature maps. Biol. Cybern. **43**(1), 59–69 (1982)
15. Liu, X., Cheng, G., Wu, J.: Noise and uncertainty management in intelligent data modelling. In: Proceedings 12th National Conference on Artificial Intelligence (AAAI-94), American Association for Artificial Intelligence, Seattle, pp. 263–268 (1994)
16. Vapnik, V.: The Nature of Statistical Learning Theory. Springer, New York (1995)
17. Liu, X.: Intelligent data analysis: issues and challenges. Knowl. Eng. Rev. **11**(4), 365–371 (1996)

18. Hand, D.J.: Intelligent data analysis: issues and opportunities. In: Liu, X., Cohen, P., Berthold, M. (eds.) Advances in Intelligent Data Analysis: Reasoning about Data. LNCS, vol. 1280, pp. 1–14. Springer-Verlag (1997)
19. Berthold, M., Kruse, R., Liu, X., Szczerbicka, H. (eds.) Intelligent Data Analysis, Dagstuhl Report No 00331. In: International Conference and Research Center for Computer Science, Schloss Dagstuhl, Germany (2000)
20. Rosaria, S., Di Fatta, G.: Learning to reason about data: the spring school on intelligent data analysis. Intell. Data Anal. **5**, 439–442 (2001)
21. Berthold, M., Hand, D.J. (eds.) Intelligent Data Analysis: An Introduction, 2nd edn. Springer-Verlag (2003)
22. Nouretdinov, I., et al.: Machine learning classification with confidence: application of transductive conformal predictors to MRI-based diagnostic and prognostic markers in depression. Neuroimage **56**, 809–813 (2010)
23. Blixt, O., et al.: Autoantibodies to aberrantly glycosylated MUC1 in early stage breast cancer are associated with a better prognosis. Breast Cancer Res. **13**(2), R25 (2011)
24. Thornton, C., Hutter, F., Hoos, H., Leyton-Brown, K.: Auto-WEKA: combined selection and hyperparameter optimization of classification algorithms. In: Proceedings International Conference on Knowledge Discovery and Data Mining (KDD 2013), pp. 847–855 (2013)
25. Feurer, M., Klein, A., Eggensperger, K., Springenberg, J.T., Blum, M., Hutter, F.: Efficient and robust automated machine learning. In: Advances in Neural Information Processing Systems (NeurIPS), vol. 28, pp. 2962–2970 (2015)
26. Olson, R.S., Bartley, N., Urbanowicz, R.J., Moore, J.H.: Evaluation of a tree-based pipeline optimization tool for automating data science. In: Proceedings of the Genetic and Evolutionary Computation Conference (GECCO 2016), pp. 485–492 (2016)
27. Zoph, B., Le, Q.V.: Neural architecture search with reinforcement learning. In: Proceedings of the International Conference on Learning Representations (ICLR) (2017)
28. Liu, H., Simonyan, K., Yang, Y.: DARTS: differentiable architecture search. In: Proceedings of the International Conference on Learning Representations (ICLR) (2019)
29. Gammerman, A., Vovk, V., Vapnik, V.: Learning by transduction. In: Proceedings of the Fourteenth Conference on Uncertainty in Artificial Intelligence (UAI 1998), pp. 148–155. Morgan Kaufmann, San Francisco (1998)
30. Vovk, V., Gammerman, A., Shafer, G.: Algorithmic Learning in a Random World. Springer, (2005)
31. Gammerman, A., Vovk, V.: Hedging predictions in machine learning. Comput. J. **50**(2), 151–163 (2007)
32. Ribeiro, M.T., Singh, S., Guestrin, C.: Why should I Trust You? Explaining the predictions of any classifier. In: Proceedings of the 22nd ACM SIGKDD International Conference on Knowledge Discovery and Data Mining (KDD 2016), pp. 1135–1144. ACM, New York (2016)
33. Lundberg, S.M., Lee, S.I.: A unified approach to interpreting model predictions. In: Advances in Neural Information Processing Systems 30 (NeurIPS 2017), pp. 4765–4774. Curran Associates, Inc. (2017)
34. Sharma, A., Kiciman, E.: DoWhy: an end-to-end library for causal inference. In: Proceedings of the 34th Conference on Neural Information Processing Systems (NeurIPS 2020), pp. 1–10. Curran Associates, Inc. (2020)
35. Selbst, A.D., Boyd, D., Friedler, S.A., Venkatasubramanian, S., Vertesi, J.: Fairness and abstraction in sociotechnical systems. In: Proceedings of the 2019 Conference on Fairness, Accountability, and Transparency (FAT*), pp. 59–68. ACM, New York (2019)
36. Szegedy, C., et al.: Intriguing properties of neural networks. In: Proceedings of the International Conference on Learning Representations (ICLR 2014). arXiv:1312.6199 (2013)

37. Settles, B.: Active Learning. Morgan & Claypool (2012)
38. Christiano, P.F., Leike, J., Brown, T.B., Martic, M., Legg, S., Amodei, D.: Deep reinforcement learning from human preferences. In: Advances in Neural Information Processing Systems 30 (NeurIPS 2017), pp. 4299–4307. Curran Associates, Inc. (2017)
39. Miettinen, K.: Nonlinear Multiobjective Optimization. Springer, Boston (1999)
40. Deb, K.: Multi-Objective Optimization Using Evolutionary Algorithms. John Wiley, New York (2001)
41. Li, M., Yang, S., Liu, X.: Shift-based density estimation for pareto-based algorithms in many-objective optimization. IEEE Trans. Evol. Comput. **18**(3), 348–365 (2014)
42. Li, M., Yang, S., Liu, X.: Bi-goal evolution for many-objective optimization problems. Artif. Intell. **228**, 45–65 (2015)
43. Hierons, R.M., Li, M., Liu, X., Segura, S., Zheng, W.: SIP: optimal product selection from feature models using many-objective evolutionary optimisation. ACM Trans. Softw. Eng. Methodol. **25**(2), 1–39 (2016)
44. Xue, Y., et al.: Many-objective simulation optimization for camp location problems in humanitarian logistics. Int. J. Netw. Dyn. Intell. **13**(3), 1–14 (2024)
45. Mienye, I.D., Swart, T.G.: A comprehensive review of deep learning: architectures. Recent Adv. Appl. Inf. **15**, 755 (2024)
46. Liu, W., Wang, Z., Liu, X., Zeng, N., Liu, Y., Alsaadi, F.E.: A survey of deep neural network architectures and their applications. Neurocomputing **234**, 11–26 (2017)
47. LeCun, Y., Bottou, L., Bengio, Y., Haffner, P.: Gradient-based learning applied to document recognition. Proc. IEEE **86**(11), 2278–2324 (1998)
48. Krizhevsky, A., Sutskever, I., Hinton, G.E.: ImageNet classification with deep convolutional neural networks. In: Pereira, F., Burges, C.J.C., Bottou, L., Weinberger, K.Q. (eds.) Advances in Neural Information Processing Systems, vol. 25, pp. 1097–1105. Curran Associates, Inc. (2012)
49. Vaswani, A., et al.: Attention is all you need. In: Guyon, I., et al. (eds.) Advances in Neural Information Processing Systems, vol. 30, pp. 5998–6008. Curran Associates, Inc. (2017)
50. Hochreiter, S., Schmidhuber, J.: Long short-term memory. Neural Comput. **9**(8), 1735–1780 (1997)
51. Liu, Y., Wang, Z., Liu, X.: Global exponential stability of generalized recurrent neural networks with discrete and distributed delays. Neural Netw. **19**(5), 667–675 (2006)
52. Cho, K., et al.: Learning phrase representations using RNN encoder-decoder for statistical machine translation. In: Proceedings of the 2014 Conference on Empirical Methods in Natural Language Processing (EMNLP 2014), pp. 1724–1734 (2014)
53. Goodfellow, I.J., et al.: Generative adversarial networks. In: Ghahramani, Z., Welling, M., Cortes, C., Lawrence, N., Weinberger, K.Q. (eds.) Advances in Neural Information Processing Systems, vol. 27, pp. 2672–2680. Curran Associates, Inc. (2014)
54. Han, B.: Trustworthy machine learning under imperfect data. In: Proceedings of the Thirty-Third International Joint Conference on Artificial Intelligence (IJCAI-24), Early Career Track (2024)
55. Wang, C., et al.: Fusionformer: a novel adversarial transformer utilizing fusion attention for multivariate anomaly detection. IEEE Trans. Neural Netw. Learn. Syst. (2025). https://doi.org/10.1109/TNNLS.2025.3542719
56. Shorten, C., Khoshgoftaar, T.M.: A survey on image data augmentation for deep learning. J. Big Data **6**(1), 1–48 (2019)
57. Wen, Q., Sun, L., Song, X., Gao, J., Wang, X., Xu, H.: Time series data augmentation for deep learning: a survey. Int. J. Comput. Vision **130**(6), 1367–1405 (2021)
58. Fraser, K., Wang, Z., Liu, X.: Microarray Image Analysis: An Algorithmic Approach. Chapman & Hall/CRC, December 2009

59. Song, H., Kim, M., Park, D., Shin, Y., Lee, J.-G.: Learning from noisy labels with deep neural networks: a survey. IEEE Trans. Neural Netw. Learn. Syst. **34**(12), 8135–8153 (2023)
60. Wen, C., Xue, Y., Liu, W., Chen, G., Liu, X.: Bearing fault diagnosis via fusing small samples and training multi-state Siamese neural networks. Neurocomputing **576** (2024). https://doi.org/10.1016/j.neucom.2024.127355
61. Kim, J.Y., Patel, S.B., Zhang, R., Lee, K., Johnson, A.E.W.: Limitations of large language models in clinical problem-solving arising from inflexible reasoning. Nat. Digit. Med. (2025). https://doi.org/10.1038/s41746-025-XXXXX
62. Shojaee, P., Mirzadeh, I., Alizadeh, K., Horton, M., Bengio, S., Farajtabar, M.: The illusion of thinking: understanding the strengths and limitations of reasoning models via the lens of problem complexity. Apple Mach. Learn. Res. Also available as arXiv:2501.12948, June 2025

Statistical Modelling in Specific Case Analysis with Bayesian Belief Networks

Colin G. G. Aitken$^{(\boxtimes)}$ [ID]

School of Mathematics and Maxwell Institute of Mathematical Sciences,
The University of Edinburgh, Edinburgh EH9 3FD, UK
cgga@ed.ac.uk
https://www.maths.ed.ac.uk/~cgga/

Abstract. Work from the 1990's is described of the role in criminal investigations of statistical modelling in the form of Bayesian belief networks for specific case analysis, also known as offender profiling. The benefits for research and investigations of a collaboration amongst statisticians, computer scientists and detectives are described. The work is illustrated with a Home Office Police Research Group project motivated by a criminal case involving sexually motivated child homicides from the 1980's.

Keywords: Bayesian belief network · CATCHEM · Forensic science · Specific case analysis

1 Introduction

I first met Alex when we were joint co-investigators on a European Community (as it then was) Development of Statistical Expert Systems (DOSES) research project *Linking Informal Knowledge and Expertise to Forecasting Models* which ran from the late 1980's to the early 1990's. Amongst all the publications from this project, we had only one joint publication ([1]), published in 1992. This described the application of stochastic simulation to mixed graphical association models to enable the estimation of marginal probabilities, means and variances of variables and the marginal densities of continuous variables.

Our first joint publication was [2], published in 1989, which concerned the development of a Bayesian belief network for a hypothetical example in forensic science. Further work related to the investigation of crimes is reported in [3–5]. This work concerns the analysis of a dataset of sexually motivated child homicides with a view to profiling offenders, a method known more formally as specific case analysis, in future cases. Two models were considered, logistic regression and Bayesian belief networks. This chapter is concerned only with Bayesian belief networks.

These Bayesian networks were developed following the methodology of [10]. The work described in [3] is a summary of consultancy work done in the early

K. An Nguyen and Z. Luo (Eds.): Alexander Gammerman Festschrift, LNCS 16290, pp. 75–86, 2026.
https://doi.org/10.1007/978-3-032-15120-9_5

1990's for the Derbyshire Police for the development and evaluation of statistical models for the prediction of offender characteristics. The paper was awarded the Philip W. Allen award of the then Forensic Science Society (now Chartered Society of Forensic Sciences) for the best paper in 1996 in their journal *Science & Justice*. The models were developed using a database of sexually-oriented child homicides and abductions that had occurred in the United Kingdom since 1960. The project to collect the data and code the details for entry into a database was named CATCHEM (an acronym for something like Centralised Analytical Team Collating Homicide Expertise) and was initiated by the Derbyshire Police in 1986 in support of a joint investigation into the murder of three young girls.

Alex and I were fortunate to be awarded support through the CATCHEM project to develop methods such that advice could be developed from an analysis of cases that had been collected previously. The work was supported by the crime sub-committee on offender profiling of what was then the Association of Chief Police Officers (ACPO, now National Police Chiefs' Council). This sub-committee was chaired by John Stevens, then Chief Constable of Northumbria Police, now Baron Stevens of Kirkwhelpington. Alex and I were able to employ two research assistants (Tom Connolly and Guanghua Zhang) and a final report [5] was published as Paper 4 of the Special Interest Series of the Home Office Police Research Group. In a foreword to the final report, John Stevens wrote that '[t]he CATCHEM project is now routinely used as a source of advice by detectives investigating child homicides; both in the UK and abroad'. In the context of the statistical models developed by Alex, me and our colleagues, John Stevens wrote that '[t]he benefits and limitations identified with this approach are of general interest to 'offender profiling' research as they apply to *any* approach that is based on the analysis of past cases'.

It is important to note that the use of the research described in the work Alex and I did on offender profiling is as an investigative tool. The tool is designed to help police officers in their investigation and provide guidance as to the characteristics of the offender. It is most definitely not to be used as evidence. The work cannot be used to help in the identification of an offender and then to be used to secure a conviction. It is not permissible for it to be said that the defendant was identified through the use of offender profiling and then to argue in court that the defendant fits the profile and is thus the offender.

A search on Google Scholar shows that [3] has been cited 62 times and [2] has been cited 80 times at the time of writing (January 2026). It was commented ([7]) in 2024 that [2] is 'amongst the first proponents for their application in forensic science as a visual aid for evidence evaluation in the late 1980s'.

Other joint publications of Alex and me are [8,9].

2   Bayesian Belief Networks

In a criminal investigation, investigators gather evidence and formulate propositions. There are several types of evidence. For example, the evidence may be trace evidence, such as a blood sample, or testimonial, such as that of an eye-witness, or documentary, such as a ransom document. There are three levels of proposition.

Consider an alleged assault at which a blood stain (trace evidence) is found at the scene of the alleged assault. The level of the proposition may be about

  (i) the source of the trace evidence, such as a DNA profile from the blood sample (stain) whose source may be a person of interest or a victim for example;

 (ii) the activity involving the evidence, the person of interest hit the victim, or

(iii) an offence related to the evidence, the person of interest criminally assaulted the victim.

A Bayesian belief network is a graphical representation, in the form of nodes and edges, of the relationships amongst propositions and evidence.

In the example just given, the evidence has two parts, the DNA profile of the blood sample and the DNA profile of the person of interest; denote this evidence by E. The proposition of current interest in the investigation relates to the source of the blood sample; the source either is, or is not, the person of interest; denote these possibilities by H. The relationship between the evidence and the proposition may be represented by the network in Fig. 1.

Fig. 1. Basic network to illustrate that evidence E is related to the proposition H.

Figure 1 has two nodes, one (H) representing the proposition and one (E) representing the evidence. The directed arrow, known as a directed edge, joining the nodes indicates the relationship between the proposition and the evidence, that E depends on H.

The proposition of interest has two possible outcomes. First, the source of the blood sample is the person of interest, denote this outcome h . Use of the Bayesian belief network requires there to be an alternative proposition, mutually exclusive, but not necessarily exhaustive, to the proposition of interest. For example, this could be that the source of the blood sample is a person unrelated to the person of interest. This outcome is conventionally denoted $\bar{h}$ though this may be thought misleading as these two propositions are not complementary. People related to the person of interest are not considered though could be included with the addition of a third proposition about the source of the blood sample.

The evidence E of the two DNA profiles has two possible outcomes. First, the profile of the blood sample and the profile of the person of interest match (in some sense which is not relevant to this discourse) and, second, the two profiles do not match. Denote the first outcome e and the second $\bar{e}$.

The structure of the network is completed with a table of conditional probabilities. Notationally, this is represented in Table 1.

Table 1. Mathematical expressions for the conditional probabilities for the two-node network of Fig. 1.

Proposition	Evidence E	
H	e	$\bar{e}$
h	$\Pr(e \mid h)$	$\Pr(\bar{e} \mid h)$
$\bar{h}$	$\Pr(e \mid \bar{h})$	$\Pr(\bar{e} \mid \bar{h})$

In practice, numerical values have to be substituted for these expressions and a value given for the initial (prior) probability for h, prior to consideration of E. Illustrative values for the tabular expressions are given in Table 2.

Table 2. Possible numerical values for the conditional probabilities for the two-node network of Fig. 1.

Proposition	Evidence E	
H	e	$\bar{e}$
h	1	0
$\bar{h}$	10^{-6}	$1 - 10^{-6}$

The choice of a value for $\Pr(h)$ is problematic. One candidate is suggested by the idea that a person is innocent until proven guilty. This suggests that the person is as guilty as anyone else in some relevant population which may be very large, where there has to be an assumption as to what the relevant population is. With a relevant population of size N, a simplistic assumption of independence, for example, would suggest a value for $Pr(h)$ of $1/N$. An alternative may be provided by other evidence to suggest that the person of interest is one of only a small number of people that could have committed the assault.

The concept of a relevant population begs the question as to what may be thought of as a relevant population. In a rape case, this may be thought to be the population of all males of appropriate age in some neighbourhood of the crime. These thoughts in turn lead on to considerations of what is meant by 'appropriate age' and the possibilities of the rapist being a visitor to, not a resident of, the neighbourhood. It is easy to see that the definition of a relevant population is difficult!

The first row of Table 2 has values such that the probability of a match in DNA profiles between the sample at the scene of the assault and the person of interest (e) if the person of interest is the source of the profile at the scene h is taken to be 1. There is no consideration for a false positive. As a consequence, $\Pr(\bar{e} \mid h)$ is assumed to be 0; no relatives are assumed to be involved. The second row has values such that the probability of a match in DNA profiles between the sample at the scene of the assault and the person of interest (e) if the person of interest is not the source of the profile at the scene h is taken to be 10^{-6}. The DNA profile is such that the probability of a random match between the profile of the person of interest and the profile of the scene sample is 1 in a million. This probability could be estimated from some relevant population[1] from which the two profiles are taken. As a consequence, $\Pr(\bar{e} \mid \bar{h})$ is assumed to be $1 - 10^{-6}$.

The probability of interest to the investigator is the probability, $Pr(h \mid e)$, that the person of interest is the source of the profile of the scene sample. Assume for the sake of argument, that it has been established that the person of interest is one of only ten people who could have committed the assault. Then, $\Pr(h) = 0.1$. For this two-node network, the numerical calculation is a simple application of Bayes' theorem.

$$
\begin{aligned}
\Pr(h \mid e) &= \frac{\Pr(e \mid h) \times \Pr(h)}{\Pr(e)} \\
&= \frac{\Pr(e \mid h) \times \Pr(h)}{\Pr(e \mid h)\,\Pr(h) + \Pr(e \mid \bar{h})\,\Pr(\bar{h})} \\
&= \frac{1 \times 0.1}{(1 \times 0.1) + (10^{-6} \times 0.9)} \\
&= 0.999991.
\end{aligned}
$$

This result is intuitively very reasonable. The pool of possible perpetrators has only ten people in it. The probability of a random match of two unrelated people has been taken to be 1 in a million so the probability of there being such a match amongst ten, unrelated, randomly selected, people is extremely low.

Of course, the calculations become very much more complicated once the number of nodes and edges increase. The work that Alex and I did about thirty years ago provides examples of how the complications may be resolved with appropriate software. The software used in these examples was created by Alex and based on the fundamental work described in [10].

3   Probabilistic Reasoning in Evidential Assessment

In 1988, the fundamental work described in [10] showed that it was possible to adopt a sound probabilistic approach to uncertain inference using Bayesian belief networks. That work was illustrated with a medical example. An analogy in the administration of criminal justice was described in [2]. Five stages

---

[1] Note this population is unconnected to the relevant population mentioned above that is taken to be the one to which the criminal is thought to belong.

of the analysis were considered: qualitative representations of relationships as a directed graph, quantitative expression of subjective beliefs as probabilities, efficient storage as an undirected graph, coherent evidence propagation through the graph and uncertain evidence. Alex and I wrote then that '[t]he ideas discussed in this paper are at a very tentative stage of development and are published to provide a basis for discussion and in the hope of directing thoughts along useful lines of enquiry'.

Figure 2 illustrates factors considered in a hypothetical case of the investigation of a murder of victim V in which there are two suspects, X and Y. There is an eyewitness. There is also a possible transfer of fibres which may have occurred during the commission of the crime or may have occurred innocently because Y drives X's car regularly. The structure of the network may be developed in a collaboration amongst statisticians, computer scientists and investigators.

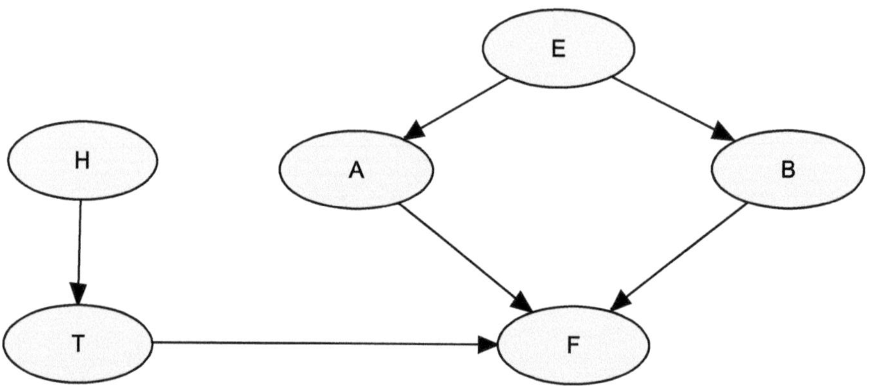

Fig. 2. Bayesian belief network of nodes in a fictitious example where each node can take one of two possible values. A: X did or did not commit the murder; B: Y did or did not commit the murder; E: Eyewitness evidence (or its absence) of a row between X, Y and the victim V sometime before the commission of the crime; F: Fibres from a jacket similar (or dissimilar) to one found in the possession of X are found at the crime scene; H: Y does or does not drive X's car regularly; T: Y does or does not pick up fibres from X's jacket. Reproduced from [2].

Previous case histories existed for the medical example in [10] and from these probabilities could be assigned with reasonable confidence. Such assignations are difficult to imagine in a forensic context because each crime is unique. For the Bayesian belief network analysis to work, probability assignments are needed; in a forensic context these assignments are, of necessity, often subjective. In practice, these can be provided by investigators based on their personal experiences.

The structure presented in Fig. 2 requires many conditional probabilities to represent the uncertain relationships amongst the nodes. The interpretations given in [2] are listed below. Notice the last item is particularly complicated as the evidence about the fibres F is dependent on three other factors, A, B

and T, each of which has two possible outcomes; this evidence thus requires the assignment of eight subjective probabilities.

(i) Eyewitness evidence of a row between X and V or between Y and V is unlikely.
(ii) X and Y are unlikely *a priori* to commit the crime but possibly might have, given eyewitness evidence.
(iii) It is quite likely that Y drives X's car.
(iv) If Y drives X's car frequently, there is a possibility that Y will pick up fibres from X's jacket; otherwise it is unlikely.
(v) Fibres from X's jacket are quite likely to be found if X committed the crime, regardless of whether Y was involved or not. If Y did not pick up fibres from X's jacket and X did not commit the crime, it is unlikely that fibres from X's jacket will be found at the scene.

Notice the use of verbal measures of uncertainty such as 'unlikely', 'possibly might have', etc. There has been much discussion of the assignation of numerical values to these verbal measures. In practice, each case has to be treated on its own merits and the numerical assignation made from collaboration amongst those working on the network. One advantage of this network process is the ease with which the outcome of a different choice of probabilities may be investigated.

Once the structure of the network and the conditional probabilities have been agreed, it is possible to determine changes in the probabilities of certain nodes once the outcome of other nodes has been accepted. For example, using the conditional probabilities from [2] it is possible to determine the marginal probabilities for each node. Then, assume the fibre evidence has been accepted: fibres from a jacket similar to one found in possession of X have been found at the crime scene. All the other initial conditional probabilities are updated, a process known as *propagation*. For full details of the initial probabilities, refer to [2]. From these, the initial probability that X committed the murder is 0.035. After acceptance of the fibre evidence, the revised probability that X committed the murder is 0.696.

The process of the development of a network with possible contributory items of evidence, the consideration of the uncertainties amongst the various components of the network, and then the implications for the investigation of information about particulars of possible contributory items of evidence should prove helpful in the prioritisation of resources in an investigation through knowledge of which items of evidence will contribute most to increasing or decreasing the probability a person of interest committed the crime under investigation.

4   Specific Case Analysis

Section 3 discussed a hypothetical and early illustration of the use of Bayesian belief networks in a criminal investigation. A few years later, Alex and I were approached for help with a research project, CATCHEM, for the development of a system to provide advice for detectives investigating child homicides both in the UK and abroad.

In the foreword to [5], John Stevens, then Chief Constable of Northumbria Police wrote:

> The CATCHEM project was initiated by Derbyshire Police in 1986 in support of a joint investigation into the murder of three young girls. The project team collected and coded details of *all* sexually-oriented child homicides and abductions that had occurred in the United Kingdom since 1960. As a result, sound advice was developed for the strategic management of the thousands of lines of potential lines of enquiry that were available to the investigation. The offender[2] is now serving a life sentence for the murder of the three girls and for other serious crimes.

The work we did, reported in [5], was said by John Stevens in the foreword to 'take the CATCHEM project one step further with the development and evaluation of statistical models for the prediction of offender characteristics'.

There are several benefits to a criminal investigation in the use of a Bayesian belief network. First, the network can be constructed in discussion amongst investigating officers, statisticians and computer scientists. The relationships amongst the characteristics are made explicit.

The second benefit is in the assignation of the conditional probabilities. The work on specific case analysis was helped considerably by the existence of the CATCHEM database. At the time of our work, it contained data for all child sexual murders in Great Britain from 1960 to 1991 inclusive. We restricted our research to those cases in which there was a single offender and a single victim. This gave us 320 cases with which to work, a depressingly high figure given the nature of the database. Characteristics of the offender, the victim and the crime were recorded. Those used in the analysis discussed here and their levels are given in Table 3, note that the victim's age has been divided into three, not two, levels.

The network of seven nodes shown in Fig. 3 was constructed in three stages in discussion with a senior detective. The first stage involved the specification of the characteristics to be included; the choice of the nodes. The second stage constructed the relationship amongst the characteristics, i.e., which characteristics (nodes) were to be joined by arcs as shown in Fig. 3. The third stage was the determination of the associated conditional probabilities. These can be determined from a database if available, as it was for this study, or from discussions with investigating officers.

Given the existence of a database, the conditional probabilities relating the characteristics can be determined automatically, for example through the calculation of relative frequencies. In the absence of a database but the presence of a senior detective, these probabilities can be assigned as a measure of belief as provided by the detective. We had the advantage of both a database and a

---

[2] Robert Black; he was sentenced to life imprisonment in 1994 with a recommendation he serve at least 35 years. He died of a heart attack in prison in 2016.

Table 3. Description of the nodes, possible levels and node identifier in the seven-node network of Fig. 3.

Node	Description	Level
1.	Age of victim	$0-7, 8-12, 13+$
2.	Sex of victim	Male, Female
3.	Location of last sighting	Home, Other
4.	Method of killing	Strangulation, Other
5.	Marital status of offender	Living with partner, Other
6.	Relationship of offender to victim	Known, Stranger
7.	Preconviction status of offender	Yes, No

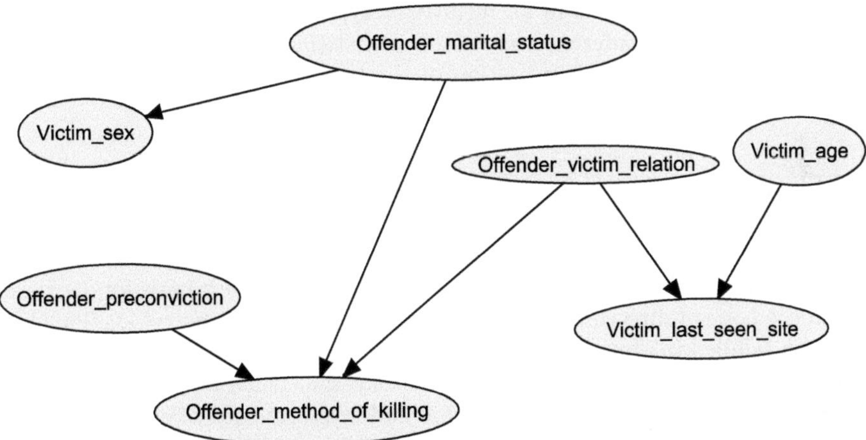

Fig. 3. Seven node network showing the relationships amongst three offender characteristics and four victim and crime scene characteristics. Reproduced from [3].

senior detective so had a check on the consistency between the measure of belief of the detective and the conditional probability estimated from the database. For example, the detective suggested about 70% of offenders had previous convictions, the database had 73% of offenders with previous convictions. The detective suggested about 60% of offenders were known to the victim, the database had 57% of offenders known to the victim.

A more detailed report of the work Alex, I and the other co-authors of [3, 4] is given in [5], a report of the Police Research Group of the Home Office Police Department in 1995. As well as Bayesian belief networks, it contains details of other models whose performances were investigated. These models included logistic regression models with binary and ordinal variables for the prediction of individual offender characteristics and categorical data models for the prediction of a combination of offender characteristics.

In the discussion section of [3] it was noted that

'In theory, it should be possible to construct a network with probability tables for a particular type of crime which may be regarded as a standard network for that type of crime. The expertise of senior police officers, which could be used to construct the initial network, could then be made available to all and could be used in operational conditions.'

In practice, there will, of course, be difficulties associated with special characteristics of a particular crime within a general type of crime. However, if investigators are unhappy with the generic network it is straightforward to amend the network and use that for analysis. There is also an underlying resistance to statistics for situations in which much is subjective, a resistance which was mentioned explicitly in [12] '[r]esistance to the use of statistics ... is often derived from a feeling that the practical person has knowledge that can be incorporated into the inferential process in an intuitive way, but which is not allowed for in the formal scheme of inference used by a statistician'. Bayesian belief networks allow for the 'intuitive' incorporation of the knowledge of the practical person 'into the inferential process' that they represent.

5   Conclusion

In 1989, Alex and I noted that 'The ideas discussed in this paper are at a very tentative stage of development and are published to provide a basis for discussion and in the hope of directing thoughts along useful lines of enquiry' [2]. This hope has been fulfilled in abundance. There have been many, many papers and books (e.g., [6]) published since then which use Bayesian belief networks in research related to forensic science, the evaluation and interpretation of evidence and the administration of criminal justice. This claim is easily verified with an on-line search. For those wishing to use Bayesian belief networks, an on-line search for 'Bayesian belief network software' provides a list of many options.

The importance of the incorporation of the data and the expertise of detectives in the analytical process surrounding the use of Bayesian belief networks has to be emphasised. The success of the methods depended crucially on the accuracy with which the CATCHEM database was constructed and this construction was done by the Derbyshire Police. The expertise of the detectives was also very helpful in the construction of the belief networks to ensure the resulting relationships were sensible.

We acknowledged the referees and editor of [2] for helpful comments on a possible area of application other than the administration of criminal justice. This area was that of databases of intelligence information with possible links between individuals, something that is now of great concern. Another area in which Bayesian belief networks have been proposed is that of international criminal trials, see [11]. There, it was argued that Bayesian belief networks are potentially useful in both the examination of international criminal judgements and the processes of trial preparation and fact-finding before international criminal tribunals. With the use of a practical case study based on a completed case from the International Criminal Tribunal for the former Yugoslavia (ICTY), it was illustrated how Bayesian belief networks could be used by international criminal tribunals to strengthen judges' confidence in their findings, to assist lawyers in preparing for trial, and to provide a tool for the assessment of international criminal tribunals' factual findings.

In the first half of the 2010's, the Statistics and Law Section of the Royal Statistical Society published a series of four reports on the subject of 'Communicating and Interpreting Statistical Evidence in the Administration of Criminal Justice'. The third report, written by myself and Paul Roberts (Professor of Criminal Jurisprudence at The University of Nottingham), concerns 'The logic of forensic proof: inferential reasoning in criminal evidence and forensic science' [13]. The application of Bayesian belief networks (and other methods of inferential reasoning) is described there, in a manner 'shorn of dispensable technology', for those interested in matters relating to the administration of criminal justice and forensic science, such as 'judges, lawyers and forensic scientists who need to concern themselves with fact-finding, case building and the generation, presentation and evaluation of evidence in criminal proceedings'.

Acknowledgments. I am very grateful to Alex for his contributions to my work on evidence evaluation and interpretation in forensic science. His support and computational work are very much appreciated.

I acknowledge helpful discussions with Gloria Laycock and Dick Oldfield with whom I worked on the CATCHEM project. They were able to provide advice during the writing of this chapter on future uses of CATCHEM. Alas, it appears there has been very little future use.

Disclosure of Interests. The author has no competing interests to declare that are relevant to the content of this article.

References

1. Brewer, M.J., Aitken, C.G.G., Luo, Z., Gammerman, A.: Stochastic simulation in mixed graphical association models. In: Dodge, Y., Whittaker, J. (eds.) Computational Statistics, vol. 1, pp. 257–262. Physica-Verlag, Heidelberg, Germany (1992)
2. Aitken, C.G.G., Gammerman, A.: Probabilistic reasoning in evidential assessment. J. Forensic Sci. Soc. **29**, 303–316 (1989)
3. Aitken, C.G.G., et al.: Statistical modelling in specific case analysis. Sci. Justice **36**(4), 245–255 (1996)
4. Aitken,C.G.G., et al.: Bayesian belief networks with an application in specific case analysis. In: Gammerman,A. (ed.) Computational Learning and Probabilistic Reasoning, pp. 169–184. John Wiley & Sons Ltd., Chichester, U.K.(1996)
5. Aitken,C.G.G., Connolly,T., Gammerman,A., Zhang,G., Oldfield,D.: Predicting an offender's characteristics: an evaluation of statistical modelling. Home Office Police Department, Police Research Group, Special Interest Series – Paper 4 (1995)
6. Taroni, F., Biedermann, A., Bozza, S., Garbolino, P., Aitken, C.G.G.: Bayesian Networks for Probabilistic Inference and Decision Analysis in Forensic Science, 2nd edn. John Wiley and Sons Ltd., Chichester (2014)

7. Lau,V.: Revisiting Textile Fibre Transfer and Persistence to Improve the Evaluation of Forensic Evidence. University of Technology Sydney (Australia) ProQuest Dissertations & Theses, (2024)

8. Gammerman,A., Luo,Z., Aitken,C.G.G., Brewer,M.J.:. Computational systems for mixed graphical models. Adaptive Computing and Information Processing. In: UNICOM Seminars Ltd, Brunel Conference Centre, London. (1994)

9. Gammerman,A., Luo,Z., Aitken,C.G.G., Brewer,M.J.: Exact and approximate algorithms and their implementations in mixed graphical models. In: Gammerman,A. (ed.) Probabilistic Reasoning and Bayesian Belief Networks, pp. 33–53. Alfred Waller Ltd., Henley-on-Thames, U.K. (1995)

10. Lauritzen, S.L., Spiegelhalter, D.J.: Local computations with probabilities on graphical structures and their application to expert systems (with Discussion). J. Royal Stat. Soc. Series B **50**, 157–224 (1988)

11. McDermott, Y., Aitken, C.G.G.: Analysis of evidence in criminal trials using Bayesian belief networks. Law, Probability and Risk **16**, 111–129 (2017)

12. Holgate, P.: Contribution to discussion of 'Statistical inference of phylogenies', by Felsenstein. J. Royal Stat. Soc. Series A **146**, 264 (1983)

13. Roberts,P. and Aitken,C.G.G.: The logic of forensic proof: inferential reasoning in criminal evidence and forensic science. Royal Stat. Soc. London (2014). https://rss.org.uk/news-publication/publications/law-guides/

Randomness, Exchangeability, and Conformal Prediction

Vladimir Vovk[(✉)]

Department of Computer Science, Royal Holloway, University of London, London, UK
`v.vovk@rhul.ac.uk`

Abstract. This paper argues for a wider use of the functional theory of randomness, a modification of the algorithmic theory of randomness getting rid of unspecified additive constants. Both theories are useful for understanding relations between the assumptions of IID data and data exchangeability. While the assumption of IID data is standard in machine learning, conformal prediction relies on the weaker assumption of data exchangeability. Nouretdinov, V'yugin, and Gammerman showed, using the language of the algorithmic theory of randomness, that conformal prediction is a universal method under the assumption of IID data. In this paper, I will selectively review connections between exchangeability and the property of being IID, early history of conformal prediction, my encounters and collaboration with Alex and other interesting people, and a translation of Nouretdinov et al.'s results into the language of the functional theory of randomness, which moves it closer to practice. Namely, the translation says that every confidence predictor that is valid for IID data can be converted into a conformal predictor without losing much in predictive efficiency.

Keywords: Conformal prediction · universality of conformal prediction · fundamental limitation of conformal prediction · functional theory of randomness · IID · exchangeability · p-values · e-values

1 Introduction

The functional theory of randomness was proposed in [50] under the name of non-algorithmic theory of randomness. The algorithmic theory of randomness originated with Kolmogorov in the 1960s [25] and has been extensively developed in numerous papers and books (see, e.g., [40]). It has been a powerful source of intuition, but its weakness is the dependence on the choice of a specific universal partial computable function. This dependence leads to the presence of unspecified additive (sometimes multiplicative) constants in its mathematical results. Kolmogorov [24, Sect. 3] speculated that for natural universal partial computable functions the additive constants will be in hundreds rather than in tens of thousands of bits, but this accuracy is very far from being sufficient in machine-learning and statistical applications (an additive constant of 100 in the

K. An Nguyen and Z. Luo (Eds.): Alexander Gammerman Festschrift, LNCS 16290, pp. 87–117, 2026.
https://doi.org/10.1007/978-3-032-15120-9_6

definition of Kolmogorov complexity translates into the astronomical multiplicative constant of 2^{100} in the corresponding p-value).

The way of dealing with unspecified constants proposed in [50] is to express statements of the algorithmic theory of randomness as relations between various function classes. It will be introduced in Sect. 6. In this paper we call this approach the functional theory of randomness. While it loses somewhat in intuitive simplicity, it is closer to practical machine learning and statistics.

The main message of this paper is that the functional theory of randomness can be useful in the foundations of machine learning in general and conformal prediction in particular. The most standard assumption in machine learning is that the data are generated in the IID manner (are independent and identically distributed). An *a priori* weaker assumption is that of exchangeability, although for infinite data sequences being generated in the IID manner and exchangeability turn out to be essentially equivalent by the celebrated de Finetti representation theorem. The classical work on relations between the two assumptions, being IID and being exchangeable, will be the topic of Sect. 2.

The word "random" is often used in two very different senses: in the sense of statistical randomness referring to IID data (as in the title of [56]) and in the sense of algorithmic randomness (as in the title of [40]). In this paper I will try not to use "random" and its derivative "randomness" often, apart from expressions such as "algorithmic theory of randomness" and "functional theory of randomness". For the former sense, I will usually replace derivatives of "random" by compounds containing "IID", such as the *IID assumption* for the assumption that the data are generated in the IID manner (so that simply replacing "IID" by "independent identically distributed" becomes impossible, as in [57]). For the latter sense, I will often use the word "typical" and its derivative "typicalness", which was endorsed by Kolmogorov [40, Appendix 2, footnote 1] in its Russian form типичность and used in [27] (written by Uspensky [27, Introduction]).

Sections 3 and 4 contain personal elements. In Sect. 3 I recount meeting Andrei Kolmogorov and working under his supervision on the relation between IID and exchangeability. In Kolmogorov's frequentist philosophy of probability, the IID property was at the very basis of the notion of probability. In Sect. 4 I recount meeting Alex and then Vladimir Vapnik a decade later. Vapnik was a second person who impressed me with his wholehearted acceptance of the IID assumption, which quickly led to the development of conformal prediction.

In my work under Kolmogorov, I realized that for finite data sequences the difference between IID and exchangeability is important. However, conformal prediction uses only exchangeability. This raises the question whether it is possible to improve on conformal prediction by using the stronger IID assumption. The topic of Sect. 5 is the fundamental result by Nouretdinov, V'yugin, and Gammerman saying that only limited improvement is possible. The result is stated in terms of the algorithmic theory of randomness, making it very intuitive, though the intuition can sometimes be obscured by dense notation.

Despite Nouretdinov et al.'s result being fundamental, it inherits the conceptual weakness of the algorithmic theory of randomness discussed earlier. As it involves unspecified constants, it cannot, strictly speaking, have any practical implications. In Sect. 6 I will state this and several related results in terms of the functional theory of randomness.

In this paper italic e may stand for an e-value. Euler's number (the base of the natural logarithms) is roman e ≈ 2.72. The notation for binary logarithm is lb [9, Sect. 10.1.2]. No detailed knowledge of the algorithmic theory of randomness is assumed on the part of the reader, but knowing the basics would be useful.

2 IID and Exchangeability

Modelling data as IID observations is an ancient notion. Already Jacob Bernoulli [7] was using the IID assumption to state his weak law of large numbers. As Glenn Shafer reminds us in [18, Shafer's comment], the IID case has been central to probability and statistics ever since, "but its inadequacy was always obvious, and Leibniz made the point in his letters to Bernoulli: the world is in constant flux; causes do not remain constant, and so probabilities do not remain constant". There is no doubt the IID assumption is highly restrictive.

The real question is whether the IID assumption is a good *starting point*. It can be fundamental without being all-encompassing; many other, perhaps much more realistic, scenarios may reduce in some way to the IID case. For example, when dealing with prediction or decision making, is it a good strategy first to explore in detail what can be achieved under the IID assumption and then to try and relax it? Or is the assumption so restrictive that it is best to start elsewhere? My views about this have been drifting over time, and even now I remain uncertain.

One relatively modest extension of the IID assumption is the assumption of exchangeability; for potentially infinite data sequences one may even argue that it is not an extension at all. Philip Dawid says in [60, Sect. 7], "For so long, and it's still true of 97% of everything done in statistics and machine learning and everything, the fundamental assumption is basically we have just a bag of exchangeable goodies. And I thought that was just so limiting; how boring. There's a big wide world beyond that." This has been my feeling as well since I started thinking about such things (and among my colleagues, Shafer's and Dawid's views on the philosophy of probability are perhaps closest to mine). However, I have also been impressed by the power of algorithms developed under IID and exchangeability and by many ingenuous ways of greatly relaxing these assumptions.

It is not clear who introduced exchangeability (see [11] for the complicated history), but the most well-known theorem about exchangeability is de Finetti's (generalized in later papers by other people), which connects it with IID. Let $\mathbf{Z}$ be an *observation space* (formally, a measurable space), and suppose that we observe its elements $z_i \in \mathbf{Z}$, $i = 1, 2, \ldots$, sequentially. The IID assumption is that the observations z_i are generated from an *IID probability measure Q^∞*, Q being a probability measure on $\mathbf{Z}$. The assumption of exchangeability is that they are generated from an *exchangeable* probability measure on $\mathbf{Z}^\infty$, i.e., a probability measure that is invariant w.r. to permutations of finitely many observations.

Remark 1. In [56] we referred to probability measures of the form Q^n, with $n = \infty$ allowed, as "power probability measures". In this paper I am using the expression "IID probability measures" instead to simplify terminology. Exchangeability of a probability measure on $\mathbf{Z}^n$ for $n < \infty$ still means invariance w.r. to permutations of observations.

According to de Finetti's theorem (see, e.g., [35, Theorem 1.49]), for infinite sequences, the IID and exchangeability assumptions are equivalent. Namely, each exchangeable probability measure R on $\mathbf{Z}^\infty$ is a convex mixture of IID probability measures: there exists a probability measure μ on the family $\mathfrak{P}(\mathbf{Z})$ of all probability measures on $\mathbf{Z}$ such that

$$R = \int_{\mathfrak{P}(\mathbf{Z})} Q^\infty \, \mu(Q).$$

The theorem makes the weak assumption that $\mathbf{Z}$ is a standard Borel space (and then $\mathfrak{P}(\mathbf{Z})$ is equipped with the smallest σ-algebra making all evaluation functionals measurable).

I find it intuitively compelling that the assumption that the data are generated from a statistical model M (i.e., a family probability measures) is equivalent to the assumption that the data are generated from the convex hull $\bar{M}$ of that statistical model. However, there are people who do not share this intuition (e.g., a friendly reviewer for [58]), so let me try to make it more explicit. The most basic way of testing a statistical model M ("Cournot's principle") is to select in advance a *critical region* A of a small probability under any probability measure in M and reject M if the actual data happens to be in A. Since

$$\sup_{R \in M} R(A) = \sup_{R \in \bar{M}} R(A),$$

rejecting M and rejecting $\bar{M}$ are equivalent. This conclusion is not affected if Cournot's principle is replaced by more sophisticated ways of hypothesis testing, such as using p-variables and e-variables, to be discussed starting from the next section. In particular, the IID and exchangeability assumptions are equivalent. This is far from being true for finite sequences, as will be discussed in detail in Sects. 3–4.

The IID picture is fundamental in the frequentist theory of probability and statistics, at least as it was presented and developed by Richard von Mises [29, 30] and Andrei Kolmogorov [21, Sect. I.2], who was following von Mises. It is well known that Kolmogorov was the first to put the mathematical theory of probability on a firm axiomatic basis in his 1933 book [21]. However, while the axioms of probability introduced in this book eventually (albeit slowly) gained universal acceptance (see, e.g., [38]), the way in which Kolmogorov proposed to connect his axioms with reality [21, Sect. I.2] was informal and has never become widely accepted. According to Kolmogorov's frequentist Principle A, introduced in [21, Sect. I.2], we can say that an event A has a probability $\mathbb{P}(A)$ under a system of conditions $\mathfrak{S}$ if

> One can be practically certain that if the system of conditions $\mathfrak{S}$ is repeated a large number of times, n, and the event A occurs m times, then the ratio m/n will differ only slightly from $\mathbb{P}(A)$.

(When quoting [21, Sect. I.2] I am using the translation given in [38, Sect. 5.2.1].) This principle, which Kolmogorov traces back to von Mises [21, Sect. I.2, footnote 1], gives us a way of measuring $\mathbb{P}(A)$. Presumably the repetitions in Principle A are independent, and so IID observations are at the heart of frequentist probability.

Remark 2. Kolmogorov's approach was not purely frequentist. Alongside his frequentist Principle A he also had a non-frequentist Principle B, namely Cournot's principle:

> If $\mathbb{P}(A)$ is very small, then one can be practically certain that the event A will not occur on a single realization of the conditions $\mathfrak{S}$.

Kolmogorov [21, Sect. I.2] postulated both principles, but it can be argued that Principle B renders Principle A redundant [38, Sect. 5.2]. All the books that I have co-authored so far can be traced back either to Kolmogorov's Principle A [56] or to his Principle B [37,39].

For many years after the publication of his book [21], Kolmogorov talked about connections of his axioms with reality only informally, believing that von Mises's approach, which was based on a flawed definition of an individual infinite sequence of IID observations, could not be cleanly applied to the real world. (See, e.g., [22].) A breakthrough came when Kolmogorov visited India in 1962 [23]. He realized that von Mises's picture can be made applicable to finite sequences (albeit in a way that looks awkward to me, no doubt in hindsight, in view of his later elegant algorithmic approach, which was much more typical of Kolmogorov).

Soon afterwards Kolmogorov came up with his algorithmic theory of randomness subsuming his theory developed in India. In particular, he formalized what it means for a finite binary sequence to be a typical IID sequence. Since Kolmogorov was only dealing with binary sequences, he referred to typical IID sequences as "Bernoulli sequences". The key to his definition was a notion of algorithmic complexity ("Kolmogorov complexity"), and typicalness was defined as maximal complexity in a finite set. This theory was described in his papers [24–26].

Martin-Löf [28] translated Kolmogorov's definition of typicalness into a more standard statistical language defining universal p-values. Later Levin and Gács modified Martin-Löf's definition in an important way, which I will discuss in the next section.

3 Meeting Kolmogorov; IID Vs Exchangeability for Finite Sequences

In 1980 Kolmogorov became Head of the Department of Mathematical Logic at Moscow University, and in the same year I became his student. This happened after I attended his talk aimed at undergraduate students and afterwards spoke to Alexei Semenov, who in his role of the departmental scientific secretary (учёный секретарь кафедры) took care of the administrative side. First I did the specialized part of a Soviet combined BSc/MSc degree programme under Kolmogorov's supervision (the specialized part covering the last three years of the 5-year degree programme), and then I did a PhD under the joint supervision of Kolmogorov and Semenov.

One of the problems that Kolmogorov offered to me was to quantify the qualitative (and intuitively obvious) statement that Bernoulli sequences satisfy his frequentist definition as given in [23]. This was a difficult problem that did not look particularly appealing to me (some results in this direction were obtained later by Kolmogorov himself and his other student Eugene Asarin; see [5, Theorem 3] for Kolmogorov's result and [4, Theorem 1] for Asarin's). Instead, I chose to investigate the relation between IID and exchangeability for finite sequences.

As I mentioned in the previous section, Kolmogorov was only interested in finite sequences in his work on the foundations of probability and believed that infinite sequences, being empirically non-existent, are irrelevant when discussing connections between the mathematical theory of probability and reality. At one point during a walk to a train station Kolmogorov told me that we can only see finite sequences around us, but in the quote from Kolmogorov given in [2, Chap. 7, bottom of p. 57] the word "only" is misplaced (Kolmogorov could not see any infinite sequences, and neither could I).

In my first journal paper [45] I explored Kolmogorov's definition of Bernoulli sequences and argued that it was a formalization of a different kind of typicalness, not typicalness under IID. The term that I used, on Alexander Zvonkin's advice, for Kolmogorov's Bernoulli sequences was von Mises's "collectives", but in hindsight I meant exchangeability (and I talk about exchangeability in the technical report [45] containing the proofs). After that I introduced a definition of typicalness under IID for binary sequences and characterized the difference between the two definitions. To state these results, let me give the relevant (standard) definitions.

I will use the notion of an aggregate of constructive objects, as in [43, Sect. 1.0.6]. This is an infinitely countable set whose elements can be effectively numbered, such as the set $\mathbb{Z}$ of all integer numbers or the set $\{0, 1\}^*$ of all finite binary sequences. Let the observation space $\mathbf{Z}$ range over the finite non-empty subsets of a fixed aggregate of constructive objects. In Kolmogorov's work in this area and in my work reported in this section, $\mathbf{Z} = \{0, 1\}$, but let me give more general definitions for later use (e.g., in the context of conformal prediction). Let $\mathbb{N}$ be the set of natural numbers; by default we do not include 0 in $\mathbb{N}$, so that $\mathbb{N} = \{1, 2, \dots\}$.

A real-valued function f defined on an aggregate of constructive objects is *lower semicomputable* if there is an algorithm that, when fed with v in the domain of f and $r \in \mathbb{Q}$ ($\mathbb{Q}$ being the set of rational numbers), eventually stops if and only if $f(v) > r$. Similarly, it is *upper semicomputable* if this condition holds with $f(v) > r$ replaced by $f(v) < r$. (And computability is the conjunction of lower and upper semicomputability.)

Let me start from a definition equivalent to Kolmogorov's, which is closest to conformal prediction. A *p-test for exchangeability* is a $[0,1]$-valued upper semicomputable function P that takes as input $\mathbf{Z}$, $N \in \mathbb{N}$, and a sequence $z_1, \ldots, z_N$ in $\mathbf{Z}^N$ and that satisfies, for all $\mathbf{Z}$, all N, and all exchangeable probability measures R on $\mathbf{Z}^N$,

$$\forall \epsilon \in (0,1) : R(\{\zeta : P(\zeta) \leq \epsilon\}) \leq \epsilon \tag{1}$$

(omitting, here and later, mentioning $\mathbf{Z}$ and N as arguments; remember that $\mathbf{Z}$ always ranges over the finite subsets of a fixed aggregate of constructive objects). The requirement (1) is usually expressed by saying that P, for fixed $\mathbf{Z}$ and N, is a *p-variable* (and its values are *p-values*). There exists a smallest, to within a constant factor, p-test for exchangeability, which is then called *universal*. Let us fix a universal p-test for exchangeability and let $\mathrm{D^{pX}}$ stand for its minus binary logarithm. We call $\mathrm{D^{pX}}(z_1, \ldots, z_N)$ the *exchangeability p-deficiency* of the sequence $(z_1, \ldots, z_N)$. We can also allow P to depend on a *condition* (an integer number) (and then P is required to be lower semicomputable as function of all its arguments, including the condition); the full notation for the exchangeability p-deficiency is then $\mathrm{D^{pX}}(z_1, \ldots, z_N \mid k)$, where k is the condition.

The function $\mathrm{D^{pX}}$ can be defined to some degree arbitrarily; different choices of $\mathrm{D^{pX}}$, however, will coincide to within an additive constant. This renders results of the algorithmic theory of randomness inapplicable in practice. When we say that two functions (such as $\mathrm{D^{pX}}$) that are only defined to within an additive constant coincide, we mean, of course, that their difference is bounded. In general, when discussing relations (such as inequalities) between such functions, we will always ignore additive constants. Without loss of generality, we will assume that the function $\mathrm{D^{pX}}$, and similar functions introduced later in this paper, are integer-valued.

In the case $\mathbf{Z} = \{0,1\}$, $\mathrm{D^{pX}}(z_1, \ldots, z_N)$ coincides with Kolmogorov's definition of Bernoulliness, as follows from [53, Proposition 11]. Kolmogorov's expression for "$\mathrm{D^{pX}}(z_1, \ldots, z_N) \leq m$" was "$z_1, \ldots, z_N$ is m-Bernoulli" ("m-бернуллиевская").

In a similar way, we can define the *IID p-deficiency* $\mathrm{D^{pR}}(z_1, \ldots, z_N)$ of a sequence $(z_1, \ldots, z_N)$. The only difference is that we replace p-tests for exchangeability by *p-tests for IID* defined by letting R in the definition (1) range over the IID probability measures Q^N, Q being a probability measure on $\mathbf{Z}$ generating one observation. A universal p-test for IID also exists; we let $\mathrm{D^{pR}}$ stands for its minus binary logarithm and call $\mathrm{D^{pR}}(z_1, \ldots, z_N)$ the *IID p-deficiency* of $z_1, \ldots, z_N$. This is equivalent to my definition proposed in [45].

To establish connections between $\mathrm{D^{pX}}$ and $\mathrm{D^{pR}}$, I followed a strategy that is standard in the algorithmic theory of randomness. While hypothesis testing in classical statistics is based on p-values, e-values are mathematically much more convenient and often serve as a useful tool. Universal e-values were introduced in the algorithmic theory of randomness by Levin and then simplified by Gács

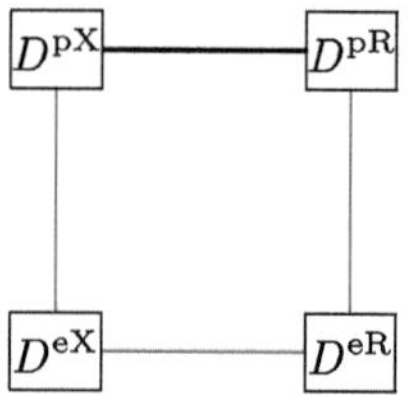

Fig. 1. Connections between 4 deficiencies of typicalness: The connection between D^{pX} and D^{pR} (shown as thick black line) is established via the chain D^{pX}–D^{eX}–D^{eR}–D^{pR} (shown as thin red lines). (Color figure online)

[15] (see also [61]), without using this expression, and nowadays non-universal e-values are gaining popularity in statistics (see, e.g., [20,33,62]). Therefore, the Martin-Löf-style functions D^{pX} and D^{pR} were connected in [45] by connecting D^{pX} and D^{eX}, then D^{eX} and D^{eR}, and finally D^{eR} and D^{pR}, where D^{eX} and D^{eR} are Levin-style analogues of D^{pX} and D^{pR}, to be defined momentarily. (The connections are shown in Fig. 1.)

An *e-test for exchangeability* is a nonnegative lower semicomputable function E on the same domain as a p-test for exchangeability, but it is required to satisfy

$$\sum_{\zeta \in \mathbf{Z}^N} E(\zeta)R(\{\zeta\}) \leq 1 \tag{2}$$

in place of (1) for all $\mathbf{Z}$, N, and exchangeable R. Both upper semicomputability for P (in (1)) and lower semicomputability for E (in (2)) are natural requirements: we reject the null hypothesis of exchangeability (or IID) when a p-value is small (say, below some threshold such as 1% or 5%) or an e-value is large, and the decision to reject should be taken in finite time. The condition (2) means that, for fixed $\mathbf{Z}$ and N, E is an *e-variable*, and then its values are *e-values*. We fix a universal (this time meaning largest to within a constant factor) e-test for exchangeability and call its binary logarithm D^{eX} *exchangeability e-deficiency*. Replacing exchangeable R by $R := Q^N$, we get the definition of D^{eR}, *IID e-deficiency*.

For any data sequence $z_1, \ldots, z_N$, let us define the IID e-deficiency of the corresponding *configuration* $\{z_1, \ldots, z_N\}$, i.e., of the bag (multiset) of its elements, as

$$\mathrm{D}^{\mathrm{eR}}(\{z_1, \ldots, z_N\}) := \min_{\pi} \mathrm{D}^{\mathrm{eR}}(z_{\pi(1)}, \ldots, z_{\pi(N)}), \tag{3}$$

π ranging over all permutations of $\{1, \ldots, N\}$. In other words, a bag is IID (compatible with the IID assumption) if it can arise from an IID data sequence. If ζ is a data sequence, we let $\{\zeta\}$ stand for the bag of its elements.

The following relation between IID and exchangeability is stated in [45, Theorem 1] for $\mathbf{Z} = \{0,1\}$ and in [55, Theorem 3] in general. It uses "$=^{+}$" to mean coincidence of two functions to within an additive constant; $\mathbf{Z}^{+}$ is the family of all non-empty finite sequences of observations. Remember that $\mathbf{Z}$ varies over the finite non-empty subsets of a fixed aggregate of constructive objects.

Theorem 1. *Let ζ range over $\mathbf{Z}^+$ for a variable $\mathbf{Z}$. Then*

$$D^{eR}(\zeta) =^+ D^{eR}(\mathfrak{z}\zeta\mathfrak{f}) + D^{eX}(\zeta \mid D^{eR}(\mathfrak{z}\zeta\mathfrak{f})). \tag{4}$$

Theorem 1 clarifies the relation between exchangeability and IID in the case of finite sequences: a sequence is IID if and only if it is exchangeable and its configuration is IID. The difference between the deficiencies of IID and exchangeability of a data sequence is, roughly, the IID deficiency of its configuration; the condition " $\mid D^{eR}(\mathfrak{z}\zeta\mathfrak{f})$" in (4) slightly obscures this, but (4) implies, e.g.,

$$D^{eR}(\mathfrak{z}\zeta\mathfrak{f}) + D^{eX}(\zeta) \leq^+ D^{eR}(\zeta) \leq^+ 1.01\, D^{eR}(\mathfrak{z}\zeta\mathfrak{f}) + D^{eX}(\zeta)$$

(with "$\leq^+$" denoting the inequality to within an additive constant).

Another result that I obtained in [45] (Theorem 2) was about how big the difference given by (4) between Kolmogorov's and my definitions can be. In the binary case considered in that paper, the configuration $\mathfrak{z}\zeta\mathfrak{f}$ carries the same information as the number of 1 s in ζ given its length N. Let k be the number of 1 s. Then $D^{eR}(k)$ can be characterized as the typicalness deficiency of k in its neighbourhood of size $\sqrt{k(N-k)/N}$ (approximately $\sqrt{N}$ if k is neither very small nor very large). Informally, this is a requirement of local typicalness; e.g., $k = \lfloor N/2 \rfloor$ is untypical for a large N since it is described in such a simple way (given N, which the definition assumes). This characterization implies that the difference between D^{eX} and D^{eR} can be as large as $\frac{1}{2} \operatorname{lb} N$ on data sequences of length N, but not larger.

In the binary case the difference between IID and exchangeability appears small, of the order of magnitude $O(\log N)$, which is much less than the attainable upper bound of $N + o(N)$ on D^{eX} and D^{eR}. In the algorithmic theory of randomness, coincidence to within a logarithmic term is often considered as being sufficiently close to disregard the difference.

These statements are also true about the definitions D^{pX} and D^{pR} in terms of p-values, as these inequalities show:

$$\begin{aligned} D^{eX} &\leq^+ D^{pX} \leq^+ D^{eX} + 2\operatorname{lb} D^{eX}, \\ D^{eR} &\leq^+ D^{pR} \leq^+ D^{eR} + 2\operatorname{lb} D^{eR} \end{aligned} \tag{5}$$

(they can be proved in the same way as Proposition 1 in [32]). We can see that D^{eX} and D^{pX}, as well as D^{eR} and D^{pR}, coincide to within logarithmic terms. Therefore, D^{pX} and D^{pR} also coincide to within a logarithmic term. This may be the reason why Kolmogorov used D^{pX} rather than D^{pR} as formalization, in the binary case, of "a result of independent tests with a probability p of getting a one during each test" [25, Sect. 2].

Remark 3. In [25, Sect. 2] Kolmogorov says (in translation) about his proposed definition, "We view, approximately, in this manner 'Bernoulli sequences' where separate signs are 'independent' and appear with a certain probability p." It is natural to assume, which I did, that the word "approximately" here means that he is ignoring the $O(\log N)$ difference between the two deficiencies (IID vs exchangeability). However, Kolmogorov told me that this was not what he meant (and he did not elaborate further).

From the vantage point of conformal prediction, however, the difference of $\frac{1}{2}\operatorname{lb} N$ between D^{eX} and D^{eR} is not small at all. Before discussing this, let us check that this difference persists when we move to the p-versions, D^{pX} and D^{pR}. Indeed, let us consider a data sequence $\zeta \in \{0,1\}^N$ which is a typical element of the set of all binary sequences of length N containing exactly $\lfloor N/2 \rfloor$ 1 s, for a large N. Then $\mathrm{D}^{\mathrm{pX}}(\zeta)$ will be close to 0, while, by the local limit theorem [41, Sect. 1.6], $\mathrm{D}^{\mathrm{pR}}(\zeta)$ will be close to $\frac{1}{2}\operatorname{lb} N$. Therefore, the difference between D^{pX} and D^{pR} can also be as large as $\frac{1}{2}\operatorname{lb} N$.

An ideal picture of conformal prediction will be introduced in the next section, but what is important for us now is that the largest p-deficiency at which an ideal conformal predictor can reject a false label for a test object is $\operatorname{lb} N$, where N is the length of the "augmented training sequence" (for details, see (10) below). Another manifestation of this phenomenon, which we will call the "fundamental limitation of conformal prediction", is the fact that the smallest possible conformal p-value is $1/N$. Now $\frac{1}{2}\operatorname{lb} N$ does not look small anymore. Even in the binary case, the difference between D^{pX} and D^{pR} can eat up half of the largest p-deficiency achievable by an ideal conformal predictor. In the non-binary case, the difference between IID and exchangeability becomes even more substantial; see inequality (8) below and its discussion.

The paper [45] did not contain any proofs. Full proofs were first published only in 2016, but the main components appeared in [48], as indicated in [45, Appendix C].

Remark 4. There are versions of de Finetti's theorem for finite sequences that assert near equivalence between a finite sequence being IID and being a prefix of a much longer finite sequence that is exchangeable (see, e.g., [13,14] for much stronger results). This idea of using exchangeable extensions makes it possible to adapt de Finetti's theorem to finite sequences, but in this paper we are only interested in basic exchangeability, with a fixed length of the data sequence.

4   Meeting Alex and Vapnik; Emergence of Conformal Prediction

I first met Alex in Barcelona at EuroCOLT 1995, the Second European Conference on Computational Learning Theory (Barcelona, Spain, 13–15 March 1995). Shortly before that, Norman Gowar, the Principal of Royal Holloway, University of London, had suggested that Alex become the next Head of Department of Computer Science. Despite some initial misgivings, Alex agreed. He proposed establishing a machine-learning group in the department, and the Principal enthusiastically supported him. One of Alex's goals in attending EuroCOLT

1995 was to meet, as new Head of Department, active researchers in machine learning.

It is interesting that the First European Conference on Computational Learning Theory had been held at Royal Holloway, University of London, on 20–22 December 1993, yet neither Alex nor I attended it (even though Alex had started teaching at Royal Holloway in September of that year).

Apart from discussing research at EuroCOLT 1995 (in particular, I learned about Alex's interest in Kolmogorov complexity), I remember an enjoyable walk past the Columbus monument at the bottom of La Rambla, Barcelona's iconic pedestrian street. My paper presented at the conference (and published in the proceedings as [47]) elaborated on the key element of the connection between IID and exchangeability found in [45] (I talked about it at length in the previous section). Later Paul Vitányi invited me to submit an extended version of [47] to a Special Issue of the *Journal of Computer and System Sciences* devoted to EuroCOLT 1995; the extended journal version appeared as [48] and later led to the publication of the proofs in [45].

In the summer of 1995 I moved to Stanford to spend a year at the Centre for Advanced Studies in the Behavioral Sciences (now part of Stanford University but then an independent institution). It had been difficult to survive doing science in Russia (I even attended a Business School in Moscow for a year, with an internship in the USA in the summer of 1992), and when Alex invited me to apply for a lectureship position at Royal Holloway to join the emerging machine-learning group (later called CLRC, Computer Learning Research Centre), I saw it as an exciting opportunity. In December 1995 I had an interview there, and at the same time my friend Philip Dawid arranged a backup interview at UCL in case I was unsuccessful. As it turned out, I was successful at Royal Holloway— and I have no idea how I fared at UCL. Even though I was appointed as lecturer, Alex told me that I would be promoted to Professorship within three years—and indeed, I was.

Alex's first two hires were Vladimir Vapnik (part-time) and, shortly afterwards, me. My family and I moved permanently to the UK in June 1996. That summer, Alex, Vapnik, and I had very fruitful discussions which later led, among other things, to the development of conformal prediction. Vapnik was working (mainly or even exclusively) on support vector machines and writing his 1998 book [44], we discussed them repeatedly, and I was eager to contribute.

Before meeting Vapnik, I had not taken the IID assumption particularly seriously. My philosophy was affected by my work on what Shafer and I later called "game-theoretic probability" ([12, 46], with later books [37, 39] joint with Shafer), and as I mentioned earlier, this assumption appeared narrow to me. But Vapnik was taking it very seriously and in many cases did not even mention explicitly that he was making it (which at first even made it difficult for me to follow his arguments). It was a live demonstration of its importance, and indeed I soon realized that it was the most fundamental assumption in machine learning. And the problem of prediction under the IID assumption looked much more down-to-earth and less philosophical than providing frequentist foundations of probability (my preferred approach to the foundations of probability being based on Kolmogorov's Principle B rather than Principle A).

It was very natural to apply what I knew about typicalness deficiency to Vapnik's IID picture, as described briefly in [56, Sect. 2.9.2]. The ideal picture of prediction under IID or exchangeability is straightforward (and described in [55]). Let us suppose that each observation z consists of two components, an object x and its label y, and our task is to predict the label of a test object. Suppose the observation space $\mathbf{Z} := \mathbf{X} \times \mathbf{Y}$ is finite, where $\mathbf{X}$ is the object space and $\mathbf{Y}$ is the label space, both non-empty. Let $|\mathbf{Y}| > 1$. The possibility of the decomposition $\mathbf{Z} := \mathbf{X} \times \mathbf{Y}$ does not restrict generality since we allow $|\mathbf{X}| = 1$. The upper or lower semicomputable functions producing IID and exchangeability deficiencies are given both $\mathbf{X}$ and $\mathbf{Y}$ as inputs (which are subsets of fixed aggregates of constructive objects). Given a training sequence $z_1, \ldots, z_n$ and a test object x_{n+1}, our task is to predict the label y_{n+1} of x_{n+1}. We say that y_{n+1} is the *true label* of the test object x_{n+1} while labels $y \neq y_{n+1}$ are *false*. The number $N := n + 1$ can be interpreted, as is often done in conformal prediction, as the length of the "augmented training sequence" (the training sequence extended by the test object x_{n+1} with a possible label y).

In "universal prediction" we can use typicalness deficiency for evaluating the plausibility of various potential labels for the test object x_{n+1}. For that we can use any of D^{pX}, D^{pR}, D^{eX}, or D^{eR}, but for concreteness, let us concentrate on Kolmogorov's D^{pX} (which is particularly close to conformal prediction).

Remark 5. In the rest of this paper I will avoid using the expression "universal prediction" because of another unfortunate terminological clash (in addition to that between statistical randomness and algorithmic randomness discussed in Sect. 1). On one hand, the adjective "universal" may mean "related to universal partial computable functions", as in "universal p-test". On the other hand, Nouretdinov et al. [31] applied it to conformal prediction meaning that, under IID, it does not lose much in efficiency as compared with any other prediction method that is valid in the same sense. The two meanings are very different, so I will usually say "ideal prediction" rather than "universal prediction" when talking about prediction in the ideal picture based on universal partial computable functions.

Our prediction and how confident we can be in it can be figured out by looking at the exchangeability deficiencies

$$f(y) := \mathrm{D}^{\mathrm{pX}}(z_1, \ldots, z_n, x_{n+1}, y) \tag{6}$$

for various potential labels $y \in \mathbf{Y}$ for the test object. (I am omitting parentheses in expressions such as $\mathrm{D}^{\mathrm{pX}}(z_1, \ldots, z_n, (x_{n+1}, y))$ if this is unlikely to lead to a misunderstanding.) For example, we can use

$$\hat{y}_{n+1} \in \arg\min_{y \in \mathbf{Y}} \mathrm{D}^{\mathrm{pX}}(z_1, \ldots, z_n, x_{n+1}, y)$$

(let us assume, for simplicity, that the arg min is attained at one point only) as the *point prediction* for the true test label y_{n+1}. However, the full *prediction* *function f* defined by (6) contains a lot of other useful information. For example, we can be confident that our point prediction is correct, $\hat{y}_{n+1} = y_{n+1}$, if the second smallest value $f(y)$ is large (presumably the smallest value is $f(y_{n+1})$ under exchangeability).

The point prediction $\hat{y}_{n+1}$ complemented by the second smallest value of f is a useful summary of the full prediction function f. Another way to summarize the prediction function (6) is to fix a significance level $\epsilon \in \mathbb{Q}$ (such as 5% or 1%) and output a prediction set using $-\operatorname{lb}\epsilon$ as threshold,

$$\Gamma^\epsilon := \{y \in \mathbf{Y} : \mathrm{D}^{\mathrm{p^X}}(z_1, \ldots, z_n, x_{n+1}, y) < -\operatorname{lb}\epsilon\} \tag{7}$$

(as in [56, Sect. 2.2.4] but with p-values measured on the logarithmic scale). Notice that in this ideal picture the prediction sets are constructively closed (i.e., their indicator functions are upper semicomputable), which is natural: when computing the ideal prediction sets we keep making them narrower and narrower (i.e., better and better) as time passes.

The next question is how to make this ideal picture computable, so that we could use, e.g., support vector machines to find some practical approximations to the ideal prediction sets (7). This was a well-rehearsed step, which I had done earlier in, e.g., [46] when developing game-theoretic probability. The idea is to use the algorithmic theory of randomness for getting a clear intuitive picture of some area of probability or statistics, and then to strip the picture of its algorithmic content. This makes results more precise and, in particular, eliminates unspecified additive constants. The process is described in detail in [59, Sect. 6], where Shafer and I present the algorithmic theory of randomness as a tool of discovery.

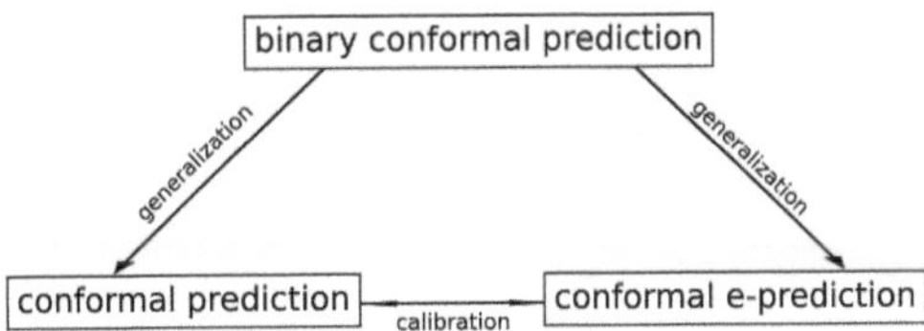

Fig. 2. Binary conformal prediction as special case of both conformal prediction and conformal e-prediction. Calibration will be discussed in Sect. 6.4.

Conformal prediction in its primitive binary form, which is a special case of both conformal prediction and conformal e-prediction, was introduced in [16]. It was applicable only to binary classification, since it was based on support vector machines. The nonconformity measure used in that paper assigns nonconformity scores of 1 to support vectors and nonconformity scores of 0 to all other observations. Let us call conformal prediction based on binary conformity measures *binary conformal prediction*. See Fig. 2 for a pictorial representation, and see [63, Sect. 2] for a more general comparison of hypothesis testing based on p-values and e-values with binary testing based on Cournot's principle. The exposition in [16], however, emphasized conformal e-prediction much more prominently than conformal prediction.

Remark 6. In [16] we apply binary conformal prediction to a binary classification problem. This is a random coincidence, and binary conformal prediction is applicable to a wide range of prediction problems (see [52, Sect. 3] for an example).

Perhaps the first public announcement of conformal prediction in a wide sense (namely, of binary conformal prediction) happened in Alex's inaugural lecture in December 1996, which was later published locally as [17]. He and I regarded it as a public report about the work carried out at CLRC and worked on it together. My main contribution was to the section "Transduction" describing the binary conformal predictor based on support vector machines and connecting it to Vapnik's idea of transductive inference. Conformal prediction proper was introduced in [55] (and soon afterwards in [34], which mainly concentrated on support vector machines).

Untypically for literature on conformal prediction, the paper [55] that introduced it paid some attention to the ideal picture based on (6). One remark that it makes [55, Remark 5] is that, in the binary classification problem ($\mathbf{Y} = \{-1, 1\}$), if the true data sequence $z_1, \ldots, z_N$, where $N = n + 1$ and $z_N = (x_N, y_N)$ is the true test observation, is typical under IID, then the maximal value of the prediction function (6) will be $\operatorname{lb} N$. This is again a manifestation of the fundamental limitation of conformal prediction already mentioned in the previous section.

Two other results stated in [55] extended relations between IID and exchangeability discussed in the previous section to the case of a general observation space $\mathbf{Z}$; it turned out that Theorems 1 and 2 of [45] behave very differently when the assumption $\mathbf{Z} = \{0, 1\}$ is dropped. Theorem 1 carries over to the case of general $\mathbf{Z}$ without any problems. On the other hand, a chasm between IID and exchangeability opens up when $\mathbf{Z}$ is large (or even infinite, which is allowed in [55]); namely,

$$\sup_{\zeta \in \mathbf{Z}^N} \mathrm{D}^{\mathrm{pR}}(\wr \zeta \smallint) \geq^+ \sup_{\zeta \in \mathbf{Z}^N} \mathrm{D}^{\mathrm{eR}}(\wr \zeta \smallint) \geq^+ N \operatorname{lb} \mathrm{e} - \frac{1}{2} \operatorname{lb} N \tag{8}$$

[55, Theorem 4], where $\mathrm{D}^{\mathrm{pR}}(\wr \cdot \smallint)$ is defined analogously to (3). The difference between IID and exchangeability deficiencies now dwarfs the best p-deficiency of $\operatorname{lb} N$ that can be used for prediction (the fundamental limitation of conformal prediction; cf. (10) below). Formally, we did not allow $|\mathbf{Z}| = \infty$ (just for simplicity of definitions), but it is sufficient to assume that $|\mathbf{Z}| \geq N$. There are no proofs in [55], but the argument in the proof of [50, Theorem 2, (16)] also proves (8).

In principle, the vast difference (8) per se does not necessarily imply that IID and exchangeability are so very different: *a priori*, both can be very large, much greater than the difference. However, we can complement (8) by

$$\sup_{\zeta \in \mathbf{Z}^N} \mathrm{D}^{\mathrm{pX}}(\wr \zeta \smallint) =^+ \sup_{\zeta \in \mathbf{Z}^N} \mathrm{D}^{\mathrm{eX}}(\wr \zeta \smallint) =^+ 0,$$

where $\mathrm{D}^{\mathrm{pX}}(\wr \cdots \wr)$ and $\mathrm{D}^{\mathrm{eX}}(\wr \cdots \wr)$ are also defined analogously to (3). Therefore, exchangeability deficiency can be small while IID deficiency is large. A specific example of a data sequence demonstrating this is an algorithmically random permutation of $1, \ldots, N$; while it is perfectly exchangeable, it does not look IID at all: given N, its IID p- and e-deficiency is

$$\mathrm{lb}\,\frac{N^N}{N!} \sim N\,\mathrm{lb}\,\mathrm{e} - \frac{1}{2}\,\mathrm{lb}\,N.$$

5 Nouretdinov et al.'s Discovery: Universality of Conformal Prediction

In the previous section, we saw that the difference between IID and exchangeability deficiencies can be very large. Does it mean that, under the IID assumption, we can achieve much more than what can be achieved by conformal prediction, which only relies on exchangeability? An important discovery by Nouretdinov, V'yugin, and Alex [31] was that conformal prediction is universal: we do not lose much even under IID when using conformal prediction. (I was among the authors of early versions of this paper, but at some point Volodya V'yugin's exposition became too technical for me, and I switched to other projects. The final version of the paper is still very generous about my contribution.)

To discuss the universality of conformal prediction under IID, it is useful to distinguish between two sides of our prediction problem. For concreteness, let us talk about IID p-deficiency D^{pR}.

- If the true data sequence $z_1, \ldots, z_n, x_{n+1}, y_{n+1}$ looks IID, i.e.,

$$\mathrm{D}^{\mathrm{pR}}(z_1, \ldots, z_n, x_{n+1}, y_{n+1})$$

  is small, we are in the situation of *prediction proper*; we can output y_{n+1} as a confident prediction for the label of the test object x_{n+1} if

$$\mathrm{D}^{\mathrm{pR}}(z_1, \ldots, z_n, x_{n+1}, y)$$

  is large for all false labels y.
- If the true data sequence $z_1, \ldots, z_n, x_{n+1}, y_{n+1}$ does not look IID, i.e., $\mathrm{D}^{\mathrm{pR}}(z_1, \ldots, z_n, x_{n+1}, y_{n+1})$ is large, we are in the situation of *anomaly detection*; in this case all of $\mathrm{D}^{\mathrm{pR}}(z_1, \ldots, z_n, x_{n+1}, y)$, $y \in \mathbf{Y}$, can be expected to be large.

Nouretdinov et al. were interested in prediction proper, which is the most natural setting of the prediction problem. While the vast difference between IID and exchangeability might well show in anomaly detection, it does not have to show in prediction proper. In this terminology, the remark in [55, Remark 5] mentioned in the previous section says that, in the situation of prediction proper, the maximal value of the prediction function is $\mathrm{lb}\,N$ (in the case of binary classification; remember that $N := n + 1$). Now at least we have a rough coincidence of the upper bounds: what can be achieved under IID (namely, deficiency of $\operatorname{lb} N$) can also be achieved already by conformal prediction ($\operatorname{lb} N$ is allowed by its fundamental limitation). This coincidence hints at the universality of conformal prediction, but Nouretdinov et al. [31] paint a much fuller picture.

An e-test E for exchangeability or IID is said to be *train-invariant* if, for all n and for all data sequences $(z_1, \ldots, z_n, z_{n+1}) \in \mathbf{Z}^{n+1}$,

$$E(z_1, \ldots, z_n, z_{n+1}) = E(z_{\sigma(1)}, \ldots, z_{\sigma(n)}, z_{n+1})$$

for all permutations σ of $\{1, \ldots, n\}$. In this definition, $z_1, \ldots, z_n$ is interpreted as training sequence and z_{n+1} as test observation. If such an E is used as predictor (e.g., replacing D^{pX} in (6) by E), we can refer to $\langle z_1, \ldots, z_n \rangle$ as *training bag*, or, colloquially, as *training set*, which is a standard expression in machine learning; E does not depend on the ordering of the bag. In the same way we define train-invariant p-tests for exchangeability. (Nouretdinov et al. used the expression "invariant" for our "train-invariant", but in this paper we will use "invariant" in a different, much narrower, sense.)

The first result reported in [31], their Proposition 1, is Ilia Nouretdinov's observation that the class of conformal predictors (understood to be functions producing conformal p-values for all possible labels $y \in \mathbf{Y}$ for the test object) essentially coincides with the class of train-invariant p-tests for exchangeability. Namely, each function in the former class is dominated (in the sense of being less than or equal to) by some function in the latter class, and vice versa.

It is also easy to check that the class of train-invariant e-tests for exchangeability essentially coincides, in the same sense, with the class of conformal e-predictors as defined in [51]. (The only difference is that "dominates" means "is greater than or equal to" in the case of e-tests.)

There exist a universal train-invariant p-test for exchangeability, a universal train-invariant e-test for exchangeability, a universal train-invariant p-test for IID, and a universal train-invariant e-test for IID. We fix such tests and denote their binary logarithms (with the sign reversed in the case of the p-tests) by $\mathrm{D}^{\mathrm{ptX}}$, $\mathrm{D}^{\mathrm{etX}}$, $\mathrm{D}^{\mathrm{ptR}}$, and $\mathrm{D}^{\mathrm{etR}}$, respectively.

Remark 7. Nouretdinov et al. [31, Sect. 4.2] used the expression "i-test" rather than "e-test", and I had used "i-values" for "e-values" earlier in [49, Sect. 5]. When working on [62], I misremembered "i-" as "e-". In hindsight, "e-" (standing for "expectation") appears to be a better counterpart of "p-" (standing for "probability" in the context of p-values) than "i-" (standing for "integral").

The following theorem is the main result of [31] (Theorem 2, slightly simplified). In it, n ranges over $\mathbb{N}$, $\mathbf{Z} = \mathbf{X} \times \mathbf{Y}$ as before, $(z_1, \ldots, z_n)$ (training sequence) ranges over $\mathbf{Z}^n$, (x_{n+1}, y_{n+1}) (test observation) over $\mathbf{Z}$, and y (possible labels of x_{n+1}) over $\mathbf{Y}$.

Theorem 2. *Letting $\zeta := (z_1, \ldots, z_n)$ stand for the training sequence, we have*

$$\mathrm{D^{pR}}(\zeta, x_{n+1}, y) - 2\,\mathrm{lb}\,\mathrm{D^{pR}}(\zeta, x_{n+1}, y) - 4\,\mathrm{D^{pR}}(\zeta, x_{n+1}, y_{n+1}) - 4\,\mathrm{lb}\,|\mathbf{Y}|$$
$$\leq^{+} \mathrm{D^{ptX}}(\zeta, x_{n+1}, y) \leq^{+} \mathrm{D^{pR}}(\zeta, x_{n+1}, y). \quad (9)$$

Theorem 2 is Nouretdinov et al.'s statement of universality for conformal prediction in classification problems. By classification I mean, informally, prediction with a small number $|\mathbf{Y}|$ of classes. In this case and in the situation of prediction proper (i.e., $\mathrm{D^{pR}}(\zeta, x_{n+1}, y_{n+1})$ also being small), (9) implies that

$$\mathrm{D^{ptX}}(\zeta, x_{n+1}, y) \approx \mathrm{D^{pR}}(\zeta, x_{n+1}, y),$$

i.e., ideal conformal prediction is almost as efficient as ideal prediction under IID.

Since $\mathrm{D^{pR}} \approx \mathrm{D^{eR}}$ and $\mathrm{D^{ptX}} \approx \mathrm{D^{etX}}$ (with the approximate equalities holding to within logarithmic terms), (9) also holds, perhaps with the coefficients 2 and 4 replaced by larger ones, for $\mathrm{D^{eR}}$ and $\mathrm{D^{etX}}$ in place of $\mathrm{D^{pR}}$ and $\mathrm{D^{ptX}}$. In other words, conformal e-prediction is also universal in classification problems.

Theorem 1 connecting IID and exchangeability was an important component of the proof of Theorem 2 in [31] (that component was stated there as Proposition 7). The role of this connection will be clearly seen in the functional version of Nouretdinov et al.'s result stated in the following section.

Now we can discuss properly the fundamental limitation of conformal prediction and its significance. Because of the upper limit of $1/(n+1)$ on conformal p-values, we have

$$\mathrm{D^{etX}}(\zeta, x_{n+1}, y) \leq^{+} \mathrm{D^{ptX}}(\zeta, x_{n+1}, y) \leq^{+} \mathrm{lb}(n+1) = \mathrm{lb}\,N, \quad (10)$$

where $\zeta := (z_1, \ldots, z_n)$. In the situation of classification and prediction proper, $\mathrm{D^{pR}}(\zeta, x_{n+1}, y_{n+1}) =^{+} 0$, (9) implies

$$\mathrm{D^{pR}}(\zeta, x_{n+1}, y) \leq^{+} \mathrm{lb}\,N + O(\mathrm{lb}\,\mathrm{lb}\,N).$$

This continues to hold with $\mathrm{D^{eR}}$ in place of $\mathrm{D^{pR}}$. Therefore, the fundamental limitation of conformal prediction is also a limitation of prediction proper under IID in general classification problems (not necessarily binary classification, as in [55, Remark 5]).

Remark 8. The inequalities "$\leq^{+}$" in (10) are, of course, tight. The most confident prediction can be made where, e.g., the training sequence is $(0, \ldots, 0)$ (n zeros for a large n); then the confidence with which we can predict that the test label is 0 as well is reflected in the large value of

$$\mathrm{D^{etX}}(0, \ldots, 0, 1) =^{+} \mathrm{D^{ptX}}(0, \ldots, 0, 1) =^{+} \mathrm{lb}(n+1).$$

6   Perspective of the Functional Theory of Randomness

As already mentioned, most of the groundbreaking results in [31] (all but Proposition 1) involve unspecified constants. The goal of this section is to explain how the functional theory of randomness makes those results more practical: instead of dealing with functions defined to within an additive constant, now we are dealing with inclusions and other relations between various function classes. Very few proofs will be given, and most of them can be found in [54] (which also covers the case of regression, while this paper is constrained to classification, similarly to [31]). Otherwise, this section is more detailed and self-contained than the previous ones.

In one respect, the setting of this section is simpler than our setting so far; since unspecified constants are gone, the observation space $\mathbf{Z}$ and the length of the training sequence n do not need to vary explicitly. Our setting can also be made more general for free; since the theory of algorithms is also gone, now we just assume that the object space $\mathbf{X}$ is a non-empty measurable space. However, we are still interested in the classification problem, where the label space $\mathbf{Y}$ is finite with $|\mathbf{Y}| \geq 2$ and equipped with the discrete σ-algebra. The observation space $\mathbf{Z} = \mathbf{X} \times \mathbf{Y}$ is then also a measurable space. In informal explanations, I will assume that $\mathbf{Y}$ is a small set, such as in the case of binary classification $|\mathbf{Y}| = 2$ (it might be a good idea for the reader to concentrate on this case, at least at first). Now both $\mathbf{Z}$ and the length n of the training sequence $z_1, \ldots, z_n$ can be fixed throughout the section. Given a new test object x_{n+1}, our task is to predict x_{n+1}'s label y_{n+1}. We will be interested in "confidence predictors", i.e., algorithms for this prediction problem producing valid measures of confidence, such as p-values or e-values. While the notion of confidence predictor is informal, later I will give formal definitions of several classes of confidence predictors.

6.1   Eight Function Classes

To translate Nouretdinov et al.'s results into the functional theory of randomness, it is useful to introduce eight function classes representing eight kinds of confidence predictors based on three dichotomies:

- the assumption about the data-generating mechanism can be IID (R) or exchangeability (X);
- with each potential label of a test object we can associate its p-value or e-value;
- optionally, we can require the train-invariance (abbreviated to "t") of the confidence predictor.

The combination X/p/t corresponds to the conformal predictors, while the combinations R/p and R/e correspond to the most general confidence predictors under the IID assumption. Following [31], one of our goals will be to establish the closeness of the conformal predictors (i.e., X/p/t predictors) to the R/p predictors; this goal is attractive since confidence predictors based on p-values enjoy a more intuitive property of validity than those based on e-values. Our argument will also establish the closeness of the conformal e-predictors (i.e., X/e/t predictors) to the R/e predictors. Such closeness can be interpreted as the universality of conformal prediction and conformal e-prediction. We will also consider simplified versions of these goals.

In the rest of this section we will explore

- the difference between IID and exchangeability predictors in Sect. 6.2 (and these results will be summarized in Corollary 1 and simplified in Corollary 2),
- the effect of imposing the requirement of train-invariance in Sect. 6.3 (summarized in Theorem 6),
- and the difference between confidence predictors based on p-values and those based on e-values in Sect. 6.4.

The overall picture will be summarized in Corollary 3 and simplified in Corollary 4, both in Sect. 6.4.

My informal explanations will sometimes be couched in the language of the "naive theory of randomness" postulating the existence of the largest or smallest, as appropriate, element in each function class; this element will be called "universal". Even though formally self-contradictory, this postulate makes some intuitive sense along the lines of the algorithmic theory of randomness. (To make statements of the naive theory of randomness more palatable, it may sometimes be helpful to qualify them using words such as "almost", but not in this paper.) In this way, instead of using the algorithmic theory of randomness for both formal analysis and intuitive considerations, we may use the functional theory of randomness for the former and the naive theory of randomness for the latter.

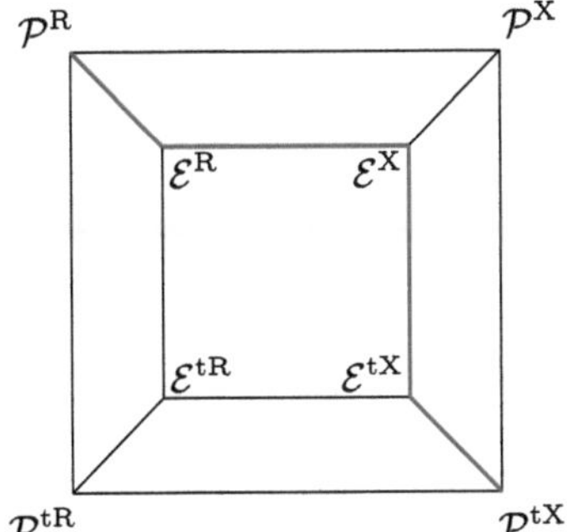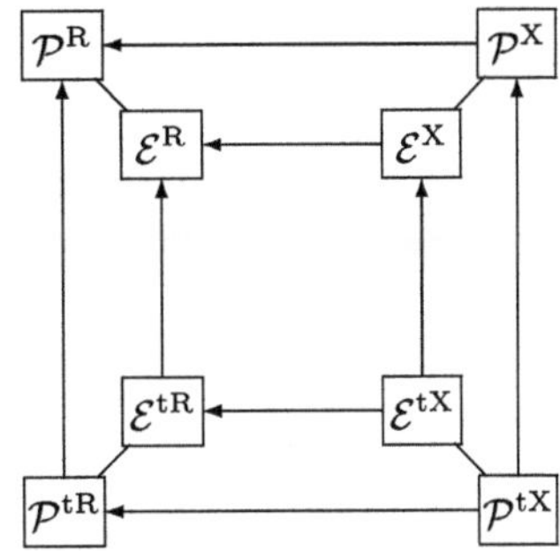

Fig. 3. A cube representing eight function classes. The polygonal chain $\mathcal{P}^{\mathrm{R}}$–$\mathcal{E}^{\mathrm{R}}$–$\mathcal{E}^{\mathrm{X}}$–$\mathcal{E}^{\mathrm{tX}}$–$\mathcal{P}^{\mathrm{tX}}$ is shown in red in the left panel. (Color figure online)

Figure 3 shows the eight function classes as a cube in each panel. Let us concentrate on its left panel for now ignoring the right one. We start from the function class in the top left corner (of the exterior square), $\mathcal{P}^{\mathrm{R}}$. It consists of all *IID p-variables* on $\mathbf{Z}^{n+1}$, i.e., functions $P : \mathbf{Z}^{n+1} \to [0, 1]$ such that, for all IID probability measures $R = Q^{n+1}$ on $\mathbf{Z}^{n+1}$, we have (1). (By default all functions referred to as "variables" are assumed to be measurable.) The importance of the class $\mathcal{P}^{\mathrm{R}}$ stems from IID being the standard assumption of machine learning.

The p-variable P can be used as a "confidence transducer", in the terminology of [56, Sect. 2.7.1]. Given a training sequence $z_1, \ldots, z_n$ and a test object x_{n+1}, we can compute the p-value $P(z_1, \ldots, z_n, x_{n+1}, y)$ for each possible label y for x_{n+1} (as before, p-values and e-values are just values taken by p-variables and e-variables, respectively). We can regard $P(z_1, \ldots, z_n, x_{n+1}, \cdot)$ to be a fuzzy set predictor for y_{n+1}. To obtain a crisp set predictor, we can choose a *significance level* $\epsilon \in (0, 1)$ and define the prediction set

$$\Gamma^\epsilon(z_1, \ldots, z_n, x_{n+1}) := \{y \in \mathbf{Y} : P(z_1, \ldots, z_n, x_{n+1}, y) > \epsilon\} \qquad (11)$$

by thresholding (cf. (7)). By the definition of p-variables, the probability of error (meaning $y_{n+1} \notin \Gamma^\epsilon(z_1, \ldots, z_n, x_{n+1})$) for this crisp set predictor will not exceed ϵ.

In [56, Sect. 2.1.6] confidence predictors were defined as nested families Γ^ϵ, $\epsilon \in (0, 1)$, and were called *conservatively valid* if Γ^ϵ makes an error with probability at most ϵ. This includes the families defined by (11) as a subclass, and in general the inclusion is proper. However, the difference is not essential, as spelled out in Propositions 2.14 and 2.15 of [56]. We will refer to the IID p-variables $P : \mathbf{Z}^{n+1} \to [0, 1]$ as *IID p-predictors* (or, more fully, IID confidence p-predictors).

The top right corner ("Kolmogorov's corner") in Fig. 3, $\mathcal{P}^{\mathrm{X}}$, is the class that consists of all *exchangeability p-variables* on $\mathbf{Z}^{n+1}$, i.e., functions $P : \mathbf{Z}^{n+1} \to [0, 1]$ satisfying (1) for all $\epsilon \in (0, 1)$ and all exchangeable probability measures R on $\mathbf{Z}^{n+1}$. Such p-variables serve as *exchangeability p-predictors*. Naively, a data sequence $\zeta \in \mathbf{Z}^{n+1}$ is exchangeable (resp. IID) if $U(\zeta)$ is not small, U being the universal exchangeability (resp. IID) p-variable.

The bottom left corner of Fig. 3, $\mathcal{P}^{\mathrm{tR}}$, is the class of the train-invariant IID p-variables (elements of $\mathcal{P}^{\mathrm{R}}$), and its bottom right corner $\mathcal{P}^{\mathrm{tX}}$ is the class of the train-invariant exchangeability p-variables.

The top left corner $\mathcal{E}^{\mathrm{R}}$ of the interior square in Fig. 3 consists of all *IID e-variables* on $\mathbf{Z}^{n+1}$, i.e., functions $E : \mathbf{Z}^{n+1} \to [0, \infty]$ such that, for all IID probability measures R on $\mathbf{Z}^{n+1}$,

$$\int E \, dR \leq 1 \qquad (12)$$

(which generalizes (2)). By Markov's inequality, $1/\mathcal{E}^{\mathrm{R}} \subseteq \mathcal{P}^{\mathrm{R}}$ (where $1/\mathcal{E}^{\mathrm{R}}$ consists of all $1/E$, $E \in \mathcal{E}^{\mathrm{R}}$). We will also refer to e-variables $E \in \mathcal{E}^{\mathrm{R}}$ as *IID e-predictors*. For a given training sequence $z_1, \ldots, z_n$ and test object x_{n+1}, the e-value $E(z_1, \ldots, z_n, x_{n+1}, y)$ for each $y \in \mathbf{Y}$ tells us how unlikely y is as label y for x_{n+1}. Therefore, $E(z_1, \ldots, z_n, x_{n+1}, \cdot)$ is again a soft set predictor for y_{n+1}.

The other function classes in Fig. 3 are defined in a similar way; $\mathcal{E}^{\mathrm{X}}$ consists of all *exchangeability e-variables* on $\mathbf{Z}^{n+1}$, i.e., functions $E : \mathbf{Z}^{n+1} \to [0, \infty]$ satisfying (12) for all exchangeable R. Finally, $\mathcal{E}^{\mathrm{tR}}$ and $\mathcal{E}^{\mathrm{tX}}$ consist of all train-invariant functions in $\mathcal{E}^{\mathrm{R}}$ and $\mathcal{E}^{\mathrm{X}}$, respectively. As before, these e-variables may be referred to as e-predictors, depending on context. The right panel of Fig. 3 shows all inclusions between our eight classes, with an arrow $A \to B$ from A to B meaning $A \subseteq B$.

It is interesting that all four confidence predictors on the right of the cubes in Fig. 3 have names (either existing or trivial modifications of existing) containing the word "conformal":

- $\mathcal{P}^{\mathrm{X}}$ (the exchangeability p-predictors) are the weak conformal predictors [56, Sect. 2.2.8 and Proposition 2.9];
- $\mathcal{E}^{\mathrm{X}}$ (the exchangeability e-predictors) are the weak conformal e-predictors;
- $\mathcal{P}^{\mathrm{tX}}$ (the train-invariant exchangeability p-predictors) are the conformal predictors [56, Proposition 2.9];
- $\mathcal{E}^{\mathrm{tX}}$ (the train-invariant exchangeability e-predictors) are the conformal e-predictors (see Sect. 6.3 below).

As already mentioned, the equivalence of $\mathcal{P}^{\mathrm{tX}}$ and conformal prediction was first established in [31, Proposition 1].

Returning to the very informal language of the naive theory of randomness, our discussion of calibration in Sect. 6.4 will show that IID data sequences $\zeta \in \mathbf{Z}^{n+1}$ can be defined as those for which $U(\zeta)$ is not large, U being the universal (i.e., largest in this context) IID e-variable. Similarly, exchangeable data sequences ζ can be defined as those for which $U(\zeta)$ is not large, U being the universal (largest) exchangeability e-variable.

Following [31], we will connect two opposite vertices of the cube in the left panel of Fig. 3, $\mathcal{P}^{\mathrm{R}}$ (IID p-prediction) and $\mathcal{P}^{\mathrm{tX}}$ (conformal prediction). These vertices are important since $\mathcal{P}^{\mathrm{R}}$ corresponds to most general confidence prediction under the standard assumption of machine learning and $\mathcal{P}^{\mathrm{tX}}$ is understood very well and has been widely implemented (see, e.g., [8] and [10]).

A convenient path connecting $\mathcal{P}^{\mathrm{R}}$ and $\mathcal{P}^{\mathrm{tX}}$ is shown as the bold red polygonal chain in the left panel of Fig. 3. It will be used in stating the closeness of $\mathcal{P}^{\mathrm{R}}$ and $\mathcal{P}^{\mathrm{tX}}$ considered as predictors (Corollary 3 below, analogous to Nouretdinov et al.'s main result). Namely, we will establish the closeness for each step in the path separately:

- The step from $\mathcal{P}^{\mathrm{R}}$ to $\mathcal{E}^{\mathrm{R}}$ (from p-values to e-values for IID) is the calibration step, to be discussed in Sect. 6.4.
- The step from $\mathcal{E}^{\mathrm{R}}$ to $\mathcal{E}^{\mathrm{X}}$ (from IID to exchangeability) is the key one; we will call it *Kolmogorov's step*. It is the topic of Sect. 6.2.
- The step from $\mathcal{E}^{\mathrm{X}}$ to $\mathcal{E}^{\mathrm{tX}}$ (adding train-invariance) is easier (if we do not worry about its optimality). We will call it the *train-invariance step*. It is discussed in Sect. 6.3.
- The step from $\mathcal{E}^{\mathrm{tX}}$ (conformal e-prediction) to $\mathcal{P}^{\mathrm{tX}}$ (conformal prediction) is the e-to-p calibration step, and it is also one of the topics of Sect. 6.4.

However, we will also be interested in other edges of the cube in the left panel of Fig. 3, first of all the edge connecting $\mathcal{P}^{\mathrm{tR}}$ and $\mathcal{P}^{\mathrm{tX}}$. A useful connection between these two classes will be obtained as a by-product (Corollary 4).

6.2 Kolmogorov's Step

In principle, besides the eight function classes shown in Fig. 3, we are also interested in the following two:

- the class $\mathcal{E}^{\mathrm{iR}}$ consisting of *invariant IID e-variables*: $E \in \mathcal{E}^{\mathrm{iR}}$ if $E \in \mathcal{E}^{\mathrm{R}}$ and E is invariant w.r. to all permutations of its arguments;
- the analogous class $\mathcal{P}^{\mathrm{iR}}$ consisting of *invariant IID p-variables*, which is the class of all $P \in \mathcal{P}^{\mathrm{R}}$ that are invariant w.r. to all permutations of their arguments.

In this paper we will only use $\mathcal{E}^{\mathrm{iR}}$. When $E \in \mathcal{E}^{\mathrm{iR}}$ is chosen in advance and $E(z_1, \ldots, z_{n+1})$ is large for the realized data sequence $z_1, \ldots, z_{n+1}$, we are entitled to reject the hypothesis that its configuration $\lfloor z_1, \ldots, z_{n+1} \rfloor$ was generated in the IID manner. Therefore, $\mathcal{E}^{\mathrm{iR}}$ is the analogue of $\mathrm{D}^{\mathrm{eR}}(\lfloor \cdot \rfloor)$ in the functional theory of randomness (and $\mathcal{P}^{\mathrm{iR}}$ is the analogue of $\mathrm{D}^{\mathrm{pR}}(\lfloor \cdot \rfloor)$).

The following theorem is the functional version of the relation (4) between IID and exchangeability.

Theorem 3. *The class $\mathcal{E}^{\mathrm{R}}$ is the pointwise product of $\mathcal{E}^{\mathrm{X}}$ and $\mathcal{E}^{\mathrm{iR}}$:*

$$\mathcal{E}^{\mathrm{R}} = \mathcal{E}^{\mathrm{X}} \, \mathcal{E}^{\mathrm{iR}} \, . \tag{13}$$

The pointwise product of function classes $\mathcal{E}_1$ and $\mathcal{E}_2$ is defined as the class of all products $E_1 E_2$ for $E_1 \in \mathcal{E}_1$ and $E_2 \in \mathcal{E}_2$, where $E_1 E_2$ is the pointwise product of functions, $(E_1 E_2)(\zeta) := E_1(\zeta) E_2(\zeta)$. For a proof of Theorem 3, see [50, Corollary 3]. However, since this result is derived in [50] as a corollary of a much more general statement [50, Theorem 1] and in order to make the exposition more self-contained, let me give a simple independent derivation.

Proof (of Theorem 3). We consider the probability space $\mathbf{Z}^{n+1}$ equipped with an IID probability measure R and consider Z_i, $i = 1, \ldots, n+1$, to be z_i regarded as a random observation (formally, Z_i is the random element on that probability space defined by $Z_i(z_1, \ldots, z_{n+1}) := z_i$). The expectation symbol $\mathbb{E} = \mathbb{E}_R$ refers to this probability space.

To prove the inclusion "$\subseteq$" in (13), let $E \in \mathcal{E}^{\mathrm{R}}$. Set

$$F(z_1, \ldots, z_{n+1}) := \frac{1}{(n+1)!} \sum_{\pi} E(z_{\pi(1)}, \ldots, z_{\pi(n+1)}),$$

$$E'(z_1, \ldots, z_{n+1}) := \frac{E(z_1, \ldots, z_{n+1})}{F(z_1, \ldots, z_{n+1})}$$

(with $0/0 := 1$), π ranging over the permutations of $\{1, \ldots, n+1\}$. It is obvious that $E' \in \mathcal{E}^{\mathrm{X}}$, and it is also easy to check that $F \in \mathcal{E}^{\mathrm{iR}}$:

$$\mathbb{E}(F(Z_1, \ldots, Z_{n+1})) = \frac{1}{(n+1)!} \sum_{\pi} \mathbb{E}(E(Z_{\pi(1)}, \ldots, Z_{\pi(n+1)}))$$

$$\leq \frac{1}{(n+1)!} \sum_{\pi} 1 = 1$$

(the inequality uses the fact that $Z_{\pi(1)}, \ldots, Z_{\pi(n+1)}$ are IID).

To prove the inclusion "$\supseteq$" in (13), let $E \in \mathcal{E}^{\mathrm{X}}$ and $F \in \mathcal{E}^{\mathrm{iR}}$. Let us check that their product is in $\mathcal{E}^{\mathrm{R}}$:

$$
\begin{aligned}
&\mathbb{E}\left(E(Z_1, \ldots, Z_{n+1})F(Z_1, \ldots, Z_{n+1})\right) \\
&\quad = \mathbb{E}\left(\mathbb{E}\left(E(Z_1, \ldots, Z_{n+1})F(Z_1, \ldots, Z_{n+1}) \mid \mathcal{G}\right)\right) \\
&\quad = \mathbb{E}\left(F(Z_1, \ldots, Z_{n+1})\mathbb{E}\left(E(Z_1, \ldots, Z_{n+1}) \mid \mathcal{G}\right)\right) \leq \mathbb{E}\left(F(Z_1, \ldots, Z_{n+1})\right) \leq 1,
\end{aligned}
$$

where $\mathcal{G}$ is the bag σ-algebra as defined in [56, Sect. A.5.2]; the first inequality follows from [56, Lemma A.3].

$\square$

In terms of the naive theory of randomness, (13) implies that the universal IID e-variable is the product of the universal exchangeability e-variable and the universal invariant IID e-variable. This shows that the difference between being IID and exchangeability lies in the configuration being IID. Therefore, the following theorem establishes a connection between IID e-predictors and exchangeability e-predictors.

Theorem 4. *For each invariant IID e-variable F there exists an IID e-variable G such that, for all $z_1, \ldots, z_n$, $z_{n+1} = (x_{n+1}, y_{n+1})$, and $y \neq y_{n+1}$,*

$$
G(z_1, \ldots, z_n, z_{n+1}) \geq \frac{1}{\mathrm{e}(|\mathbf{Y}| - 1)}F(z_1, \ldots, z_n, x_{n+1}, y). \tag{14}
$$

We apply this theorem (see Corollary 1 below) in the context where $z_1, \ldots, z_n, z_{n+1}$ is the true data sequence with z_{n+1} being the test observation and y is a false label; ideally such y should be excluded by our confidence predictor. If $z_1, \ldots, z_{n+1}$ is IID and F is the universal invariant IID e-variable, $F(z_1, \ldots, x_{n+1}, y)$ will be small, and so there will be little difference between the degrees to which a false label for the test object will be rejected by the universal e-predictors under IID and exchangeability.

Let me give an informal argument why $\lfloor z_1, \ldots, z_n, x_{n+1}, y \rfloor$ not being IID for $y \neq y_{n+1}$ implies the true data sequence $(z_1, \ldots, z_n, x_{n+1}, y_{n+1})$ not being IID either. Consider, for simplicity, the case of binary labels. If after flipping the last label in the true data sequence $(z_1, \ldots, z_n, x_{n+1}, y_{n+1})$ the bag of its elements becomes non-IID, then either already the original bag $\lfloor z_1, \ldots, z_n, x_{n+1}, y_{n+1} \rfloor$ was non-IID or the last element (x_{n+1}, y_{n+1}) was special in the true data sequence, and in any case already the original data sequence was non-IID. A formal proof is given in [54].

The following asymptotic result says that the $|\mathbf{Y}|$ in the denominator of (14) is in some sense optimal (provided it is large enough).

Theorem 5. *For each constant $c > 1$ the following statement holds true for a sufficiently large $|\mathbf{Y}|$ and a sufficiently large n. There exists an invariant IID* *e/variable F such that for each IID e-variable G there exist $z_1, \ldots, z_n$, $z_{n+1} = (x_{n+1}, y_{n+1})$, and $y \neq y_{n+1}$ such that*

$$G(z_1, \ldots, z_n, z_{n+1}) < \frac{c}{e\,|\mathbf{Y}|} F(z_1, \ldots, z_n, x_{n+1}, y).$$

Theorem 5 is proved in [54]. The idea of the proof can be explained informally using the algorithmic theory of randomness (or even more informally using the naive theory of randomness): we can make the label y in the bag $\{z_1, \ldots, z_n, x_{n+1}, y\}$ encode the bag $\{y_1, \ldots, y_n\}$ of the other labels; if we also make y easily distinguishable from the other labels, the value $F(z_1, \ldots, z_n, x_{n+1}, y)$ of the universal invariant IID e-variable will be large.

Let us now state explicitly a corollary of Theorems 3 and 4 that expresses the universality of weak conformal e-prediction.

Corollary 1. *For each IID e-predictor E there exist an exchangeability e-/predictor E' and an IID e-variable G such that, for all $z_1, \ldots, z_n$, $z_{n+1} = (x_{n+1}, y_{n+1})$, and $y \neq y_{n+1}$,*

$$E'(z_1, \ldots, z_n, x_{n+1}, y) \geq \frac{1}{e(|\mathbf{Y}| - 1)} \frac{E(z_1, \ldots, z_n, x_{n+1}, y)}{G(z_1, \ldots, z_n, z_{n+1})}. \tag{15}$$

The informal interpretation of (15) is that, in classification, every false label y for the test object is excluded by an exchangeability e-predictor once it is excluded by an IID e-predictor, unless the true data sequence $(z_1, \ldots, z_{n+1})$ is not IID. For this interpretation, there is no need to take E, E', and G universal.

Proof (of Corollary 1). Let E be an IID e-variable. By Theorem 3, there exist an exchangeability e-variable E' and an invariant IID e-variable E'' such that

$$E(z_1, \ldots, z_n, x_{n+1}, y) = E'(z_1, \ldots, z_n, x_{n+1}, y)E''(z_1, \ldots, z_n, x_{n+1}, y) \tag{16}$$

for all $z_1, \ldots, z_n, x_{n+1}, y$. By Theorem 4 there exists an IID e-variable G such that

$$G(z_1, \ldots, z_n, z_{n+1}) \geq \frac{1}{e(|\mathbf{Y}| - 1)} E''(z_1, \ldots, z_n, x_{n+1}, y) \tag{17}$$

for all $z_1, \ldots, z_n$, $z_{n+1} = (x_{n+1}, y_{n+1})$, and $y \neq y_{n+1}$. It remains to combine (16) and (17).

$\qquad\qquad\qquad\qquad\qquad\qquad\qquad\qquad\qquad\qquad\qquad\qquad\qquad\qquad\qquad\square$

Remark 9 In Corollary 1 it is possible to have, in principle, 0 in the denominator in (15). Our interpretation of an inequality $A \geq c\frac{B}{C}$, where A, c, B, C are all nonnegative, covering the possibility of $C = 0$ is that it is equivalent, by definition, to $AC \geq cB$. Similar remarks can be made about other results, such as Theorem 6 below.

An important variation on Corollary 1 is where the original IID e-predictor E is already train-invariant. Under the IID assumption, it seems useless to consider predictors that are not train-invariant, and indeed the requirement of train-invariance follows [54, Sect. 2] from fundamental statistical principles, such as the sufficiency principle and the invariance principle. In this case the resulting predictor E' will also be train-invariant and, therefore, a conformal predictor.

Corollary 2. *For each train-invariant IID e-predictor E there exist a conformal e/predictor E' and an IID e-variable G such that, for all $z_1, \ldots, z_n$, $z_{n+1} = (x_{n+1}, y_{n+1})$, and $y \neq y_{n+1}$, we have (15).*

Corollary 2 is a statement of universality of conformal e-prediction under train-invariance.

6.3 Train-Invariance Step

Let us say that $G = G(z_1, \ldots, z_n \mid z_{n+1})$ is a *test-conditional exchangeability e-variable* (given the test observation) if

$$\forall (z_1, \ldots, z_{n+1}) : \frac{1}{n!} \sum_{\sigma} G\left(z_{\sigma(1)}, \ldots, z_{\sigma(n)} \mid z_{n+1}\right) \leq 1,$$

σ ranging over the permutations of $\{1, \ldots, n\}$. This property implies $G \in \mathcal{E}^{\mathrm{X}}$. If $G = G(z_1, \ldots, z_n \mid z_{n+1})$ is large for the universal test-conditional exchangeability e-variable G, the sequence $z_1, \ldots, z_n$ is not exchangeable given z_{n+1}. (And G being large is a stronger property than $z_1, \ldots, z_{n+1}$ not being exchangeable.)

For any exchangeability e-predictor E, define the corresponding train-invariant exchangeability e-predictor $\bar{E}$ by

$$\bar{E}(z_1, \ldots, z_n, z_{n+1}) := \frac{1}{n!} \sum_{\sigma} E\left(z_{\sigma(1)}, \ldots, z_{\sigma(n)}, z_{n+1}\right),$$

σ again ranging over the permutations of $\{1, \ldots, n\}$. This is the train-invariance step. The following theorem says that $\bar{E}$ is almost as good as E in our prediction problem unless $z_1, \ldots, z_{n+1}$ is not exchangeable.

Theorem 6. *For each exchangeability e-predictor E there exists a test-/conditional exchangeability e-variable G such that, for all $z_1, \ldots, z_n$, all $z_{n+1} = (x_{n+1}, y_{n+1})$, and all $y \neq y_{n+1}$,*

$$\bar{E}(z_1, \ldots, z_n, x_{n+1}, y) \geq \frac{1}{|\mathbf{Y}| - 1} \frac{E(z_1, \ldots, z_n, x_{n+1}, y)}{G(z_1, \ldots, z_n \mid z_{n+1})}.$$

The intuition behind Theorem 6 is that each exchangeability e-predictor can be made train-invariant without significant loss of efficiency in classification proper. For a simple proof, see [54].

6.4   Other Steps

In previous sections we discussed the two interior red steps shown in the left panel of Fig. 3. Here we will discuss the two other red steps and summarize the overall picture obtaining a version of [31, Theorem 2] in the functional theory of randomness. These two steps are analogues of the inequalities (5) in the functional theory of randomness.

Conversion from p-values to e-values (*calibration*) and vice versa (*e-to-p calibration*) is understood very well: see, e.g., [62, Sect. 2]. E-to-p calibration is particularly simple: there is one optimal e-to-p-calibrator, $e \mapsto \min(1/e, 1)$ [62, Proposition 2.2]. As for calibration, a decreasing function $f : [0,1] \to [0,\infty]$ is a *calibrator* (transforms p-values into e-values) if and only if $\int_0^1 f \leq 1$ [62, Proposition 2.1]. We will use the calibrator

$$f(p) := \delta p^{\delta-1} \tag{18}$$

for a fixed value $\delta \in (0,1)$. If δ is small, $f(p)$ will be close to $1/p$ if we ignore the multiplicative constant (as customary in the algorithmic theory of randomness). Other popular calibrators are

$$f(p) := \begin{cases} \infty & \text{if } p = 0 \\ \kappa(1+\kappa)^\kappa p^{-1}(-\ln p)^{-1-\kappa} & \text{if } p \in (0, \exp(-1-\kappa)] \\ 0 & \text{if } p \in (\exp(-1-\kappa), 1] \end{cases}$$

for a constant $\kappa > 0$ (see [62, Appendix B]; this calibrator is even closer to $1/p$ than (18) with a small δ) and Shafer's [36, Sect. 3, (6)] calibrator

$$f(p) := p^{-1/2} - 1.$$

The lines between the corresponding $\mathcal{P}$ and $\mathcal{E}$ vertices in the right panel of Fig. 3 stand for the possibility of calibration or e-to-p calibration (similarly to the double-headed arrow in Fig. 2).

The following result combines all the previous statements in this section.

Corollary 3. *Let $\delta \in (0,1)$. For all $P \in \mathcal{P}^{\mathrm{R}}$ there exist $P' \in \mathcal{P}^{\mathrm{tX}}$ and $G \in \mathcal{E}^{\mathrm{R}}$ such that, for all observations $z_1, \ldots, z_n$, $z_{n+1} = (x_{n+1}, y_{n+1})$, and labels $y \neq y_{n+1}$,*

$$P'(z_1, \ldots, z_n, x_{n+1}, y)$$
$$\leq \frac{e(|\mathbf{Y}|-1)^2}{\delta} G(z_1, \ldots, z_{n+1})^2 P(z_1, \ldots, z_n, x_{n+1}, y)^{1-\delta}. \tag{19}$$

Corollary 3 reduces (as usual, imperfectly) IID p-predictors to conformal predictors. It says that in classification problems every false label excluded by an IID p-predictor is excluded by a conformal predictor (perhaps less strongly) unless the true data sequence is non-IID. It is an analogue of [31, Theorem 2].

Proof. (of Corollary3). Let $P \in \mathcal{P}^{\mathrm{R}}$ and $\delta \in (0,1)$. We will construct $P' \in \mathcal{P}^{\mathrm{tX}}$ and $G \in \mathcal{E}^{\mathrm{R}}$ satisfying (19) in several steps. Since (18) is a calibrator, there is $E \in \mathcal{E}^{\mathrm{R}}$ satisfying

$$E(z_1, \ldots, z_n, x_{n+1}, y) \geq \delta P(z_1, \ldots, z_n, x_{n+1}, y)^{\delta-1} \tag{20}$$

(in fact, with "=" in place of "$\geq$"); here and in the rest of the proof we will leave "for all $z_1, \ldots, z_n$, $z_{n+1} = (x_{n+1}, y_{n+1})$, and $y \neq y_{n+1}$" implicit. By Corollary 1, there exist $E' \in \mathcal{E}^{\mathrm{X}}$ and $G_1 \in \mathcal{E}^{\mathrm{R}}$ such that

$$E'(z_1, \ldots, z_n, x_{n+1}, y) \geq \frac{1}{\mathrm{e}(|\mathbf{Y}|-1)} \frac{E(z_1, \ldots, z_n, x_{n+1}, y)}{G_1(z_1, \ldots, z_n, z_{n+1})}. \tag{21}$$

By Theorem 6, there exist $E'' \in \mathcal{E}^{\mathrm{tX}}$ and $G_2 \in \mathcal{E}^{\mathrm{R}}$ such that

$$E''(z_1, \ldots, z_n, x_{n+1}, y) \geq \frac{1}{|\mathbf{Y}|-1} \frac{E'(z_1, \ldots, z_n, x_{n+1}, y)}{G_2(z_1, \ldots, z_n, z_{n+1})}. \tag{22}$$

Finally, since $e \mapsto \min(1/e, 1)$ is an e-to-p calibrator, there is $P' \in \mathcal{P}^{\mathrm{tX}}$ satisfying

$$P'(z_1, \ldots, z_n, x_{n+1}, y) \leq 1/E''(z_1, \ldots, z_n, x_{n+1}, y). \tag{23}$$

It remains to combine (20)–(23) and set $G := \sqrt{G_1 G_2}$. (By the inequality between the geometric and arithmetic means, $G \in \mathcal{E}^{\mathrm{R}}$.)

$$\square$$

Of course, Corollary 3 continues to hold if the condition $P \in \mathcal{P}^{\mathrm{R}}$ is replaced by $P \in \mathcal{P}^{\mathrm{X}}$. In this case, however, we can drop step (21) and replace (19) by

$$P'(z_1, \ldots, z_n, x_{n+1}, y) \leq \frac{|\mathbf{Y}|-1}{\delta} G(z_1, \ldots, z_{n+1}) P(z_1, \ldots, z_n, x_{n+1}, y)^{1-\delta}.$$

The most important case, however, is where $P \in \mathcal{P}^{\mathrm{tR}}$. Now we can drop step (22), as spelled out in the following corollary.

Corollary 4. *Let $\delta \in (0,1)$. For each $P \in \mathcal{P}^{\mathrm{tR}}$ there exist $P' \in \mathcal{P}^{\mathrm{tX}}$ and $G \in \mathcal{E}^{\mathrm{R}}$ such that, for all $z_1, \ldots, z_n$, $z_{n+1} = (x_{n+1}, y_{n+1})$, and $y \neq y_{n+1}$,*

$$P'(z_1, \ldots, z_n, x_{n+1}, y) \leq \frac{\mathrm{e}(|\mathbf{Y}|-1)}{\delta} G(z_1, \ldots, z_{n+1}) P(z_1, \ldots, z_n, x_{n+1}, y)^{1-\delta}.$$

Corollary 4 reduces train-invariant IID p-prediction to conformal prediction without significant loss in efficiency. Since the condition of train-invariance is so natural under the IID assumption, Corollaries 2 and 4 may be the most useful statements in this section.

7   Conclusion

This paper further develops the functional theory of randomness in the direction of Nouretdinov, V'yugin, and Gammerman's work on universality of conformal prediction. While the statements of the functional theory of randomness may be less intuitive than those of the algorithmic theory of randomness, they are more precise avoiding unspecified constants and are simpler in an important respect: e.g., the analogue of (13) in the algorithmic theory of randomness, (4), involves the condition "$\mid \mathrm{D}^{\mathrm{eR}}(\lfloor\zeta\rfloor)$", which disappears in (13). In the naive theory of randomness we could write, instead,

$$\mathrm{D}^{\mathrm{eR}}(\zeta) = \mathrm{D}^{\mathrm{eX}}(\zeta) + \mathrm{D}^{\mathrm{eR}}(\lfloor\zeta\rfloor).$$

For the reader familiar with the algorithmic theory of complexity, the condition "$\mid \mathrm{D}^{\mathrm{eR}}(\lfloor\zeta\rfloor)$" is analogous to the second entry of "$K(x)$" in the Levin–Chaitin formula

$$K(x,y) =^{+} K(x) + K(y \mid x, K(x))$$

[40, Theorem 67]. Conditions of this type can be removed in the functional theory of randomness and functional theory of complexity (the latter introduced in [50, Appendix]).

In Sect. 2 I emphasized the narrowness of both IID and exchangeability assumptions, whereas this paper concentrates on what can be achieved under IID. There have been many developments in conformal prediction beyond exchangeability, such as those in [3,6,19,42]; see also the review [1, Chap. 7]. For all of them, exchangeability serves as a starting point.

Acknowledgments. Many thanks to Alex for sharing his recollections. I am also grateful to Ruodu Wang, Irina Shevtsova, Nicolo Colombo, and Alexander Shen for their advice. As always, the Stack Exchange $T_{\!E}X - -\!\mathit{L\!\!^{A}\!T_{\!E}X}$ community have been ready to help.

References

1. Angelopoulos, A.N., Barber, R.F., Bates, S.: Theoretical foundations of conformal prediction. Tech. Rep. arXiv:2411.11824 [math.ST], arXiv.org e-Print archive (2025). Pre-publication version of a book to be published by Cambridge University Press

2. Angelopoulos, A.N., Bates, S.: Conformal prediction: A gentle introduction. Found. Trends Mach. Learn. **16**(4), 494–591 (2023)

3. Angelopoulos, A.N., Candès, E.J., Tibshirani, R.J.: Conformal PID control for time series prediction. In: Advances in Neural Information Processing Systems 36 (NeurIPS 2023) (2023)

4. Asarin, E.A.: Some properties of Kolmogorov Δ-random finite sequences. Theory of Probability and its Applications **32**, 507–508 (1987), Russian original: О некоторых свойствах Δ-случайных по Колмогорову конечных последовательностей

5. Asarin, E.A.: On some properties of finite objects random in the algorithmic sense. Soviet Math. Doklady **36**, 109–112 (1988). Russian original: О некоторых свойствах случайных в алгоритмическом смысле конечных объектов, published in 1987

6. Barber, R.F., Candès, E.J., Ramdas, A., Tibshirani, R.J.: Conformal prediction beyond exchangeability. Ann. Stat. **51**, 816–845 (2023)

7. Bernoulli, J.: Ars Conjectandi. Thurnisius, Basel (1713)

8. Boström, H.: Conformal prediction in Python with crepes. Proc. Mach. Learn. Res. **230**, 236–249 (2024). COPA 2024

9. British Standards Institution: Quantities and units, Part 2: Mathematical signs and symbols to be used in the natural sciences and technology (2010). BS ISO 80000-2:2009

10. Cordier, T., Blot, V., Lacombe, L., Morzadec, T., Capitaine, A., Brunel, N.: Flexible and systematic uncertainty estimation with conformal prediction via the MAPIE library. Proc. Mach. Learn. Res. **204**, 549–581 (2023). COPA 2023

11. Dale, A.I.: A study of some early investigations into exchangeability. Hist. Math. **12**, 323–336 (1985)

12. Dawid, A.P., Vovk, V.: Prequential probability: principles and properties. Bernoulli **5**, 125–162 (1999)

13. Diaconis, P.: Finite forms of de Finetti's theorem on exchangeability. Synthese **36**, 271–281 (1977)

14. Diaconis, P., Freedman, D.A.: Finite exchangeable sequences. Ann. Probab. **8**, 745–764 (1980)

15. Gács, P.: Exact expressions for some randomness tests. Zeitschrift für Mathematische Logik und Grundlagen der Mathematik **26**, 385–394 (1980)

16. Gammerman, A., Vovk, V., Vapnik, V.: Learning by transduction. In: Proceedings of the Fourteenth Conference on Uncertainty in Artificial Intelligence, pp. 148–155. Morgan Kaufmann, San Francisco, CA (1998)

17. Gammerman, A.: Machine Learning: Progress and Prospects. An Inaugural Lecture by Alexander Gammerman, Professor of Computer Science. Presented at Royal Holloway, University of London, on 11th December 1996. Royal Holloway, University of London, Egham, Surrey (1997)

18. Gammerman, A., Vovk, V.: Hedging predictions in machine learning (with discussion). Comput. J. **50**, 151–177 (2007)

19. Gibbs, I., Candès, E.J.: Adaptive conformal inference under distribution shift. Adv. Neural. Inf. Process. Syst. **34**, 1660–1672 (2021)

20. Grünwald, P., de Heide, R., Koolen, W.M.: Safe testing (with discussion). J. Roy. Stat. Soc. B **86**, 1091–1171 (2024)

21. Kolmogorov, A.N.: Grundbegriffe der Wahrscheinlichkeitsrechnung. English translation: Foundations of the Theory of Probability, 1950. Springer, Berlin. Chelsea, New York (1933)

22. Kolmogorov, A.N.: Теория вероятностей. In: Математика, ее содержание, методы и значение, vol. 2, pp. 252–284. Издательство АН СССР, Moscow (1956)

23. Kolmogorov, A.N.: On tables of random numbers. Sankhyā. Indian J. Stat. A **25**, 369–376 (1963)

24. Kolmogorov, A.N.: Three approaches to the quantitative definition of information. Probl. Inf. Trans. **1**, 1–7 (1965). Russian original: Три подхода к определению понятия количество информации

25. Kolmogorov, A.N.: Logical basis for information theory and probability theory. IEEE Trans. Inf. Theory IT-14, 662–664 (1968). Russian original: К логическим основам теории информации и теории вероятностеЙ, published in Проблемы передачи информации, 1969
26. Kolmogorov, A.N.: Combinatorial foundations of information theory and the calculus of probabilities. Russian Math. Surv. **38**, 29–40 (1983). Russian original: Комбинаторные основания теории информации и исчисления вероятностеЙ
27. Kolmogorov, A.N., Uspensky, V.A.: Algorithms and randomness. Theory Probability Appl. **32**, 389–412 (1987). Russian original: Алгоритмы и случаЙность
28. Martin-Löf, P.: The definition of random sequences. Inf. Control **9**, 602–619 (1966)
29. von Mises, R.: Grundlagen der Wahrscheinlichkeitsrechnung. Math. Z. **5**, 52–99 (1919)
30. von Mises, R.: Wahrscheinlichkeit, Statistik, und Wahrheit. Springer, Berlin (1928). English translation: Probability, Statistics and Truth. William Hodge, London (1939)
31. Nouretdinov, I., V'yugin, V., Gammerman, A.: Transductive Confidence Machine Is Universal. In: Gavaldá, R., Jantke, K.P., Takimoto, E. (eds.) ALT 2003. LNCS (LNAI), vol. 2842, pp. 283–297. Springer, Heidelberg (2003). https://doi.org/10.1007/978-3-540-39624-6_23
32. Novikov, G.: Relations between randomness deficiencies. Tech. Rep. arXiv:1608.08246 [math.LO], arXiv.org e-Print archive (2016). Published in Lecture Notes in Computer Science 10307:338–350 (2017)
33. Ramdas, A., Wang, R.: Hypothesis testing with e-values. Technical Report arXiv:2410.23614 [math.ST], arXiv.org e-Print archive (2025). Book draft
34. Saunders, C., Gammerman, A., Vovk, V.: Transduction with confidence and credibility. In: Dean, T. (ed.). In: Proceedings of the Sixteenth International Joint Conference on Artificial Intelligence. vol. 2, pp. 722–726. Morgan Kaufmann, San Francisco, CA (1999)
35. Schervish: Theory of Statistics. Springer, New York (1995). https://doi.org/10.1007/978-1-4612-4250-5_10
36. Shafer, G.: The language of betting as a strategy for statistical and scientific communication (with discussion). J. Roy. Stat. Soc. A **184**, 407–478 (2021)
37. Shafer, G., Vovk, V.: Probability and Finance: It's Only a Game! Wiley, New York (2001)
38. Shafer, G., Vovk, V.: The sources of Kolmogorov's Grundbegriffe. Stat. Sci. **21**, 70–98 (2006). Extended version: arXiv:1802.06071 [math.HO]
39. Shafer, G., Vovk, V.: Game-Theoretic Foundations for Probability and Finance. Wiley, Hoboken, NJ (2019)
40. Shen, A., Uspensky, V.A., Vereshchagin, N.: Kolmogorov Complexity and Algorithmic Randomness. American Mathematical Society, Providence, RI (2017)
41. Shiryaev: Probability-1, vol. 95. Springer, New York (2016). https://doi.org/10.1007/978-0-387-72206-1_3
42. Tibshirani, R.J., Barber, R.F., Candès, E.J., Ramdas, A.: Conformal prediction under covariate shift. In: Advances in Neural Information Processing Systems 32 (NeurIPS 2019) (2019)
43. Uspensky, V.A., Semenov, A.L.: Algorithms: Main Ideas and Applications. Kluwer, Dordrecht (1993)
44. Vapnik, V.N.: Statistical Learning Theory. Wiley, New York (1998)
45. Vovk, V.: On the concept of the Bernoulli property. Russian Mathematical Surveys 41, 247–248 (1986), Russian original: О понятии бернуллиевости, another English translation with proofs: arXiv:1612.08859 (math.ST)

46. Vovk, V.: A logic of probability, with application to the foundations of statistics (with discussion). J. Roy. Stat. Soc. B **55**, 317–351 (1993)
47. Vovk, V.: Minimum description length estimators under the optimal coding scheme. In: Vitányi, P. (ed.) Computational Learning Theory. Lecture Notes in Computer Science, vol. 904, pp. 237–251. Springer, Berlin (1995). EuroCOLT 1995
48. Vovk, V.: Learning about the parameter of the Bernoulli model. J. Comput. Syst. Sci. **55**, 96–104 (1997). EuroCOLT 1995 Special Issue, this paper is the journal version of [47]
49. Vovk, V.: Kolmogorov's complexity conception of probability. In: Hendricks, V.F., Pedersen, S.A., Jørgensen, K.F. (eds.) Probability Theory: Philosophy, Recent History and Relations to Science, pp. 51–69. Kluwer, Dordrecht (2001)
50. Vovk, V.: Non-Algorithmic Theory of Randomness. In: Blass, A., Cégielski, P., Dershowitz, N., Droste, M., Finkbeiner, B. (eds.) Fields of Logic and Computation III. LNCS, vol. 12180, pp. 323–340. Springer, Cham (2020). https://doi.org/10.1007/978-3-030-48006-6_21
51. Vovk, V.: Conformal e-prediction. Pattern Recogn. **166**, article 111674 (2025). Special Issue on Conformal Prediction and Distribution-Free Uncertainty Quantification
52. Vovk, V.: Inductive randomness predictors: beyond randomness. In: Proceedings of Machine Learning Research, vol. 266, pp. 6–33 (2025)
53. Vovk, V.: Testing exchangeability in the batch mode with e-values and Markov alternatives. Mach. Learn. **114**, article 99 (2025). Special Issue on Conformal Prediction and Distribution-Free Uncertainty Quantification
54. Vovk, V.: Universality of conformal prediction under the assumption of randomness. Technical Report arXiv:2502.19254 [cs.LG], arXiv.org e-Print archive (2025). Published in the Proceedings of ALT 2026
55. Vovk, V., Gammerman, A., Saunders, C.: Machine-learning applications of algorithmic randomness. In: Proceedings of the Sixteenth International Conference on Machine Learning, pp. 444–453. Morgan Kaufmann, San Francisco, CA (1999)
56. Vovk, V., Gammerman, A., Shafer, G.: Algorithmic Learning in a Random World. Springer, Cham, second edn (2022)
57. Vovk, V., Nouretdinov, I., Gammerman, A.: On-line predictive linear regression. Ann. Stat. **37**, 1566–1590 (2009)
58. Vovk, V., Nouretdinov, I., Gammerman, A.: Conformal e-testing. Pattern Recogn. **168**, article 111841 (2025). Special Issue on Conformal Prediction and Distribution-Free Uncertainty Quantification
59. Vovk, V., Shafer, G.: Kolmogorov's contributions to the foundations of probability. Probl. Inf. Transm. **39**, 21–31 (2003)
60. Vovk, V., Shafer, G.: A conversation with A. Philip Dawid. Statistical Science **40**, 148–166 (2025)
61. Vovk, V., V'yugin, V.V.J.: On the empirical validity of the Bayesian method. Roy. Stat. Soc. B **55**, 253–266 (1993)
62. Vovk, V., Wang, R.: E-values: Calibration, combination, and applications. Ann. Stat. **49**, 1736–1754 (2021)
63. Vovk, V., Wang, R.: Confidence and discoveries with e-values. Stat. Sci. **38**, 329–354 (2023)

An Investigation of
Conformal Test Martingales

Henrik Boström$^{(\boxtimes)}$ (iD)

KTH Royal Institute of Technology, Stockholm, Sweden
`bostromh@kth.se`
`https://www.kth.se/profile/henbos`

Abstract. Conformal test martingales are used for testing the assumption of exchangeability, which is central for the validity guarantees of conformal predictors. All algorithms for computing conformal test martingales share the fundamental property of allowing the false alarm rate to be controlled by a user-specified significance level. However, the ability to detect violations of the exchangeability assumption may vary largely across algorithms and distributions. Three algorithms for computing conformal test martingales, i.e., Simple Jumper, Sleeper/Stayer and Sleeper/Drifter, are analyzed and three central components are identified; the investment strategy, the betting function and the wake-up scheme. By combining the options of these components, a total of nine novel algorithms are obtained, complementary to the three previously proposed algorithms. Results from a large-scale investigation using both synthetic and real-world datasets are presented. The results on the former show that while all combinations successfully keep the false alarm rate below the requested significance level, which follows from the theoretical guarantees, the relative effectiveness of the algorithms is shown to depend on the type of distribution shift, i.e., whether there is a change in mean or variance (or both) and whether the change is abrupt or continuous. While the original Sleeper/Stayer algorithm is shown to be the most effective of the twelve considered combinations in a majority of cases, the results also show that the novel combinations are competitive in some cases.

Keywords: Conformal test martingales · exchangeability

1 Introduction

As the assumption of exchangeability is central for the validity guarantees of many approaches, e.g., conformal predictors [7], the ability to test it in practice is of crucial importance. Conformal test martingales have been proposed for this purpose in online scenarios, i.e., where p-values are observed over time [5, 6, 9, 10]. The false alarm rate, i.e., the probability that the martingale incorrectly signals that the exchangeability assumption is violated, is in these approaches controlled by a user-specified significance level. However, the effectiveness of the algorithms,

K. An Nguyen and Z. Luo (Eds.): Alexander Gammerman Festschrift, LNCS 16290, pp. 118–143, 2026.
https://doi.org/10.1007/978-3-032-15120-9_7

i.e., their ability to detect violations of the exchangeability assumption, may vary to a large extent across algorithms and tasks.

In this paper, we will take a closer look at three algorithms for computing conformal test martingales that are described in [7]; Simple Jumper [7, p. 232], Sleeper/Stayer [7, p. 283], and Sleeper/Drifter [7, p. 283]. We will discuss the main components of these algorithms and show how they may be combined in new ways, leading to novel algorithms for computing conformal test martingales. The original algorithms are all based on the idea of betting against the null hypothesis of the p-values being uniformly distributed, and for this purpose, multiple accounts are maintained, containing the earnings from betting on various discrete outcomes. Three different betting strategies are employed in the algorithms; i) betting on that the next p-value in the sequence is less (or greater) than 0.5 (Simple Jumper), ii) betting on that the p-value falls above (or below) a fixed, account specific, threshold (Sleeper/Stayer), and iii) betting on that the p-value falls above (or below) a threshold that varies over time (Sleeper/Drifter). Moreover, the algorithms vary in how the original and earned capital are (re-)invested. Simple Jumper initially distributes the original capital over available accounts and employs taxation to redistribute the capital that is accumulated over time. Sleeper/Stayer and Sleeper/Drifter instead keep separate investment accounts that are gradually consumed. Finally, Sleeper/Drifter allows for dynamically waking up new accounts, in contrast to Simple Jumper and Sleeper/Stayer. Hence, three betting strategies, two investment strategies and two wake-up strategies have been employed in the previously described algorithms. The described algorithms thus represent only three out of twelve possible combinations, and we will in this work investigate the relative effectiveness of all these combinations.

In the next section, we provide terminology and the previous three algorithms for computing conformal test martingales. In Sect. 3, we will formalize nine novel combinations, obtained by combining the betting, investment and wake-up strategies in new ways. We will evaluate both the original and new combinations experimentally in Sect. 4. Finally, we summarize the main findings and outline some directions for future work in Sect. 5.

2   Conformal Test Martingales

Given a sequence of examples $z_1, z_2, \ldots$, where each $z_i \in Z$, for some example space Z, a sequence of nonconformity scores $\alpha_1 = A(z_1)$, $\alpha_2 = A(z_2)$, $\ldots$ is obtained using some nonconformity function $A : Z \to \mathbb{R}$. From the sequence of nonconformity scores $\alpha_1, \alpha_2, \ldots$ and a sequence of values $\tau_1, \tau_2, \ldots$ that are sampled IID and uniformly from the unit interval, a (smoothed) p-value for the nth nonconformity score is obtained by:

$$p_n = \frac{|i : \alpha_i > \alpha_n| + \tau_n |i : \alpha_i = \alpha_n|}{n} \tag{1}$$

where $i \in \{1, \ldots, n\}$. The above p-values are computed in the *semi-online* mode, according to the terminology in [7], and if the examples $z_1, z_2, \ldots$ are IID, or if the examples are exchangeable and we consider a finite horizon, then the p-values are IID and distributed uniformly on $[0, 1]$ [6].

For each p-value in the sequence, a *conformal test martingale* is computed by:

$$s_n = f_1(p_1) \cdots f_n(p_n) \tag{2}$$

where each f_i is a *betting function* $f : [0, 1] \rightarrow [0, \infty]$ satisfying:

$$\int_0^1 f(u)du = 1 \tag{3}$$

If the sequence of examples $z_1, z_2, \ldots$ is exchangeable, then Ville's inequality implies [7]:

$$P(\{(z_1, z_2, \ldots) : \exists n : s_n \geq c\}) \leq 1/c \tag{4}$$

where c is some positive constant.

The conformal test martingale can hence be used for testing the assumption of exchangeability, where the false alarm rate is controlled by choosing a suitable c. A common choice for the latter is $c = 100$, implying that the probability of observing a conformal test martingale greater than or equal to this is less than or equal to 1%, if the sequence of observations is exchangeable.

In Algorithm 1, a slight reformulation[1] of the original Simple Jumper algorithm [7, p. 232] is given.

Algorithm 1. Simple Jumper

Input: $p_1, p_2, \ldots$ (p-values), $t \in (0, 1)$ (tax rate), E (epsilons)
Output: $s_1, s_2, \ldots$
1: **for** $\epsilon \in E$ **do** $c_\epsilon \leftarrow 1/|E|$
2: **for** $i = 1, 2, \ldots$ **do**
3:      **for** $\epsilon \in E$ **do** $c_\epsilon \leftarrow c_\epsilon \, f_\epsilon(p_i)$
4:      $s_i \leftarrow \sum_{\epsilon \in E} c_\epsilon$
5:      **for** $\epsilon \in E$ **do** $c_\epsilon \leftarrow (1 - t)\, c_\epsilon + (t/|E|)\, s_i$

Simple Jumper employs betting functions of the following form:

$$f_\epsilon(p) = 1 + \epsilon(p - 0.5) \tag{5}$$

where $\epsilon \in [-2, 2]$ is a parameter determining the maximum gain (and loss) and whether a p-value below 0.5 gives a gain or not. In [7, p. 232], the following values for ϵ is considered: $E = \{-1, -0.5, 0, 0.5, 1\}$ and the output of the corresponding betting functions are displayed in Fig. 1.

---

[1] The variable C for storing the sum of all capital has been removed and the variable J (*jumping rate*) has been renamed to t (*tax rate*).

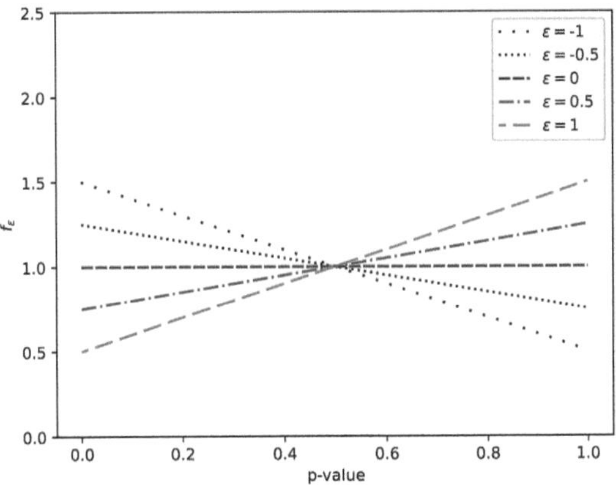

Fig. 1. The output of Simple Jumper's betting function for various ϵ.

Simple Jumper keeps one account (c_ϵ) for each $\epsilon \in E$, and the accounts are initialized with an equal amount summing up to 1 (l. 1 in Algorithm 1). For each observed p-value, the amount on each account is multiplied with the output of the corresponding betting function (l. 3), and the output conformal test martingale is the sum of the accumulated capital in all accounts (l. 4). The accounts are finally taxed (at the rate t, which is a parameter of the algorithm), and the tax is distributed uniformly across all the accounts (l. 5). Values for t considered in [7, p. 232], are $t \in \{10^{-4}, 10^{-3}, 10^{-2}, 10^{-1}, 1\}$.

In Algorithm 2, we present a slight modification[2] of the original Sleeper/Stayer algorithm [7, p. 283].

Algorithm 2. Sleeper/Stayer

Input: $p_1, p_2, \ldots$ (p-values), $r \in (0, 1)$ (investment rate), G (grid)
Output: $s_1, s_2, \ldots$
1: **for** $(a, b) \in G$ **do** $c_{a,b} \leftarrow r/|G|$
2: $c \leftarrow 1 - r$
3: **for** $i = 1, 2, \ldots$ **do**
4:     **for** $(a, b) \in G$ **do** $c_{a,b} \leftarrow c_{a,b}\, f_{a,b}(p_i)$
5:     $s_i \leftarrow c + \sum_{(a,b) \in G} c_{a,b}$
6:     **for** $(a, b) \in G$ **do** $c_{a,b} \leftarrow c_{a,b} + r\, c/|G|$
7:     $c \leftarrow (1 - r)\, c$

Sleeper/Stayer employs the following betting function:

$$f_{a,b}(p) = \begin{cases} \frac{b}{a} & \text{if } p \le a \\ \frac{1-b}{1-a} & \text{otherwise} \end{cases} \tag{6}$$

---

[2] Investments are here distributed before computing the first martingale, which otherwise always will be 1, effectively ignoring the first p-value.

where the parameter a determines the threshold value (to compare the observed p-value p with) and b determines the gain (if $b > a$) or loss (if $b < a$), if $p \leq a$. The output of the betting function for the set of pairs $G = \{(0.1, 0.8), (0.4, 0.8), (0.5, 0.5), (0.6, 0.4), (0.9, 0.4)\}$ and all possible p-values are displayed in Fig. 2.

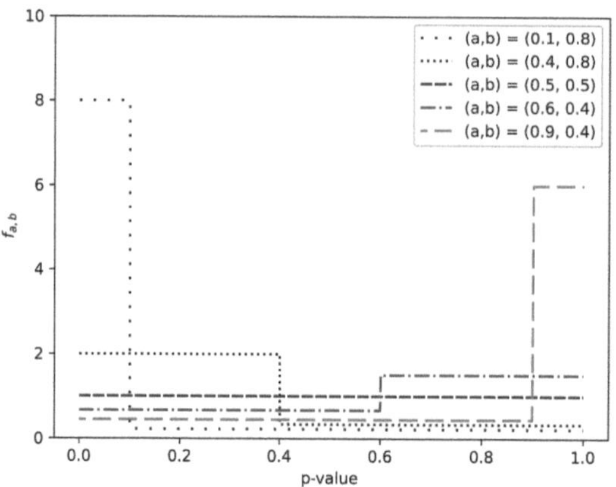

Fig. 2. Output of Sleeper/Stayer's betting function for various a and b.

The grid G is in [7, p. 282] defined by:

$$G = \{(i/g, j/g) : (i, j) \in \{1, \ldots, g-1\} \times \{1, \ldots, g-1\}\} \tag{7}$$

where $g \in \mathbb{N}^+$ is the grid size.

In contrast to Simple Jumper, which employs taxation of all accounts to obtain capital to invest, Sleeper/Stayer keeps an investment account (c), from which capital is transferred to each betting account $(c_{a,b})$. All betting accounts are initialized with an equal amount summing up to r (l. 1), where r (the investment rate) is a parameter of the method, while the investment account is initialized to 1 minus the invested amount (l. 2). For each observed p-value, the amount on each betting account is multiplied with the output of the corresponding betting function (l. 4), and the output conformal test martingale is the sum of the accumulated capital in all accounts, including the investment account (l. 5). Investments are finally transferred from the investment account to the betting accounts (l. 6 and 7) according to the rate r. Parameter values considered in [7, p. 283], are $g = 10$ and $r = 0.001$.

In Algorithm 3, a slight modification[3] of the original Sleeper/Drifter algorithm [7, p. 284] is presented.

Similarly to Sleeper/Stayer, Sleeper/Drifter maintains accounts for pairs $(a, b) \in G$. The set of accounts are initialized similarly to Sleeper/Stayer by

---

[3] Accounts are allowed to be updated before observing m p-values.

Algorithm 3. Sleeper/Drifter

Input: $p_1, p_2, \ldots$ (p-values), $r \in (0, 1)$ (investment rate), G (grid), $m \in \mathbb{N}^+$ (wakeup frequency)

Output: $s_1, s_2, \ldots$

1: **for** $(a, b) \in G$ **do** $c_{0,a,b} \leftarrow r/|G|$
2: $c \leftarrow 1 - r$
3: **for** $i = 1, 2, \ldots$ **do**
4:      **for** $j = 0, \ldots, i//m$ **do**
5:          **for** $(a, b) \in G$ **do**
6:              $w \leftarrow (j + 1)\, m/(m + i)$
7:              $a' \leftarrow a\,w + b\,(1 - w)$
8:              $c_{j,a,b} \leftarrow c_{j,a,b}\, f_{a',b}(p_i)$
9:      $s_i \leftarrow c + \sum_{j=0,\ldots,i//m} \sum_{(a,b)\in G} c_{j,a,b}$
10:     **if** $(i + 1) \bmod m = 0$ **then**
11:         **for** $(a, b) \in G$ **do** $c_{(i+1)//m,a,b} \leftarrow r\,c/|G|$
12:         $c \leftarrow (1 - r)\,c$

---

transferring from an investment account (l. 1 and 2). However, in contrast to Sleeper/Stayer (and Simple Jumper), the number of accounts is not fixed, but just before every mth time-step, a new set of accounts are created and initialized, where m is a parameter of the algorithm (l. 10–12). For each observed p-value, the active accounts are updated using the same betting function as employed by Sleeper/Stayer (l. 8), where the employed threshold value however is no longer the value a of each pair (a, b) in G, but a weighted average $a\,w + b\,(1 - w)$ (l. 7), where w decreases with the number of observed p-values. It should be noted that in the original Sleeper/Drifter algorithm [7, p. 284], the employed weight w corresponds to the wake-up time of the account divided by the number observed p-values, while in the modified version, the weight is computed by adding m to both the numerator and denominator. This results in that the same set of weights are employed by both the original and modified algorithm, while the modified version allows accounts to be updated already before observing m p-values.

In Fig. 3, the output of the betting function is shown for the pair ($a = 0.1, b = 0.8$) when observing a p-value less than a, for different number of p-values observed after creating the accounts at various time steps ($s \in \{100, 200, 300\}$) and when using the original threshold a (no drift). As can be observed, the potential gains (as well as the potential losses) are reduced over time when allowing the thresholds to drift, compared to using the original threshold. It can also be seen that of the accounts with drifting thresholds, the ones introduced earlier will give relatively lower gains as more weight is given to b when computing the thresholds.

Parameter values considered in [7, p. 284], are $g = 10$, $m = 100$ and $r = 0.001$.

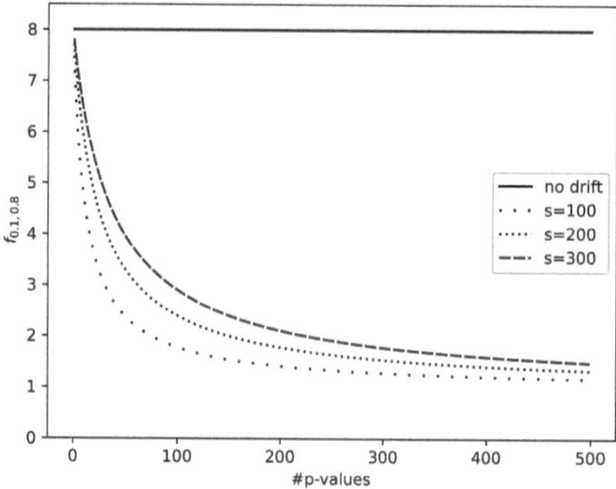

Fig. 3. Output of betting function with time-dependent thresholds.

3   Novel Combinations

3.1   Summary of Strategies

The algorithms described in the previous section employ two investment strategies; using taxation (I1), and using an investment account (I2). Three betting functions are also used; the epsilon betting function in Eq. 5 (B1), the step betting function in Eq. 6 with fixed thresholds (B2), and the step betting function in Eq. 8 with time-varying thresholds (B3). Finally, two wake-up strategies are employed; using a fixed set of active accounts, i.e., no wake-up of additional accounts (W1), and allowing the set of active accounts to grow, i.e., wake-up of additional accounts is allowed (W2).

In Table 1, the three original algorithms are characterized in terms of the above strategies. The original algorithms hence only represent a minor part of the possible algorithms that we may obtain by combining the above strategies. In the following subsections, we formalize the remaining nine possible combinations.

Table 1. The three original algorithms

Algorithm	Investment	Betting	Wake-Up
Simple Jumper	I1	B1	W1
Sleeper/Stayer	I2	B2	W1
Sleeper/Drifter	I2	B3	W2

3.2   Generalizing Simple Jumper

In the following, we will assume betting functions to be provided as input to the algorithms, rather than being specified by parameters for a specific type of function. For example, rather than providing $E = \{-1, -0.5, 0, 0.5, 1\}$ as parameter values for the epsilon betting function, the input set of betting functions is $B = \{f_{\epsilon=-1}, f_{\epsilon=-0.5}, f_{\epsilon=0}, f_{\epsilon=0.5}, f_{\epsilon=1}\}$. Similarly, a set of step betting functions may be declared by $B = \{f_{a=0.1,b=0.1}, f_{a=0.2,b=0.1}, \ldots\}$, i.e., specifying the parameters a and b. When applied, both types of betting function require a single argument only, i.e., a p-value. Step betting functions with time-varying thresholds of the following form may also be provided, requiring three additional arguments; account index (j), wake-up frequency (m) and index of the observed p-value (i):

$$f'_{a,b}(p, j, m, i) = f_{aw_{j,m,i}+b(1-w_{j,m,i}),b}(p) \tag{8}$$

where $w_{j,m,i} = (j+1)m/(m+i)$.

When the betting functions are evaluated in the algorithms below, provided arguments that are not used will simply be ignored. This means that the arguments following the first have an effect only on step betting functions with time-varying thresholds (Eq. 8).

In Algorithm 4, we modify Algorithm 1 to assume the betting functions to be provided as input, rather than to be of a specific form. This allows us to employ Simple Jumper not only with betting functions in accordance with Eq. 5, but also Eq. 6 and Eq. 8 (step betting functions with and without time-varying thresholds).

The original investment strategy of Simple Jumper is still used, i.e., taxation (I1) is employed. Generalized Simple Jumper hence generalizes Simple Jumper by not only providing an implementation of the original combination of strategies, i.e., I1+B1+W1, but also I1+B2+W1 and I1+B3+W1.

Algorithm 4. Generalized Simple Jumper

Input: $p_1, p_2, \ldots$ (p-values), $t \in (0, 1)$ (tax rate), B (betting functions), m (used for step betting functions with drift)

Output: $s_1, s_2, \ldots$

1: **for** $b \in B$ **do** $c_b \leftarrow 1/|B|$
2: **for** $i = 1, 2, \ldots$ **do**
3:     **for** $b \in B$ **do** $c_b \leftarrow c_b\, b(p_i, 0, m, i)$
4:     $s_i \leftarrow \sum_{b \in B} c_b$
5:     **for** $b \in B$ **do** $c_b \leftarrow (1 - t)\, c_b + t\, s_i/|B|$

We may now go one step further and allow the number of accounts to grow over time, i.e., employ the wake-up strategy W2, in Algorithm 4. The new algorithm, presented in Algorithm 5, hence allows for the implementation of three additional combinations, i.e., I1+B1+W2, I1+B2+W2 and I1+B3+W2.

The main addition to Generalized Simple Jumper is that a new set of accounts is activated the step before every mth observed p-value (1. 8–12). In contrast to Sleeper/Drifter, there is no investment account from which capital can be withdrawn, so all current accounts are instead taxed, using a tax rate (t') that gives the new set of accounts the same amount as available on average for the existing sets of accounts (l. 9).

It should be noted that in case a fixed set of parameters are used for the betting functions, e.g., when employing B1 and B2, the newly created accounts can be merged with the existing accounts (thus saving space); the main effect of the wake-up step (l. 8–12) in this case, is just a re-distribution of capital among the existing accounts.

Algorithm 5. Generalized Simple Jumper with Wake-Up

Input: $p_1, p_2, \ldots$ (p-values), $t \in (0,1)$ (tax rate), B (betting functions), m (wake-up frequency)

Output: $s_1, s_2, \ldots$

 1: **for** $b \in B$ **do** $c_{0,b} \leftarrow 1/|B|$
 2: **for** $i = 1, 2, \ldots$ **do**
 3:      **for** $j = 0, \ldots, i//m$ **do**
 4:          **for** $b \in B$ **do** $c_{j,b} \leftarrow c_{j,b}\, b(p_i, j, m, i)$
 5:      $s_i \leftarrow \sum_{j=0,\ldots,i//m} \sum_{b \in B} c_{j,b}$
 6:      **for** $j = 0, \ldots, i//m$ **do**
 7:          **for** $b \in B$ **do** $c_{j,b} \leftarrow (1-t)\, c_{j,b} + t\, s_i/(|B| i//m)$
 8:      **if** $(i+1) \bmod m = 0$ **then**
 9:          $t' \leftarrow 1/((i+1)//m)$
10:          **for** $b \in B$ **do** $c_{(i+1)//m,b} \leftarrow t'\, s_i/|B|$
11:          **for** $j = 0, \ldots, i//m$ **do**
12:             **for** $b \in B$ **do** $c_{j,b} \leftarrow (1-t')\, c_{j,b}$

3.3 Generalizing Sleeper/Stayer

In Algorithm 6, we present a straightforward extension of Sleeper/Stayer (Algorithm 2), now allowing any of the three considered betting functions to be used, i.e., implementing I2+B1+W1, I2+B2+W1 and I2+B3+W1.

Algorithm 6. Generalized Sleeper/Stayer

Input: $p_1, p_2, \ldots$ (p-values), $r \in (0,1)$ (investment rate), B (betting functions), m (used for step betting functions with drift)

Output: $s_1, s_2, \ldots$

 1: **for** $b \in B$ **do** $c_b \leftarrow r/|B|$
 2: $c \leftarrow 1 - r$
 3: **for** $i = 1, 2, \ldots$ **do**
 4:      **for** $b \in B$ **do** $c_b \leftarrow c_b\, b(p_i, 0, m, i)$
 5:      $s_i \leftarrow c + \sum_{b \in B} c_b$
 6:      **for** $b \in B$ **do** $c_b \leftarrow c_b + r\, c/|B|$
 7:      $c \leftarrow (1-r)\, c$

3.4 Generalizing Sleeper/Drifter

Finally, a generalized version of Sleeper/Drifter (Algorithm 3) is presented in Algorithm 7. The only modification of the original algorithm is that it now allows any of the three betting functions to be used, effectively implementing the combined strategies I2+B1+W2, I2+B2+W2 and I2+B3+W2.

Similarly to Generalized Simple Jumper with Wake-Up, the newly created accounts can be merged with the existing accounts if a fixed set of parameters are used for the betting functions. The main effect of the wake-up step (l. 7–9) when using time-independent betting functions is then just an injection of a fixed amount to all existing accounts.

Algorithm 7. Generalized Sleeper/Drifter

Input: $p_1, p_2, \ldots$ (p-values), $r \in (0, 1)$ (investment rate), B (betting functions), $m \in \mathbb{N}^+$ (wakeup frequency)

Output: $s_1, s_2, \ldots$

1: **for** $b \in B$ **do** $c_{0,b} \leftarrow r/|G|$
2: $c \leftarrow 1 - r$
3: **for** $i = 1, 2, \ldots$ **do**
4:      **for** $j = 0, \ldots, i//m$ **do**
5:          **for** $b \in B$ **do** $c_{j,b} \leftarrow c_{j,b}\, b(p_i, j, m, i)$
6:      $s_i \leftarrow c + \sum_{j=0,\ldots,i//m} \sum_{b \in B} c_{j,b}$
7:      **if** $(i + 1) \bmod m = 0$ **then**
8:          **for** $b \in B$ **do** $c_{(i+1)//m,b} \leftarrow r\, c/|B|$
9:          $c \leftarrow (1 - r)\, c$

3.5 Illustration

To illustrate the workings of the algorithms, we will consider two synthetic cases. The first concerns observing p-values obtained using Eq. 1 from the two concatenated sequences $X_1 = \{x_{1,i}\}_{i=1}^k \sim \mathcal{N}(\mu_1, \sigma_1^2)$ and $X_2 = \{x_{2,i}\}_{i=1}^k \sim \mathcal{N}(\mu_2, \sigma_1^2)$, where $k = 1000$, $\mu_1 = 0$, $\sigma_1 = 1$ and $\mu_2 = 0.5$, i.e., the mean of the underlying normal distribution changes after 1000 observations. The second case concerns concatenating X_1 with $X_3 = \{x_{3,i}\}_{i=1}^k \sim \mathcal{N}(\mu_1, \sigma_2^2)$, where $\sigma_2 = 2$, i.e., the standard deviation of the underlying distribution changes after 1000 observations.

We will consider all twelve algorithms with the following (default) parameter settings: $t = 0.01$ (used with the investment strategy I1), $r = 0.001$ (used with the investment strategy I2), and $m = 100$ (used in conjunction with betting function B3 and wake-up strategy W2). The considered epsilons (for betting function B1) are $E = \{-1, -0.5, 0, 0.5, 1\}$ and the grid (for betting function B2) is defined by Eq. 7, using grid size $g = 10$.

Figure 4 shows the martingales generated by the twelve combinations in the first case. It can be seen that all algorithms are able to detect a violation of the exchangeability assumption, but at various time steps as indicated by the dotted vertical lines, showing when the corresponding martingale reaches a value greater than or equal to 100. The most effective algorithm in this case is the original Simple Jumper (I1+B1+W1), signaling at index 1122, closely followed by Simple Jumper with wake-up (I1+B1+W2), signaling at index 1123. The least effective are two of the modifications of Simple Jumper and Sleeper/Stayer, using a drifting step betting function (B3), signaling at index 1371 and 1368, respectively.

In Fig. 5, the results from applying the twelve combinations in the case of a changing standard deviation. In this case, it turns out that not all algorithms are able to detect a violation of the exchangeability assumption. In particular, none of the combinations using the epsilon betting function (B1) manage to detect data drift. This may not be very surprising as the p-values even after the change point will still be uniformly distributed on each side of the 0.5 threshold, which is employed by the epsilon betting function. In contrast, the step betting functions, which employ multiple thresholds, are able to detect deviations from the expected at a more fine-grained scale.

Hence, the original Simple Jumper is not effective in this case. However, by modifying the algorithm to use a step betting function (B2), either without or with wake-up (W2), data drift can be detected (at index 1134 and 1136, respectively). Also modifying the algorithm to use a drifting betting function (B3) together with wake-up (W2) results in that data drift can be detected (at index 1139). The most effective combination in this case is Sleeper/Stayer (without modification), signaling already at index 1099.

4   Experimental Investigation

In this section, we will present a large-scale evaluation of the algorithms for generating conformal test martingales. We will compare the three original algorithms with the nine novel combinations. We will consider both synthetic data (with known change points) and real-world datasets (for which data drift can be expected but is not known) to investigate the effectiveness of the algorithms.

4.1   Experimental Setup

Synthetic Data Drift. We will consider sequences of a specific length (l) with values sampled randomly from the normal distribution $\mathcal{N}(\mu, \sigma)$, where the parameters change either abruptly (at some randomly chosen time point $1 < k < l$) or continuously (between two randomly chosen time points k_1 and k_2, where $1 < k_1 < l - 1$ and $k_1 < k_2 < l$). A sequence obtained by abrupt change is hence defined by concatenating two sequences[4]:

$$\begin{aligned} X_{l,k,\mu_1,\sigma_1,\mu_2,\sigma_2} = \{x_i\}_{i=1}^{k-1} &\sim \mathcal{N}(\mu_1, \sigma_1^2) \quad \oplus \\ \{x_i\}_{i=k}^{l} &\sim \mathcal{N}(\mu_2, \sigma_2^2) \end{aligned} \tag{9}$$

---

[4] where $\oplus$ denotes the concatenation operator.

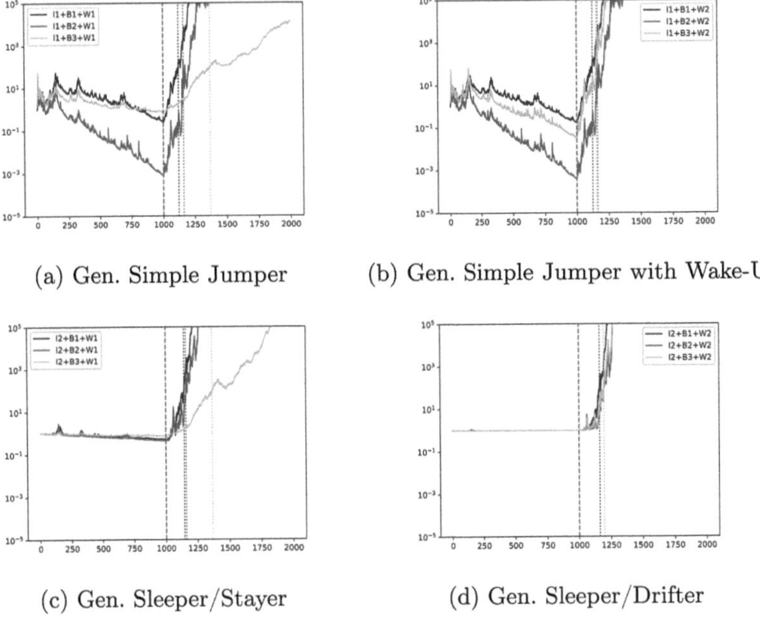

(a) Gen. Simple Jumper

(b) Gen. Simple Jumper with Wake-Up

(c) Gen. Sleeper/Stayer

(d) Gen. Sleeper/Drifter

Fig. 4. Change of mean.

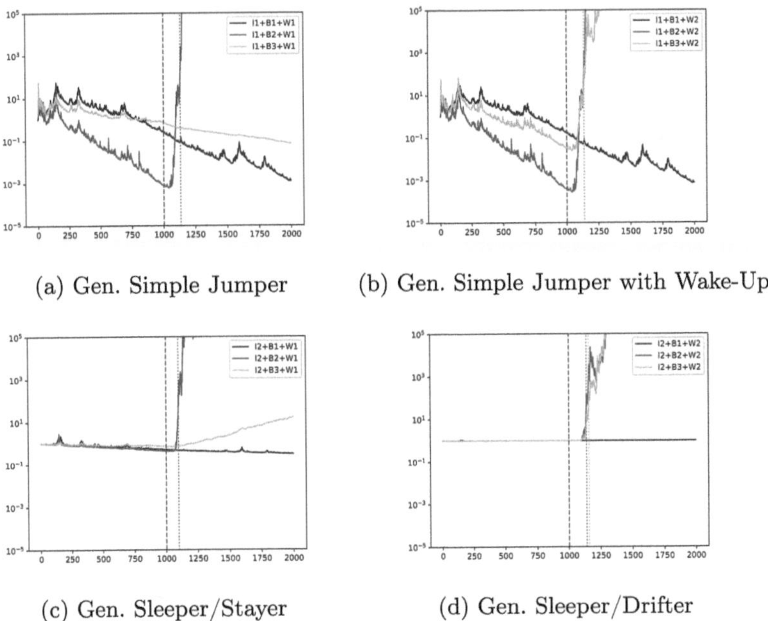

(a) Gen. Simple Jumper

(b) Gen. Simple Jumper with Wake-Up

(c) Gen. Sleeper/Stayer

(d) Gen. Sleeper/Drifter

Fig. 5. Change of standard deviation.

while a sequence obtained by continuous change is defined by:

$$X_{l,k_1,k_2,\mu_1,\sigma_1,\mu_2,\sigma_2} = \{x_i\}_{i=1}^{k_1-1} \sim \mathcal{N}(\mu_1,\sigma_1^2) \quad \oplus$$
$$\{x_i\}_{i=k_1}^{k_2-1} \sim \mathcal{N}(w(i,k_1,k_2,\mu_1,\mu_2), w(i,k_1,k_2,\sigma_1,\sigma_2) \quad \oplus$$
$$\{x_i\}_{i=k_2}^{l} \sim \mathcal{N}(\mu_2,\sigma_2^2)$$

$$(10)$$

where $w(i,k_1,k_2,a,b)$ gives a weighted average of a and b using the weight $1 - \frac{i-k_1}{k_2-k_1} - 1$.

We will investigate both abrupt and continuous changes of the mean and standard deviation individually and in combination, where the change points (defined by k or k_1) are sampled uniformly over the range of integers $(1,l)$ and $(1,l-2)$, respectively, where for the latter k_2 is sampled uniformly from the range $(k_1+1,l-1)$. In all scenarios, $\mu_1 = 0$ and $\sigma_1 = 1$, i.e., the values occurring before a change-point are sampled from the standard normal distribution. In Table 2, we summarize the changes that are considered in the experiments.

Table 2. Scenarios for synthetic data drift

Type	μ_2	σ_2	Description
abrupt	$\sim \mathcal{N}(0,1)$	1	change of mean
abrupt	0	$\sim \mathcal{U}(0,2)$	change of standard deviation
abrupt	$\sim \mathcal{N}(0,1)$	$\sim \mathcal{U}(0,2)$	change of mean and standard deviation
continuous	$\sim \mathcal{N}(0,1)$	1	change of mean
continuous	0	$\sim \mathcal{U}(0,2)$	change of standard deviation
continuous	$\sim \mathcal{N}(0,1)$	$\sim \mathcal{U}(0,2)$	change of mean and standard deviation

Both shorter sequences of length $l = 1\,000$ and longer sequences of length $l = 10\,000$ will be considered in conjunction with the six scenarios.

Parameter tuning was first conducted for the four introduced algorithms in the previous section, when using the betting functions proposed for the original (non-generalized) versions, i.e., the epsilon betting functions for Generalized Simple Jumper with and without wake-up, the step betting functions without and with drift, respectively, for Generalized Sleeper/Stayer and Generalized Sleeper/Drifter. For the two former, the same set of values for t (tax rate) were considered as in [7, p. 232], i.e., $t \in \{10^{-4}, 10^{-3}, 10^{-2}, 10^{-1}, 1\}$, together with a composite martingale formed from all values. For the parameter r (investment rate) of the latter two algorithms, the following values were considered: $r \in \{10^{-4}, 10^{-3}, 10^{-2}, 10^{-1}\}$, where again also a composite martingale was considered. By repeating the above scenarios 1000 times (with unique random seeds), the following parameter values were found to be more effective than the alternatives, including the composite martingales: $t = 0.01$ for Generalized Simple Jumper with and without wake-up for short sequences and $t = 0.0001$ for long sequences. For continuous change of mean and change of mean and standard deviation for short sequences, the parameter value $t = 0.001$ did not result in a significant difference from $t = 0.01$; in all other cases the latter was significantly more effective. For Generalized Sleeper/Stayer the best performing setting was $r = 0.001$ and $r = 0.0001$ for short and long sequences, respectively. Finally, for Generalized Sleeper/Drifter the best values for r were $r = 0.1$ and $r = 0.01$, respectively.

The above set of parameters were used in 1000 additional repetitions of the above experiments (using novel random seeds). To determine whether the observed difference in performance was significant or not, a Friedman test followed by a post-hoc Nemenyi test was employed [2], i.e., comparing all twelve algorithms over the 1000 observations for each scenario, where performance is evaluated by the time-point at which the conformal test martingale reaches $s = 100$ (an earlier time-point is better).

House Sales. The first real-world dataset that will be used in the evaluation concerns predicting house sale prices for the King County, Washington, between May 2014 and May 2015. The dataset, named `house_sales` in the OpenML repository[5], consists of 21 613 instances represented with 19 features. The instances were sorted according to year, month and day; these three features were subsequently removed and the 10 000 first instances were used to train a random forest regressor (with default parameter settings as provided in `scikit-learn`[6]). The remaining 11 613 test instances were used to obtain p-values according to Eq. 1 from a standard conformal predictive system, as implemented in `crepes`[7] [1] in a semi-online setting, where an initially empty calibration set was updated sequentially with each instance in the test set.

For the twelve conformal test martingale algorithms, the parameter values obtained from tuning using long sequences in the experiment with synthetic data were employed.

Lipophilicity. A second real-world dataset used in the evaluation is Lipophilicity from MoleculeNet [11], which contains measurements of the octanol/water distribution coefficient for 4200 chemical compounds, represented by the simplified molecular-input line-entry system (SMILES). The Python package `RDKit`[8] was used to generate features from the SMILES strings, more specifically, *Morgan fingerprints* (binary vectors, all of length 1024).

The dataset was sorted according to the molecular identifier (ChEMBLID) under the assumption that the number correlates to the date of inclusion. The first 2000 labeled compounds were used to form a random forest regressor, again using the default parameter settings in `scikit-learn`, while the remaining 2200

---

[5] www.openml.org.

[6] www.scikit-learn.org.

[7] http://www.github.com/henrikbostrom/crepes/.

[8] https://www.rdkit.org.

test instances were used to obtain p-values using a standard conformal predictive system generated in the semi-online setting, again using the **crepes** package.

For the twelve conformal test martingale algorithms, the parameter values obtained from tuning using short sequences in the experiment with synthetic data were employed.

PCBA. The third real-world case concerns a set of binary classification tasks; predicting properties of chemical compounds as extracted from the **pcba** dataset from **DeepChem**[9]. The chemical compounds are again represented by SMILES strings and the **RDKit** package was again employed to generate features from the SMILES strings, in the same way as for the Lipophilicity dataset. The complete dataset includes 128 binary targets in total, but labels are not available for all compounds; tasks for which at least 30 000 labeled instances (for which also finger-prints could be generated) were included, leading to a set of 120 datasets.

The datasets were sorted according to the molecular identifier (a number), again under the assumption that the number correlates to the date of inclusion. The first 20 000 labeled compounds were used to form a binary classifier (a random forest classifier, generated using the default parameter settings in **scikit-learn**), while the last 10 000 instances were used to obtain p-values using a conformal classifier generated in the semi-online setting, again using the **crepes** package.

For the twelve conformal test martingale algorithms, the parameter values obtained from tuning using long sequences in the experiment with synthetic data were employed.

4.2   Experimental Results

Synthetic Data. In Fig. 6, the average ranks (over 1000 repeated experiments) are displayed for the twelve algorithms when comparing their ability to signal early for abrupt changes in long sequences ($l = 10\ 000$), where a low rank indicates high performance. Methods for which their corresponding lines are not connected perform significantly different (at the level $\alpha = 0.05$) according to the Nemenyi test. Hence, when it comes to detecting abrupt data drift for long sequences, the combination I2+B2+W1, i.e., the original Sleeper/Stayer algorithm, is significantly outperforming all competing approaches, for all three types of change, i.e., change of mean and standard deviation either alone or in combination.

More details from the evaluation are shown in Appendix A (Tables 4, 5, 6, 7, 8 and 9), where in addition to the ranks, also the true detection rate, i.e., fraction of changes detected after the change point, the false error rate, and the average detection time points. It can be seen that all observed false error rates are well below the requested level (0.01).

In Fig. 7, the average ranks are shown for the twelve algorithms when comparing their ability to signal early for continuous changes in long sequences

---

[9] https://github.com/deepchem/deepchem.

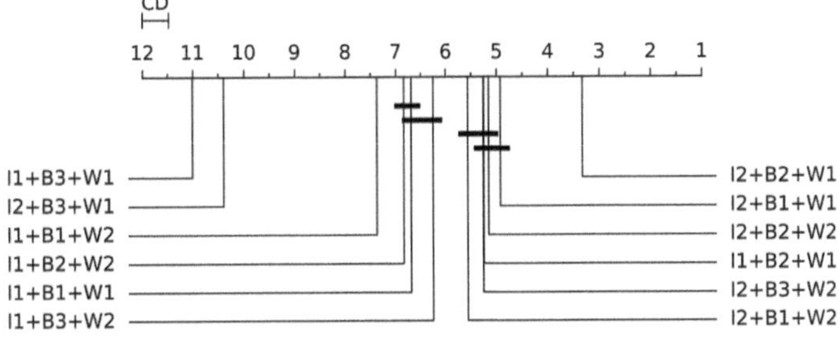

(a) Change of mean

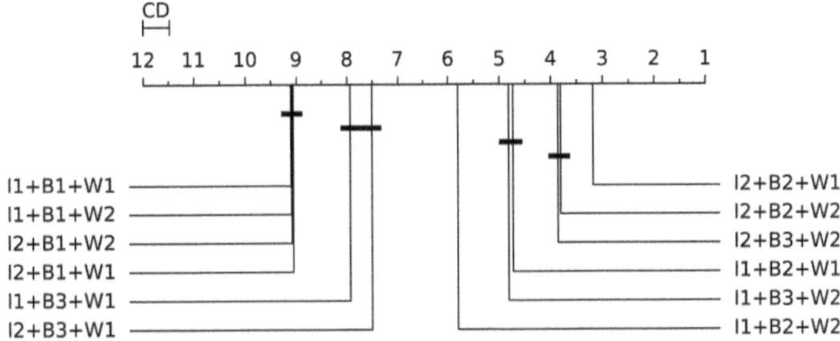

(b) Change of standard deviation

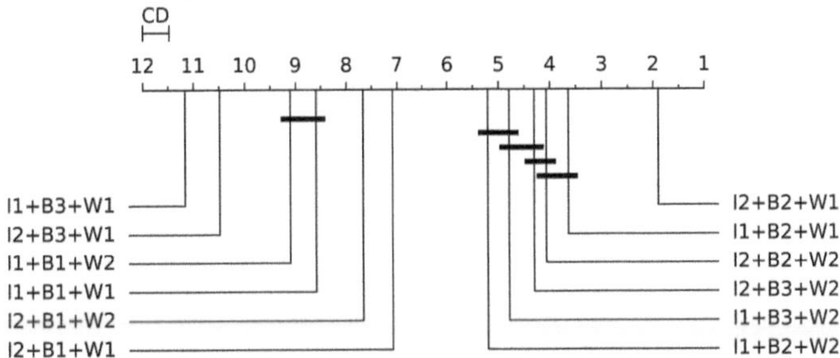

(c) Change of mean and standard deviation

Fig. 6. Abrupt change; long sequences.

($l = 10\ 000$). In this case, I2+B1+W1 and I2+B1+W2, i.e., Generalized Sleeper/Stayer and Sleeper/Drifter with epsilon betting functions, are the most effective for detecting change of mean. This type of betting function is however not effective for detecting change of standard deviation and is consequently employed by the worst performing approaches in the second scenario. The two top-ranked methods are in this case the original Sleeper/Drifter and Sleeper/Stayer. Finally, for continuous drift of both mean and standard devi-

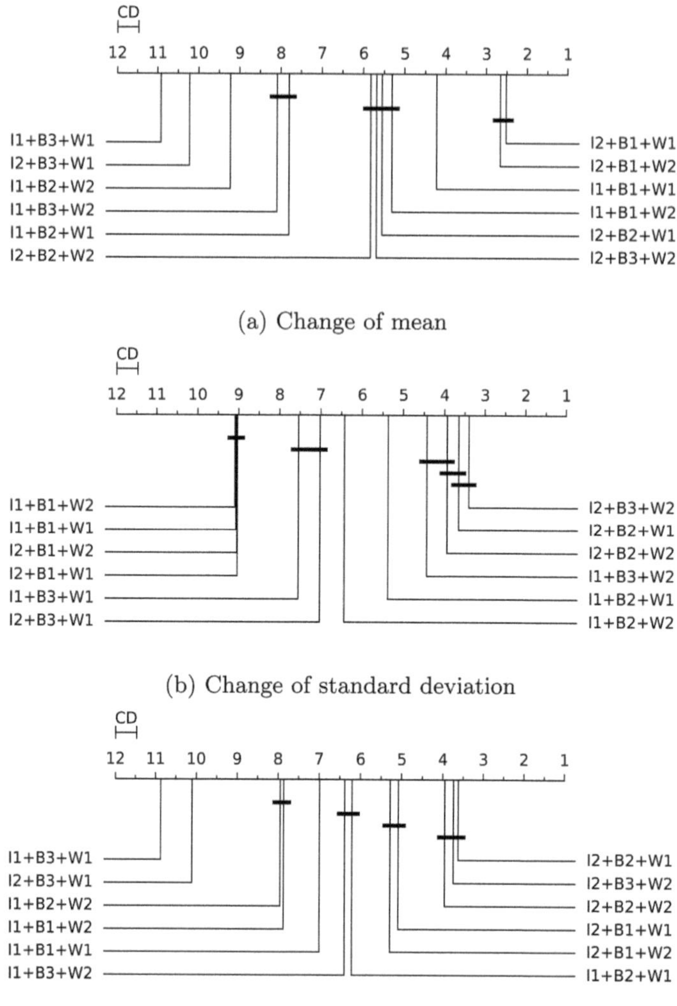

(a) Change of mean

(b) Change of standard deviation

(c) Change of mean and standard deviation

Fig. 7. Continuous change; long sequences.

ation, the three top-ranked combinations are the original Sleeper/Stayer and Sleeper/Drifter (using a step betting function with or without drift).

Results using short sequences ($l = 1000$) are shown in Appendix B (See Figs. 8 and 9). The results are fairly similar to those obtained from long sequences. For abrupt changes, the original Sleeper/Stayer is again significantly outperforming all competing approaches, while for continuous changes, the results are again more mixed, with the approaches using epsilon betting functions being ranked higher than approaches using step betting functions (with or without drift) in the first scenario (change of mean). The three best performing methods in this case are Generalized Sleeper/Stayer, Generalized Sleeper/Drifter and Generalized Simple Jumper, all using epsilon betting functions. In contrast, for the second and third scenarios, Sleeper/Stayer is significantly outperforming all other algorithms, except for Sleeper/Drifter in the second scenario.

Real-World Data. In Table 3, the rank and detection time are shown for all methods in all three experiments with real-world data. For the house sales dataset, the two best-performing methods are, similar to the continuous change-of-mean scenario in the experiment with synthetic data, Generalized Sleeper/Stayer and Sleeper/Drifter, both using epsilon betting functions (as shown in column 2 and 3 of the table). It can be seen that they are very closely followed by the original Sleeper/Stayer and Generalized Sleeper/Drifter, both using time-independent step betting functions; the latter two signal for a change just one time-step after the former two.

As can be seen in Table 3 (column 4 and 5), the most effective methods for the Lipophilicity dataset are Generalized Simple Jumper with time-independent step betting functions, with or without wake-up. They are very closely followed by Generalized Simple Jumper using time-dependent step betting functions, again with or without wake-up. All four combinations using epsilon betting functions are among the worst-performing half, very similar to the synthetic scenarios with (abrupt or continuous) change of standard deviation.

The results for PCBA in Table 3 (column 6 and 7), are averaged over 120 datasets, as described above. The algorithm with the highest ranking on average is the original Sleeper/Stayer. However, it is not significantly outperforming five of the competing approaches, as shown in Appendix C (Fig. 10).

Table 3. Results on real-world datasets

Method	House sales		Lipophilicity		PCBA	
	Rank	Time	Rank	Time	Rank	Time
I1+B1+W1	6	6508	8	347	6.8	5717.1
I1+B2+W1	7	6650	1.5	90	6.3	5582.6
I1+B3+W1	12	9491	3.5	92	10.0	7664.1
I1+B1+W2	8	6690	6	168	6.6	5679.3
I1+B2+W2	9	6702	1.5	90	7.3	5737.0
I1+B3+W2	10	6705	3.5	92	5.8	5433.7
I2+B1+W1	1.5	6180	10.5	770	5.3	5479.2
I2+B2+W1	3.5	6181	5	94	4.4	5360.0
I2+B3+W1	11	8886	12	1264	9.7	7537.0
I2+B1+W2	1.5	6180	10.5	770	5.8	5640.9
I2+B2+W2	3.5	6181	7	170	5.0	5383.2
I2+B3+W2	5	6371	9	648	4.9	5305.4

4.3 Discussion

In the following, we discuss the main observations from the empirical investigation regarding the original three algorithms, the investment strategies, the betting functions and the wake-up strategies.

Observations Regarding the Original Algorithms. The original Sleeper/Stayer (I2+B2+W1) is clearly the most effective algorithm for the synthetic scenarios; it is top-ranked among the twelve considered combinations in five out of six cases for short sequences, and four out of six cases for long sequences, and it is significantly outperformed by some other method in only one case each for short and long sequences. It is furthermore top-ranked for PCBA, although it is in that case not significantly outperforming a group of five competing approaches. It is very close in performance to the top-ranked methods also on House sales.

The original Sleeper/Drifter (I2+B3+W2) is ranked first in the synthetic scenario involving long sequences with a continuous change of the standard deviation and ranked second in the synthetic scenario involving long sequences with a continuous change of both the mean and the standard deviation, as well as on PCBA.

The original Simple Jumper (I1+B1+W1) is not providing top-performance for any of the synthetic or real-world cases. It is not effective for detecting change of standard deviation (either abruptly or continuously) due to the use of the epsilon betting function (B1).

Observations Regarding the Investment Strategies. The use of taxation (I1) does not lead to top performance for any of the synthetic scenarios, but the strategy is employed by the two top-ranked algorithms on the Lipophilicity dataset.

The use of an investment account (I2), which is the strategy employed by all variants of Sleeper/Stayer and Sleeper/Drifter, is the most effective in all twelve synthetic scenarios, and also in two out of the three real-world cases.

Observations Regarding the Betting Functions. The use of epsilon betting functions (B1) is the most effective strategy for detecting continuous change of the mean in long synthetic sequences, and also gives a slight advantage on House sales, compared to the other betting functions.

The use of time-independent step betting functions (B2) is frequently employed by the most effective methods; the top-ranked approach is using B2 in 9 out of 12 synthetic cases. It is also employed by the most effective approach for the Lipophilicity and PCBA datasets.

Time-dependent step betting functions (B3) are used by two of the most effective methods on long synthetic sequences with continuous change of either the standard deviation alone or in combination with a change of the mean. The use of B3 without the wake-up strategy W2 is significantly outperformed by all other approaches for PCBA and all synthetic (both short and long) where not only the standard deviation change; the combination of B3 and W1 also gives the worst performance for House sales (using either I1 or I2) and for Lipophilicity when used with I2.

Observations Regarding the Wake-up Strategies. In only one of the twelve synthetic cases, the most effective method is employing wake-up (W2), where I2+B3+W2 is the most effective for continuous change of the standard deviation in long sequences. The top-ranked methods on each real-world case include variants with and without wakeup; I2+B1+W1 and I2+B1+W2 for House sales, I1+B2+W1 and I1+B2+W2 for Lipophilicity. For PCBA, the top-ranked method I2+B2+W1 is not significantly outperforming I2+B2+W2.

5   Concluding Remarks

The main components of Simple Jumper, Sleeper/Stayer and Sleeper/ Drifter have been analyzed and combined in novel ways, resulting in twelve unique combinations. Results from a large-scale empirical investigation have been presented, showing that the ability to signal early for data drift varies across distributions and algorithms. The original Sleeper/Stayer algorithm is shown to perform consistently relatively well, but the results also show that the novel combinations are competitive in some cases.

There are several directions for future work. First of all, the empirical investigations may be extended with additional existing approaches for generating conformal test martingales, e.g., using alternative betting functions as proposed in [3,4]. Other directions concern exploring more refined approaches to the investment and wake-up strategies, e.g., using time-varying or non-uniform tax rates and allowing accounts to fall asleep (temporary or permanently). Finally, the effectiveness of the algorithms for generating conformal test martingales may be compared to other frameworks for testing the exchangeability assumption, most notably conformal e-testing [8].

Acknowledgments. HB was partly funded by Vinnova (RAPIDS, grant no. 2021-02522) and Digital Futures.

Disclosure of Interests. The author has no competing interests to declare that are relevant to the content of this article.

Appendix A: Results on Long Synthetic Sequences

Table 4. Abrupt change of mean

Method	Rank	True	False	Time
I1+B1+W1	6.688	0.871	0.005	5770.927
I1+B2+W1	5.246	0.832	0.007	5937.425
I1+B3+W1	11.003	0.682	0.004	6719.652
I1+B1+W2	7.370	0.868	0.004	5766.657
I1+B2+W2	6.838	0.826	0.006	6005.186
I1+B3+W2	6.244	0.852	0.007	5870.039
I2+B1+W1	4.924	0.878	0.003	5757.825
I2+B2+W1	3.326	0.850	0.003	5879.993
I2+B3+W1	10.393	0.714	0.000	6596.872
I2+B1+W2	5.558	0.878	0.003	5756.900
I2+B2+W2	5.154	0.846	0.004	5902.072
I2+B3+W2	5.256	0.859	0.003	5866.648

Table 5. Abrupt change of standard deviation

Method	Rank	True	False	Time
I1+B1+W1	9.104	0.017	0.004	9880.603
I1+B2+W1	4.734	0.714	0.007	6532.851
I1+B3+W1	7.938	0.380	0.006	8108.458
I1+B1+W2	9.096	0.014	0.006	9875.320
I1+B2+W2	5.800	0.708	0.006	6561.221
I1+B3+W2	4.818	0.767	0.008	6291.646
I2+B1+W1	9.062	0.030	0.003	9850.647
I2+B2+W1	3.184	0.724	0.007	6485.451
I2+B3+W1	7.507	0.397	0.000	8102.434
I2+B1+W2	9.084	0.029	0.001	9877.841
I2+B2+W2	3.810	0.721	0.006	6498.719
I2+B3+W2	3.862	0.756	0.005	6306.539

Table 6. Abrupt change of mean and standard deviation

Method	Rank	True	False	Time
I1+B1+W1	8.593	0.861	0.002	5799.294
I1+B2+W1	3.642	0.953	0.006	5246.048
I1+B3+W1	11.160	0.695	0.005	6500.949
I1+B1+W2	9.106	0.855	0.005	5802.876
I1+B2+W2	5.203	0.949	0.006	5265.744
I1+B3+W2	4.790	0.961	0.005	5197.592
I2+B1+W1	7.085	0.871	0.000	5773.713
I2+B2+W1	1.896	0.959	0.003	5227.509
I2+B3+W1	10.486	0.737	0.001	6374.680
I2+B1+W2	7.665	0.870	0.000	5784.554
I2+B2+W2	4.069	0.957	0.003	5257.127
I2+B3+W2	4.304	0.964	0.002	5222.380

Table 7. Continuous change of mean

Method	Rank	True	False	Time
I1+B1+W1	4.226	0.855	0.011	6523.837
I1+B2+W1	7.798	0.819	0.007	6818.321
I1+B3+W1	10.918	0.625	0.004	7612.732
I1+B1+W2	5.306	0.837	0.016	6544.911
I1+B2+W2	9.224	0.799	0.008	6883.441
I1+B3+W2	8.084	0.833	0.006	6781.863
I2+B1+W1	2.512	0.877	0.001	6496.233
I2+B2+W1	5.550	0.833	0.003	6755.364
I2+B3+W1	10.218	0.676	0.001	7471.693
I2+B1+W2	2.656	0.876	0.001	6497.460
I2+B2+W2	5.823	0.833	0.002	6762.698
I2+B3+W2	5.684	0.843	0.002	6707.939

Table 8. Continuous change of standard deviation

Method	Rank	True	False	Time
I1+B1+W1	9.067	0.017	0.005	9916.323
I1+B2+W1	5.372	0.710	0.003	7587.600
I1+B3+W1	7.543	0.371	0.003	8642.488
I1+B1+W2	9.088	0.008	0.005	9932.623
I1+B2+W2	6.428	0.683	0.003	7667.625
I1+B3+W2	4.432	0.785	0.002	7263.375
I2+B1+W1	9.030	0.022	0.004	9922.977
I2+B2+W1	3.652	0.733	0.002	7466.639
I2+B3+W1	7.018	0.383	0.001	8593.943
I2+B1+W2	9.039	0.023	0.003	9924.837
I2+B2+W2	3.934	0.729	0.002	7469.080
I2+B3+W2	3.397	0.764	0.002	7289.790

Table 9. Continuous change of mean and standard deviation

Method	Rank	True	False	Time
I1+B1+W1	6.984	0.853	0.008	6474.302
I1+B2+W1	6.201	0.953	0.003	6131.255
I1+B3+W1	10.868	0.668	0.004	7349.708
I1+B1+W2	7.862	0.849	0.006	6505.246
I1+B2+W2	7.944	0.947	0.004	6177.332
I1+B3+W2	6.378	0.960	0.003	6073.269
I2+B1+W1	5.077	0.873	0.000	6428.811
I2+B2+W1	3.620	0.964	0.000	6071.820
I2+B3+W1	10.103	0.715	0.001	7193.724
I2+B1+W2	5.278	0.873	0.001	6421.142
I2+B2+W2	3.948	0.960	0.003	6059.413
I2+B3+W2	3.736	0.961	0.003	6010.003

Appendix B: Results on Short Synthetic Sequences

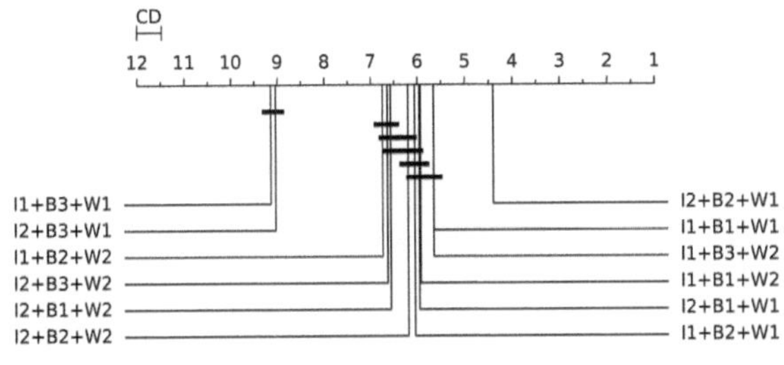

(a) Change of mean

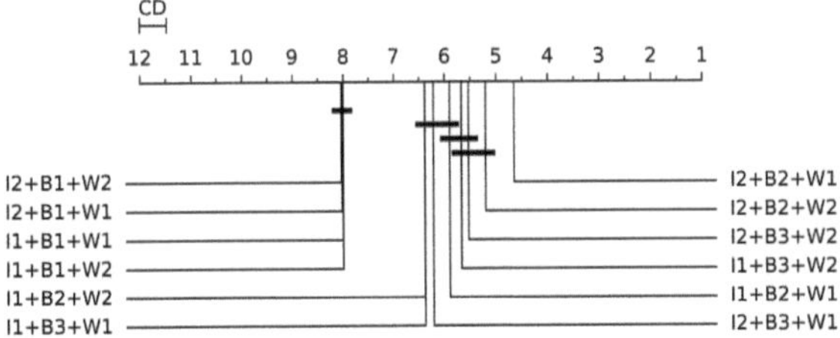

(b) Change of standard deviation

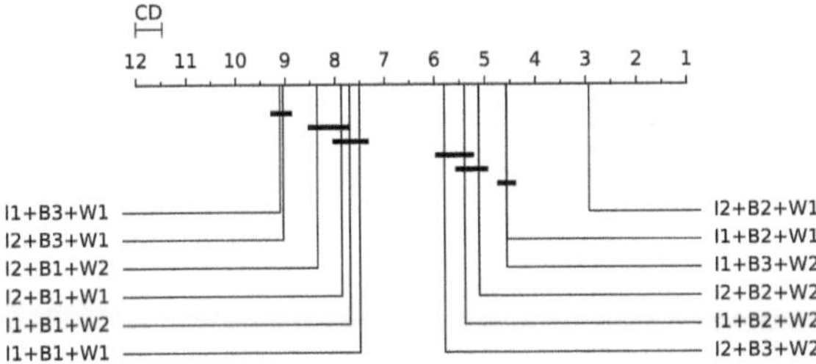

(c) Change of mean and standard deviation

Fig. 8. Abrupt change; short sequences.

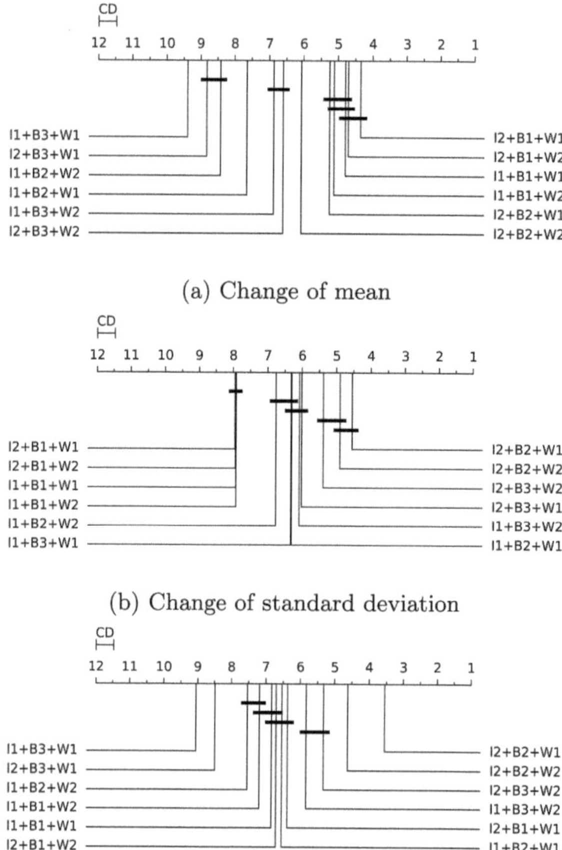

(a) Change of mean

(b) Change of standard deviation

(c) Change of mean and standard deviation

Fig. 9. Continuous change; short sequences.

Appendix C: Results on PCBA

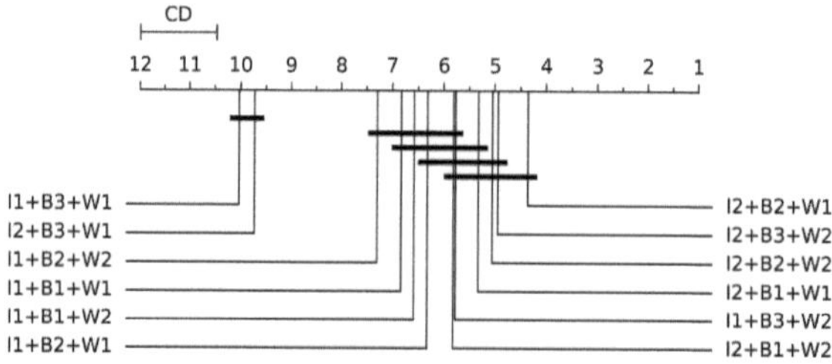

Fig. 10. Average ranks on 120 PCBA datasets.

References

1. Boström, H.: Conformal prediction in python with crepes. In: Proc. of the 13th Symposium on Conformal and Probabilistic Prediction with Applications, pp. 236–249. PMLR (2024)
2. Demšar, J.: Statistical comparisons of classifiers over multiple data sets. J. Mach. Learn. Res. **7**, 1–30 (2006)
3. Eliades, C., Papadopoulos, H.: ICM ensemble with novel betting functions for concept drift. Mach. Learn. **113**(9), 6911–6944 (2024)
4. Fedorova, V., Gammerman, A., Nouretdinov, I., Vovk, V.: Plug-in martingales for testing exchangeability on-line. In: Proceedings of the 29th International Conference on Machine Learning, pp. 923–930 (2012)
5. Volkhonskiy, D., Burnaev, E., Nouretdinov, I., Gammerman, A., Vovk, V.: Inductive conformal martingales for change-point detection. In: Gammerman, A., Vovk, V., Luo, Z., Papadopoulos, H. (eds.) Proceedings of the Sixth Workshop on Conformal and Probabilistic Prediction and Applications. Proceedings of Machine Learning Research, vol. 60, pp. 132–153. PMLR (2017)
6. Vovk, V.: Testing randomness online. Stat. Sci. **36**(4), 595–611 (2021)
7. Vovk, V., Gammerman, A., Shafer, G.: Algorithmic Learning in a Random World, 2nd edn. Springer, New York (2022)
8. Vovk, V., Nouretdinov, I., Gammerman, A.: Conformal e-testing. Pattern Recogn. **168**, 111841 (2025)
9. Vovk, V., Nouretdinov, I., Gammerman, A.: Testing exchangeability on-line. In: Proceedings of the 20th International Conference on Machine Learning (ICML 2003), pp. 768–775 (2003)
10. Vovk, V., Petej, I., Nouretdinov, I., Ahlberg, E., Carlsson, L., Gammerman, A.: Retrain or not retrain: conformal test martingales for change-point detection. In: Carlsson, L., Luo, Z., Cherubin, G., An Nguyen, K. (eds.) Proceedings of the Tenth Symposium on Conformal and Probabilistic Prediction and Applications. Proceedings of Machine Learning Research, vol. 152, pp. 191–210. PMLR (2021)
11. Wu, Z., et al.: MoleculeNet: a benchmark for molecular machine learning. arXiv 1703.00564 (2018). https://arxiv.org/abs/1703.00564

Bootstrapped Chains for Learning Using Privileged Information

Evgueni Smirnov[(✉)][iD] and Filip Schlembach[iD]

Department of Advanced Computing Sciences, Maastricht University,
Maastricht, The Netherlands
{smirnov,filip.schlembach}@maastrichtuniversity.nl

Abstract. This paper introduces Bootstrapped Chains for Learning Using Privileged Information (BC-LUPI), a model-independent method designed to reconstruct privileged variables at test time. BC-LUPI builds an ensemble of predictive chains, where each chain sequentially predicts privileged variables from standard inputs and previously predicted privileged variables. The chains differ in training data via bootstrapping, and in prediction order via random permutations to provide diversity and reduce correlation between individual errors. The final output is obtained by aggregating predictions across chains which allows modeling privileged-variable dependencies and lowering variance, and, thus, improving predictive performance. Experiments on UCI datasets show that BC-LUPI is capable of outperforming baselines, particularly when the privileged information is informative and partially predictable.

Keywords: Learning Using Privileged Information · Ensemble Learning · Classifier Chains

1 Introduction

In many real-world scenarios, additional information is available during training that cannot be accessed at test time. For example, consider predicting at the end of a student's first year whether she will complete her three-year Bachelor's program on time [9]. While training, we can employ historical data from alumni, including grades from all three years. The grades from the second and third years are the most informative but unavailable at prediction time for our first-year student. This scenario exemplifies *Learning Using Privileged Information* (LUPI). We use privileged features such as grades from the second and third years during training to enhance learning, but the final model predicts based on information available at test time, in this case our student's first-year grades.

The LUPI framework was first introduced by Vapnik and Vashist [10]. As stated above, the core idea is to utilize additional informative input variables available during training time to guide the learning process, even though these variables are not accessible at test time. There exist several model-dependent [10,12] and model-independent [1,2] methods for LUPI. In this paper we focus mainly on model-independent methods.

K. An Nguyen and Z. Luo (Eds.): Alexander Gammerman Festschrift, LNCS 16290, pp. 144–156, 2026.
https://doi.org/10.1007/978-3-032-15120-9_8

Model-independent LUPI methods do not rely on specific underlying models such as SVMs or neural networks [1,2]. Instead, they can be applied on almost any base learner which broadens their applicability. We consider three main types of model-independent LUPI methods below.

Feature-transformation methods for LUPI use privileged information during training to learn a new representation that enables predictions at test time using only standard inputs [1,8]. They usually avoid explicitly reconstructing privileged variables, and, thus, can be applied for tasks for which the predicting the privileged variables is difficult [2]. However, these methods may suffer from complexity, reduced interpretability, and reliance on the quality of the learned transformation [3].

Knowledge distillation methods for LUPI treat the model trained with privileged information as a "teacher" and transfer its knowledge to a simpler "student" model that uses only standard inputs at test time. The student mimics the teacher's soft predictions or internal representations [2]. While often effective, the success of these methods depends on the quality of the teacher model and the fidelity of the knowledge transfer [3].

A class of LUPI methods focuses on explicitly predicting the privileged variables from standard variables at test time [7]. The privileged variables are modeled individually using regression or classification models, typically one per privileged variable. At test time, the predicted privileged variables are added to the standard inputs to improve the final prediction for the output variable. While conceptually simple and practically appealing, these methods often assume independence between privileged variables and do not explicitly model their dependencies during reconstruction. Despite this limitation, they have shown promising results across various domains.

In this paper we propose *Bootstrapped Chains for Learning Using Privileged Information* (BC-LUPI), a simple model-independent method for LUPI, that explicitly predicts the privileged variables taking into account their mutual dependencies. The method has its roots in classifier chains [6]. It builds chains of models that sequentially reconstruct privileged variables from standard inputs and already predicted privileged variables. Training data bootstrapping and random ordering of privileged variables create diverse chains, whose aggregated predictions take into account the privileged variables' dependencies, reduce variance, and, thus, enhance predictive performance.

The remainder of the paper is structured as follows. Section 2 defines formally the LUPI problem. BC-LUPI is presented in Sect. 3. Section 4 provides the experimental setup and results for BC-LUPI and baseline models. The experiments are analysed in Sect. 5. Section 6 concludes the paper.

2   Problem of Learning Using Privileged Information

Let $\mathcal{X}$ be the *standard instance space* and $\mathcal{X}^*$ be the *privileged instance space*. $\mathcal{X}$ is given with a vector X of P input variables $X_1, \ldots, X_P$ available at both training time and test time, and $\mathcal{X}^*$ is given with a vector X^* of Q input variables

$X_1^*, \ldots, X_Q^*$ available at training time only. The *output space* $\mathcal{Y}$ can be discrete or continuous for classification or regression tasks, respectively, and it is given by the variable Y. The triplet (X, X^*, Y) follows an unknown joint distribution $\mathbb{P}$ defined over the product space $\mathcal{X} \times \mathcal{X}^* \times \mathcal{Y}$.

During training, a labeled dataset D_N is provided consisting of N instances $(x_n, x_n^*, y_n) \in (\mathcal{X} \times \mathcal{X}^* \times \mathcal{Y})$ drawn from $\mathbb{P}$. At test time, for any test instance $x \in \mathcal{X}$, the value $\hat{y}$ is estimated for the output variable Y. Note that the privileged inputs $x^* \in \mathcal{X}^*$ for the test instance are not observed. This implies that the prediction function h is a mapping $\mathcal{X} \to \mathcal{Y}$ that exploits the privileged information using the triplets (x, x^*, y) at training time.

3   Bootstrapped Chains for Learning Using Privileged Information

Bootstrapped Chains for Learning Using Privileged Information (BC-LUPI) is a model-independent LUPI method. The central idea is to train an ensemble of predictive chains, where each chain sequentially reconstructs the privileged variables. In each chain, one model predicts a single privileged variable using the standard input variables along with the previously predicted privileged variables. Once all privileged variables have been predicted for a test instance, i.e. the instance has been reconstructed, the final model in the chain predicts the output variable. We note that BC-LUPI creates diverse chains by bootstrapping the training data and using different orders to predict privileged variables. Therefore, it aggregates predictions from all chains for the test instance (e.g., via majority voting or averaging), which helps capture privileged-variable dependencies and reduce variance of the chain models.

The pseudocode of BC-LUPI for the training time is given in Algorithm 1. The input consists of the training dataset D_N and the number of chains T. BC-LUPI starts by initializing the ensemble of chains $\mathcal{C}$ equal to $\emptyset$. Then it independently trains the chains C_t for t from 1 to T. To train chain C_t BC-LUPI first draws the bootstrapped dataset D_N' from the training data D_N and generates a random permutation π of $\{1, \ldots, Q\}$. The permutation determines the order in which privileged variables are predicted in C_t; i.e. it determines the order in which we train models in C_t. The q-th model h_q in C_t is trained on the bootstrapped dataset D_N' to predict privileged variable $X_{\pi(q)}^*$ from the concatenated input vector $X \| (X_{\pi(1)}^*, \ldots, X_{\pi(q-1)}^*)$. After the models h_q for all the privileged variables have been trained, they are stored in the list $\mathcal{H}_t$ and the final model h_t for chain C_t is trained to predict the output variable Y using the standard and privileged variables. The final trained chain C_t is consists of the permutation π, the list $\mathcal{H}_t$ of privileged models, and the final output model h_t. C_t is added to the ensemble $\mathcal{C}$. This process is repeated for t from 1 to T and the final ensemble $\mathcal{C}$ of size T is returned.

The pseudocode of BC-LUPI for the test time is given in Algorithm 2. The input consists of a test instance $x \in \mathcal{X}$ and the trained ensemble $\mathcal{C}$. BC-LUPI processes the test instance x independently across all chains C_t, for $t = 1, \ldots, T$.

Algorithm 1. BC-LUPI: Training Time

Require: Training dataset D_N.
Require: Number T of chains.
Ensure: Ensemble of chains $\mathcal{C} = \{C_1, \ldots, C_T\}$.
 1: Set ensemble of chains $\mathcal{C}$ equal to $\emptyset$.
 2: **for** $t = 1$ to T **do**
 3:     Generate bootstrap dataset D'_N from D_N.
 4:     Generate random permutation π of $\{1, \ldots, Q\}$.
 5:     Set list $\mathcal{H}_t$ of privileged prediction models equal to empty list [].
 6:     **for** $q = 1$ to Q **do**
 7:         Train model h_q on D'_N to predict variable $X^*_{\pi(q)}$ from input
            vector $X \,\|\, (X^*_{\pi(1)}, \ldots, X^*_{\pi(q-1)})$
 8:         Add h_q to $\mathcal{H}_t$.
 9:     **end for**
10:     Train final model h_t on D'_N to predict output variable Y from input
        vector $X \,\|\, X^*$.
11:     Set chain C_t equal to $(\pi, \mathcal{H}_t, h_t)$.
12:     $\mathcal{C} \leftarrow \mathcal{C} \cup \{C_t\}$.
13: **end for**
14: **return** $\mathcal{C}$.

For each chain C_t, the algorithm first retrieves the permutation π of privileged variables, the corresponding sequence $\mathcal{H}_t$ of privileged-variable predictors, and the final output model h_t. It then reconstructs the privileged vector $\hat{x}^*$ sequentially by predicting each privileged variable $X^*_{\pi(q)}$ in the order defined by π. The estimate $\hat{x}^*_{\pi(q)}$ is obtained by applying model h_q to the concatenated vector $x \,\|\, \hat{x}^{*1}$, and is subsequently appended to $\hat{x}^*$. Once the full privileged vector $\hat{x}^*$ has received the estimates for all the privileged variables, we obtain a complete reconstruction of the test instance given by the concatenated vector $x \,\|\, \hat{x}^*$. The final model h_t is then applied on $x \,\|\, \hat{x}^*$ to obtain the chain's prediction $\hat{y}_t$. This process is repeated independently for all T chains in the ensemble. Finally, the set of predictions $\{\hat{y}_t\}_{t=1}^{T}$ is aggregated to produce the final prediction $\hat{y}$ for the test instance x.

We note three important details related to BC-LUPI:

- BC-LUPI can be applied for regression and classification LUPI problems. If the output variable Y is continuous, the final model h_t in any chain C_t is a regression model and the aggregation mechanism in line 11 of Algorithm 2 is simple averaging. If the output variable Y is discrete, then the final model h_t is a classification model and the aggregation mechanism is majority voting.
- BC-LUPI can handle data represented with both continuous and discrete variables. If a privileged variable $X^*_{\pi(q)}$ is continuous or discrete, h_q is a regression or classification model, respectively.

---

[1] We note that $\hat{x}^* = (\hat{x}^*_{\pi(1)}, \ldots, \hat{x}^*_{\pi(q-1)})$.

– Different orders in chains and aggregating chain predictions model implicitly the dependencies between the privileged variables [6]. This contrasts with the previous methods from the similar cohort [7].

Algorithm 2. BC-LUPI: Test Time

Require: Test instance $x \in \mathcal{X}$.
Require: Trained ensemble $\mathcal{C}$ from Algorithm 1.
Ensure: Prediction $\hat{y}$
1: **for** $t = 1$ to T **do**
2:     Take chain $C_t = (\pi, \mathcal{H}_t, h_t) \in \mathcal{C}$.
3:     Initialize $\hat{x}^*$ as an empty vector to store the privileged-variable predictions.
4:     **for** $q = 1$ to Q **do**
5:         Get model $h_q \in \mathcal{H}_t$ for privileged variable $X^*_{\pi(q)}$.
6:         Apply model h_q on $x \,||\, \hat{x}^*$ to get prediction $\hat{x}^*_{\pi(q)}$ for $X^*_{\pi(q)}$.
7:         $\hat{x}^* = \hat{x}^* \,||\, (\hat{x}^*_{\pi(q)})$.
8:     **end for**
9:     Apply model h_t on $x \,||\, \hat{x}^*$ to get prediction $\hat{y}_t$ for output variable Y.
10: **end for**
11: Aggregate prediction set $\{\hat{y}_t\}_{t=1}^T$ into final prediction $\hat{y}$.
12: **return** $\hat{y}$.

The prediction error e of BC-LUPI arises from the individual chains and the way their predictions are aggregated at the ensemble level. As a result, e depends primarily on three components:

– **Per-chain biases** b_t: comprising the model bias b_t^m of the final predictor h_t and the reconstruction bias b_t^r from using predicted privileged inputs.
– **Per-chain variances** v_t: comprising the model variance v_t^m of h_t and the reconstruction variance v_t^r, plus their interaction.
– **Error correlations** ρ_{st}: the correlations between the prediction errors of chains C_s and C_t for $s \neq t$.

Small model biases b_t^m and variances v_t^m can be achieved when the privileged variables X^* are *informative* for the output variable Y. This motivates the first factor[2] contributing to the predictive performance of BC-LUPI:

(F1) The privileged variables X^* should carry substantial information about the output variable Y.

The better (F1) is satisfied, the lower the per-chain biases b_t^m and variances v_t^m, and, hence, the lower the prediction error e of BC-LUPI.

The correlation terms ρ_{st} grow with the predictability of the privileged variables X^*. When the *predictability is high*, the reconstructions of X^* become

---

[2] This is not a necessary condition, as noted by [2].

very similar across the chains which results in highly correlated errors. Although reconstruction biases b_t^r and variances v_t^r may be small, BC-LUPI derives little benefit from averaging/majority vote because $\rho_{st} \approx 1$.[3]

At the opposite extreme, if the privileged variables X^* are *completely unpredictable*, each chain guesses the privileged variables, leading to large reconstruction biases b_t^r and variances v_t^r. Correlations ρ_{st} are then low, but the high noise level leaves little room for variance reduction and the prediction error e of BC-LUPI remains large.

BC-LUPI is expected to be the most effective in the intermediate case when the privileged variables X^* are *partially* predictable. In this case, the reconstruction errors are of moderate magnitude, resulting in moderate levels of per-chain reconstruction biases b_t^r, reconstruction variances v_t^r, and inter-chain error correlations ρ_{st}. This enables BC-LUPI to reduce the variance component of the prediction error e through aggregation over diverse chains.

We recall that within each BC-LUPI chain, the privileged variables in X^* are predicted sequentially using all standard variables X and the previously reconstructed privileged variables. Therefore, when we need the privileged variables to be *partially predictable*, we need two additional factors for effectiveness of BC-LUPI to hold:

(F2) The privileged variables X^* should be *partially* predictable from the standard variables X.

(F3) The privileged variables X^* should be *partially* predictable from one another.

Modeling dependencies among privileged variables in BC-LUPI reduces both reconstruction biases b_t^r and variances v_t^r. By allowing each privileged variable to be predicted not only from the standard inputs but also from previously reconstructed privileged variables, the model can better correct systematic errors, lowering the bias, and leverage inter-variable dependencies to stabilize predictions, thereby lowering the variance. Although error propagation through the chain can amplify early mistakes, BC-LUPI mitigates this through bootstrapping and ensemble averaging. In this context, we note that when the factor (F3) is observed, multiple chains may reconstruct the same privileged variables similarly, increasing the error correlations ρ_{st} between their final predictions $\hat{y}_t$. To counteract this, BC-LUPI uses random orderings of privileged variables in each chain, ensuring that every variable is predicted in a different context. This decorrelates reconstruction paths across chains and helps reduce BC-LUPI's the final prediction error e further.

BC-LUPI is most effective when applied to high-variance base learners, as it reduces the variance through bootstrapping and random permutation of privileged variables across chains. For stable, low-variance models, this variance reduction is minimal, while the added reconstruction steps introduce extra bias that cannot be averaged out, potentially worsening performance.

---

[3] In fact, if X^* can be fully determined, the problem ceases to be a genuine LUPI task.

4 Experiments

This section describes our experiments with BC-LUPI. It starts with the experimental setup in Subsect. 4.1 and continues with the experimental results in Subsect. 4.2.

4.1 Experimental Setup

The performance of BC-LUPI is evaluated on classification and regression datasets from the UCI repository provided via scikit-learn [4]. The datasets are summarized in Table 1.

Table 1. Summary of the six datasets used in the experiments.

Dataset	# Instances	# Variables	Type
Breast Cancer	569	31	Classification
Wine	178	14	Classification
Digits	1,797	65	Classification
Diabetes	442	11	Regression
California Housing	20,640	9	Regression
Auto MPG	398	8	Regression

To simulate the LUPI setting, we partition the input variables into normal and privileged subsets based on their relevance to the output variable. For this purpose, we compute the mutual information (MI) between each input variable and the output variable. Input variables are ranked in descending order of MI. For each privileged-information proportion $p \in \{10\%, 20\%, \ldots, 90\%\}$, the top $\lceil p\% \cdot d \rceil$ variables (where d is the total number of input variables) are selected as privileged, while the remaining variables are treated as normal. This ranking-based partitioning ensures a fair and consistent evaluation across different scenarios by systematically increasing the number of privileged variables while decreasing the number of normal variables. As a result, we can assess the effect of varying privileged information availability on the performance of predictors.

We experiment with BC-LUPI for $T = 20$ predictive chains. The base predictors employed within BC-LUPI's chains are decision trees. Following [2], we compare BC-LUPI with three baseline decision trees:

- (1) XX^* decision trees trained on all the input variables (normal + privileged).
- (2) X decision trees trained on the normal input variables only.
- (3) X^* decision trees trained on the privileged input variables only.

All decision trees, both the baseline ones, and those used within BC-LUPI, are trained using the default configuration of scikit-learn [4].

The performance of the classification models is measured using the accuracy rate. The performance of the regression models is measured using the root-mean-squared error (RMSE). We employ 5-fold cross-validation repeated 5 times. For each method and each privileged percentage, the average score (accuracy or RMSE) across all 25 evaluation folds is reported.

We assess whether BC-LUPI significantly outperforms the X decision trees using a paired t-test at the significance level $\alpha = 0.05$, separately for each proportion of privileged variables. We note that for classification, we perform a one-tailed test for higher accuracy rate and for regression, we perform a one-tailed test for lower RMSE.

4.2   Experimental Results

The experimental results for each dataset are provided in Figs. 1, 2, 3, 4, 5 and 6. Each figure contains a *performance plot* and a *mutual information (MI) plot*.

The performance plots visualize how different learning models behave as the proportion of privileged variables increases. These plots display the predictive performance (accuracy rate for classification datasets and RMSE for regression datasets) for: XX^* decision trees, X decision trees, X^* decision trees, and BC-LUPI. Statistically significant improvements of BC-LUPI over X decision trees are highlighted using circle markers based on paired t-tests at significance level $\alpha = 0.05$.

In addition to the performance plots, we also include mutual information (MI) plots that characterize the statistical relationships between the input, privileged, and output variables. For each privileged variable proportion, we report the following four metrics:

- The average mutual information $\mathrm{MI}(X \leftrightarrow Y)$ between the normal variables and the output variable. This serves as a baseline indicator of how informative the standard inputs are on their own.
- The average mutual information $\mathrm{MI}(X^* \leftrightarrow Y)$ between the privileged variables and the output variable. This reflects the informativeness of the privileged variables and directly corresponds to factor (F1).
- The average mutual information $\mathrm{MI}(X \leftrightarrow X^*)$ between the normal and privileged variables. This quantifies the extent to which privileged variables can be reconstructed from standard variables, and is related to factor (F2).
- The average mutual information $\mathrm{MI}(X^* \leftrightarrow X^*)$ between the privileged variables themselves. This captures the internal dependencies among privileged variables and connects to factor (F3).

The mutual information plots allow us to assess how the quality and structure of the privileged information influence the effectiveness of the baseline predictors and BC-LUPI in each experiment.

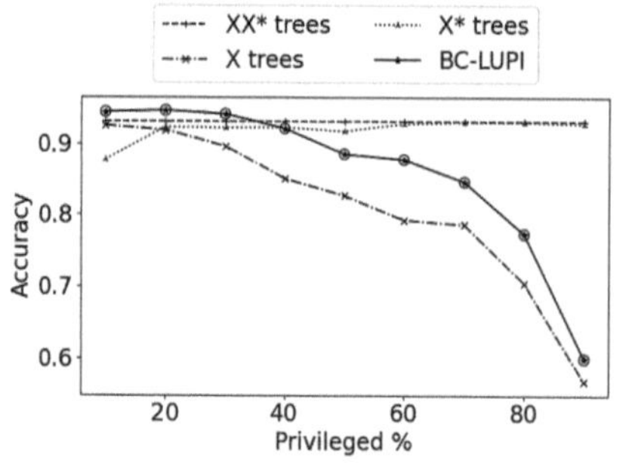

(a) Accuracy vs. % privileged for XX* trees, X trees, X* trees, and BC-LUPI.

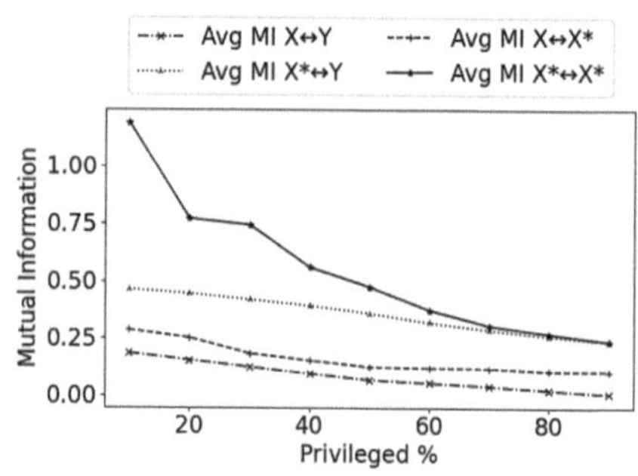

(b) Mutual information metrics.

Fig. 1. Breast Cancer data: experimental results.

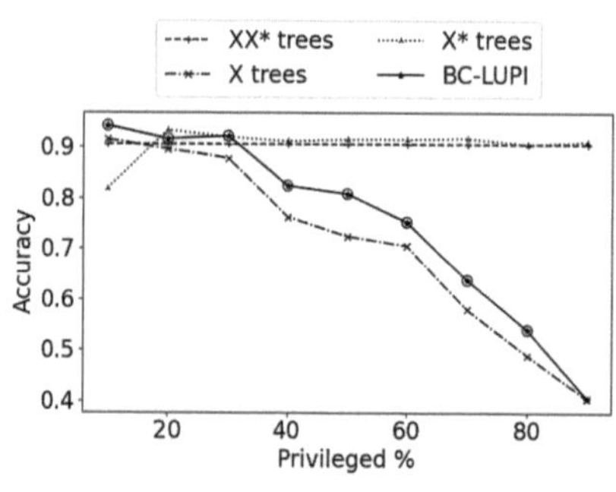

(a) Accuracy vs. % privileged for XX* trees, X trees, X* trees, and BC-LUPI.

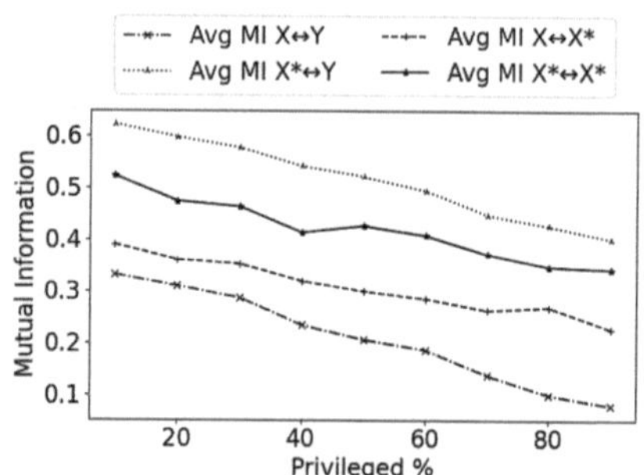

(b) Mutual information metrics.

Fig. 2. Wine data: experimental results.

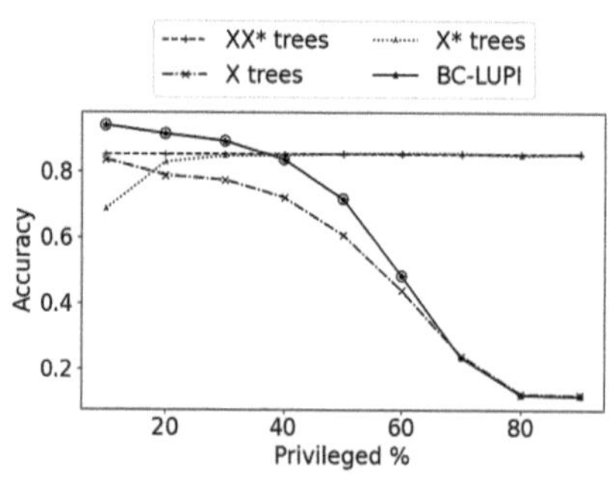

(a) Accuracy vs. % privileged for XX* trees, X trees, X* trees, and BC-LUPI.

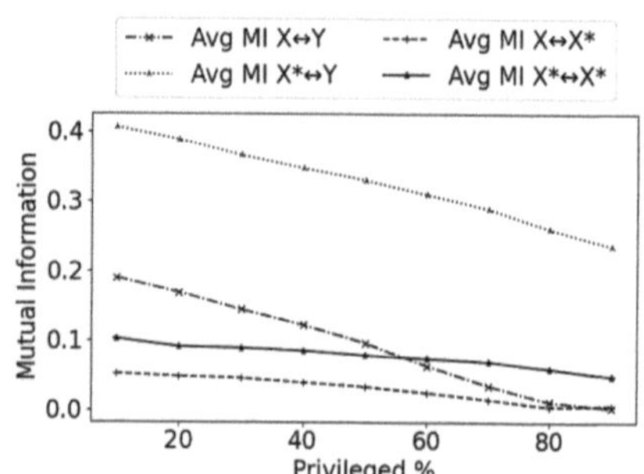

(b) Mutual information metrics.

Fig. 3. Digits data: experimental results.

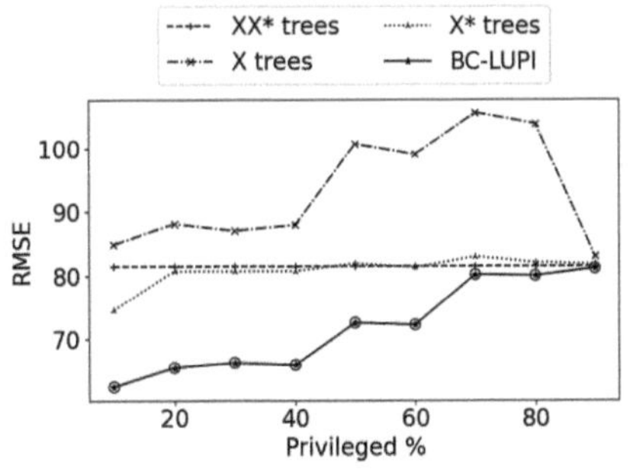
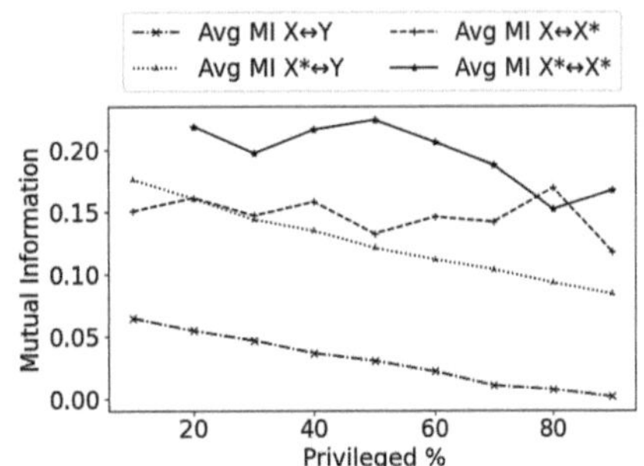

(a) RMSE vs. % privileged for XX* trees, X trees, X* trees, and BC-LUPI.

(b) Mutual information metrics.

Fig. 4. Diabetes data: experimental results.

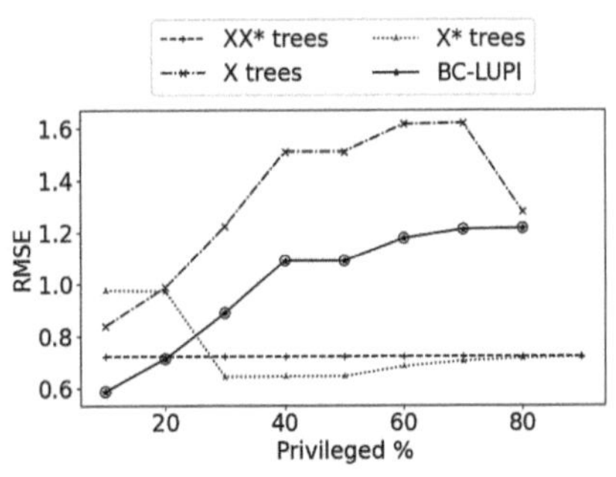
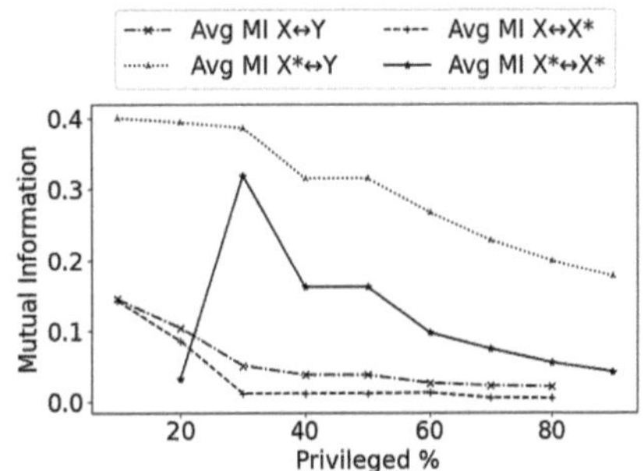

(a) RMSE vs. % privileged for XX* trees, X trees, X* trees, and BC-LUPI.

(b) Mutual information metrics.

Fig. 5. California Housing data: experimental results.

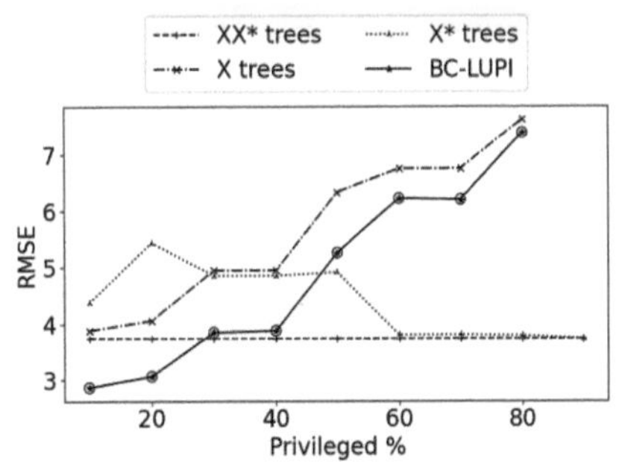

(a) RMSE vs. % privileged for XX* trees, X trees, X* trees, and BC-LUPI.

(b) Mutual information metrics.

Fig. 6. Auto MPG data: experimental results.

5   Discussion

The plots in Figs. 1, 2, 3, 4, 5 and 6 support several observations for the baseline models. The performance of the XX^* decision trees is constant across different privileged-information proportions p as it shown in Figs. 1(a)–6(a) since these models have access to the full set of input variables. The performance of the X decision trees consistently decreases with p since these models have access to sets of input that decrease in size and become less informative with p (see Figs. 1(b)–6(b)). The performance of the X^* decision trees increases with p and converges to that of the XX^* decision trees since these models have access to the full set of input variables for $p = 100\%$.

BC-LUPI consistently outperforms all baseline models at low privileged-information proportions p, particularly when $p < 20\%$, across all datasets. For these low values of p, the mutual information metrics: $\mathrm{MI}(X^* \leftrightarrow Y)$, $\mathrm{MI}(X \leftrightarrow X^*)$, and $\mathrm{MI}(X^* \leftrightarrow X^*)$ tend to be relatively high (see Figs. 1(b)–6(b)) which implies that factors (F1), (F2), and (F3) are contributing. This favors the BC-LUPI mechanism: privileged variables are informative, partially predictable from standard inputs, and exhibit internal dependencies. A notable example in this context is the diabetes dataset (see Fig. 4), for which all three mutual information metrics are particularly high. This indicates strong presence of factors (F1), (F2), and (F3) for this dataset that enables BC-LUPI to effectively reconstruct privileged information and achieve lower RMSE than all baseline models across all levels of privileged-information proportion p.

As the proportion of privileged variables p exceeds 30–40%, the performance of BC-LUPI begins to decline. This decline is associated with decreasing mutual information between normal and privileged variables $\mathrm{MI}(X \leftrightarrow X^*)$, as well as among privileged variables $\mathrm{MI}(X^* \leftrightarrow X^*)$ (see Figs. 1(b)–6(b)). Lower $\mathrm{MI}(X \leftrightarrow X^*)$ implies that privileged variables become harder to reconstruct from the reduced set of normal variables, while lower $\mathrm{MI}(X^* \leftrightarrow X^*)$ weakens the effectiveness of dependency modeling within chains. These two effects correspond to a weaker observation of factors (F2) and (F3), respectively. As a result, the predictors for privileged variables in the chains suffer from increased bias and reconstruction noise, which cannot be sufficiently mitigated by the ensemble structure of BC-LUPI.

In contrast, the X^* decision trees, due to direct access to privileged variables at test time, are unaffected by reconstruction errors and thus outperform BC-LUPI at high values of p. This advantage does not apply to the X decision trees: their bias and performance continue to deteriorate with p which leads to consistently worse results compared to BC-LUPI.

6   Conclusion

We introduced *Bootstrapped Chains for Learning Using Privileged Information* (BC-LUPI), a model-independent method that reconstructs privileged variables by exploiting their mutual dependencies. By combining sequential prediction over randomly ordered privileged variables with diversity introduced through bootstrap resampling, BC-LUPI transforms high-variance base learners into a low-variance ensemble. Theoretical analyses in Sect. 3 identify three key factors contributing to BC-LUPI's effectiveness: when privileged variables are (F1) informative for the output, (F2) partially predictable from the standard inputs, and (F3) mutually dependent on one another.

Experimental results confirm the effectiveness of BC-LUPI and support these theoretical expectations. Across multiple controlled scenarios that varied the informativeness and predictability of privileged variables, BC-LUPI consistently outperformed baseline models whenever the identified factors were present.

Future research will proceed along two main directions: (1) exploring alternative architectures for bootstrapped chains to enhance predictive performance and reduce computational complexity, and (2) developing conformal versions of bootstrapped chains to provide reliable uncertainty quantification to control reconstruction error [5, 11].

References

1. Fouad, S., Tiño, P., Raychaudhury, S., Schneider, P.: Incorporating privileged information through metric learning. IEEE Trans. Neural Netw. Learn. Syst. **24**(7), 1086–1098 (2013). https://doi.org/10.1109/TNNLS.2013.2251470
2. Lopez-Paz, D., Bottou, L., Schölkopf, B., Vapnik, V.: Unifying distillation and privileged information. In: Bengio, Y., LeCun, Y. (eds.) 4th International Conference on Learning Representations, ICLR 2016, San Juan, Puerto Rico, 2–4 May 2016, Conference Track Proceedings (2016). http://arxiv.org/abs/1511.03643
3. Pechyony, D., Vapnik, V.: On the theory of learning with privileged information. In: Lafferty, J.D., Williams, C.K.I., Shawe-Taylor, J., Zemel, R.S., Culotta, A. (eds.) Advances in Neural Information Processing Systems 23: 24th Annual Conference on Neural Information Processing Systems 2010. Proceedings of a Meeting Held 6–9 December 2010, Vancouver, British Columbia, Canada, pp. 1894–1902. Curran Associates, Inc. (2010). https://proceedings.neurips.cc/paper/2010/hash/c73dfe6c630edb4c1692db67c510f65c-Abstract.html
4. Pedregosa, F., et al.: Scikit-learn: machine learning in Python. J. Mach. Learn. Res. **12**, 2825–2830 (2011). https://doi.org/10.5555/1953048.2078195, https://dl.acm.org/doi/10.5555/1953048.2078195
5. Provodin, D., van den Akker, B., Katsimerou, C., Kaptein, M., Pechenizkiy, M.: Rethinking knowledge transfer in learning using privileged information. CoRR abs/2408.14319 (2024). https://doi.org/10.48550/ARXIV.2408.14319
6. Read, J., Pfahringer, B., Holmes, G., Frank, E.: Classifier chains: a review and perspectives. J. Artif. Intell. Res. **70**, 683–718 (2021). https://doi.org/10.1613/JAIR.1.12376
7. Sarafianos, N., Nikou, C., Kakadiaris, I.A.: Predicting privileged information for height estimation. In: 23rd International Conference on Pattern Recognition, ICPR 2016, Cancún, Mexico, 4–8 December 2016, pp. 3115–3120. IEEE (2016). https://doi.org/10.1109/ICPR.2016.7900113
8. Sharmanska, V., Quadrianto, N., Lampert, C.H.: Learning to transfer privileged information. CoRR abs/1410.0389 (2014). http://arxiv.org/abs/1410.0389

9. Smirnov, E.N., Delava, R., Diris, R., Nikolaev, N.I.: Multi-view semi-supervised learning using privileged information. In: Iliadis, L., Maglogiannis, I., Alonso, S., Jayne, C., Pimenidis, E. (eds.) Engineering Applications of Neural Networks - 24th International Conference, EAAAI/EANN 2023, León, Spain, 14–17 June 2023, Proceedings. Communications in Computer and Information Science, vol. 1826, pp. 144–152. Springer, Cham (2023). https://doi.org/10.1007/978-3-031-34204-2_13
10. Vapnik, V., Vashist, A.: A new learning paradigm: Learning using privileged information. Neural Netw. **22**(5-6), 544–557 (2009). https://doi.org/10.1016/J.NEUNET.2009.06.042
11. Vovk, V., Gammerman, A., Shafer, G.: Algorithmic Learning in a Random World, 2 edn. Springer, Cham (2022). https://doi.org/10.1007/978-3-031-06649-8
12. Yan, S., Odom, P., Pasunuri, R., Kersting, K., Natarajan, S.: Learning with privileged and sensitive information: a gradient-boosting approach. Front. Artif. Intell. **6** (2023). https://doi.org/10.3389/FRAI.2023.1260583

VC-Theoretical Explanation of Double Descent

Vladimir Cherkassky$^{(\boxtimes)}$ [iD] and Eng Hock Lee [iD]

Department of Electrical and Computer Engineering, University of Minnesota Twin Cities,
Minneapolis, MN 55455, USA
`{cherk001,leex7132}@umn.edu`

Abstract. Despite remarkable practical success of Deep Learning (DL) networks, their theoretical understanding is still limited. Recently, in DL community there has been many new theories that explain generalization properties of heuristic learning methods. These new theories appear to be unaware of scientific understanding of inductive inference, dating back to Fisher's (1935) paper. Notably, phenomenon called 'double descent', discovered by DL practitioners, seems to contradict known statistical theories. However, we show that double descent phenomenon is in perfect agreement with classical VC-theory. In particular, VC-theory explains both first and second descent regime, corresponding to under and over-parameterized estimators. Proposed theoretical explanation is supported by empirical modeling of double descent curves using analytic VC-bounds, under standard classification setting. Finally, we discuss methodological reasons for misunderstanding of VC-theoretical concepts, such as VC-dimension and Structural Risk Minimization, in machine learning research community.

Keywords: Deep Learning · Double Descent · Complexity Control · Structural Risk Minimization · VC Dimension · VC Generalization Bounds

1 Introduction and Motivation

Over past 40 years, the field of Machine Learning (ML) has been driven by the growth of data and digital technology, focusing on development of powerful learning algorithms, such as Deep Learning (DL). However, theoretical understanding of Deep Learning technology is still limited. Popular DL algorithms are usually motivated by heuristic arguments, and their mathematical justification often appears later. This happened with artificial neural networks in mid-1990's, and then later with AdaBoost, Random Forest, and DL networks. In the absence of a common mathematical framework, ML researchers usually propose many different theoretical explanations. Often, these new theories represent old theories in disguise. This disconnects between mathematical theory and applications prevents methodological understanding of ML by practitioners.

There is a clear difference between classical scientific and modern data-driven approaches to knowledge discovery. Science is based on hypothetico-deductive approach:

$$Hypothesis \rightarrow Experiment \rightarrow Theory\ (Knowledge)$$

© The Author(s), under exclusive license to Springer Nature Switzerland AG 2026
K. An Nguyen and Z. Luo (Eds.): Alexander Gammerman Festschrift, LNCS 16290, pp. 157–176, 2026.
https://doi.org/10.1007/978-3-032-15120-9_9

Under this approach, empirical data is used to validate scientific theory proposed by intelligent humans. In this case, empirical data is obtained from designed experiments.

Modern data-centric approach derives predictive models from observational data:

$$ML\,software\ +\ Data\ \rightarrow\ Data - Driven\,Model\ (Knowledge)$$

Under this view, human intelligence is replaced by machine learning software, enhanced by powerful computers and Big Data. This approach relies mostly on complex observational data, and it yields large data-analytic models that are not intelligible, brittle and often non-reproducible. This leads to poor understanding of its limitations and capabilities. Methodologically, modern data-centric approach emphasizes good engineering, rather than science, where the goal is to develop new ML algorithms for various applications. For this reason, DL methods are often regarded as magic tools for generating knowledge from data, aka inductive inference process.

Discussion about mathematical foundations of inductive inference can be traced back to classical statistics and philosophy of science [1, 2]. Fisher [1] discussed the role of statistics in the process of inductive inference. He argued that 'uncertain inferences' from the particular to the general can be made rigorous only for properly defined mathematical quantities. Then, he introduced one such formalization, known as parametric estimation using mathematical likelihood, 'where we start with a parametric form of unknown distribution, describing the population sampled, but without knowledge of the values of one or more parameters which enter into this form.' At the same time, Fisher was fully aware that there may be different formalizations of inductive inference [1].

In the discussion part of Fisher's paper [1], J. Neyman pointed out the recurring pattern observed in the development of statistics: 'Starting with several problems of purely practical character, the branch of mathematics is being developed primarily as a set of methods useful for the solution of similar problems. These methods are then compiled in theories already forming what could be called a science. However, the early theories contain usually many gaps and inaccuracies. When these are noticed a new period comes, in which a marked effort to penetrate into the depth rather than into the width of the science may be observed'. Of course, the history repeats itself in the case of DL, where currently the period of "solving problems" is over. However, nowadays it is common to introduce multiple mathematical theories for DL. These theories are often based on different formalizations and assumptions.

This chapter presents a case study, using generalization properties of DL networks, known as "double descent", originally discovered by practitioners [3–5]. Currently, there is a consensus view that double descent cannot be explained by existing statistical learning theories. This motivates development of various new theories [4, 6]. However, this chapter shows that double descent can be indeed well explained by classical VC-theory that applies to all methods based on minimization of training error. In particular, we show how to apply VC-theoretical results for modeling double descent, under classification setting.

This chapter is organized as follows. Section 2 presents an overview of double descent and relevant VC-theoretical results. Section 3 presents empirical modeling of double descent using VC-bounds. Section 4 extends the proposed VC-theoretical approach to understanding generalization in DL networks. Section 5 presents VC-theoretical modeling for low-dimensional data. Section 6 presents summary and discussion.

2 Overview VC-Theory and Double Descent

Despite many (often exaggerated) claims about capabilities of DL, they usually select a function (or model) from an infinite number of possible models. This selection is made using finite training data, based on the notion that a model that explains well the training data, is likely to generalize well for future data. Usually, good explanation (of training data) and generalization (for test data) correspond to minimization of a pre-defined loss function, i.e., training and test errors. Therefore, it is possible to formalize learning from data as a mathematical problem. Formal framework for methods based on minimization of training error, aka Empirical Risk Minimization (ERM), is given by VC-theory [7, 8]. It provides mathematical conditions for generalization of models estimated from training data, and states general conditions and assumptions for all methods based on ERM.

Here, we also point out the difference between classical statistical view of inductive inference (parameter estimation via maximum likelihood) and ERM. That is, classical statistics developed theory for methods for low-dimensional problems, such as estimating parameters of distributions and hypothesis testing. Even though it is not explicitly stated by [1], classical statistics deals with under-parameterized settings, when the number of data samples is larger than the number of parameters being estimated. In contrast, VC-theoretical framework is more general and applies to both under and over-parameterized settings.

Many DL application studies use large networks trained to fit perfectly available training data, that also achieve good generalization for test data. This appears to be at odds with classical bias-variance trade-off view that overfitting leads to poor generalization [3–5]. Belkin et al. [6] systematically described this phenomenon as 'double descent' and pointed out the difference between classical regime (first descent) and modern one (second descent). Graphical representation of double descent is shown in Fig. 1, following [6]. This figure shows training and test error curves, as a function of 'capacity' of possible models, or hypotheses $\mathcal{H}$. In DL literature [6, 9, 10], capacity of $\mathcal{H}$ usually corresponds to network size, or the number of hidden units. Researchers proposed several theoretical explanations of generalization capability of DL networks and double descent, such as:

- special properties of multilayer network parameterization [11].
- choosing proper inductive bias implemented during second descent [6].
- properties of Stochastic Gradient Descent (SGD) training algorithm [4, 5, 12].
- the effect of various heuristics (used for training) on generalization [13].
- he effect of margin on generalization in DL networks [14].
- efficient quantization (encoding) of large DL network models [9].

The current consensus view on 'generalization paradox' in DL networks is that:

- Existing indices for model complexity, such as VC-dimension and Rademacher complexity, cannot explain generalization performance of DL networks. Therefore, researchers often introduce new complexity indices designed for particular parameterization (e.g., nets with ReLU units) and particular training method (such as SGD).
- 'Classical' theories developed in ML and Statistics cannot explain generalization performance of large DL networks, such as 'double descent' phenomenon.

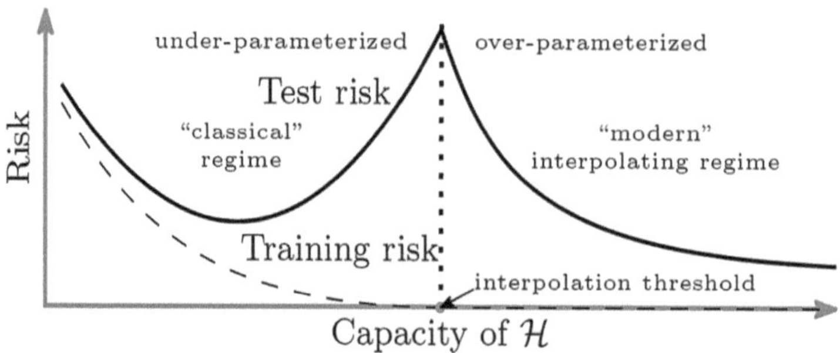

Fig. 1. Double Descent generalization curve.

On the other hand, classical VC-theory [7, 8] provides clear conditions for understanding generalization of all learning methods based on minimization of training error, such as:

- Finite VC-dimension provides *necessary* and *sufficient* conditions for good generalization.
- Analytic VC-bounds on test error, as a function of training error and VC-dimension.

This clear disagreement between the consensus view and VC-theory can be explained primarily by misunderstanding of VC-theoretical results in DL research community. For example, there is a common view that VC-dimension grows with network size, or the number of weights, and so, 'traditional measures of model complexity struggle to explain the generalization ability of large artificial neural networks' [5]. This view is reflected in double descent curve in Fig. 1, where 'capacity' grows with network size [5, 6]. In fact, VC-dimension can be equal, larger, or smaller than the number of parameters [7, 8]. Referring to Fig. 1, during second descent VC-dimension is decreasing with increasing network size (as shown later in Sect. 3). Another misconception is that 'VC-dimension depends only on the model family and data distribution, and not on the training procedure used to find models' [3]. In fact, VC-dimension does not depend on data distribution, but depends on a training procedure, i.e., SGD [15].

VC theory provides analytic bounds for (unknown) test error, as a function of Empirical Risk (or training error) and VC-dimension (h) of a set of admissible models. For classification problems, given training data set (of size n), there are two types of VC-bounds [7, 8, 15]:

Uniform convergence bound:

$$R_{tst} \leq R_{trn} + \tfrac{1}{2}\sqrt{\varepsilon} \tag{1}$$

Uniform relative convergence bound:

$$R_{tst} \leq R_{trn} + \tfrac{\varepsilon}{2}\left(1 + \sqrt{1 + \tfrac{4R_{trn}}{\varepsilon}}\right) \tag{2}$$

where $\varepsilon = \tfrac{a_1}{n}\left(h\left[\ln\left(\tfrac{a_2 n}{h}\right) + 1\right] - \ln\tfrac{\eta}{4}\right)$ and $\eta = \min\left(\tfrac{4}{\sqrt{n}}, 1\right)$. When $R_{trn} = 0$, the uniform relative convergence bound is reduced to:

$$R_{tst} \leq \varepsilon \tag{2a}$$

VC-bounds (1) and (2) hold with probability $1 - \eta$ (aka confidence level) for all admissible models (functions) including the one minimizing the training error (R_{trn}). According to (1) and (2), test error is bounded by the sum of training error and second additive term, called the confidence interval, a.k.a excess risk, that depends on both the training error and VC-dimension (h). These bounds were originally intended for conceptual understanding of complexity control, i.e., the effect of VC-dimension on controlling test error. According to [7, 8], uniform relative convergence bound (2) is more accurate than uniform convergence bound (1), especially during second descent. That is, consider the bound on excess error $R_{tst} - R_{trn}$ when $R_{trn} = 0$. . In this case, the uniform convergence bound (1) gives an estimate of excess error of the order $\mathcal{O}(\sqrt{h/n})$, , whereas bound (2a) gives an estimate of the order $\mathcal{O}(h/n)$. Since h/n is under 0.5, the value of $\mathcal{O}(h/n)$ is much smaller than $\mathcal{O}(\sqrt{h/n})$.

Note that uniform relative convergence bound (2) has no analogy outside VC-theory [8], and it remains virtually unknown in DL community. All technical arguments, suggesting that VC-theory cannot explain double descent, are based on analysis of uniform convergence bound (1) [5, 6, 16–19], where the excess error is of the order $\mathcal{O}(\sqrt{h/n})$.

Application of VC-bound (2) for quantitative modeling of double descent for various data sets requires (a) selecting proper values of positive constants a_1 and a_2, and (b) analytic estimation of VC-dimension. Classical VC-bounds describe general form of analytic relationship between test error, training error, and VC-dimension. However, the values $a_1 = 4$ and $a_2 = 2$, provided in VC-theory for the worst-case distributions, yield bounds that are too loose for real-life data [15, 20]. Recently, we proposed using 'practical' values $a_1 = 3$ and $a_2 = 1$ for real-life data sets [21, 22]. These values are used for all empirical results presented in this chapter. For over-parameterized settings, the training error in bound (2) equals zero, so using these 'practical' values $a_1 = 3$, $a_2 = 1$ results in a simplified bound for test error:

$$R_{tst} \leq \tfrac{3}{n}\left(h\left[\ln\left(\tfrac{n}{h}\right) + 1\right] - \ln \tfrac{\eta}{4}\right) \tag{3}$$

Clearly, bound (3) is minimized when VC-dimension is minimized during 2nd descent. Later in Sect. 3, we consider a simplified two-layer networks, where during 2nd descent VC-dimension is given by the norm of weights $\|\mathbf{w}\|^2$, so minimization of VC-bound (3) is achieved by minimization of $\|\mathbf{w}\|^2$.

Application of VC-bounds to modeling generalization requires analytic estimates of VC-dimension. For many learning methods, such estimates are not known. For example, for SGD algorithms, the combined effect of various heuristics (e.g., weight initialization, learning rate schedule) on VC-dimension is impossible to estimate analytically.

Next, we discuss VC-theoretical framework for conceptual understanding of double descent. Recall the consensus view that generalization ability of large (over-parameterized) DL networks cannot be explained by VC-theory. This view is not correct, because according to VC-bound (3), good generalization is possible if zero training error can be achieved using a set of models with small VC-dimension.

According to VC-bound (2), generalization depends only on two factors, the training error and VC-dimension. A formal procedure for minimizing this bound is known as Structural Risk Minimization (SRM) inductive principle [7, 8]. Under SRM, a set of admissible models has a nested structure, where each element of a structure S_k has finite

VC-dimension h_k. Hence, SRM structure provides ordering of all admissible models according to their complexity (~VC-dimension). There are two different constructive strategies for minimizing VC-bound (2) using SRM [7, 8, 15, 20]:

- Strategy 1: Consider nested structure $\{S_k\}$ that provides ordering of admissible models according to VC-dimension $h_1 < h_2 < h_3 < \cdots$. Assume that fitting training data using admissible models has non-zero training error, for each element of a structure S_k. Then minimize training error for each element of a structure S_k having VC-dimension h_k. The goal is to find optimal element (~optimal VC-dimension), providing optimal trade-off between the training error and confidence interval in bound (2). This corresponds to the first descent regime, implemented in 'classical' statistical methods;
- Strategy 2: For a set of models providing fixed training error, find a model with the smallest VC-dimension. This corresponds to over-parameterized set of models during second descent, when training error is fixed at zero (or very small value).

These two strategies for controlling VC-dimension have been known long before DL [7, 8, 15]. Strategy 2 has been used in SVM where larger margin corresponds to smaller VC-dimension. Notably, these two strategies implement two different VC-theoretical structures, or complexity orderings. Traditional learning methods typically implement a single strategy, while DL practitioners observed the effect of both strategies on generalization, by varying a single network design parameter, such as the network size or the number of epochs. According to this VC-theoretical explanation, there is no magic in double descent phenomenon. The real issue is to understand how and why one strategy performs better than the other.

As shown later in this chapter, applying classical VC-theoretical concepts for modeling double descent helps to 'demystify' generalization performance of DL networks. In particular, VC-theoretical framework helps to answer important practical questions, such as:

- How double descent is affected by statistical properties of application data?
- Can double descent be observed for low-dimensional data?
- Characterization of specific DL network settings, where analytic VC-bounds can be rigorously applied.

We also briefly mention several alternative complexity indices (besides VC-dimension) in theoretical machine learning. These indices reflect additional assumptions about data distributions and/or specific methods used for training [14, 23, 24]. For practitioners, an obvious question is: why do we need so many different complexity indices? What is the 'best' complexity index for real-life data sets? In many cases, different complexity indices (and related theories) provide similar explanations. For example, Rademacher complexity index measures the capacity of a set of functions to fit random labels for a given training sample. Rademacher complexity index depends on data distribution and therefore is considered to be better than VC-dimension. However, Rademacher bounds on test error [23, 25, 26] are conceptually similar to VC-bound (1), since both provide estimates of excess error of the order $\mathcal{O}\left(\sqrt{h/n}\right)$. As noted earlier, VC-bound (2) is more accurate than (1), but there are no bounds similar to (2) in Rademacher theory.

3   Modeling Double Descent Using VC-Bounds

This section presents VC-theoretical explanation of double descent for classification. Several papers [6, 27, 28] discussed double descent for two-layer network shown in Fig. 2. This simplified network setting enables theoretical analysis of double descent. In this network, a learning model is estimated in two steps:

- first, input vector x is encoded using N features, formed by nonlinear mapping $z = G(x)$. Usually, random features (~weak features) are used, such as random ReLU features or random Fourier features;
- second, a linear model is estimated in this N-dimensional feature space.

Here, we adopt the same setting and experimental set up, when the number of features N is gradually increased. Then, generalization curve (test error) exhibits double descent shape, i.e., 1st descent can be observed for small N, and 2nd descent for N larger than interpolation threshold (as shown in Fig. 1).

In order to apply VC-bound (2) for modeling double descent, we need an analytic estimate of VC-dimension. Since the network output is formed as linear combination of N features in Z-space, analytic estimate of VC-dimension for linear hyperplanes $f(z, \mathbf{w}) = (\mathbf{w}z) + b$ is available [7, 8, 20]:

$$h \leq \min\big(\|\mathbf{w}\|^2, N\big) + 1 \qquad (4)$$

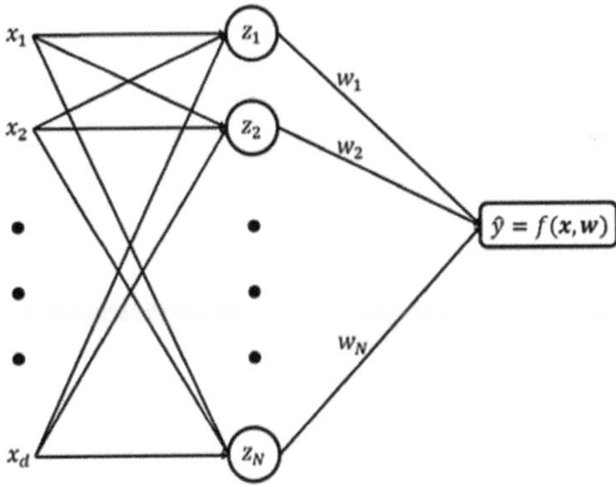

Fig. 2. Two-layer network for estimating a linear classifier using N nonlinear features $z = G(x)$.

This bound holds under assumption that all training samples are enclosed within a sphere of radius 1, in Z-space. In practice, this constraint is usually enforced by pre-scaling input variables to the same range, or by rescaling heuristics during training. According to (4), VC-dimension is bounded either by input dimension N, or by the *norm_squared* of weights $\|\mathbf{w}\|^2$. Note that the norm of weights is independent of dimensionality. Bound (4) has been originally introduced in VC-theory for Δ-margin hyperplanes (for SVM), where large separation margin corresponds to minimization of the norm_squared $\|\mathbf{w}\|^2$. The same bound holds for (over-parameterized) classifiers minimizing least-squares (LS) loss [29, 30]. In addition, a bound similar to (4) holds for Rademacher complexity of a set of linear functions with bounded weights [23, 25, 31]. Separately, [12] show that SGD training results in large-margin models for separable data.

VC-theoretical bound (2) holds for all training methods based on minimization of training error (using different loss functions). Hence, this bound is used for modeling double descent when network weights are estimated using LS and SVM loss. In both cases, minimization of the *norm_squared* corresponds to VC-theoretical 'structure' where admissible models are ordered according to the norm $\|\mathbf{w}\|^2$. For over-parameterized two-layer network in Fig. 2, training using least-squares loss finds minimum-norm solution known as pseudo-inverse, thus minimizing VC-dimension during 2nd descent.

When network size N is increasing, VC-bounds (2) and (3) can explain double descent, because:

- for small network size (N), the VC-dimension grows linearly with N, and VC-bound (2) is minimized using Strategy 1 for under-parameterized setting, a.k.a 1st descent.
- for overparameterized networks, bound (2) is minimized using Strategy 2, when VC-dimension is given by the smallest *norm_squared* of weights. This norm is decreasing with increasing network size N, leading to 2nd descent.

Next, we present empirical results showing that double descent can be accurately predicted using analytic VC-bounds (2) and (3). All experiments presented in this section use:

- SVM and least-squares (LS) loss for training linear classifier in N-dimensional feature space. All SVM modeling results use the same value of regularization parameter C = 64, to ensure large margin.
- MNIST digits data adapted for binary classification (odd digits 1, 3, 5, 7, 9 vs even digits 0, 2, 4, 6, 8). The digits are grey-scale images of size 28×28 pixels. The training set size n = 1000, and test set size is 2,000.

We use two distinct types of random nonlinear features, ReLU and Fourier:

- Random ReLU features formed as: $z_i = \max(\langle v_i, x \rangle, 0), i = 1, \ldots, N$ where random vectors $v_1, \ldots, v_N$ are sampled uniformly from the range $[-1, 1]$.
- Random Fourier features formed as: $z_i = \exp(\sqrt{-1}\langle v_i, x \rangle), i = 1, \ldots, N$ where random $v_1, \ldots, v_N$ are sampled from Gaussian distribution with standard deviation $\sigma = 0.05$.

In all experiments, input (x) values were pre-scaled to [0, 1] range, for training and test data. Following nonlinear mapping $\mathbf{X} \rightarrow \mathbf{Z}$, all z-values are re-scaled to $[-1, 1]$ range. Such re-scaling is performed to satisfy the condition for bound (4), stating that all training samples in $\mathbf{Z}$-space are within a sphere of radius 1.

Training samples (z, y) are used to estimate linear decision function $f(z,\mathbf{w}) = (\mathbf{w} \cdot z)+b$ in $\mathbf{Z}$-space, via minimization of training error using LS and SVM loss functions. VC-bound (2) is used to predict double descent for both methods and for both types of features. Empirical test error is estimated using test data set.

Figure 3 shows VC-theoretical modeling for MNIST data, using random ReLU and Fourier features. These results show empirical training and test error curves, as a function of the number of features (N), along with VC-bound on test error obtained via bound (2). Additional figures show the 'norm_squared' $\|\mathbf{w}\|^2$ of estimated model, and the number of support vectors (for SVM models). These results are summarized next:

- *for small* N-*values*, VC-dimension grows linearly with N for both SVM and LS estimators. Empirical results for LS in Fig. 3(a) show that 1st descent error curve can be explained by VC-bound (2). Results for SVM classifier in Fig. 3(b) show that VC-bound (2) is too loose (inaccurate) in the region of 1st descent. This can be explained by the fact that class distributions (for odd vs even digits) are well-separable, so there is no visible 1st descent for SVM. This will change for noisy data, presented later in Fig. 4.
- *for large* N, VC-dimension is controlled by the *norm_squared*, according to bound (2). Empirical results show that 2nd descent can be explained by VC-bounds (2) and (3), for both SVM and LS methods.
- double descent curves for LS and SVM in Fig. 3 different have *interpolation threshold* $N*$ (when the training error reaches zero). That is, for SVM, the value $N* \sim 300$ is achieved when the number of features equals the number of support vectors. For LS classifier, the interpolation threshold $N* \sim 1000$ is achieved when the number of features equals the number of training samples.

The dependence of test error on the norm of weights in large networks has been recently discovered by DL researchers [6, 32, 33]. However, these papers do not mention the bound (4) that clearly relates VC-dimension to the norm of weights and generalization performance. Note that the same VC-bounds explain double descent for different classifiers, using SVM and LS loss, as shown in Fig. 3.

Dependence of interpolation threshold $N*$ on the training sample size has been also observed in DL literature. Without sound theoretical framework for double descent, interpretation of this empirical dependency often leads to convoluted explanations. For example, [3] investigated how the training set size may adversely affect generalization performance. They analyzed double descent curves, for the same DL network trained using smaller and larger-size training set, showing that during 2nd descent, near interpolation threshold, the test error of a network trained on a smaller training set is smaller than when it is trained on larger data set. [3] referred to this phenomenon as 'sample-wise non-monotonicity' in DL networks and proposed a new theory for explaining why 'increasing the number of training samples actually hurts test performance'.

However, this phenomenon has simple VC-theoretical explanation. Consider LS modeling results in Fig. 3(a) showing that the shape of the *'norm_squared'* during 2nd descent closely follows the shape of test error, according to VC-bound (2a). Note that interpolation threshold $N*$ equals the training set size. Therefore, for larger training set, the test error for N-values near its interpolation threshold will be larger than test error of a network trained on a smaller training set.

There is a common view among DL researchers that models estimated during 2nd descent generalize better than during 1st descent [3, 5, 6, 34]. That is, it is always better to use large (over-parameterized) networks. However, empirical evidence alone cannot justify this view, because prediction performance obviously depends on statistical properties of unknown distributions.

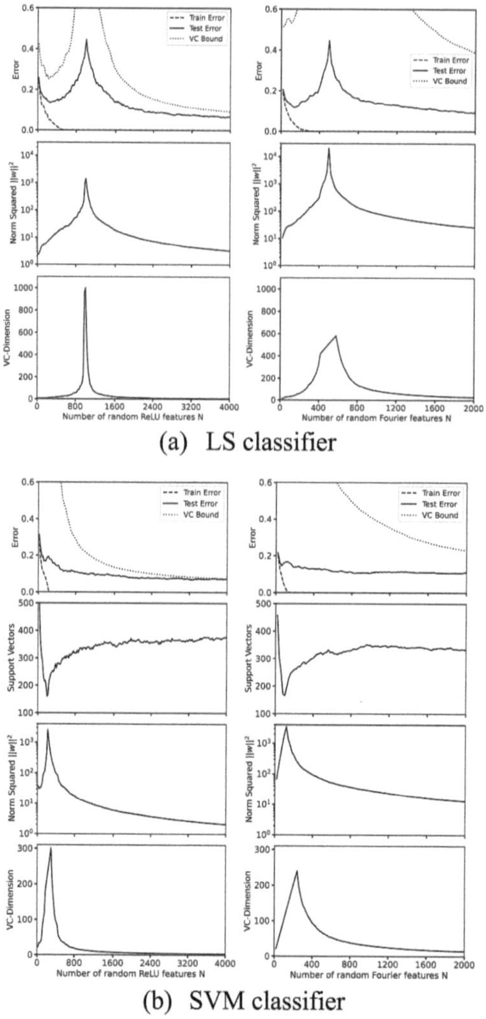

(a) LS classifier

(b) SVM classifier

Fig. 3. Double descent curves for odd vs even MNIST data set for (a) LS classifiers and (b) SVM classifiers, using random ReLU features (left) and random Fourier features (right).

Next, we show how VC-theoretical framework can explain the effect on noisy data on double descent curves. These results are obtained for two-layer network in Fig. 2 with ReLU features, trained using SVM and LS classifiers. We use odd vs even digits data with randomly corrupted class labels, i.e. 10% noise means that class labels are randomly assigned for 10% of data samples. Original clean data corresponds to 0% noise level. Training set size is 1,000, and test set size is 2,000. Figure 4 shows the effect of noise level on the shape of double descent curves. Note that the shape of double descent curves is different for LS and SVM classifiers. For LS models, we observe first and second descent for both clean and noisy data, and the interpolation threshold is the same for different noise levels (it equals training size 1,000). But for SVM, interpolation threshold is larger

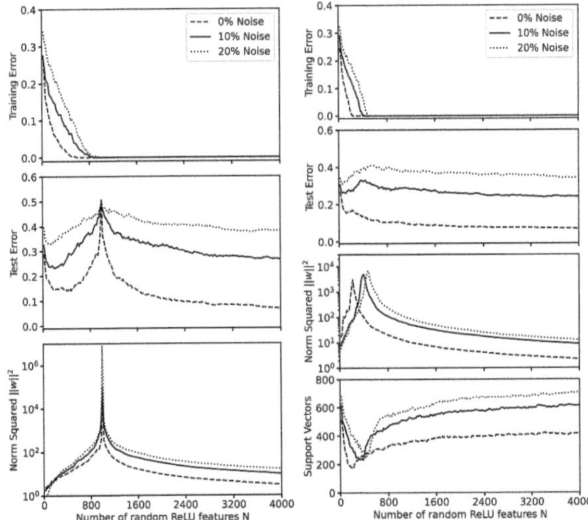

Fig. 4. Effect of noise on double descent for odd vs even MNIST digit data, for LS (left) and SVM (right) classifiers using random ReLU features.

for noisy data. This can be explained by noting that for SVM interpolation threshold is determined by the number of support vectors (SV's), which is increasing with noise level.

During second descent, test error for noisy data is increasing with noise level. This is explained by noting that VC-bound (2a) depends only on VC-dimension (~the norm squared of weights). During second descent, fitting noisy training data requires larger VC-dimension (i.e., larger *norm_squared*) than fitting clean training data (from separable class distributions). Hence, increasing noise level results in larger test error and flattening of second descent curve (as shown in Fig. 4, for both SVM and LS).

Further, results in Fig. 4 show that for very noisy data, generalization during 1st descent may be better than 2nd descent. This can be seen in Fig. 4, for test error curves with 20% label noise. This phenomenon has not been known in DL, but it has VC-theoretical explanation. That is, memorizing very noisy data (during 2nd descent) requires larger VC-dimension than modeling during 1st descent (when training error is nonzero). This phenomenon can be also modeled using VC-bound (2).

In summary, generalization ability of over-parameterized networks for high-dimensional digits data can be explained by well-known VC-bounds. This conclusion contradicts [5, 32, 35] who argue that generalization in large networks cannot be explained by VC-theory.

4 Modeling Generalization in Deep Learning Networks

Empirical results in Sect. 3 show that double descent can be explained using VC-theoretical bounds. However, these results are shown for simplified two-layer network in Fig. 2, where only second layer weights are learned during training, and nonlinear random features are not adapted to training data. In this section, we try to address two questions:

1) why analytic results for two-layer networks with random features are relevant for understanding generalization of DL networks?
2) is it possible to apply analytic VC-bounds for modeling generalization of multi-layer networks?

Regarding the first question, we note that network setting in Fig. 2 has been used in many recent studies [6, 27, 28], because theoretical analysis of simple networks precedes understanding generalization of complex networks. VC-theoretical analysis for simple networks presented in Sect. 2 clearly suggests that using wide networks is important for good generalization, in agreement with earlier studies [32, 35, 36]. Further, generalization of DL networks depends on the norm of weights, rather than network size [12, 16, 32, 35, 36], which is attributed to 'inductive bias' of SGD training, that favors small norm of weights in over-parameterized networks. However, controlling generalization using the norm of weights has been known in VC-theory since 1990's.

The second question, about VC-theoretical modeling of multilayer networks, is challenging because:

- analytic estimates of VC-dimension do not exist for general multilayer networks. Yet, it may be possible to identify restricted DL network settings where analytic estimates can be found.
- typical double descent curves show dependence of test error on a *single factor* that controls generalization, with all other network hyper parameters preset to some 'good' values. However, modern DL networks have many tuning parameters, and their combined effect is difficult to quantify analytically. For example, various hyper parameters of SGD training directly affect VC-dimension and generalization.

As explained in Sect. 2, VC-theoretical understanding of double descent is based on two different strategies for minimizing the same VC-bound (2) on test error. These two strategies implement Structural Risk Minimization (SRM) inductive principle [7, 8, 20], where 'structure' refers to complexity ordering of admissible models.

The choice of the 'best' SRM structure for modeling particular data set is outside the scope of VC-theory. For example, classical statistical methods use small number of 'strong' informative input features, and implement 'dictionary' structure [15], where the number of features (or degrees-of-freedom) control model complexity. This corresponds to 1st descent regime.

Section 3 presented VC-theoretical explanation of double descent for two-layer network in Fig. 2, where inputs $\mathbf{x}$ are encoded as N nonlinear features (z-features), and then a linear classifier is estimated in this feature space. In this case, non-linear mapping $z = G(\mathbf{x})$ *does not depend* on the training data, and analytic estimates of VC-dimension can be used for modeling double descent. For general DL networks, nonlinear features depend on the training data, so analytic estimates of VC-dimension are not known.

However, rigorous application of VC-bounds may still be possible for certain restricted DL network settings. Next, we apply VC-bounds for *transfer learning* setting commonly used in DL. Transfer learning is used for many applications, where practitioners use pre-trained network for modeling their own application data. The idea is to leverage informative features extracted from pre-trained network, in order to use them for training another (simple) network using domain-specific data. For example, existing pre-trained networks, such as VGGNet [37] or ResNet [38], trained on vast amounts of natural image data, are commonly used to fine-tune models for specific applications [39].

VC-theoretical modeling can be used for transfer learning setting implemented as a two-step procedure:

1) *Pre-training.* A network is trained using general application data, in order to learn N nonlinear features. That is, pre-trained network effectively learns nonlinear mapping $z = G(x)$.

2) *Fine tuning,* or training a simple network with N input features (found in Step 1), using domain-specific training data. For fine tuning, we will use simple linear classifier, as shown in Fig. 2.

Since non-linear mapping (estimated in Step 1) *does not depend* on domain-specific data used for fine tuning (in Step 2), we can apply VC-theoretical bounds for linear classifier estimated in Step 2. That is, we can model generalization following the same procedure as described in Sect. 3.

Next, we present VC-theoretical modeling of generalization curves, empirically observed for transfer learning, for several pre-trained networks and several data sets with different statistical properties. For example, consider a small CNN network [21], pre-trained using handwritten letters data set for 26 upper case letters [40]. This data set is large ~ 370K letter images. The final layer of CNN represents $N = 4608$ features learned for 'generic' handwritten images. These nonlinear features are used for training a binary classifier for handwritten odd vs even digits. That is, in *Step 2*, a single-layer linear network is trained on a small domain-specific data set (~ 1,000 digits). During the pre-training step, training is performed by minimizing cross-entropy loss using Adam optimizer with the number of epochs set to 50 (or when the test error converges to a minimum value). During the fine-tuning step, MSE loss with SGD optimizer is selected during the model training, with the number of epochs set to 10,000. For both steps, the learning rate is set to 0.001, with the learning rate monotonically decreasing by a factor of 0.8 at every 500 epochs.

During fine tuning, the number of nonlinear features $N = 4608$ is fixed, so generalization curves show the dependence of training/test error on the number of training epochs of SGD training. Figure 5 shows modeling results for fine-tuned linear network trained using digits data, where training set size is 1,000 and test size is 2000 [21]. Figures 5(a)–(c) show results for 3 data sets: original digits data, and noisy digits with 5% and 10% label noise. For each data set, the top row of Fig. 5 displays the VC-bound (2) on test error, that follows empirical test error curve. The bottom row shows VC-dimension, estimated as the norm squared of weights, that is increasing during training (since the initial random weights are very small).

Results in Fig. 5 for fully trained networks (at 10,000 epochs) at various noise levels provide insights on the effect of noisy data on transfer learning:

- Transfer learning can memorize noise-free training data in Fig 5(a) (~ zero training error). However, it cannot memorize noisy data in Figs. 5(b), (c). That is, for noisy

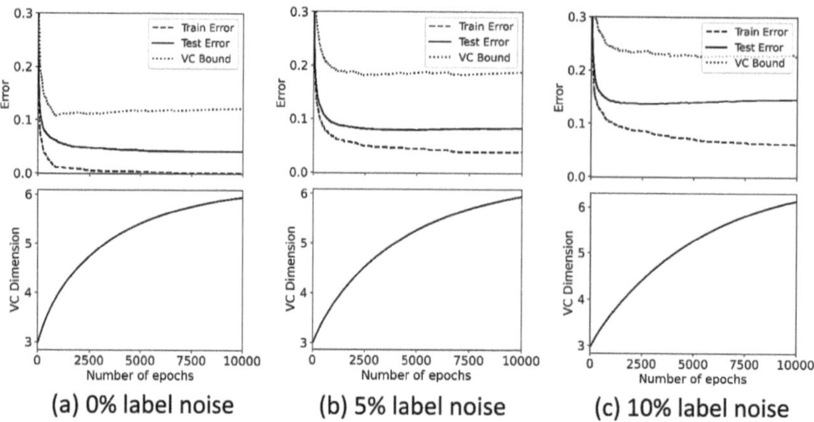

(a) 0% label noise (b) 5% label noise (c) 10% label noise

Fig. 5. Modeling results for linear classifier (with nonlinear features) trained on odd vs even MNIST digit data, for various levels of label noise.

data, the fine-tuned network *does not overfit*, so its error curves cannot be regarded as second descent, even though their shape resembles second descent.

- VC-dimension of networks trained on clean and noisy data has roughly the *same value ~ 6.*

Note that earlier results in Fig. 4 for a network with random ReLU features show different effect of noise on generalization. That is, a trained network in Fig. 4 can memorize noisy training data, and its VC-dimension *increases* with increasing noise level.

We observed similar effect of noisy data on generalization when a fine-tuned network was trained using SVM. That is,

- SVM models trained on clean and noisy digits data had the same VC-dimension;
- SVM model trained using clean digits data had zero training error, but models trained on noisy data had non-zero training error.

Modeling results in Fig. 5 explain generalization curves observed in transfer learning, i.e. that VC-bounds explain the effect of label noise in the training data on the performance of a fine-tuned model. Arguably, similar procedure can be used for evaluating the effect of different distributions (used for pre-training and fine-tuning) on generalization performance of transfer learning. In particular, this effect can be modeled by observing the VC-dimension and training error of fine-tuned models.

In summary, VC-theoretical analysis of transfer learning can help:

- to understand the effect of statistical properties of application data on generalization performance,
- to perform objective comparison of different DL networks used for pre-training. That is, different pre-trained networks generate different sets of nonlinear features. Then, the effect of these nonlinear features on generalization of fine-tuned models can be evaluated and explained by VC-theoretical arguments.

5 Modeling Generalization for Low-Dimensional Data

For high-dimensional data, generalization is typically controlled by the norm of weights. This is well-known in DL research community and also evident in empirical results presented in Sects. 3 and 4. However, for *low-dimensional data*, the norm-based VC bounds, such as (3) and (4), are too loose for modeling generalization [15], because the margin size is usually too small, even for separable low-dimensional data. Therefore, in this section, we discuss generalization curves for low-dimensional data, using the same two-layer network shown in Fig. 2.

There exists another approach to inductive inference, developed outside VC-theoretical framework, called Minimum Description Length (MDL), based on the information-theoretic analysis of the randomness concept. Under MDL, learning amounts to finding the most efficient encoding of the training data using the smallest number of bits [41]. In the case of binary classification, consider n labeled training samples (x_i, y_i), where x and y represent inputs and binary class labels, respectively. Then, binary string $\{y_i\}$ is described by n bits. However, this binary string is not completely random, since the binary values y_i depend on x_i. So, according to MDL principle, it is possible to describe this binary sequence using smaller number of bits. If such efficient encoding of the training data can be found, then the coefficient of compression determines an upper bound on test error [8]. Even though VC-theory itself is based on the notion of minimization of training error (ERM), which is different from MDL, both approaches yield similar-style generalization bounds [8]. Conceptually, it means that for a given finite training set, good prediction is achieved by significant compression of this data set (in MDL), or by a 'simple' model found by minimization of training error in VC-theory (where 'simple' means small VC-dimension).

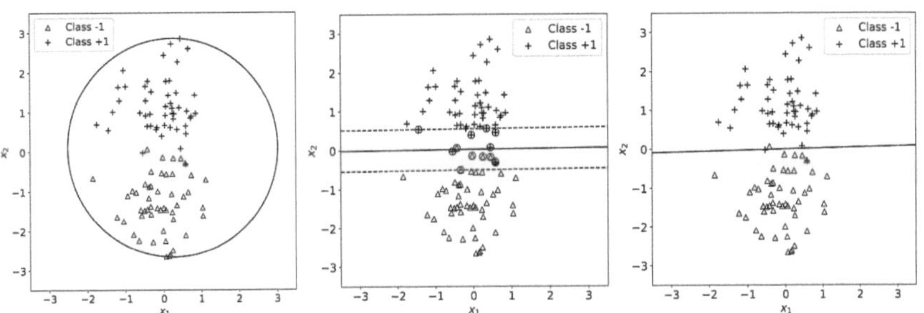

(a) Training data set (100 samples)
(b) Estimated linear SVM model (SV's are indicated by circles)
(c) Estimated LS model

Fig. 6. Linear SVM and LS modeling for low-dimensional data set.

Most machine learning methods implement minimization of training error, so they do not provide constructive procedure for implementing MDL approach. However, SVM methods implicitly implement data compression, because a trained SVM classifier depends *only* on support vectors (SV's). So, the data compression coefficient equals the ratio of a small number of SV's to the total number of training samples. This leads to MDL-style theoretical bound, where SVM test error is bounded by the fraction of the number of support vectors in the training data set (of size n):

$$R_{tst} \leq \frac{Number\ of\ SVs}{n} \tag{5}$$

This bound is also known as the Leave-One-Out (LOO) bound for SVM [15]. It can be readily seen that for low-dimensional data sets, the number of SV's is a better measure of complexity than margin size. For example, consider two-dimensional data set in Fig. 6, where class distributions are two Gaussians with the same standard deviation 0.5, and centers separated by the distance 2.5 in x_2-dimension. Fig. 6(a) shows 100 training samples and the minimal sphere enclosing all samples. For this data set, we estimate a linear SVM classifier (since a Bayes optimal solution is linear). For SVM modeling, optimal C = 1 is selected using separate validation data set. Fig. 6(b) shows optimal linear SVM model (solid line) along with margins (dotted line) and SV's (circled samples). This SVM model has 3% training error and 3.5% test error, where test error is estimated using 200 test samples. Using bound (4) gives VC-dimension $h = 3$ based on the input dimensionality N = 2, which is smaller than VC-dimension based on the norm squared $h = 28.5$. Using $h = 3$ in VC-bound (2) on test error yields 53.3%, but using LOO SVM bound (5) gives a better bound of 13%.

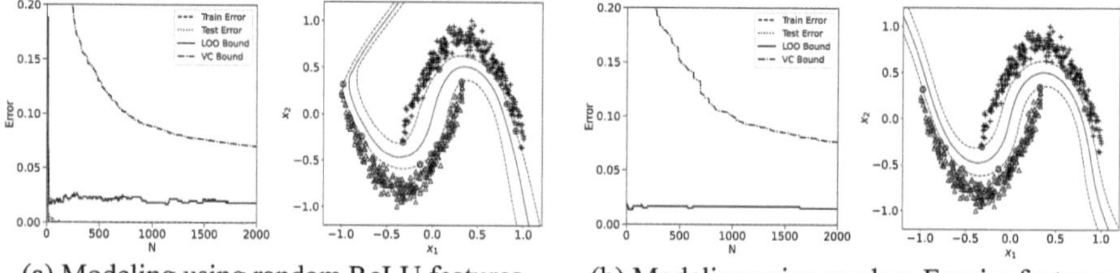

(a) Modeling using random ReLU features.   (b) Modeling using random Fourier features.

Fig. 7. Modeling results for non-overlapping Hyperbolas data set, using SVM with N nonlinear features.

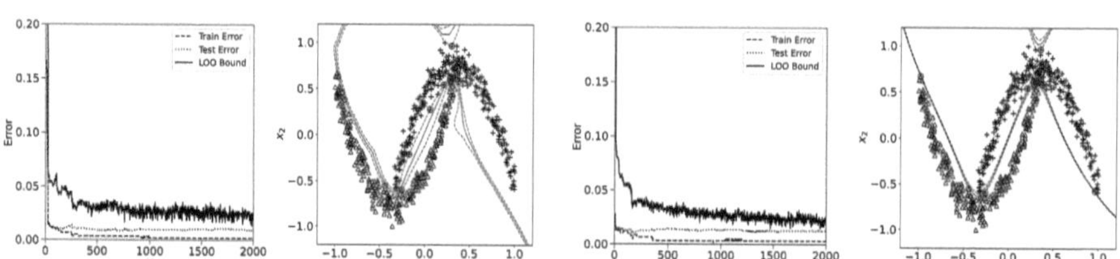

Fig. 8. Modeling results for overlapping Hyperbolas data set, using SVM with N nonlinear features.

For comparison, for the same training data set, Fig. 6(c) shows a linear classifier estimated by minimizing LS error. In this case, an optimal LS model achieves 3% training error and 3.5% test error. For this case, VC-dimension based on the *norm_squared* is h = 4.0. Using $h = 4.0$ in VC-bound (2) yields the bound 63%. For LS models, there are no known MDL-style bounds similar to (5).

Next, we present modeling for two non-overlapping Hyperbolas data set in Fig. 7, to illustrate a nonlinear problem. This data set has 600 training and 2,000 test samples. We follow the same experimental set up as in Fig. 2, where inputs are mapped onto N -dimensional feature space, using random ReLU and Fourier features, and then a linear SVM model (with regularization parameter C fixed to 64) is estimated in this feature space. Modeling results in Fig. 7 display an empirical test error along with bound (5) on test error, when the number of nonlinear features N is increasing. These results in Fig. 7 suggest that bound (5) can accurately predict test error curves, for both random ReLU and Fourier features.

These modeling results show that the norm-based VC-bounds (2) and (3) do not work well for low-dimensional data. In order to understand why the norm-based VC-bounds are too loose, we can visually interpret SVM models, estimated for large network size N = 1000. These models (shown in figures on the right) have very small margin size, relative to the size of a sphere enclosing all training samples. This implies large estimates of VC-dimension based on the *norm of weights*, via bound (3), and explains why norm-based complexity control does not work well for low-dimensional data.

Modeling results for overlapping low-dimensional distributions are presented in Fig. 8, using the Overlapping Hyperbolas data set. These modeling results follow the same procedure as earlier results in Fig. 7 (for non-overlapping data). Modeling results for overlapping and non-overlapping data sets are remarkably similar, and in both cases MDL-style LOO bounds provide accurate estimates of the test error, whereas the norm-based bounds are very loose (and they are not shown in the figure for that reason).

In summary, for low-dimensional data, model complexity is controlled by the number of SV's, which is different from the norm-based complexity control for high-dimensional data settings. Further, as shown in Figs. 7 and 8, generalization curves for low-dimensional data *do not* follow familiar double descent shape (as in Fig. 1), when N is increasing. These curves cannot be interpreted as either first or second descent. This is because bound (5) describes different mechanism for controlling generalization.

6 Summary and Conclusions

We discussed generalization phenomenon known as 'double descent', under classical VC-theoretical framework. The main message is that double descent curves can be fully explained in terms of fundamental mechanisms for generalization, and there is no need to invent new theories. In DL community, generalization is controlled by various computational heuristics that effectively implement these mechanisms. The main source of confusion is that VC-theoretical concepts are not well understood in DL, and so explanation of successful heuristics (developed for various applications) usually leads to new theories.

In this chapter, we showed that double descent can be explained using three different mechanisms for controlling generalization [8], i.e.

- small number of input features (relative to training sample size); small VC-dimension;
- large data compression (for encoding training data set).

The first mechanism is adopted in classical statistical methods [1], that effectively implement 'first descent'. Second approach is implemented in over-parameterized DL networks for high-dimensional data, and it allows to achieve zero training error during second descent. Third approach, also known as MDL, is not commonly used in machine learning methods, except for SVM. However, it is distinctly different from the first two approaches, and it achieves the best explanation (~most accurate bounds) for low-dimensional data. Based on this understanding, it is possible to observe different generalization curves showing a single descent or double descent, depending on statistical properties of the training data (such as dimensionality, noise level etc.) and chosen experimental set up.

Even though conceptual understanding of three different mechanisms for generalization is well-known [8], this chapter shows how these mechanisms can accurately explain quantitative modeling of double descent. In addition, we show that selecting the best approach for generalization depends on statistical characteristics of the training data, such as the number of training samples, input data dimensionality, the number of nonlinear features, and separability of class distributions.

Application of analytic VC-bounds for modeling real-life data is possible only for simple learning methods and simple settings. However, *methodological application* of VC-theoretical framework can be still quite useful for understanding generalization curves observed for analytically intractable methods such as DL networks. In this respect, the main methodological points are:

- explaining generalization of complex learning methods in terms of 3 main mechanisms for controlling generalization
- understanding the relationship between general characteristics of application data and appropriate mechanism for generalization. Here it is important to keep in mind that statistical properties of high-dimensional and low-dimensional data are inherently different [15].

Finally, we can relate three different mechanisms for generalization to philosophical aspects of inductive data-analytic inference, presented in Sect. 1. That is, classical statistical methods use strong a priori knowledge (about parametric form of unknown distributions). Consequently, these methods use a small number of input features that reflect strong a priori knowledge about reality. In contrast, inductive mechanisms based on minimization of training error (ERM) and data compression (MDL) use weak a priori knowledge. Therefore, all modern data-analytic methods can only imitate reality, in the sense of making useful predictions. Based on this understanding, we conclude that:

- all learning methods should follow sound mathematical methodology. We favor VC-theoretical approach, because it applies to all learning methods based on ERM.
- predictive data-analytic models are not interpretable, in principle, because one can estimate several models that have similar prediction performance but completely different parameterization.

References

1. Fisher, R.A.: The logic of inductive inference. J. R. Stat. Soc. **98**, 39 (1935)
2. Russell, B.: The Problems of Philosophy. Oxford University Press (1912)
3. Nakkiran, P., Kaplun, G., Bansal, Y., Yang, T., Barak, B., Sutskever, I.: Deep double descent: where bigger models and more data hurt*. J. Stat. Mech. Theory Exp. **2021**, 124003 (2021)
4. Neyshabur, B., Bhojanapalli, S., Mcallester, D., Srebro, N.: Exploring generalization in deep learning. In: Advances in Neural Information Processing Systems. Curran Associates, Inc. (2017)
5. Zhang, C., Bengio, S., Hardt, M., Recht, B., Vinyals, O.: Understanding deep learning (still) requires rethinking generalization. Commun. ACM **64**, 107–115 (2021)
6. Belkin, M., Hsu, D., Ma, S., Mandal, S.: Reconciling modern machine-learning practice and the classical bias–variance trade-off. Proc. Natl. Acad. Sci. **116**, 15849–15854 (2019)
7. Vapnik, V.N.: Statistical Learning Theory. Wiley, New York (1998)
8. Vapnik, V.N.: The Nature of Statistical Learning Theory. Springer, New York (2000)
9. Arora, S., Ge, R., Neyshabur, B., Zhang, Y.: Stronger generalization bounds for deep nets via a compression approach. In: Proceedings of the 35th International Conference on Machine Learning, pp. 254–263. PMLR (2018)
10. Neyshabur, B., Li, Z., Bhojanapalli, S., LeCun, Y., Srebro, N.: The role of over-parametrization in generalization of neural networks. In: International Conference on Learning Representations (2019)
11. Bengio, Y.: Learning deep architectures for AI. Found. Trends® Mach. Learn. **2**, 1–127 (2009)
12. Soudry, D., Hoffer, E., Nacson, M.S., Gunasekar, S., Srebro, N.: The implicit bias of gradient descent on separable data. J. Mach. Learn. Res. **19**, 1–57 (2018)
13. Srivastava, N., Hinton, G., Krizhevsky, A., Sutskever, I., Salakhutdinov, R.: Dropout: a simple way to prevent neural networks from overfitting. J. Mach. Learn. Res. **15**, 1929–1958 (2014)
14. Bartlett, P.L., Foster, D.J., Telgarsky, M.J.: Spectrally-normalized margin bounds for neural networks. In: Advances in Neural Information Processing Systems. Curran Associates, Inc. (2017)
15. Cherkassky, V., Mulier, F.: Learning from Data: Concepts, Theory, and Methods. Wiley (2007)
16. Bartlett, P.L., Montanari, A., Rakhlin, A.: Deep learning: a statistical viewpoint. Acta Numer. **30**, 87–201 (2021)
17. Koltchinskii, V.: Rademacher penalties and structural risk minimization. IEEE Trans. Inf. Theory **47**, 1902–1914 (2001)
18. Sokolic, J., Giryes, R., Sapiro, G., Rodrigues, M.R.D.: Robust large margin deep neural networks. IEEE Trans. Signal Process. **65**, 4265–4280 (2017)
19. Tewari, A., Bartlett, P.L.: Learning theory. In: Academic Press Library in Signal Processing, pp. 775–816. Elsevier (2014)
20. Vapnik, V.: Estimation of Dependencies Based on Empirical Data. Springer, New York, NY (2006)
21. Cherkassky, V., Lee, E.H.: To understand double descent, we need to understand VC theory. Neural Netw. **169**, 242–256 (2024)
22. Lee, E.H., Cherkassky, V.: Understanding double descent using VC-theoretical framework. IEEE Trans. Neural Netw. Learn. Syst. **35**, 18838–18847 (2024)
23. Bartlett, P.L., Mendelson, S.: Rademacher and Gaussian complexities: risk bounds and structural results. J. Mach. Learn. Res. **3**, 463–482 (2003)
24. Zhou, D.-X.: The covering number in learning theory. J. Complex. **18**, 739–767 (2002)
25. Mohri, M., Rostamizadeh, A., Talwalkar, A.: Foundations of Machine Learning. The MIT Press, Cambridge, Mass. London (2012)

26. Cortes, C., Kloft, M., Mohri, M.: Learning kernels using local Rademacher complexity. In: Advances in Neural Information Processing Systems. Curran Associates, Inc. (2013)
27. Belkin, M., Hsu, D., Xu, J.: Two models of double descent for weak features. SIAM J. Math. Data Sci. **2**, 1167–1180 (2020)
28. Neal, B., et al.: A modern take on the bias-variance tradeoff in neural networks (2019). http://arxiv.org/abs/1810.08591
29. Liang, T., Recht, B.: Interpolating classifiers make few mistakes. J. Mach. Learn. Res. **24**, 1–27 (2023)
30. Ardeshir, N., Sanford, C., Hsu, D.J.: Support vector machines and linear regression coincide with very high-dimensional features. In: Advances in Neural Information Processing Systems, pp. 4907–4918. Curran Associates, Inc. (2021)
31. Koltchinskii, V., Panchenko, D.: Empirical margin distributions and bounding the generalization error of combined classifiers. Ann. Stat. **30**, 1–50 (2002)
32. Belkin, M., Ma, S., Mandal, S.: To understand deep learning we need to understand kernel learning. In: Dy, J., Krause, A. (eds.) Proceedings of the 35th International Conference on Machine Learning, pp. 541–549. PMLR (2018)
33. Gunasekar, S., Woodworth, B.E., Bhojanapalli, S., Neyshabur, B., Srebro, N.: Implicit regularization in matrix factorization. In: Advances in Neural Information Processing Systems. Curran Associates, Inc. (2017)
34. Neyshabur, B., Tomioka, R., Srebro, N.: In Search of the Real Inductive Bias: On the Role of Implicit Regularization in Deep Learning (2024)
35. Belkin, M.: Fit without fear: remarkable mathematical phenomena of deep learning through the prism of interpolation. Acta Numer. **30**, 203–248 (2021)
36. Bietti, A., Bach, F.: Deep equals shallow for ReLU networks in kernel regimes. In: International Conference on Learning Representations (2021)
37. Simonyan, K., Zisserman, A.: Very deep convolutional networks for large-scale image recognition. In: International Conference on Learning Representations (2015)
38. He, K., Zhang, X., Ren, S., Sun, J.: Deep residual learning for image recognition. In: 2016 IEEE Conference on Computer Vision and Pattern Recognition (CVPR), pp. 770–778. IEEE, Las Vegas, NV, USA (2016)
39. Aversano, L., Bernardi, M.L., Cimitile, M., Iammarino, M., Rondinella, S.: Tomato diseases classification based on VGG and transfer learning. In: 2020 IEEE International Workshop on Metrology for Agriculture and Forestry (MetroAgriFor), pp. 129–133. IEEE, Trento, Italy (2020)
40. Cohen, G., Afshar, S., Tapson, J., van Schaik, A.: EMNIST: extending MNIST to handwritten letters. In: 2017 International Joint Conference on Neural Networks (IJCNN), pp. 2921–2926 (2017)
41. Akaike, H.: A new look at the statistical model identification. IEEE Trans. Autom. Control **19**, 716–723 (1974)

Conformal Prediction and Trustworthy AI

Anthony Bellotti$^{(\boxtimes)}$ and Xindi Zhao

School of Computer Science, University of Nottingham, Ningbo, China
`anthony-graham.bellotti@nottingham.edu.cn`

Abstract. Conformal predictors are machine learning algorithms developed in the 1990's by Gammerman, Vovk, and their research team, to provide set predictions with guaranteed confidence level. Over recent years they have grown in popularity and have become a mainstream methodology for uncertainty quantification in the machine learning community. From their beginning, there was an understanding that they enable reliable machine learning with well-calibrated uncertainty quantification. This makes them extremely beneficial for developing trustworthy AI, a topic that has also risen in interest over the past few years, in both the AI community and society more widely. In this chapter, we review the potential for conformal prediction to contribute to trustworthy AI beyond its marginal validity property, addressing problems such as generalization risk and AI governance. Experiments and examples are also provided to demonstrate its use as a well-calibrated predictor and for bias identification and mitigation.

Keywords: conformal prediction · trustworthy AI · algorithmic bias

1 Introduction

Over the past ten years the increasing impact of AI in society has been extraordinary, and the impact is set to continue to increase. Although much of the excitement around AI is with generative models such as ChatGPT for text and DALL-E for images and video, many of the AI workhorses in industry and government are predictive machine learning models. Predictive machine learning can be used for decision-support or agentic tasks. Such supervised machine learning models can provide either a prediction or propose some action. They are widely used in all areas of society [36]. For example, they can be used to predict the type of cancer from gene expression data extracted from blood sample [50], or used to screen loan applicants based on characteristics of an applicant such as income, outstanding debt, and employment status; e.g. [18]. However, with the rise in the use of AI, there has been a growing rise of concern and suspicion with the use of AI. People, governments and companies wonder if AI can be trusted to perform the tasks they are set to do [27]. How do we know that they are helping, and not hindering? Human medical doctors have been trained over many years, so why should we trust a machine learning model that has been trained in just a few hours? The risks associated with using AI are broad, but most can be summarized within these seven categories:

K. An Nguyen and Z. Luo (Eds.): Alexander Gammerman Festschrift, LNCS 16290, pp. 177–197, 2026.
https://doi.org/10.1007/978-3-032-15120-9_10

Poor performance. This is the most basic risk, that the AI does not perform as well as it should to perform reliably. For example, in a classification problem, it would be typical to ensure that overall misclassification rate is low across an independent test dataset. For regression problems, it would be typical to use a measure such as mean-squared-error. Many alternative measures are available and the choice of performance measure depends on the task. For predictive machine learning projects, this type of evaluation is standard [35]. However, there remain some subtle problems that are sometimes overlooked: (i) Generalizability: is the model robust when presented with new data which may have shifted from the original training data in some way? This depends on both the task and the model; (ii) Calibration: how far can we trust each individual prediction? In particular, if the model provides an estimate of uncertainty, how well does the estimate match measured uncertainty?

Bias. Does the model give unequal predictions or performance across different subpopulations of data? This can give rise to discrimination. For example, in their audit of commercial face recognition systems, [8] found discrimination against black women. In particular, darker-skinned females had much higher misclassification rates (up to 34.7%) relative to lighter-skinned males (up to 0.8%). If the subpopulations are protected classes such as gender, race or religion, then such models that exhibit bias are not only unethical, but could be illegal if deployed.

Human/AI interaction. The relationship between humans and AI is a concern. It may be that humans will become over-reliant on AI, or alternatively, humans may irrationally under-rate the value of AI. We may wonder what the extensive use of AI by humans will have on human behaviour and happiness. As humans we may be concerned that the AI ought to be able to explain itself or its decisions (transparency). We are also concerned that AI is not *misused* by humans to cause harm. For example, criminals could use AI for fraud using DeepFake technology, or it could be used to generate disinformation to influence people politically.

Objective misalignment. Although AI is largely autonomous, humans still need to specify the task they perform to the AI system. For machine learning, typically this is done using an objective function, translating the task description into an algebraic form. This can create difficulty since not all nuances of the task may be correctly and completely specified in the required precise mathematical form. There is a risk that the objective function is misaligned with what is really required, hence exhibiting unintended consequences. For example, [2] describes how they used reinforcement learning to build a system to play CoastRunner, a boat racing video game, with the objective of maximizing the score. However, instead of learning to complete the race as intended, the AI learned to drive the boat in a circle to pick up points from peripheral tasks. In a sense, the system was doing what we tasked it to do, but it was not what was *intended*. This experiment was in a game setting, but in a real-world setting such as automated vehicles such behaviour could be dangerous [3].

Security and Privacy. Machine learning uses data for training. There is a risk that private data is used without consent, or confidential data may be compromised during the training process, or confidential data is able to be reconstructed from the output of the algorithm. These raise serious security and privacy concerns when using AI.

Ethics, Legal and Accountability. There are ethical issues for some applications of AI. For example, is it ethical to use AI to determine who should be paroled, or to predict who is most likely to commit a crime in the future? There are legal and regulatory issues regarding the fair use of AI in society. There are questions about the governance of AI systems and accountability for AI systems that need to be considered [36].

Broad social consequences. Finally, there are a broad range of social concerns regarding AI, perhaps the strongest is the *singularity*: will AI supercede human intelligence with profound consequences for human society and world power? Another important concern is the potential impact of AI on the labor markets and human employment (see e.g. [13] on post-labor economics), and also on the environment due to excessive energy or water use by large AI systems. Furthermore, how will AI affect human relationships, both personal relations (with AI and amongst humans) and in wider society?

Trustworthy AI endeavours to mitigate and control these risks in AI systems. Conformal Prediction (CP) is a technology to augment machine learning models by providing reliable measures of uncertainty [45]. In this chapter, we explore how their use can help address some of these risks and therefore support development of trustworthy AI.

The CP framework was originally developed by Gammerman, Vovk and their research group in the 1990's. One of their earliest publications, with Vapnik, introduced the exchangeability and predicting with confidence framework in the context of Support Vector Machines [17]. Later the algorithms reached maturity with the important validity results when they were first named confidence machines [33,46]. Since then, with the need to develop reliable and trustworthy AI, interest in CPs has grown considerably, especially over the past few years, as is evident in Fig. 1 which shows citation rates for the key reference book in the field [45].

AI is data-driven and most AI tools use deep neural networks which are trained on large datasets. As such, often the source of many AI risks can be found with problems in the data, and hence it makes sense for solutions to be found at the data collection and pre-processing steps of AI development [20]. Nevertheless, it is also possible to find solutions that can be implemented in the machine learning algorithms themselves. Of the risks listed above the first three are amenable to technical solutions and hence these are the ones we will focus on in this paper. The fourth one, objective misalignment, can be mitigated by better management of the process of task specification, and so align with better software engineering processes and validation in machine learning. The fifth one, security and privacy, is also amenable to a technical solution, but in the domain of computer security, rather than by applying machine learning algorithms such as

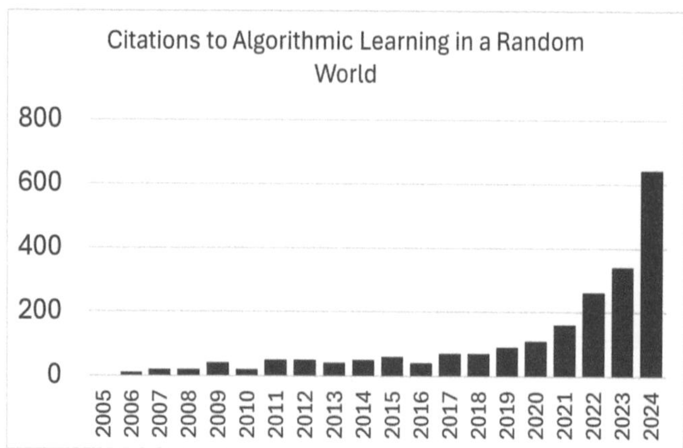

Fig. 1. Annual citations to [45], as measured by Google Scholar (accessed on 22 July 2025).

CP. Interestingly, the sixth risk for legal and regulatory responsibility, although a human, management and governance issue, may be informed by use of CP, as we will discuss later.

Predictive machine learning algorithms work by *learning* a function that takes an input example and outputs some prediction. This may not be sufficient in many domains since the end-user may want to know how confident the algorithm is. For example, if a medical diagnosis system suggests that a patient has a certain subtype of leukaemia, it would be valuable if the doctor also knew the probability that this is the case. Many machine learning algorithms are able to do this by providing probabilities for the prediction. These probability estimates are typically based on underlying distributional assumptions. With a frequentist approach, an implicit assumption of normality is often required; whereas, for a Bayesian approach, an explicit assumption of a prior distribution is required [45]. CP works in a different way by providing a range of possible predicted values as a prediction set at a *guaranteed* level of confidence. This guarantee is achieved as a mathematical property, with exchangeability being the only distributional assumption on the data. This is a minimal assumption which is even weaker than the usual independently and identically distributed (i.i.d.) assumption, which has led scholars to refer to CP as *distribution-free* [26]. This is a valuable property, especially for trustworthy machine learning, as we will see.

In Sect. 2 we give a brief overview of CP, and then in Sect. 3 describe how CP can help develop trustworthy AI systems.

2   Conformal Predictors

Conformal Predictors are set predictors. Instead of predicting one possible value, called a *point prediction*, they can predict a set of possible values for a target outcome. The set prediction is also referred to as a *region prediction* in the literature. A prediction set is correct if the true target value is in the prediction set, and wrong otherwise. Example 1 illustrates this approach.

Example 1. Suppose that the machine learning task is to model a particular disease group. For a given patient, it is required to predict whether they have disease A, B, C or D.

- A basic machine learning algorithm, given the details and symptoms of a new patient John Smith, may predict that he has disease B. But this provides no information on how confident we should be in this point prediction.
- A probabilistic machine learning algorithm may be able to predict disease B with probability 75%. This is more useful but depends on whether the assigned probability is well-calibrated, and if the underlying assumptions of the probabilistic model are correct.
- Finally, a Conformal Predictor produces the prediction set {B, D} at 95% confidence level which means that the disease is either B or D with 95% probability. This is valuable for the doctor since it means that she can take action to perform further tests to isolate which of these two it is, and the probability is reliable since CP has a guarantee of validity. The confidence level itself can be set by the doctors according to their own threshold. However, there is no "free lunch" and as confidence level increases, so the size of the prediction set tends to increase.
- If later it emerges that John Smith actually has disease B, then the CP was correct. On the other hand, if it is disease C then the prediction set was wrong.

The size of the prediction set is governed by a user-defined confidence level. A high confidence level can be set, which means we can be more certain of the prediction, but this can only be achieved with larger prediction sets: the CP will hedge its prediction by offering more options. In the extreme case, setting 100% confidence level means the prediction set will contain all possible target values, which is not any use. This demonstrates an interesting quality of the CP: in other machine learning algorithms, performance will normally be measured by its accuracy, but in CP, this is guaranteed; instead, performance is related to the size of the prediction set: the smaller the better (except for the empty set which is a special case). This is called the *predictive inefficiency* of the CP and there are several ways to measure this, although the mean set size (N criterion) is typical [44].

Since this chapter is discursive in nature, we will not dwell on the technical details. However, there is a rich mathematical foundation to CPs which the reader is invited to follow for details [45]. However, we can give a brief overview of how CP works. At the heart of CP is a *conformity measure* which is a function that takes an example (observation): both the input values and the target outcome and outputs how typical that example is. For example, if input is type of animal and the target variable is height in meters, then ("horse", 2) would receive a high conformity score (say, 10) since 2 m is a normal height for a horse, whereas ("mouse", 5) would be low (conformity score=−3) since a 5 m mouse is very unlikely. The conformity measure is typically based on a machine learning model that has learned the typical relationship between the input and the target output. The Conformal Predictor is an algorithm which takes the conformity measure to construct prediction sets for new unlabelled examples in such a way that the probability that the prediction set is correct is equal to a user-defined confidence level: this is the property of *validity* [1]. Remarkably, validity holds regardless of choice of conformity measure, so long as the data is exchangeable. However, poor choice of conformity measure will lead to high predictive inefficiency.

There are two types of CP: transductive CP (TCP) which operates in the online learning setting, and the inductive CP (ICP) which operates in the offline, batch setting. Since the batch setting, with a fixed training dataset and separate test dataset, is the usual setting in practice, we will focus on ICP. For ICP, a separate calibration dataset is required to achieve validity. Figure 2 provides a schematic for ICP.

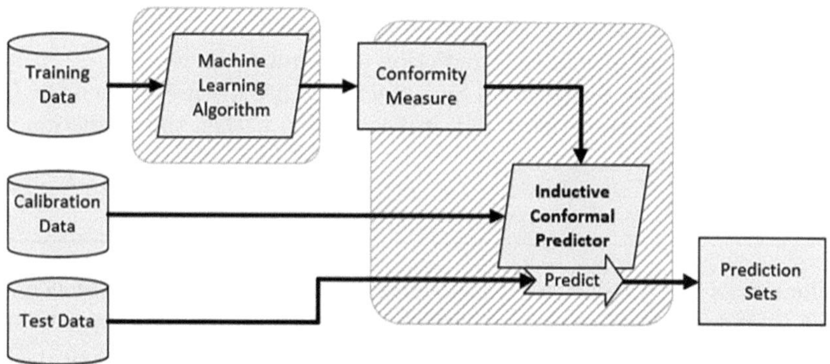

Fig. 2. Inductive Conformal Predictor (ICP) (adapted from [25]).

CP can be used for both classification and regression learning tasks. Classification involves tasks where the target variable is a finite set of labels, just two for binary classification, and so the prediction set is also finite. However, for regression the target variable is a real number. This means the prediction set

---

[1] Technically, the probability of being correct is greater or equal to the confidence level, but so long as conformity scores are mostly unique, this becomes equality. It is easy to enforce this by adding small random noise to the conformity measure.

is a subset of the real numbers and is typically infinite. It is usual to treat the special case when the prediction set is a prediction interval since this is manageable. The normalized conformity measure proposed by [33], or some variant, is typically used for regression and yields prediction intervals.

Traditionally Conformal Predictors are *wrapper* algorithms, in the sense that a machine learning model is first built and then it becomes to the input to a conformity measure which is then used by the Conformal Predictor to produce reliable prediction sets, as illustrated in Fig. 2. Therefore, all the learning is done by the machine learning algorithm and CP provides reliable prediction sets. The problem with this is that the learning is not optimized to the CP task. A new approach is to integrate the Conformal Predictor into the optimization and learning process. Studies have shown that this can lead to training Conformal Predictors which have lower predictive inefficiency for both classification [39] and regression tasks [25].

3   CP for AI Risk Mitigation

There are several aspects of AI Risk that CP can help reduce or avoid. In this section, we break down the risks in detail and describe how CP can contribute. Typically, all machine learning models are evaluated according to aggregate performance on a task. If a classification task is learnt, then performance can be measured by accuracy, area under the ROC curve (AUC), F1-score and a myriad of other measures. For a regression task, typically mean-square-error, R^2 or mean-absolute-error. Academic articles reporting machine learning results will provide such aggregate performance measures and industry projects will also report these performance measures so, as long as the evaluation has been performed correctly, this is not usually a risk [35]. However, there are several AI risks that may not typically be addressed and so deserve special attention.

3.1   Performance - Calibration Risk

When we use a machine learning algorithm to make a prediction, we would like to know how reliable this prediction is. This is different from accuracy. For example, if I consult a weather forecasting system, if it says that there is 75% chance of rain I need to trust the probability really is 75% (and not 65% or 85%). This is called *calibration* and if a model's probability estimates are measurably evident in operation we say it is *well-calibrated*. Then we can trust its uncertainty estimates are reliable. Calibration can be tested over the long run by observing its prediction against outcome. For example, if the weather forecasting system makes 50 daily forecasts of rain between 70% and 80% over the course of the year, and of those days, 38 had rain, that would be about right. If it was only 20 (40% empirical frequency) we would not trust it is well-calibrated. Statistical tests can be used to test for calibration; e.g. using a binomial test.

Conformal Predictor is a set predictor that outputs set predictions at a given confidence level. In this context, we say that a set predictor is well-calibrated if the frequency that the set predictions are correct is the same as the confidence level. As we have explained, CPs have the important property of validity which means that they are guaranteed to be well-calibrated, and this is true in the finite sample setting. This remarkable feature clearly ensures they are trustworthy for calibration, which means they are valuable tools when applying machine learning in safety-critical applications. Of course, the validity guarantee is dependent on both the data being exchangeable and a lack of coding errors when developing the prediction software (e.g. processing the data wrongly). For this reason it remains important to test the empirical validity.

Not all set predictors are well-calibrated. One study performed experiments with several prediction interval algorithms, essentially set predictors in the regression setting [25]. They found that whilst ICP was evidently well-calibrated, an alternative algorithm QD-Soft [34] was not. Figure 3 shows results for the bias correction [10] and RSNA data sets [19]. The left graph shows that both models' coverage is slightly lower than the confidence level; however, ICP is superior being closer to the calibration line, especially at higher confidence level. The right graph shows a clear difference with ICP well-calibrated, whilst QD-Soft is extremely over- and under-calibrated, especially at high confidence levels.

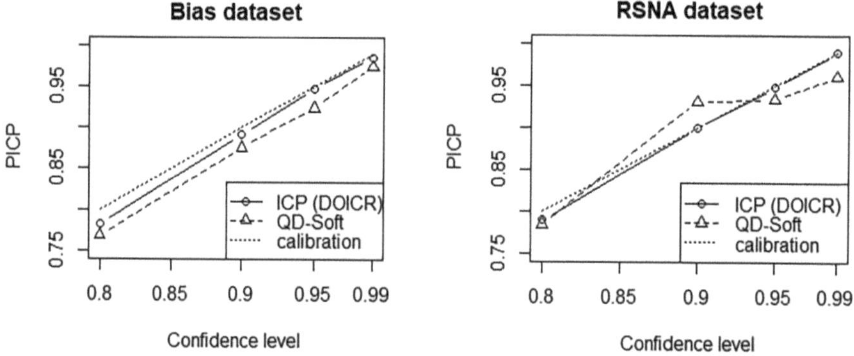

Fig. 3. Experimental results showing calibration plots of ICP and QD-Soft algorithms, from [25], Tables 4 & 5. The two experiments were building neural networks on the bias correction data set (left) and building convolutional neural network (EfficientNet) on RSNA imaging data for bone age prediction (right). DOICR is the authors directly trained ICP for regression, and PICP is prediction interval coverage percentage. The calibration line is the target when PICP is equal to confidence level (perfectly calibrated).

The validity property is the key reliability property that CPs have, hence when they are used for reliable or trustworthy machine learning, it is usually the calibration risk that is addressed. For example, CP is used to develop trustworthy deep learning in the clinical setting [16]. In [29], disease severity rating based on medical imaging is made reliable using CP. They provide experimental results to demonstrate the successful use of conformal predictors in this setting. In another study, CP is integrated into a multi-view (multi-model) deep learning framework, using the Conformalized Multi-view Deep Classification, to enable prediction set prediction with validity for trustworthy models, and provide results for medical imaging data [28]. However, calibration is not the only AI risk that can be addressed by CP and the next sections will address these issues.

3.2   Performance Generalization Risk

Although the model may be built and perform well on the development data, it may not generalize well to new data. This is important because the purpose of a model is to predict on new, previously unseen, data. There are different forms of this risk. The problem of models *overfitting* to training data is well-known. If a model overfits, it will not generalize well beyond the training data. However, this problem is largely dealt with by setting up a proper validation procedure, ensuring that the model is evaluated on data that is independent of data that it was trained on. Randomly splitting development data into training and test data is the easiest approach to do this, or cross-validation is an alternative.

More challenging versions of generalization risk are due to (i) Selection bias: the population in development data is different in some way to the data that the model will be applied to; (ii) Population drift: the time period of the development data is different from the data the model will be applied to in the future. An example of selection bias is when a credit score model is built on a historic portfolio of credit card holders who are on average old and predominantly male, but the model will be applied on a population of applicants who are expected to be younger with a higher proportion of females. Such a model may not perform so well on the target population because of difference in behaviour across age groups and sex. In the medical context, an example of population drift may occur with the development of CT scanner imaging technology over time. A machine learning algorithm built on old CT images may not work well on future ones. Other sources of drift could be change in patient demographics, change in disease prevalence and behaviour, change in disease definition and understanding, or change in physician behaviour, perhaps as a result of automation [38]. What all forms of generalization risk have in common is that there is a distributional change between development and target operational data.

We can utilize the property of CP that if data is drawn from the same exchangeable distribution then a CP built on that data will exhibit validity. By *modus tollens*, this means that if we detect that a Conformal Predictor is not exhibiting empirical validity then the data is not exchangeable and not drawn from a single homogeneous distribution. This would signal a potential generalization risk. In this way, CP can be used as a generalization risk warning system and this is the underlying idea of *conformal testing*. A framework can be set up using conformal test martingales based on the p-values output by the CP making sequential predictions, as described in [48]. Although the paper focusses on drift, it could be adapted for the selection bias setting. A more recent development of this idea is conformal e-testing for distribution change [47]. There are at least two ways this conformal testing methodology can be used to reduce generalization risk.

- Conformal testing is a warning system to determine when to rebuild a model.
- There are several methods proposed to reduce the impact of generalization risk such as resampling or weighting the development data. These often involve approaches to make development data similar to target data in some sense [6]. Conformal testing can be used as a way to test whether these approaches have worked successfully, or not.

3.3   Performance Robustness Risk

An AI system is robust if it is able to operate well even under unusual conditions; in particular, if it is given a new example unlike those it has been trained on, we would still like it to perform well. Unfortunately, many AI systems lack this type of robustness. If they are given data that is quite different, or untypical, of training data, they behave poorly (see e.g. [37]). This is related to generalization risk, but these cases are typically rarer and may lead to more extreme outcome. A real example is illustrated by the case of Tesla cars operating in autopilot on the highway with 11 accidents between 2018 and 2021. The accidents all had the same common feature: the car crashed into a stationary emergency vehicle. The accidents were investigated by the National Highway Transportation Safety Administration (USA). This particular scenario, a stationary emergency vehicle on the highway, was unusual for the car's autopilot, since it would not have emerged during training. Unfortunately the autopilot was not able to adapt to the scenario in a way we would expect a human driver to do [22]. Sometimes a very small change can cause a machine learning model to fail; e.g. [40] showed that changing just one pixel in an image can upset a deep neural network. This vulnerability to very small changes is referred to as *brittleness*.

CP can be used to avoid problems when new examples are different from the training data. This takes advantage of the conformity measure in the CP. The conformity measure scores how typical a new example is, therefore an unusual case - one not seen in training - will yield a low conformity score. It would be expected that such a case would yield a very inefficient prediction set, since it is untypical, which would be a way for the AI system to signal that it is not confidently predicting in these cases. The conformity measure can be designed especially to find unusual cases not just in the relationship between the input features and the target variable but also amongst the input features themselves. In particular, [24] proposes a nearest-neighbors-based nonconformity measure that can be used for this purpose. In their experiments using a dataset of real sea vessel trajectories, they were able to achieve high accuracy levels for identification of anomalous trajectories with low false alarm rates. Furthermore, the anomaly detection is shown to be well-calibrated.

Therefore, by using CP to detect anomalous examples, it is possible to provide a warning regarding unusual examples and situations that the AI system can act upon to ensure safety.

3.4   Bias Biassed Prediction Risk

Bias has a number of different meanings. In the context of trustworthy AI, we usually refer to decisions made by AI systems that are biassed towards some particular group of people. For example, it could be bias towards women, or black people, or Jewish or Muslim people. We would say that the AI system discriminates against those people. In many countries, there are protected characteristics by law. For example, in the UK, the Equality Act 2010 protects people in law based on age, disability, sex, sexual orientation, gender reassignment, race and several other categories. Therefore, AI systems that are biassed may break the law. The bias may be more specific, and hence more difficult to uncover, crossing two or more protected characteristics: such as discrimination toward black women [8]. This is known as intersectionalism.

Bias and discrimination are linked to notions of unfairness in algorithmic decision making. Fairness is well-understood and has been defined clearly in the context of machine learning [5]. One key insight is that fairness is not a single idea, but can be defined in many ways. Common definitions of fairness are Independence, Separation, and Sufficiency. These definitions are contradictory, in the sense that if we choose one type of fairness, we cannot achieve the others [5]. In practice, this means that the stakeholders of an AI system need to agree on the form of fairness that is required for the application, and this will be strongly related to the legal and regulatory framework. This decision will then affect how we measure whether our model is unfair.

It is worth pointing out that not all bias may be unethical or unfair. For example, a medical system may find that men are more likely than women to contract a disease, all else being equal. However, this may reflect an actual bias in reality, i.e. men are genuinely more susceptible to the disease, rather than representing a poor diagnosis in women. Therefore, a judgement needs to be made regarding whether a biassed outcome represents unfair discrimination.

Bias is well-known in AI and has been well-documented. The possible causes of bias are myriad: bias in the way data is selected, historical bias in the data, bias introduced (or amplified) by the algorithm, bias by the machine learning engineer. If the bias is detected and the cause of the bias is isolated then a solution to mitigate the bias may be proposed [30].

Predictions may be biassed within subgroup. So, overall a model may predict accurately across a whole population but for some subgroups it may be that the target variable is over- or under-predicted. For example, in 2015 Amazon ditched a project to use AI in recruitment, since the model was under-scoring women applicants, specifically penalizing applicants with female-linked descriptions in their application (e.g. "women" or "women's chess club captain") [12]. This type of bias could be fixed by either returning to the data to adjust for any under-representation, or historical bias, or, in the context of supervised learning adjusting for failures in the model by building a second level model of the errors, to adjust post-hoc to the bias. Interestingly, machine learning algorithms based on boosting can do this by design [30].

3.5 Bias Biassed Performance Risk

Another form of bias is not with the prediction itself, but with the performance of the model for different subgroups. We have already given the example of intersectional bias for facial recognition systems [8] in the Introduction. In another study looking at diagnosis based on medical imaging data and using machine learning found that imbalanced data sets lead to models that reduce performance in diagnosis for the underrepresented gender, measured using AUC [23].

For CP, the key performance measure is predictive inefficiency, and so biassed performance could manifest in changes in prediction set size between subgroups. For example, if we observe larger prediction sets for an ethnic minority, on average, then this would suggest bias that will require further investigation. Hence, CP developers would be recommended to check for this. Mitigating this will require similar techniques as other machine learning algorithms (e.g. analysis and adjustment of data). However, as mentioned in Sect. 3.2, prediction inefficiency is measured without knowing outcome, so CP for bias performance tests will be particularly important in applications with delayed outcome, such as patient survival or credit scoring, where actual outcome may not be known for several years.

Another way that CP may exhibit biassed performance is with its coverage: the proportion of times that a prediction set is correct within subgroups. This is because the validity property referred to in Sect. 2 is across the entire population, but may not be true for all subgroups, as illustrated by Example 2. This means the validity property is *marginal validity*, but to ensure even coverage between groups, we are interested in is *conditional validity*; i.e. validity conditional on the subgroups. Arguably, conditional validity is more important for CP, rather than biassed prediction set size, since the validity property is the main benefit of CP over other algorithms. Unfortunately, it has been shown that conditional validity is impossible in the general finite-sample case [42]. Approaches have been taken for special cases, such as linear models [31] or asymptotic case [26], but these cannot be extended to the general case. In terms of fairness, conditional validity fits the Independence criterion [5], since validity, meeting the confidence level, across all observations is the central benefit for CP. We could imagine situations when we might want to achieve different coverage conditional on the outcome which would lead to the Separation definition, but these are not intuitive and would be special cases.

Example 2. Suppose the CP task is to model a particular disease group, and for a given patient, predict whether they have disease A, B, C or D. The medical doctor sets a confidence level of 90%.

- We observe that in practice, the CP gives empirical coverage of 89.5%. This means that the prediction sets contain the true label (for each patient) 89.5% of the time. This is sufficiently close to 90% that we can conclude that the CP is well-calibrated.

> – If we consider just the male patients, we find that the empirical coverage is higher at 94%.
> – Conversely, the female patient group has 85% coverage.
> – If the male/female populations are the same size, then this shows that overall empirical coverage is consistent with 89.5%.
> – However, it is clear that the CP is not well-calibrated *within* these subgroups, and is under-performing for women.

Mondrian CP [45] has also been proposed and gives conditional validity within the context of a given taxonomy of subgroups. Essentially, these work by splitting the calibration data into separate calibration sets for each subgroup. This guarantees conditional validity within these subgroups. However, Mondrian CP has two problems: (i) As taxonomy gets larger, the calibration sets get smaller, hence potentially degrading their performance; (ii) Mondrian is dependent on the pre-defined taxonomy and cannot guarantee validity in other subgroups.

Another approach is to aim for approximate conditional validity instead of exact. In [4], a *distribution-free approximate conditional coverage* framework is proposed that takes conditional validity to be true for at least some pre-defined proportion of the population. This is similar to the PAC-type conditional validity proposed by [43]. In particular they further propose *restricted conditional coverage* to provide local conditional validity in the neighbourhood (ball) around some point. This is a promising approach and they show that CP satisfies this condition, in a special case, but they identify computational difficulties that require further research. Another proposal for approximate conditional validity is the *iterative feedback-adjusted conformity measure* (IFACM) algorithm that uses a boosting approach to update the conformity measure based on the performance of a conformal predictor in different subgroup regions [7]. It can be set up with a taxonomy like Mondrian, but also has the advantage of being general-purpose, discovering subgroup regions with performance bias itself, without the need of a taxonomy.

Although conditional validity is a useful goal, and Mondrian and approximation approaches are promising tools, it is clear that these approaches work by shifting predictive inefficiency to improve coverage, and there is "no free lunch". In particular, IFACM works explicitly by expanding or contracting prediction set sizes in response to deviations in validity across the population. As such, although they achieve approximate conditional validity, this will typically translate into changes in predictive inefficiency, often increasing. This is acceptable since the main goal of CP is to achieve reliability. However, since many bias issues are often data-driven [20], e.g. underrepresented sub-populations, it will still be valuable to fix the problem at the data and pre-processing level.

Experiments Using Mondrian and IFACM for Performance Bias Mitigation.
We implemented experiments using Mondrian and IFACM on two real datasets, ASCIncome dataset [14] and ISIC2019 dataset [11,21,41]. We selected three subgroups which are detailed in Table 1 for each dataset. Subgroup 1 is a majority group, subgroup 2 and subgroup 3 are minority groups.

In the ASCIncome experiment, we built a neural network model (i.e. a Multi-Layer Perceptron with 64, 32, 32 and 8 hidden units in each layer) to classify annual income levels based on individual demographic features including age, educational attainment, occupation, gender and race. We perform an expository analysis of the results to illustrate performance bias correction. Figure 4 shows the empirical coverage and inefficiency within the subgroups obtained by the basic ICP, Mondrian and IFACM, with confidence levels 0.8 and 0.9. We see that all the methods achieve marginal validity. However, ICP exhibits coverage higher than the confidence level conditional on subgroup 1 and extremely low coverage conditional on subgroup 2 and subgroup 3. By contrast, Mondrian and IFACM provide coverages that are very close to the confidence level across all subgroups, albeit at the cost of increasing inefficiency. These results demonstrate the significance for applications where policymakers express major concern about the validity conditional on age, gender and race, and do not have any preference for other demographic features such as educational attainment and occupation.

In the ISIC2019 task, we built a ResNet50 model to classify diagnostic categories given the dermoscopic images of the skin lesion. Figure 5 shows the empirical coverage and inefficiency within the subgroups obtained by ICP, Mondrian and IFACM, with confidence levels 0.8 and 0.9. All of the methods achieve marginal coverages that are close to the confidence level. ICP achieves higher coverage conditional on subgroup 1 and significantly lower coverage conditional on subgroup 2 and subgroup 3. Mondrian and IFACM provide conditional coverages that are closer to the confidence level across all subgroups, albeit with higher inefficiency. The results demonstrate that Mondrian and IFACM are potential approaches to increasing trustworthiness in medical AI by providing validity across clinically relevant subgroups.

Table 1. Subgroup descriptions

Subgroup	ASCIncome	ISIC2019
1	female, age>55, White, 2k$\leq$income<70k	female, age$\leq$40, Melanocytic nevus
2	female, 30<age$\leq$55, Black, income$\geq$140k	female, 40<age$\leq$55, Benign keratosis
3	male, 30<age$\leq$55, Asian, income$\geq$140k	male, age>70, Benign keratosis

3.6 Human/AI Reliance and Regulatory Risks

When AI systems are deployed, the response of people using the system is sometimes not what we would expect or desire. Human-AI interaction is a special

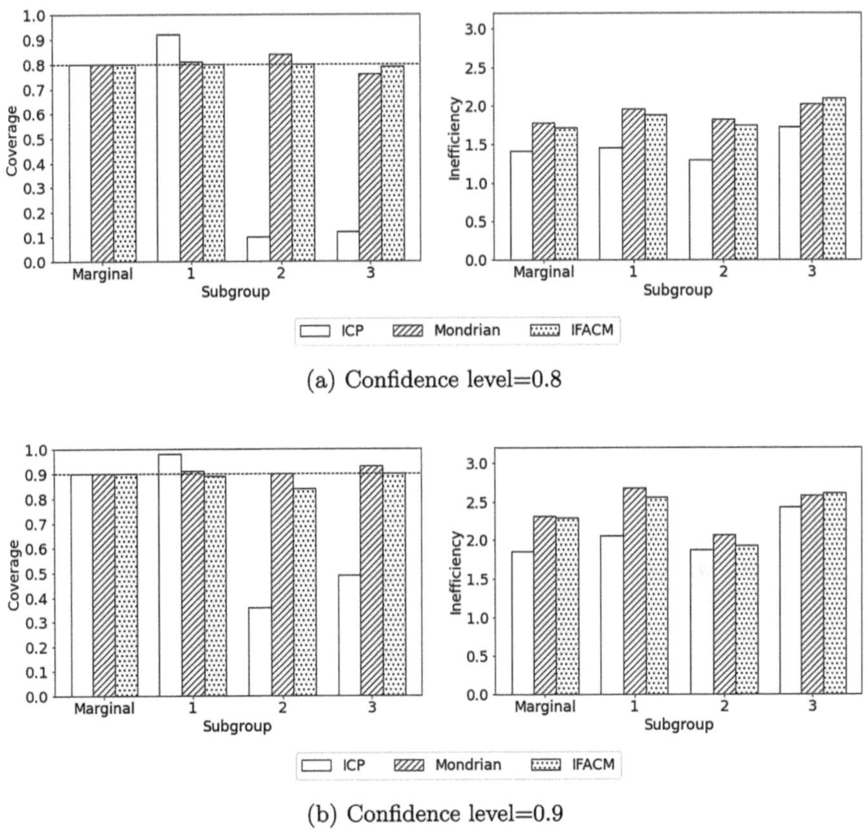

(a) Confidence level=0.8

(b) Confidence level=0.9

Fig. 4. Experimental results on ASCIncome dataset for performance bias correction with (a) confidence level=0.8 and (b) confidence level=0.9.

topic area of human-computer interaction, studying specifically the behaviour of people when working with AI, and ways we can design AI for better collaboration with humans (e.g. [1,32]). Much of the misuse or disuse of AI, which can cause failures such as railway crashes and biassed decision making comes from people over- or under-trusting the AI, and this is "often linked to the alignment between the perceived and actual performance of the system" [32]. This is particularly the case with systems that do not provide a confidence measure for their predictions or confidence measures that are not well-calibrated. In these circumstances, the user has no way to measure the AI's trustworthiness and this can lead to poor outcomes. CP can mitigate this problem since it is able to provide reliable uncertainty measures that can be trusted. In particular, if approaches that deal with conditional validity such as Mondrian and IFACM, illustrated in Sect. 3.5, are deployed, then human users will have additional confidence at subgroup and individual decision-making level.

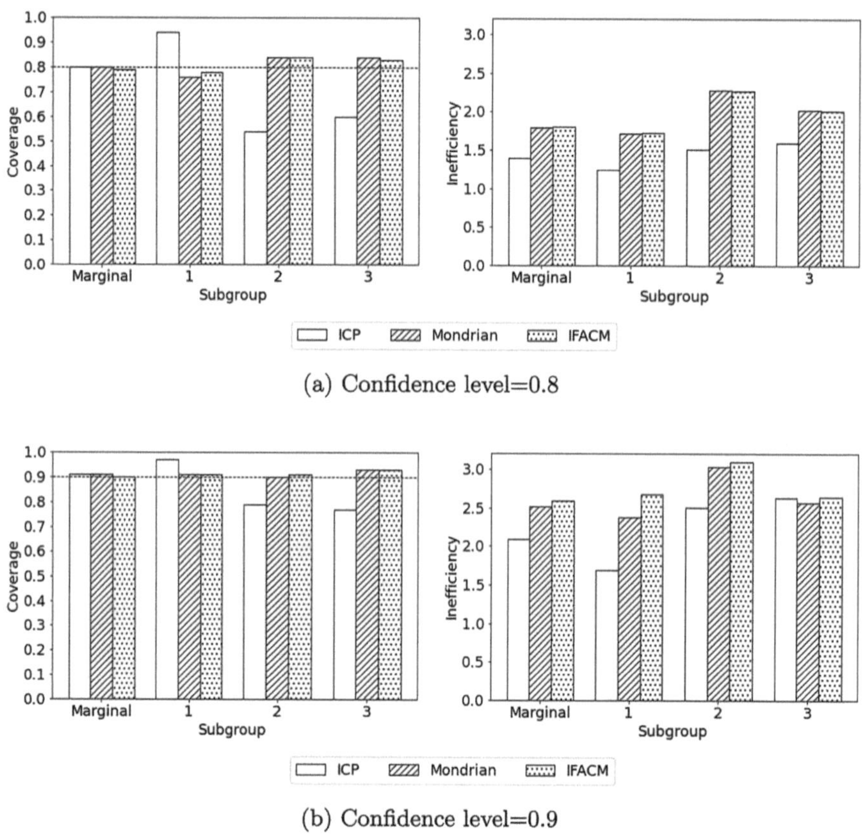

(a) Confidence level=0.8

(b) Confidence level=0.9

Fig. 5. Experimental results on ISIC2019 dataset for performance bias correction with (a) confidence level=0.8 and (b) confidence level=0.9.

With the increased adoption of regulation and laws on AI, such as the European Union AI Act, there is an emphasis on safe, trustworthy and transparent AI [15]. The qualities of CP make it a valuable tool to demonstrate the reliability and transparency of uncertainty measures in AI decision making to regulators.

Table 2. Summary of AI risks and CP mitigation

AI risk category	AI risk	CP mitigation
Poor performance	Calibration risk Generalization risk Robustness	Validity guarantee Conformal testing Conformity measure / anomaly detection
Bias	Biassed predictions Biassed performance	- Mondrian and approximate conditional validity (IFACM)
Human-AI interaction	Over- or under-reliance on AI Impact on human behaviour Explainable AI, lack of Misuse of AI	Reliable prediction sets
Objective misalignment	Difference between specification and actual task Unintended consequences	-
Security & Privacy	Data breach Copyright breach	-
Ethics, Legal & Accountability	Illegal activity by AI Regulatory noncompliance Unsafe agentic AI Lack of accountability	Reliable uncertainty measures for reporting, accountability and transparency

4   Conclusion

In this chapter, we have considered how CP can be used broadly for trustworthy AI, and given some specific experimental results to demonstrate the value of CP for calibrated uncertainty estimation and bias detection and mitigation. Since many of the risks are data-driven, we acknowledge that many solutions to enable trustworthy AI fall back to careful data collection, processing and use [20]. Nevertheless, we show that CP can provide some valuable algorithmic tools.

Table 2 provides an overview of AI risks, along with characteristics of CP that can be beneficial. Some of the AI risks have not been addressed, since they are not so obviously addressed by CP. Most previous articles that discuss CP in the context of trustworthy AI typically do so in the context of marginal validity and calibration (e.g. [16,29]). However, future work could expand the scope of CP for trustworthy AI further in areas identified in this paper, and perhaps also in novel areas of AI risk that are not addressed by CP yet according to Table 2.

Although this chapter focusses on predictive machine learning, new research is emerging applying CP to generative models such as large language models (LLMs) [9,49]. Consequently, we can foresee CP also being used for trustworthy and reliable AI in this expanding field of AI with future research. Finally, we note the value of CP to allow for better understanding and interpretation of AI, and therefore an important tool for AI governance, to demonstrate adherence to standards and regulations as these emerge throughout the world both in industry and government.

Acknowledgment. We are grateful for fruitful discussions on trustworthy AI over many years with the team at Validate AI [1], especially Shakeel Khan, David Hand and Mark Kennedy. We are especially indebted to Alex Gammerman who introduced Anthony Bellotti to CP in the early 2000's and enabled him to complete his PhD using CP in 2006. Xindi Zhao is subsequently Anthony's PhD student and will graduate with her PhD using CP this year. Many thanks to Alex, and congratulations on his 80th birthday!([1]`https://validateai.org` (accessed on 25 July 2025))

References

1. Abedin, B., Meske, C., Junglas, I., Rabhi, F., Motahari-Nezhad, H.R.: Designing and managing human-AI interactions. Inf. Syst. Front. **24**(3), 691–697 (2022)
2. Amodei, D., Clark, J.: Faulty reward functions in the wild. OpenAI Blog (2016)
3. Amodei, D., Olah, C., Steinhardt, J., Christiano, P., Schulman, J., Mané, D.: Concrete problems in AI safety. arXiv preprint arXiv:1606.06565 (2016)
4. Barber, R.F., Candès, E.J., Ramdos, A., Tibshirani, R.J.: The limits of distribution-free conditional predictive inference. Inf. Infer. A J. IMA **10**(2), 1–28 (2020)
5. Barocas, S., Hardt, M., Narayanan, A.: Fairness and Machine Learning: Limitations and Opportunities. MIT Press (2023)
6. Bellotti, A., Hand, D., Khan, S.: Predicting through a crisis, 2nd white paper. Validate AI (2020). validateai.org/white-papers
7. Bellotti, A.: Approximation to object conditional validity with inductive conformal predictors. In: Conformal and Probabilistic Prediction and Applications, pp. 4–23. PMLR (2021)
8. Buolamwini, J., Gebru, T.: Gender shades: intersectional accuracy disparities in commercial gender classification. In: Conference on Fairness, Accountability and Transparency, pp. 77–91. PMLR (2018)
9. Campos, M., Farinhas, A., Zerva, C., Figueiredo, M.A.T., Martins, A.F.T.: Conformal prediction for natural language processing: a survey. Trans. Assoc. Comput. Linguist. **12**, 1497–1516 (2024)
10. Cho, D., Yoo, C., Im, J., Cha, D.H.: Comparative assessment of various machine learning-based bias correction methods for numerical weather prediction model forecasts of extreme air temperatures in urban areas. Earth Space Sci. **7**(4), e2019EA000740 (2020)
11. Codella, N.C., et al.: Skin lesion analysis toward melanoma detection: a challenge at the 2017 international symposium on biomedical imaging (ISBI), hosted by the international skin imaging collaboration (ISIC). In: 2018 IEEE 15th International Symposium on Biomedical Imaging (ISBI 2018), pp. 168–172. IEEE (2018)
12. Dastin, J.: Amazon scraps secret AI recruiting tool that showed bias against women. In: Ethics of Data and Analytics, pp. 296–299. Auerbach Publications (2022)
13. Dehouche, N.: Post-labor economics: a systematic review. Preprints (2025)
14. Ding, F., Hardt, M., Miller, J., Schmidt, L.: Retiring adult: new datasets for fair machine learning. Adv. Neural. Inf. Process. Syst. **34**, 6478–6490 (2021)

15. European Union: Rules for trustworthy artificial intelligence in the EU. EUR-Lex (2025), https://eur-lex.europa.eu/legal-content/EN/TXT/?uri=legissum:4762484
16. Gamble, C., Faghani, S., Erickson, B.J.: Toward clinically trustworthy deep learning: applying conformal prediction to intracranial hemorrhage detection. arXiv preprint arXiv:2401.08058 (2024)
17. Gammerman, A., Vovk, V., Vapnik, V.: Learning by transduction. In: Proceedings of the Fourteenth Conference on Uncertainty in Artificial Intelligence, p. 148–155. Morgan Kaufmann Publishers Inc. (1998)
18. Gunnarsson, B.R., Broucke, S.V., Baesens, B., Óskarsdóttir, M., Lemahieu, W.: Deep learning for credit scoring: do or don't? Eur. J. Oper. Res. **295**(1), 292–305 (2021). https://doi.org/10.1016/j.ejor.2021.03.006
19. Halabi, S.S., et al.: The RSNA pediatric bone age machine learning challenge. Radiology **290**(2), 498–503 (2019)
20. Hand, D.J.: Dark Data: Why what you don't know Matters. Princeton University Press (2020)
21. Hernández-Pérez, C., et al.: Bcn20000: Dermoscopic lesions in the wild. Scientific data **11**(1), 641 (2024)
22. Isidore, C., Valdes-Dapena, P.: Tesla is under investigation because its cars keep hitting emergency vehicles. CNN Business News (2021). edition.cnn.com/2021/08/16/business/tesla-autopilot-federal-   safety-probe. Accessed 18 July 2025
23. Larrazabal, A.J., Nieto, N., Peterson, V., Milone, D.H., Ferrante, E.: Gender imbalance in medical imaging datasets produces biased classifiers for computer-aided diagnosis. Proc. Natl. Acad. Sci. **117**(23), 12592–12594 (2020)
24. Laxhammar, R., Falkman, G.: Inductive conformal anomaly detection for sequential detection of anomalous sub-trajectories. Ann. Math. Artif. Intell. **74**(1), 67–94 (2015)
25. Lei, H., Bellotti, A.: Reliable prediction intervals with directly optimized inductive conformal regression for deep learning. Neural Netw. **168**, 194–205 (2023)
26. Lei, J., Wasserman, L.: Distribution-free prediction bands for non-parametric regression. J. R. Stat. Soc. Ser. B Stat Methodol. **76**(1), 71–96 (2014)
27. Li, B., et al.: Trustworthy AI: From Principles to Practices. ACM Comput. Surv. **55**(9) (2023). https://doi.org/10.1145/3555803
28. Liu, W., Chen, Y., Yue, X.: Building trust in decision with conformalized multi-view deep classification. In: Proceedings of the 32nd ACM International Conference on Multimedia, pp. 7278–7287 (2024)
29. Lu, C., Angelopoulos, A.N., Pomerantz, S.: Improving trustworthiness of AI disease severity rating in medical imaging with ordinal conformal prediction sets. In: International Conference on Medical Image Computing and Computer-Assisted Intervention, pp. 545–554 (2022)
30. Mavrogiorgos, K., Kiourtis, A., Mavrogiorgou, A., Menychtas, A., Kyriazis, D.: Bias in machine learning: a literature review. Appl. Sci. **14**(19) (2024). https://doi.org/10.3390/app14198860
31. McCullagh, P., Vovk, V., Nouretdinov, I., Devetyarov, D., Gammerman, A.: Conditional prediction intervals for linear regression. In: Proceedings of the Eighth International Conference on Machine Learning and Applications (2009)
32. Mehrotra, S., Degachi, C., Vereschak, O., Jonker, C.M., Tielman, M.L.: A systematic review on fostering appropriate trust in human-ai interaction: trends, opportunities and challenges. ACM J. Responsible Comput. **1**(4), 1–45 (2024)

33. Papadopoulos, H., Proedrou, K., Vovk, V., Gammerman, A.: Inductive Confidence Machines for Regression. In: Elomaa, T., Mannila, H., Toivonen, H. (eds.) Machine Learning: ECML 2002. ECML 2002. Lecture Notes in Computer Science, vol. 2430, Springer, Berlin, Heidelberg (2002)
34. Pearce, T., Zaki, M., Brintrup, A., Neely, A.: High-quality prediction intervals for deep learning: a distribution-free, ensembled approach. In: Proceedings of Machine Learning Research: International Conference on Machine Learning 2018 vol. 80, 4075–4084 (2018)
35. Provost, F., Fawcett, T.: Data Science for Business: What you need to know about Data Mining and Data-Analytic Thinking. Inc, O'Reilly Media (2013)
36. Qian, Y., Siau, K.L., Nah, F.F.: Societal impacts of artificial intelligence: ethical, legal, and governance issues. Societal Impacts **3**, 100040 (2024). https://doi.org/10.1016/j.socimp.2024.100040
37. Rafiq, H., Aslam, N., Issac, B., Randhawa, R.H.: An investigation on fragility of machine learning classifiers in android malware detection. In: IEEE INFOCOM 2022 - IEEE Conference on Computer Communications Workshops (INFOCOM WKSHPS), pp. 1–6 (2022). https://doi.org/10.1109/INFOCOMWKSHPS54753.2022.9798161
38. Sahiner, B., Chen, W., Samala, R.K., Petrick, N.: Data drift in medical machine learning: implications and potential remedies. Br. J. Radiol. **96**(1150), 20220878 (2023). https://doi.org/10.1259/bjr.20220878
39. Stutz, D., Cemgil, A.T., Doucet, A.: Learning optimal conformal classifiers. In: International Conference on Learning Representations (2022). https://doi.org/10.48550/arXiv.2110.09192
40. Su, J., Vargas, D.V., Sakurai, K.: One pixel attack for fooling deep neural networks. IEEE Trans. Evol. Comput. **23**(5), 828–841 (2019). https://doi.org/10.1109/TEVC.2019.2890858
41. Tschandl, P., Rosendahl, C., Kittler, H.: The ham10000 dataset, a large collection of multi-source dermatoscopic images of common pigmented skin lesions. Scientific data **5**(1), 1–9 (2018)
42. Vovk, V.: Conditional validity of inductive conformal predictors. Mach. Learn. **92**, 349–376 (2013)
43. Vovk, V.: Beyond the basic conformal prediction framework. In: Balasubramanian, V., Ho, S.S., Vovk, V. (eds.) Conformal Predictions for Reliable Machine Learning: Theory, Adaptations and Applications. Elsevier (2014)
44. Vovk, V., Fedorova, V., Nouretdinov, I., Gammerman, A.: Criteria of efficiency for conformal prediction. In: Gammerman, A., Luo, Z., Vega, J., Vovk, V. (eds.) COPA 2016. LNCS (LNAI), vol. 9653, pp. 23–39. Springer, Cham (2016). https://doi.org/10.1007/978-3-319-33395-3_2
45. Vovk, V., Gammerman, A., Shafer, G.: Algorithmic Learning in a Random World. Springer, US (2005)
46. Vovk, V.: On-line confidence machines are well-calibrated. In: The 43rd Annual IEEE Symposium on Foundations of Computer Science, 2002. Proceedings, pp. 187–196. IEEE (2002)
47. Vovk, V., Nouretdinov, I., Gammerman, A.: Conformal e-testing. Pattern Recogn. **168**, 111841 (2025). https://doi.org/10.1016/j.patcog.2025.111841

48. Vovk, V., Petej, I., Nouretdinov, I., Ahlberg, E., Carlsson, L., Gammerman, A.: Retrain or not retrain: conformal test martingales for change-point detection. In: Carlsson, L., Luo, Z., Cherubin, G., An Nguyen, K. (eds.) Proceedings of the Tenth Symposium on Conformal and Probabilistic Prediction and Applications, Proceedings of Machine Learning Research, vol. 152, pp. 191–210. PMLR (2021). https://proceedings.mlr.press/v152/vovk21b.html
49. Wang, Z., et al.: Conu: conformal uncertainty in large language models with correctness coverage guarantees. arXiv preprint arXiv:abs/2407.00499 (2024). https://doi.org/10.48550/arXiv.2407.00499
50. Zhang, B., Shi, H., Wang, H.: Machine learning and AI in cancer prognosis, prediction, and treatment selection: a critical approach. J. Multi. Healthc. 1779–1791 (2023)

A Review of Theoretical Advances and Practical Applications of Conformal Prediction in China

Shuo Zhao[1], Zhirui Zhang[2], Yifan Liu[1], Khuong An Nguyen[3], You Wang[1,2], Zhiyuan Luo[3], and Guang Li[1(✉)]

[1] State Key Laboratory of Industrial Control Technology, Institute of Cyber-Systems and Control, Zhejiang University, Hangzhou 310027, China
`{zhaoshuo,yifanliu,king_wy,guangli}@zju.edu.cn`
[2] Rotunbot Hangzhou Co. Ltd., Hangzhou 310027, China
`zhang.zr@rotunbot.com`
[3] Department of Computer Science, Royal Holloway University of London, Surrey TW20 0EX, UK
`{Khuong.Nguyen,Zhiyuan.Luo}@rhul.ac.uk`

Abstract. Enhancing the learnability and reliability of machine learning algorithms remains a critical research objective. As a learnable framework, Conformal Prediction (CP) leverages historical data to generate statistically valid predictions for new instances at a specified confidence level. For any predictive model, CP produces prediction sets or regions that are guaranteed to contain the true label with at least the pre-specified confidence level under the assumption of exchangeability. This review highlights pioneering contributions from Chinese research communities to both theoretical advancements and practical applications of CP. Based on a systematic review of the existing literature, we analyse CP's core theoretical strengths and objectively discuss its inherent limitations, such as computational inefficiency and the curse of dimensionality. Furthermore, we evaluate CP's current potential across a range of real-world domains, including biomedicine, cybersecurity, industry, and environmental science. Notably, novel applications such as tea analysis and the identification of traditional Chinese medicine using electronic noses demonstrate CP's versatility and significant practical value.

Keywords: Conformal Prediction · Uncertainty Quantification · Prediction Regions · Machine Learning

1 Introduction

With the advancement of machine learning algorithms, remarkable results have been achieved in increasingly complex tasks. However, while pursuing higher accuracy, enhancing the learnability and reliability of algorithms has become a critical challenge that cannot be overlooked. Researchers have long recognized that as machine learning algorithms grow more complex, they increasingly resemble black-box models, making it difficult to provide meaningful explanations for

K. An Nguyen and Z. Luo (Eds.): Alexander Gammerman Festschrift, LNCS 16290, pp. 198–226, 2026.
https://doi.org/10.1007/978-3-032-15120-9_11

their learning and reasoning processes. This not only introduces reliability risks but also complicates the control of model adaptation to new environments and data. Therefore, endowing algorithms with learnability is of paramount importance.

Nowadays, many approaches have been proposed to solve this problem by building reliable prediction models. Bayesian methods [29] and Gaussian process [12] are classical statistical approaches capable of producing prediction sets, but their results heavily depend on the correctness of prior assumptions, which are often unknown in practical applications [34,44]. Other methods, such as bootstrap-based approaches [98] and quantile regression [23], while effective, cannot guarantee coverage under finite-sample conditions. Unlike these methods, Conformal Prediction (CP) [65], as a model-agnostic framework, can provide reliable prediction regions. Requiring only the exchangeability assumption of data, it guarantees finite-sample coverage. CP's validity has been extensively demonstrated both theoretically and empirically [5,52,65], with successful applications across numerous domains.

Since its introduction by Vovk, Gammerman, and Shafer, CP has gained significant attention from researchers. While its theoretical foundations continue to be refined and various extensions explored [42,43,64], researchers have also uncovered CP's potential for diverse applications. This paper synthesizes key research on Conformal Prediction, with particular emphasis on contributions from Chinese scholars to its development and applications. We quantified the number and publication periods of reviewed papers, with specific annotation of contributions from Chinese research communities. The results are presented in Fig. 1. A notable surge in research interest in CP among Chinese scholars has been observed since 2020, accompanied by an exponential increase in research output.

The following section presents a concise yet rigorous theoretical framework of CP and the remaining sections of this chapter are organized along two main dimensions:

1. **Theoretical advances** - highlighting explorations inspired by classical CP methods and improvements addressing traditional CP limitations;
2. **Practical applications** - covering not only common domains like biomedical engineering, cybersecurity, and industrial control, but also unique investigations in areas like tea analysis and traditional Chinese medicine research.

In addition to reviewing existing studies, the chapter analyses CP's strengths and limitations, along with its application prospects across various scenarios. Finally, the conclusion section elucidates CP's significance in enhancing algorithmic learnability.

In summary, this chapter's contributions to the CP research community include:

- A systematic review of contributions from the Chinese research community to CP, encompassing both theoretical advances and practical innovations.

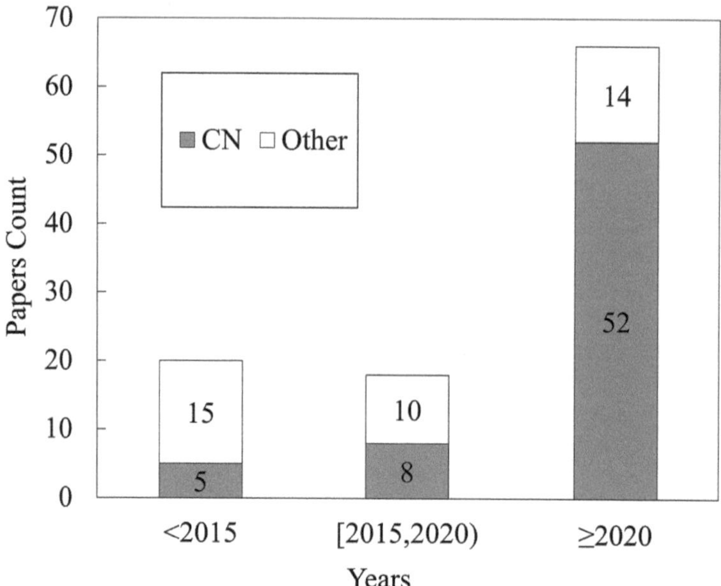

Fig. 1. Number and publication period of the reviewed papers. "CN" indicates contributions from Chinese scholars. Research output from Chinese scholars has shown consistent annual growth, with an exponential increase in recent years

– An introduction to a novel application of the electronic nose-based gas identification system, focusing on tea and traditional Chinese medicine.
– A comprehensive synthesis of applied CP implementations, including theoretical insights and performance comparison, to help propel the advancement of CP deployment.

2 Framework of Conformal Prediction

The framework of classical CP is remarkably concise. For an observed training set of $n - 1$ examples Z: $\{(x_1, y_1), ..., (x_{n-1}, y_{n-1})\}$ where $z_i = (x_i, y_i)$ consists of a feature vector x_i and its corresponding label y_i, a machine learning model learns from Z to map a new sample x_n from the sample space X to its label $\hat{y}_n$ in the label space Y. Unlike traditional machine learning algorithms, a conformal predictor can give a prediction set $\Gamma^\varepsilon \subseteq Y$ with a user defined confidence level $1 - \varepsilon$, where ε is a given additional parameter representing the significance level. The confidence level $1 - \varepsilon$ indicates how much confidence the predictor has for that the prediction set Γ^ε covers the true label. In other words, CP guarantees that the prediction set will include the true label with a probability at least of the specified confidence level, assuming the data points are exchangeable:

$$P(y_n \in \Gamma^\varepsilon) \geq 1 - \varepsilon.$$

This guarantee is known as validity and is a key advantage of conformal prediction. While validity is crucial, it is also desirable to make the prediction sets as small as possible to provide more precise and informative predictions. Efficiency refers to the size of the prediction set. CP makes it possible to avoid worrying about validity and to focus only on improving efficiency.

The foundation for CP's ability to produce prediction sets lies in its use of a nonconformity measure (or score), which quantifies how well a new sample (x_n, y) conforms to previously observed data. By extending the dataset Z to include (x_n, y), the nonconformity measurement α_i for each sample $z_i (i = 1, 2, ..., n) \in Z$ is calculated using a measurable function A:

$$\alpha_i = A(\{z_1, ..., z_{i-1}, z_{i+1}, ..., z_n\}, z_i), \quad i = 1, ..., n,$$

where A is determined by both the specific task type and the underlying algorithm, with the resulting α_i always being a real number. For example, a common nonconformity measure in K nearest neighbours (KNN) is the ratio of distances to the nearest neighbours of the same class versus those of different classes.

The nonconformity measurement enables to assess how well a new sample x_n with a hypothetical label y conforms to observations already seen, from which p-values can be derived for each potential label $y \in Y$ of x_n. For a classification problem with every potential label $y \in Y$ for x_n, the nonconformity measurement is calculated and denoted as α_n^y, then the corresponding p-value p^y is defined as:

$$p^y = \frac{|\{i = 1, ..., n \mid \alpha_i^y \geq \alpha_n^y\}|}{n}.$$

The p-value, which lies between $1/n$ and 1, indicates the fraction of the samples in the set containing both observed data and the new sample as nonconforming as $z_n = (x_n, y)$. For a potential label y for x_n, if the corresponding p-value p^y is small, then y is unlikely to be the true label of x_n. If it is large, then we can be highly confident that this label is the true label.

The significance level ε indicates the required confidence level for constructing the prediction set, defined as:

$$\Gamma^\varepsilon \left((z_1, z_2, ..., z_{n-1}), x_n\right) = \{y \mid p^y > \varepsilon\}.$$

The prediction set construction process can be summarised as follows: for each candidate label y, after incorporating (x_n, y) into the previously observed data Z to form a new set, if the proportion of examples that conform worse or the same as (x_n, y) (measured by the p-value) exceeds the predefined significance level ε, then y is included in the prediction set. In other words, for each label y in the prediction set, we are highly confident that the p-value of (x_n, y) is greater than ε, or the proportion of examples that conform equally poorly or worse than (x_n, y) will exceed ε. Moreover, the confidence level $1 - \varepsilon$ provides a quantitative reference for how confident we can be.

3 Theoretical Advances in Conformal Prediction Research

3.1 Research Inspired by Conformal Prediction's Properties

Conformal Prediction is a statistically rigorous framework that combines the theoretical guarantees of statistical theory with the flexibility of machine learning. Its core strength lies in providing mathematically sound uncertainty quantification while remaining model-agnostic, making it a powerful and reliable foundation for trustworthy AI systems. When integrating with neural network models, CP further enhances predictive reliability [55]. The unique theoretical advantages of CP not only distinguish it from other machine learning algorithms, but have also inspired extensive research aimed at enhancing its performance and broadening its range of applications.

Strict Mathematical Guarantees. CP is grounded in rigorous mathematical principles, providing non-asymptotic and distribution-free statistical guarantees. These guarantees hold regardless of the underlying data distribution or model structure, relying only on the relatively mild assumption of data exchangeability. This property ensures the framework makes no assumptions about the data distribution form, remains agnostic to model architecture, and maintains compatibility with any prediction model. Any algorithm improved upon CP that maintains its rigorous statistical guarantees can thereby establish its mathematical soundness at its core.

Wang et al. [69] conducted a theoretical analysis of the asymptotic properties of a CP variant—locally weighted jackknife prediction (LW-JP). LW-JP incorporates the nonconformity scores into the jackknife prediction [24], which is designed for regression problems with heteroscedastic errors. Unlike traditional conformal prediction, jackknife prediction does not rely on the possible y of the new example x_n; instead, the nonconformity measurements of training examples are computed solely based on the training set Z itself. Specifically, the nonconformity measurement of jackknife prediction is calculated as follows:

$$\alpha_i = A\left(\{z_1, ..., z_{i-1}, z_{i+1}, ..., z_{n-1}\}, z_i\right), \quad i = 1, ..., n - 1.$$

Thus, the nonconformity measurements of observed data can be calculated offline and reused for all candidate label y, making LW-JP more computationally efficient than the standard conformal prediction framework. LW-JP employs the absolute loss normalized by the square root of the estimated variance as its nonconformity measure, in contrast to the simple absolute loss used in conventional jackknife prediction.

The study provides a rigorous theoretical analysis of LW-JP under asymptotic conditions where the number of training samples approaches infinity. Under appropriate regularity assumptions, the asymptotic validity of LW-JP was proven for nonlinear regression scenarios with heteroscedastic errors. Building on this theoretical foundation, the authors developed two novel conformal regressors based on LW-JP, which achieved state-of-the-art performance, demonstrating empirical validity and improved informational efficiency in experimental evaluations.

An Interval Estimation Under Certain Confidence. Compared to traditional machine learning algorithms, CP-based methods can produce confidence sets or intervals for predictions with guaranteed coverage probabilities. This makes CP naturally suited not only for probabilistic inference problems, but also enables transparent model performance evaluation that surpasses conventional point-estimate metrics like accuracy and mean square error (MSE). When configured to output single predictions, CP can additionally provide credibility and confidence measures, thereby enriching the evaluation metrics for individual point estimates. Credibility is the largest p-value that indicates how reliable the best choice of prediction y in the label space Y is, based on its conformity to the training observations. Confidence is $1 -$ the second largest p-value that indicates how reliable the best prediction y is, based on the exclusion of other possible choices of prediction.

Distribution regression is a specialized regression task where the inputs are probability distributions (theoretical or empirical) rather than conventional feature vectors [9, 24, 38, 57, 58]. Typical applications include medical diagnosis [47], where the inputs are derived from multiple measurements of a patient's physiological conditions, and meteorological forecasting [11], where the inputs come from multiple numerical weather prediction models. CP-based predictive models provide statistically valid predictions at specified confidence levels, making them both theoretically sound and empirically suitable for distribution regression tasks.

Wang et al. [68] pioneered the extension of CP to distribution prediction tasks and applied it to the post-processing of ensemble predictions. The input distributions are first embedded into a Hilbert space via kernel mean embedding [8, 39], then processed by a conformal regressor to obtain the final output. Random Fourier features [49] are employed to approximate the kernel for computational efficiency. The algorithm was tested on synthetic datasets and subsequently applied to statistical post-processing of temperature and precipitation ensemble forecasts. The results of validity test and efficiency comparison shown in Fig. 2 confirm that the proposed algorithm, denoted as Algorithm 3, outperforms other algorithms and serves as a powerful interval predictor for distribution regression problems.

CP's unique capability to quantify uncertainty while guaranteeing prediction set coverage enables applications in optimizing machine learning and neural network architectures [99]. Zhao et al. [100] applied Inductive Conformal Prediction (ICP) to guide neural network pruning, specifically performing filter-level pruning for Convolutional Neural Networks (CNNs). Their approach evolved from initially estimating each filter's contribution to CP prediction efficiency in a coarse manner and removing the least contributive ones, to incorporating computational efficiency through Taylor expansion-based filter contribution

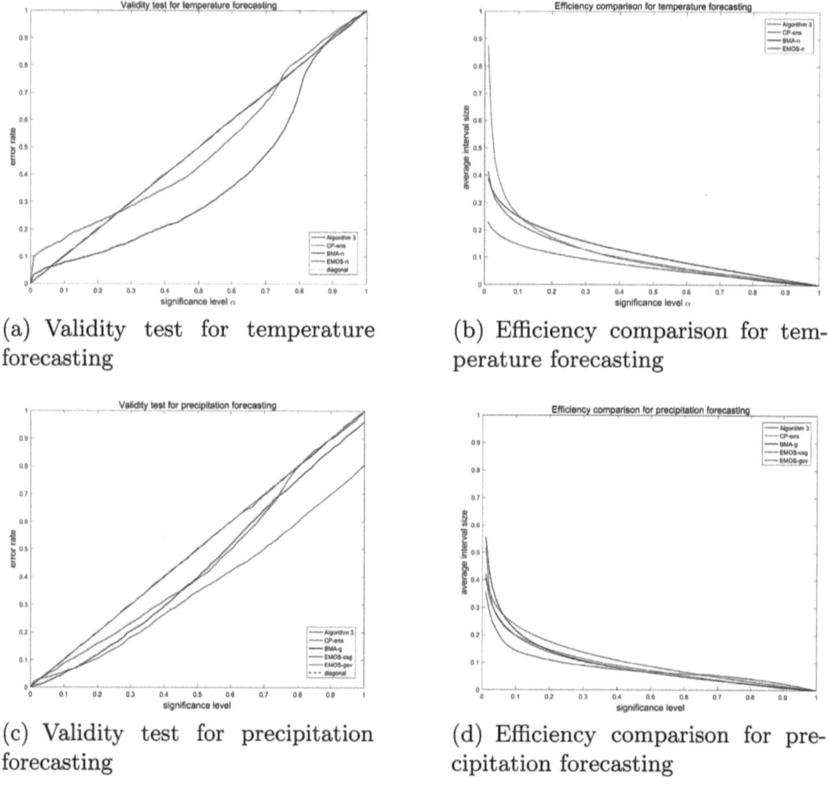

(a) Validity test for temperature forecasting

(b) Efficiency comparison for temperature forecasting

(c) Validity test for precipitation forecasting

(d) Efficiency comparison for precipitation forecasting

Fig. 2. Overall results of the validity test and efficiency comparison [68] for the post-processing of ensemble forecasts for temperature and precipitation. The proposed algorithm is denoted as Algorithm 3. CP-ens refers to a conformal regressor which is built directly on the ensemble forecasts. BMA refers to Bayesian Model Averaging. EMOS refers to Ensemble Model Output Statistics. The suffixes -n, -g, -csg, and -gev correspond to the normal, gamma, censored shifted gamma, and censored generalized extreme value distribution, respectively

estimation [37]. Further improvements were achieved by protecting the most critical filters from pruning to enhance overall pruning performance. Extensive experiments evaluated the trade-offs among prediction efficiency, computational efficiency, and network sparsity on three image datasets (SVHN, CIFAR-10, and CIFAR-100) using two network architectures (VGG16 and ResNet18). The results show that predictive efficiency does not degrade significantly, even when up to 60% of filters are pruned, although more computation time is required compared to the magnitude-based method. Empirical coverage across all experiments is approximately 99.0%, demonstrating the validity property of CP.

Extension to Online Data Augmentation. CP can quantify the nonconformity between newly observed samples and existing data. This property enables its application to online prediction tasks such as data augmentation, where high-reliability unlabeled samples are selectively added to the dataset.

Liu et al. [26] proposed an ensemble ICP-based online learning method (EICP) for data augmentation to address the labour-intensive and time-consuming challenge [90] of annotating training data in electronic nose-based traditional Chinese medicine recognition tasks. Under experimental conditions with varying training set sizes and Gaussian noise introduced to simulate sensor drift, the proposed method was compared against four alternative data augmentation strategies including standard ICP. The results, illustrated in Fig. 3, provided a systematic analysis of different augmentation approaches, demonstrating that the novel EICP method is both effective and robust in enhancing the generalization capability of classification models.

3.2   Dilemmas in the Application of Conformal Prediction

The limitations of CP fundamentally reflect a trade-off between statistical rigor and real-world complexity. While CP relies on rigorous mathematical proofs to ensure theoretical consistency, its computational process can be highly demanding. Moreover, real-world data often contain complex dependencies and noise, which can reduce efficiency and increase risks in traditional CP implementations.

Computational Inefficiency. A key limitation of existing CP-based regression methods is their computational inefficiency, despite CP's theoretical and practical effectiveness. Traditional CP employs a transductive framework, which requires computations involving all previous samples for each new prediction, resulting in substantial computational costs. Inductive CP was specifically developed to address this limitation. Additionally, optimizing the computation of nonconformity measures presents another avenue for improving CP's efficiency.

Disadvantages of the Classic CPKNN. The performance of CP is closely tied to the machine learning algorithm it incorporates. For instance, the widely-used CPKNN [46] method combines KNN's local adaptability with CP's rigorous statistical guarantees, demonstrating strong performance in nonparametric prediction tasks. The CPKNN method leverages KNN's flexibility in distance metric selection (e.g., Euclidean, Mahalanobis, or domain-specific measures like structural similarity in medical imaging) to compute nonconformity measures, while KNN's training-free properties naturally align with CP's transductive framework - particularly advantageous for online prediction scenarios. However, the KNN algorithm performs poorly with high-dimensional data and also suffers from computational inefficiency, as processing new samples requires calculations against all existing instances.

The Curse of Dimensionality. CP relies on measuring nonconformity, which becomes increasingly challenging when handling high-dimensional data such as

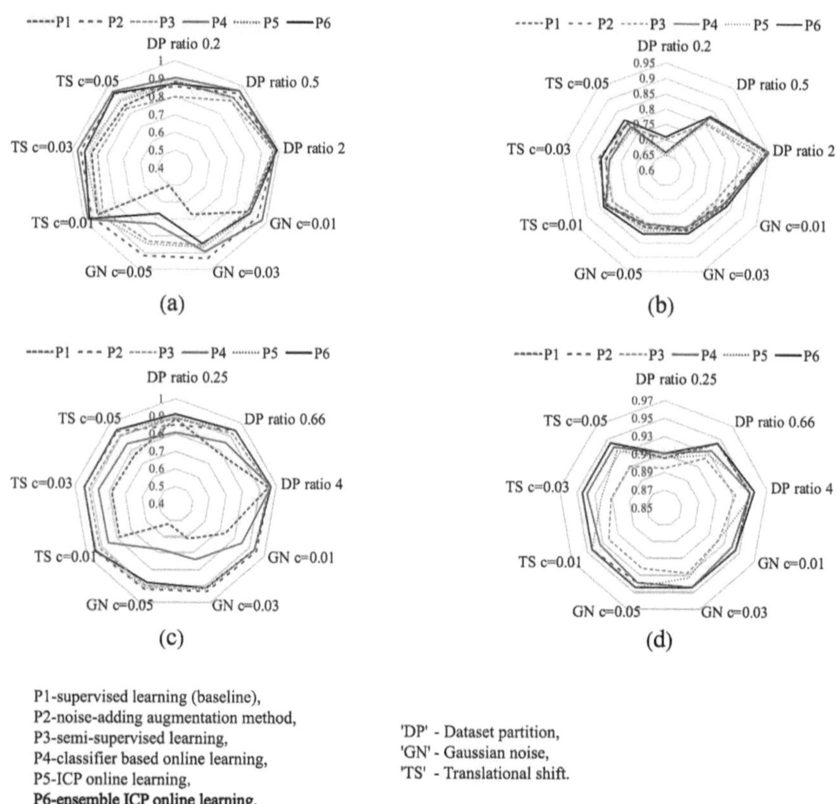

Fig. 3. Experiment results of EICP [26]. Summary of effect on classification accuracy across six processes in nine tasks on two datasets. Median classification accuracy of six processes on dataset 1 with LDA (a), dataset 1 with SVM (b), dataset 2 with LDA (c), dataset 2 with SVM (d)

images or text. In such spaces, traditional nonconformity measures struggle to reliably assess similarity between data vectors. As dimensionality increases, data points become sparse in the high-dimensional space, necessitating larger prediction sets to maintain valid coverage guarantees. This sparsity not only reduces predictive accuracy but also significantly increases memory usage and computational cost, further limiting the efficiency of CP in high-dimensional settings.

Inefficiency in Prediction Sets. CP-generated prediction sets may include excessive low-probability or non-informative candidates to ensure coverage of the true label, which can reduce decision-making efficiency. This reflects CP's inherent conservatism, rooted in the trade-off between statistical rigor and practical utility, and poses challenges for high-precision, high-risk applications such as autonomous driving and medical diagnosis. The redundancy is evident in classification tasks where prediction sets contain multiple low-probability labels, in regression tasks producing excessively wide prediction intervals that far exceed actual fluctuations, and in multi-label tasks where irrelevant labels are included in the output sets.

The Mismatch Between Real Data and Theory. Conformal Prediction theoretically requires exchangeable data, where the ordering of observations does not affect their joint distribution. However, real-world data often violate this assumption, such as strongly autocorrelated time series or spatially/temporally varying distributions. CP's limitations on non-exchangeable data highlight the sensitivity of statistical methods to their foundational assumptions.

3.3   Research on Addressing Conformal Prediction Limitations

Improving Computational Efficiency. Wang et al. [67] proposed a novel conformal regressor based on a CP variant, local-weighted jackknife prediction (LW-JP), which significantly improves computational efficiency. LW-JP computes leave-one-out errors [56] on the training set and correspondingly adjusts its predictions. The study focuses on improving the computational efficiency of CP-integrated algorithms by employing regularized extreme learning machines (RELM) [16] to accelerate LW-JP's computation. RELM enables efficient calculation of the leave-one-out prediction process, which aligns perfectly with LW-JP's requirements, thereby substantially speeding up the entire learning framework. Experiment results show that LW-JP-RELM outperforms ICP-ANN, ICP-SVR, RFNN and RFVN in computational efficiency. In addition, the average results of the validity test are illustrated in Fig. 4. The curve is close to the dashed line, indicating that LW-JP-RELM is a valid predictor in the sense of average performance.

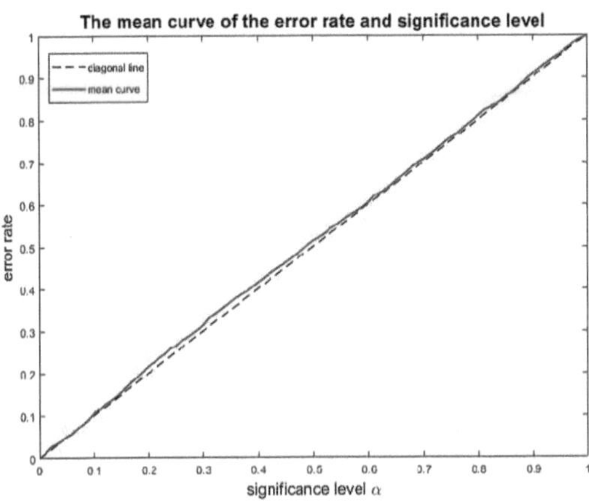

Fig. 4. Average results of the validity test of LW-JP-RELM [67]

Enhancing CP Performance via Integration with Advanced Machine Learning Algorithms. Liu et al. [27] improved the classical CP framework CPKNN by proposing CPSC, which incorporates shrunken centroids [60]. The conventional CPKNN method suffers from high bias and long computation time when processing high-dimensional data [15]. By replacing KNN with shrunken centroids, CPSC effectively regularizes class centroids to eliminate irrelevant features and reduce the sample space dimensionality, thereby yielding more reliable predictions. Both off-line prediction and online prediction with data augmentation are tested on an electronic nose dataset, with the online comparison results shown in Fig. 5.

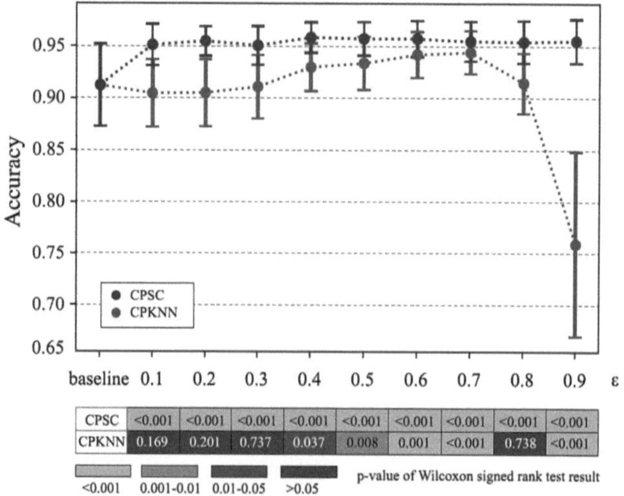

Fig. 5. Online prediction accuracy of CPKNN and CPSC [27] on the testing set with data augmentation under a series of ε settings. Accuracy is calculated using the LDA classifier after completion of the entire data augmentation process

Luo et al. [30] addressed the efficiency degradation of prediction sets caused by using validation sets for confidence calibration in CP classification tasks by proposing an entropy-based reweighted nonconformity measure. This approach simultaneously increases the probability of including the true label in prediction sets while reducing their sizes, thereby improving overall efficiency. Figure 6 shows the coverage-size plots for four datasets: AG News, CelebA Attributes (CARER), Fashion MNIST, and MNIST. The proposed ER score function maintains good coverage while yielding relatively small confidence set sizes.

Enhancing High-Dimensional Data Processing Capabilities. Qian et al. [48] proposed an ensemble CP method based on random projection [1,4] for accurate high-dimensional data classification. The study employs multiple random projection matrices to map high-dimensional data into lower-dimensional spaces and applies CP prediction separately to each low-dimensional representation, which simultaneously serves as a form of data augmentation. A voting strategy [51] is then used to aggregate the predicted classes and refine the prediction sets, achieving a balance between accuracy and prediction set efficiency.

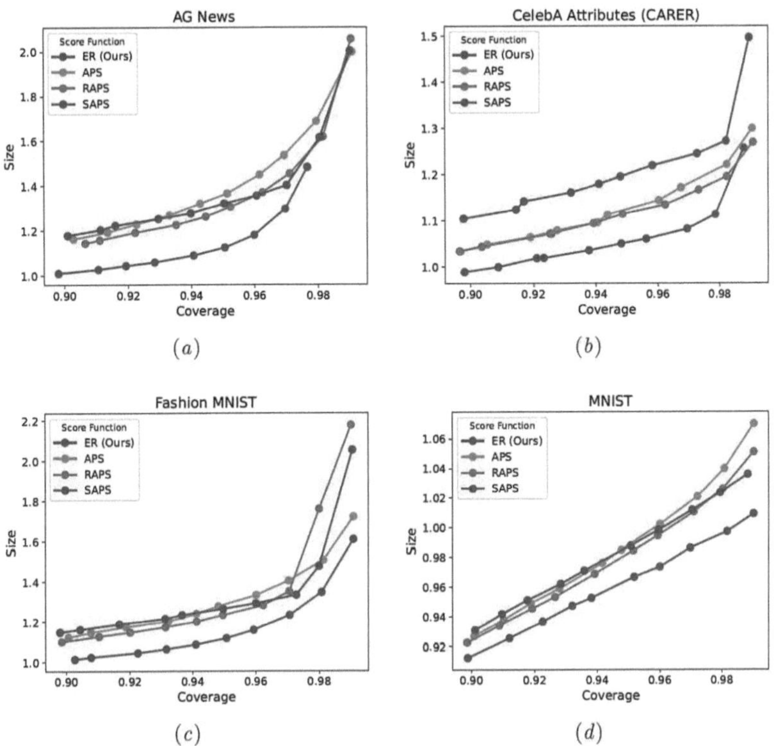

Fig. 6. Prediction set size vs. coverage plots for different datasets and score functions [30]. ER refers to the proposed Error Reweighted Conformal Prediction algorithm. APS, RAPS, and SAPS refer to Adaptive Prediction Sets, Regularized Adaptive Prediction Sets, and Sorted Adaptive Prediction Sets, respectively

Improving Prediction Efficiency by Loss Optimization. Wang et al. [66] recognized that traditional CP methods' prediction set construction only controls the coverage loss, making it difficult to constrain redundancy in prediction sets. This implies that regulating the prediction loss could potentially optimize prediction set efficiency to some extent. Moreover, in practical applications such as disease classification using MRI images [6], the specific label categories influence the associated loss, which can be viewed as a class-varying loss. Another example is tumor image segmentation [3], where controlling the false negative rate is of greater concern than merely ensuring prediction set coverage of tumor pixels. Based on this insight, the authors proposed Conformal Loss-Controlling

Prediction (CLCP), extending CP from coverage loss to generic loss control while providing finite-sample guarantees under the exchangeability assumption. By optimizing loss control, the method achieves more efficient prediction sets. The loss-controlling guarantee is empirically verified in classification with a class-varying loss and weather forecasting. Figure 7 shows the results for maximum and minimum temperature forecasting. The preset parameters α and the significance level ε ensure that the prediction loss does not exceed α with confidence $1 - \varepsilon$.

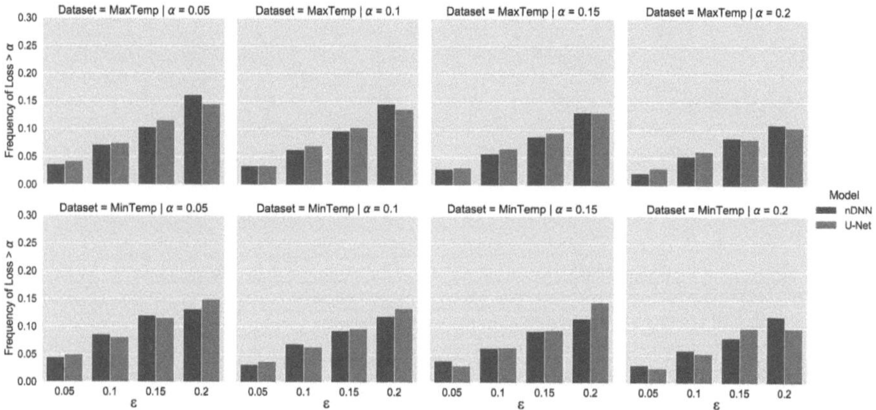

Fig. 7. Bar plots showing the frequencies of prediction losses exceeding the predefined α for $\varepsilon = 0.05, 0.1, 0.15$, and 0.2 on test data for maximum and minimum temperature forecasting. nDNN refers to the naive deep neural network. All bars are at or below the preset ε, empirically confirming the loss-controlling guarantee of CLCP

4   Applications

4.1   Biochemistry and Biomedical

The use of Artificial Intelligence (AI), including machine learning (ML), in biochemistry and biomedical fields has attracted significant attention [63]. In recent years, research in areas such as disease diagnosis, especially cancer detection, and herbal medicine classification has increasingly focused on the reliability and credibility of predictions, rather than settling for approximate results. Conformal Prediction (CP) can achieve reliable predictions by providing the credibility (i.e., uncertainty) for each instance, making it a reliable and compatible machine learning approach for applications in biomedical and biochemical fields [28].

Biomedical. In the biomedical field, CP has shown great potential in areas such as physiological condition detection, cancer detection, and disease diagnosis [33, 61], see Table 1.

Table 1. CP Applications in Biomedical in China

Application	Method	Performance
Blood Pressure Estimation	CP & DER (Deep Evidential Regression) [53]	MAD (Mean Absolute Deviation) of SBP (Systolic Blood Pressure) and DBP (Diastolic Blood Pressure) are 5.56 and 3.18 mmHg, with coverage rates of 94.8% and 95.9%. Uncertainty intervals improve reliability for hypertension diagnosis. Patented method [62].
Cancer Detection	CP & KNN [92]	Accuracies of 87.5% (1NN) and 83.33% (3NN), outperforming traditional KNN. Provides confidence and credibility for individual lung cancer predictions.
Disease Diagnosis	Mondrian CP & GBM (Gradient Boosting Machines) [88]	Predicts sepsis mortality risk; improves the credibility and interpretability of predictions; supports clinical decision-making.
Disease Diagnosis	CP & GBM, NDF (Neural Decision Forest), RF, LR (Logistic Regression) [89]	Applies CP to uncertainty estimation in patient selection for sepsis clinical trials; enables customizable confidence levels for predictions.
COVID-19 Risk Stratification	CP-based tri-light warning system using CP & SC (Soft Confidence), LGBM (Light Gradient Boosting Machine), ANN (Artificial Neural Network) [86]	Helps optimize medical resources and improve triage by stratifying hospitalized COVID-19 patients based on risk.
COVID-19 Recovery Time Prediction	CP & XGBoost (Extreme Gradient Boosting) [72]	Provides tight interval predictions for negative conversion time; MAE (Mean Absolute Error) of 3.54 days. The result is shown in Fig. 8.
Depression Diagnosis	CDP (Conformal Depression Prediction) [25]	Provides reliable, uncertainty-aware facial depression predictions with marginal coverage guarantees at any given miscoverage rate; does not require model retraining or assumptions on data distribution.

In physiological condition detection, Ding's team at the University of Electronic Science and Technology of China applied CP to cuffless blood pressure measurement by combining epistemic uncertainty with CP to generate statistically rigorous uncertainty intervals, thereby improving the reliability and accuracy of blood pressure detection [53]. Subsequently, they designed a CP-based blood pressure measurement system to enhance the reliability of non-invasive arterial blood pressure measurements in clinical settings [62].

In cancer detection, early diagnosis of lung cancer can greatly reduce its mortality. However, diagnostic misinterpretation and erroneous predictions of lung cancer can cause serious psychological and financial burden on patients and their families. Thus, the reliability and credibility of disease diagnosis are crucial. Li's team at Zhejiang University used a gas sensor array to detect the

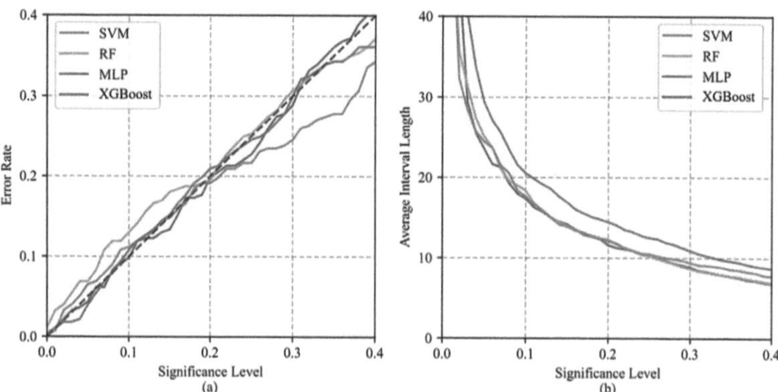

Fig. 8. Distribution of error rates and average interval lengths of the models at different significance levels [72]. (a) Curves showing the relationship between error rate and significance level. (b) Curves showing the relationship between average interval size and significance level (MLP: multilayer perceptron)

exhaled gases of lung cancer patients and healthy individuals and then applied the CP algorithm based on KNN to predict and classify lung cancer [92]. For each case, this method not only provided a diagnostic result but also quantified the confidence and credibility of the diagnosis, significantly improving the reliability of the diagnostic result and aiding doctors in making optimal decisions.

Electronic Nose. The electronic nose, also known as an artificial olfactory system, is a novel bio-inspired detection instrument designed to mimic human olfactory processes. It offers the advantages such as fast and stable performance, low cost, and non-destructive testing. With the rapid development of computational science, it has become a hot research topic to use statistics and machine learning methods to extract relevant information from the signals of electronic noses. Moreover, evaluating the credibility of the predictions is essential when making predictions based on electronic nose data, as it reflects the reliability of the prediction.

The electronic nose can be used in classification of aroma pattern. For tea classification, the primary standards are aroma, color, and taste, with aroma considered the most important. To quantify this phenomenon, Nouretdinov et al. designed a gas sensor-based electronic nose system to distinguish between four different types of tea [41]. To improve the reliability of multi-class tea classification, they proposed a new non-conformity measure for the implementation of conformal predictors based on Support Vector Machine (SVM). Empirical results showed the good performance of the implemented conformal predictor.

Table 2. CP Applications in Electronic Nose in China

Application	Method	Performance
Application	Method	Performance
Electronic Nose	CP-based SVM [41]	CP-SVM with a new non-conformity measure for multi-class classification; overall forced prediction accuracy: 89.7%; error rate controlled by selecting a suitable confidence level.
Electronic Nose	CP-based KNN [77]	CP-KNN for ginseng classification; accuracy: 84.44% (1NN), 80.63% (3NN); the result shown in Fig. 9.
Electronic Nose	CP-based KNN [91]	CP-KNN for classification of 12 herbal medicine categories; accuracy: 91.50% (1NN), 92.17% (3NN).
Electronic Nose	Online CP-based KNN [90]	Online CP enhances model adaptability and improves classification accuracy over time for 12 types of herbal medicines.
Electronic Nose	Aggregated CP-based SVM, RF [75]	Aggregated CP framework for classifying 10 dendrobium species; accuracy close to 80%, with an average improvement of 6.2%.

In China, experienced traditional Chinese medicine practitioners often identify the type of medicinal herbs based on taste and appearance. Taste is the most difficult attribute to quantify, and the classification of Chinese medicinal herbs often requires extensive experience. Because different herbs have different medicinal properties, incorrect usage can worsen the condition and cause significant adverse effects. To address this, Li's team at Zhejiang University conducted a series of studies on reliable classification of Chinese herbal medicines using electronic noses. In study [36], they discussed the classification of ginseng samples using CP and Venn prediction, both of which provided reliability estimates for the predictions. Similarly, for ginseng sample classification, they proposed a CP predictor based on KNN in [77], which not only outperformed traditional KNN but also provided reliability estimates for each prediction. Subsequently, they expanded the classification and recognition of herbal medicines. In [91] and [90], they proposed both online CP and CP-based KNN classification methods to classify 12 types of visually similar herbal medicines using the electronic nose. Both methods achieved reliable and accurate classification predictions. In a further study on valuable herbs such as Dendrobium, they developed a method to distinguish between 10 different species using the electronic nose [75]. The method applied aggregated conformal prediction and achieved nearly 80% classification accuracy while providing reliability assessments for each prediction. The CP applications on electronic nose in China are summarised in Table 2.

Bioelectronics. Neurophysiological signals, such as electroencephalogram (EEG) and electromyography (EMG), are important indicators of human physiological states and can provide insight into human cognition and perception.

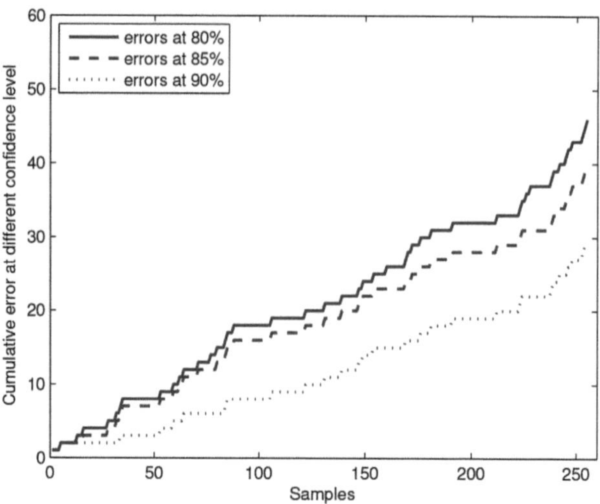

Fig. 9. Experimental results of discrimination of Ginseng [77]

These signals have significant implications for clinical diagnosis and neuroengineering. CP provides theoretical support for quantifying uncertainty in signal analysis, showing significant value in EEG and EMG applications [2] (Table 3).

Table 3. CP Applications in Bioelectronics in China

Application	Method	Performance
EEG	KIII model [93]	TCM (Transductive Confidence Machine) applied with the KIII model for recognition of brain hypoxia EEG signals; controls and assesses of risk associated with each prediction.
EMG	ICP & Bi-LSTM & RF [76,94]	Improves the reliability of decoding speech information from EMG; provides confidence regions for individual predictions with a guaranteed error rate; achieves higher accuracy (0.87) than baseline methods.
EMG	CPSC & DRCA [70]	Improves the reliability of taste sensation decoding; CPSC with DRCA significantly improves classification accuracy ($p < 0.05$) across six subjects; enhances reliability in high-dimensional sensor data while reducing computation time.

In EEG signal analysis, to improve the credibility of predictions related to brain hypoxia, Zhang et al. combined the Transductive Confidence Machine algorithm with the KIII model. They proposed a more reliable method for recognizing hypoxia EEG signals with a preset confidence level [93]. This method improved reliable prediction with confidence measures in classifying normal versus hypoxia EEGs, while also allowing for the control and assessment of risk associated with each prediction.

In the EMG domain, because muscle movement is closely linked to neural signals, it is possible to decode human behavior from neuromuscular activity. However, the decoding process is uncertain, making it essential to quantify the confidence of predictions. For example, to decode silent speech from EMG, Wang and Zhang et al. combined Inductive Conformal Prediction (ICP) with machine learning methods such as Random Forest (RF) and Bidirectional Long Short-Term Memory (Bi-LSTM) [76,94]. Their approach not only improved the performance of decoding speech information from EMG but also provided confidence regions for individual predictions with a guaranteed error rate, thereby significantly improving the reliability of the decoding process.

Furthermore, EMG signals can also be used to decode human sensations. Interpreting these sensations is particularly challenging due to the high levels of noise in EMG signals and the significant individual differences in signal strength. These factors greatly affect the reliability and accuracy of predictions, as well as the generalizability and transferability of the model. For example, human taste sensation can be qualitatively described using surface electromyography (sEMG), and improving both reliability and cross-user classification performance is crucial for effective recognition. To address this, Wang et al. proposed a method that combines Domain Regularized Component Analysis (DRCA) and Conformal Prediction with Shrunken Centroids (CPSC) [70]. Their approach aims to improve the cross-user classification performance through the following steps:

1. Build a model using training data from the source domain;
2. Predict the labels of active data from the target domain;
3. Quantify the reliability of the predictions in terms of credibility and confidence;
4. Filter predictions with high credibility;
5. Augment the training data with the selected high-confidence samples.

Ultimately, their method achieved high classification accuracy in distinguishing among the six basic taste sensations. It also effectively addressed the issue of cross-user data distribution drift and improved prediction reliability.

4.2   Cyber Security

CP has made significant progress in recent years in the field of cyber security, offering new perspectives and methods to address the complex and ever-changing nature of network threats. Recent studies have explored its applications in areas such as 6G communication systems [14,40], Android malware detection [10,45], and botnet identification [7,22]. CP has also demonstrated unique advantages in tasks such as server decision-making, intrusion detection, and anomaly behavior monitoring [5]. Table 4 summarises its applications in China.

Table 4. CP Applications in Cyber security in China

Application	Method	Performance
Wireless Federated Learning	WFCP [103]	Achieves small and informative prediction sets under noisy wireless channels, enhances reliability and communication efficiency.
Intrusion Detection	Enhanced SVM [17,21]; Multi-model CP [79]	Improves detection accuracy and provides confidence and reliability estimates; patented method [20]; Multi-model fusion for malicious code enhances robustness and prediction reliability.
Unknown Threat Intelligence	LSTM /XGBoost [97]	Maintains error rate below 2.5% and achieves F1 score > 90% for DGA (Domain Generation Algorithm) domain detection, supports secure and reliable threat propagation; patented method [78].
Log Anomaly Detection	Statistical learning method with CP [13,50]; Multi-confidence-guided anomaly detection (Multi-CAD) [84]	Achieves dynamic adaptation to changing logs and improves F1 score by 1.6–1.9% points; Multi-CAD model reaches 98.2% accuracy using feedback from CP-based non-conformity scores.
Network Traffic Prediction	MetaSTNet [31]	Provides calibrated prediction intervals with high accuracy, ensures reliable performance on real-world traffic datasets; patented method [87].

One notable study, Wireless Federated Conformal Prediction (WFCP), proposed by Zhu et al. [103], aims to enhance the reliability and efficiency of joint data processing between devices and servers over wireless channels using Conformal Prediction. In this framework, devices communicate statistical information about their local data to the server, enabling it to make more reliable predictions and ensure that each prediction falls within a predefined confidence range, even under noisy communication conditions.

Acquiring and utilizing unknown network threat intelligence remains one of the key challenges in the field of cyber security. Wang et al. [97] proposed a network threat intelligence propagation method based on CP, incorporating confidence and reliability assessment parameters into the LSTM and XGBoost algorithms. This integration allows the model to quantify prediction reliability, while controlling the error probability in propagation of unknown intelligence.

More recently, a deep learning-based model called MetaSTNet transfers the meta-knowledge to real-world environments to achieve accurate predictions [31]. The model is further deployed with Cross Conformal Prediction (CCP) to evaluate the calibrated prediction intervals, providing reliability guarantees for the prediction results as illustrated in Fig. 10. These findings highlight the advantages of CP methods in addressing network traffic prediction problems.

4.3   Industries and Environment

In industrial and environmental domains, the application of CP has been gradually gaining popularity. Table 5 summarises some of its applications in China.

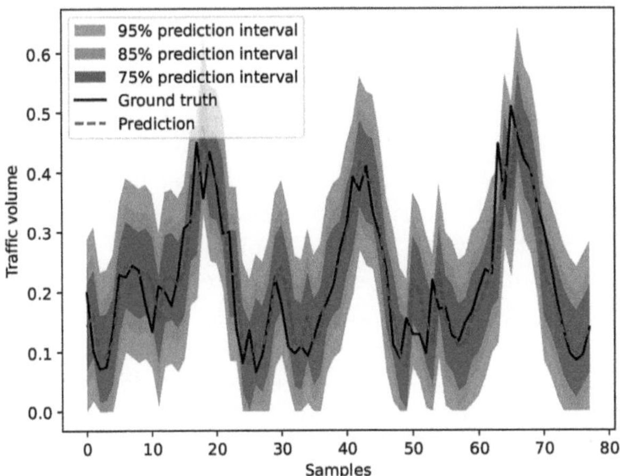

Fig. 10. The experimental results of interval prediction on the MetaSTNet [31]

Table 5. CP Applications in Industries and Environment in China

Application	Method	Performance
Mining (Iron-making)	RVFLN (Random Vector Functional Link Network) [101]; multi-output CP [96]	Faster and more accurate molten iron quality prediction in the blast furnace iron-making process; Improved efficiency of the prediction region under the guarantee of coverage.
Coal Classification	SVM [74]	Classification accuracy > 98.45%, error rate can be adjusted by confidence levels.
Industrial Time Series	Uncertainty quantification framework with CP [95]	Reduced prediction interval width (6.29%, 15.70%) while maintaining 95% coverage.
Distribution Network Topology	CNN [104]	High accuracy, especially when output size is small, with high coverage.
Wind Speed Prediction	CP-Based Support Vector Regression [18]	Improved accuracy and reliability compared to direct prediction methods; supports early warnings to reduce potential downtime and maintenance costs.
Weather Forecasting using Radar Echo	Unet [19]	Enhanced accuracy and reliability of storm forecasts; shows advantages in both point prediction and interval prediction for radar echo extrapolation.
Storm Prediction	CP-based framework [35, 71]	Comparable accuracy with operational benchmarks; better reliability via strict uncertainty intervals. Achieved RMSE (Root Mean Square Error) of 7.86 knots, outperforming other comparison algorithms.

CP can be used to optimize production processes by predicting variations in production parameters, thereby improving efficiency and product quality. Wang et al. [74] applied CP to near-infrared analysis for rapid coal classification. By combining CP with SVM through the CV-SVM method, the study achieved fast and accurate coal type identification. Furthermore, by adjusting the confidence level, the model produced results with varying error rates, playing a key role in quality control.

Zhou et al. [95] proposed a novel; uncertainty quantification framework based on CP for industrial time series analysis. The framework integrates global and local nonconformity measure information within the CP structure to dynamically adjust the confidence levels of one-sided prediction intervals. Experiments on wastewater treatment and sintering production datasets show that this method significantly reduced the width of prediction intervals while maintaining the same coverage rate. Building on this work, the team [102] proposed a patent for nonferrous metal smelting, particularly targeting the prediction of alkali liquor concentration.

CP is also gaining traction in the earth sciences [54], particularly in early warning systems for wildfires [85] and floods [32]. For tropical cyclone path prediction, Meng et al. [35] developed a CP-based system using hurricane data from 1975âĂŞ2021 for path forecasting. While maintaining competitive accuracy, the system demonstrated superior reliability in its uncertainty estimates. Similarly Wang's team [71] utilised satellite infrared imagery and CP for cyclone intensity estimation, achieving a lower RMSE of 7.86 knots, outperforming other algorithms. These results illustrate CP's strong potential in forecasting both cyclone paths and intensity.

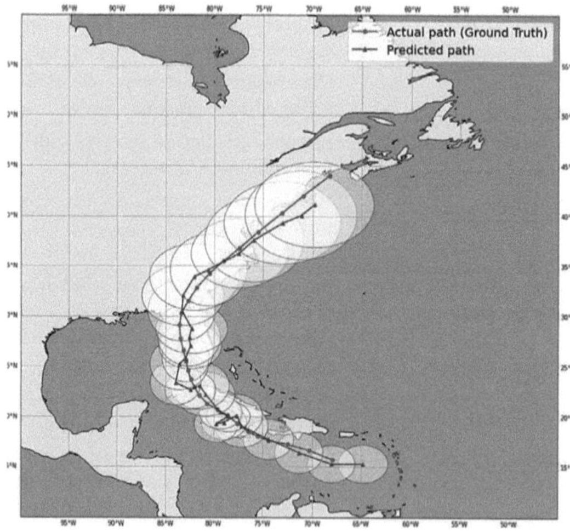

Fig. 11. The hurricane forecast results [71]: red line indicates the ground truth and green line represents the prediction path

4.4 Other Applications and the Lessons

CP has also shown significant potential across a wide range of fields, offering strong support for uncertainty quantification and decision-making in diverse applications, see Table 6.

Table 6. Other CP Applications in China

Application	Method	Performance
Transportation Planning	CQR-GAE (Conformal Quantile Regression Graph Autoencoder) [59]	Improves predictive confidence with strict coverage guarantees; outperforms baseline methods in real-world traffic flow scenarios; enhances robustness of route planning.
Behaviour Recognition	CP-based framework [80–82]	Enables effective set-based predictions for group behaviour forecasting; detects distribution shifts online even in complex environments and with limited sample conditions.
Aerospace	CP-based distribution-free uncertainty quantification method [83]	Provides finite-sample guarantees for trajectory classification and improves trajectory classification accuracy; delivers reliable prediction sets with specified confidence levels.
Finance	OSL-SCN (Online Self-learning Stochastic Configuration Network) [73]	Enhances forecast accuracy and reliability in Bitcoin price prediction; produces confidence intervals for robust financial modeling.

The cross-domain applications of Conformal Prediction (CP) demonstrate its growing importance as a reliable and robust framework for uncertainty quantification. First, CP provides strict mathematical guarantees: it ensures validity under minimal assumptions, promises high accuracy, and maintains that the error rate of predictions does not exceed a user-defined significance level. This is particularly critical in applications such as medical diagnosis, finance, and weather forecasting, where trust in the model is essential. Second, CP's ability to generate interval estimates with adjustable confidence levels enhances reliability over point predictions. It is widely applied in areas such as wind speed prediction, intrusion detection, and industrial process control. Third, CP can adapt to online, streaming, or dynamically changing data, improving system robustness. By measuring the nonconformity of new data relative to historical observations, CP allows real-time uncertainty estimation. This capability is especially valuable in behavior recognition, anomaly detection, and data augmentation for high-dimensional sensor environments. Together, these advantages highlight CP's strong theoretical foundation and practical effectiveness across diverse application domains.

5   Conclusion

This chapter briefly reviewed theoretical advances and practical applications of conformal prediction in China, highlighting key contributions from Chinese research communities.

With the rapid advancement of machine learning algorithms and computing systems, an increasing number of models have been developed to tackle complex tasks. However, the pursuit of higher accuracy has become increasingly challenging due to the growing complexity and over-parameterization of models, which often result in overfitting, reduced interpretability, and weakened generalization performance. Striking a balance between the learnability and reliability has emerged as a critical concern. Although newly developed computing systems have improved the computational process for the learning and reasoning processing of a complex system, they often fail to provide meaningful explanations for model behaviour. This lack of interpretability introduces potential reliability risks and complicates model adaptation to new environments and data.

Conformal Prediction (CP) [65], pioneered by Vovk, Gammerman and Shafer, offers a learnable framework that enhances the reliability of machine learning models. By producing valid and calibrated confidence measures, CP enables models to quantify and communicate uncertainty effectively—an essential feature for robust and trustworthy learning systems. To address computational challenges, several efficient CP variants have been developed, including adaptive CP, local-weighted jackknife prediction (LW-JP) and Loss-Controlling Prediction (CLCP). These innovations have made CP increasingly practical for large-scale and real-time applications.

The availability of well-calibrated confidence information helps learning algorithms adapt to better handle distributional shifts, manage ambiguous inputs, and make more informed updates. This ultimately improves both predictive performance and training efficiency. For this reason, CP has been successfully applied across a wide range of fields, including biomedicine, cybersecurity, industrial control, environmental science and artificial sensation - such as in tea and traditional Chinese medicine classification. Notably, many of these applications are driven by contributions from Chinese researchers. Collectively, these successes show that the CP's significance arises from both the reliability and learnability of the model predictions.

Acknowledgments. Professor Alexander Gammerman visited Professor Guang Li and his group at Zhejiang University, China, in 2010 and delivered a series of seminars and lectures on Conformal Prediction and its applications. His visit profoundly inspired much of our research in this field. We are delighted to dedicate this chapter to celebrating Professor Gammerman on the occasion of his 80th birthday.

This study was funded by the State Key Laboratory of Industrial Control Technology, China (Grant No.ICT2024A21).

References

1. Achlioptas, D.: Database-friendly random projections: Johnson-lindenstrauss with binary coins. J. Comput. Syst. Sci. **66**(4), 671–687 (2003)
2. Alvarsson, J., McShane, S.A., Norinder, U., Spjuth, O.: Predicting with confidence: using conformal prediction in drug discovery. J. Pharm. Sci. **110**(1), 42–49 (2021)
3. Angelopoulos, A.N., Bates, S., Fisch, A., Lei, L., Schuster, T.: Conformal risk control. In: The Twelfth International Conference on Learning Representations (Oct 2023)
4. Arriaga, R.I., Vempala, S.: An algorithmic theory of learning: robust concepts and random projection. Mach. Learn. **63**(2), 161–182 (2006)
5. Balasubramanian, V., Ho, S.S., Vovk, V.: Conformal Prediction for Reliable Machine Learning: Theory. Adaptations and Applications. Morgan Kaufmann Publishers Inc., San Francisco, CA, USA (2014)
6. Bates, S., Angelopoulos, A., Lei, L., Malik, J., Jordan, M.: Distribution-free, risk-controlling prediction sets. J. ACM **68**(6), 43:1–43:34 (Sep 2021)
7. Cherubin, G., et al.: Conformal clustering and its application to botnet traffic. In: Gammerman, A., Vovk, V., Papadopoulos, H. (eds.) Statistical Learning and Data Sciences, pp. 313–322. Springer International Publishing, Cham (2015)
8. Dong, X.M., Gu, Y.H., Shi, J., Xiang, K.: Random multi-scale kernel-based Bayesian distribution regression learning. Knowl.-Based Syst. **201–202**, 106073 (2020)
9. Fang, Z., Guo, Z.C., Zhou, D.X.: Optimal learning rates for distribution regression. J. Complex. **56**, 101426 (2020)
10. Georgiou, N., Konstantinidis, A., Papadopoulos, H.: Malware detection with confidence guarantees on android devices. In: Iliadis, L., Maglogiannis, I. (eds.) Artificial Intelligence Applications and Innovations, pp. 407–418. Springer International Publishing, Cham (2016)
11. Gneiting, T., Katzfuss, M.: Probabilistic forecasting. Ann. Rev. Stat. Appl. **1**(Volume 1, 2014), 125–151 (Jan 2014)
12. Gómez-Verdejo, V., Parrado-Hernández, E., Martínez-Ramón, M.: Adaptive sparse gaussian process. IEEE Trans. Neural Netw. Learn. Syst. **35**(11), 16383–16395 (2024)
13. Gu, Z., Ren, Y., Liu, C., Wang, Z.: Intranet log anomaly detection model based on conformal prediction. Netinfo Security **20**(3), 45 (2020)
14. Guo, H., Akyildiz, I.F.: Task-oriented mulsemedia communication using unified perceiver and conformal prediction in 6g wireless systems. IEEE Trans. Mobile Comput. 1–15 (2025)
15. Hastie, T., Tibshirani, R., Friedman, J.: The Elements of Statistical Learning. Springer Series in Statistics, Springer, New York, New York, NY (2009)
16. Huang, G.B., Zhou, H., Ding, X., Zhang, R.: Extreme learning machine for regression and multiclass classification. IEEE Trans. Syst., Man, Cybern., Part B (Cybernetics) **42**(2), 513–529 (Apr 2012)
17. Huo, G.: Research About Network Instrusion Detection Algorithm Based On Confidence Machine. Master's thesis, Beijing Jiaotong University (2008)
18. Ji, G.R., Dong, Z., Wang, D.F., Han, P., Xu, D.P.: Wind speed conformal prediction in wind farm based on algorithmic randomness theory. In: 2008 International Conference on Machine Learning and Cybernetics, vol. 1, pp. 131–135 (2008)

19. Jiao, X.: Research on Radar Echo Extrapolation Algorithm Based On Conformal Prediction. Master's thesis, Lanzhou University (2024)
20. Jin, H., Zhao, X.: High-reliability intrusion detection method based on conformal prediction (2021), patent, CN113947150A, 18 Jan., 2022
21. Jin, H., Zhao, X.: High-reliability intrusion detection algorithm under conformal prediction framework. Comput. Eng. **48**(7), 130–140 (2022)
22. Kiani, B.M.: A conformalized density-based clustering analysis of malicious traffic for botnet detection. In: Gammerman, A., Vovk, V., Luo, Z., Smirnov, E., Cherubin, G. (eds.) Proceedings of the Ninth Symposium on Conformal and Probabilistic Prediction and Applications. Proceedings of Machine Learning Research, vol. 128, pp. 244–256. PMLR (09–11 Sep 2020)
23. Koenker, R., Bassett, G.: Regression quantiles. Econometrica **46**(1), 33–50 (1978)
24. Lei, J., , Max, G., , Alessandro, R., , Ryan J., T., Wasserman, L.: Distribution-free predictive inference for regression. J. American Stat. Assoc. **113**(523), 1094–1111 (Jul 2018)
25. Li, Y., Qu, S., Zhou, X.: Conformal depression prediction. IEEE Trans. Affect. Comput. 1–11 (2025)
26. Liu, L., et al.: Boost ai power: data augmentation strategies with unlabeled data and conformal prediction, a case in alternative herbal medicine discrimination with electronic nose. IEEE Sens. J. **21**(20), 22995–23005 (2021)
27. Liu, L., et al.: CPSC: conformal prediction with shrunken centroids for efficient prediction reliability quantification and data augmentation, a case in alternative herbal medicine classification with electronic nose. IEEE Trans. Instrum. Meas. **71**, 1–11 (2022)
28. Lu, C., Lemay, A., Chang, K., Höbel, K., Kalpathy-Cramer, J.: Fair conformal predictors for applications in medical imaging. Proc. AAAI Conf. Artif. Intell. **36**(11), 12008–12016 (2022)
29. Lu, J., Ding, J., Liu, C., Chai, T.: Hierarchical-Bayesian-based sparse stochastic configuration networks for construction of prediction intervals. IEEE Trans. Neural Netw. Learn. Syst. **33**(8), 3560–3571 (2022)
30. Luo, R., Colombo, N.: Entropy reweighted conformal classification. In: Proceedings of the Thirteenth Symposium on Conformal and Probabilistic Prediction with Applications, pp. 264–276. PMLR (Sep 2024)
31. Ma, H., Yang, K.: Metastnet: multimodal meta-learning for cellular traffic conformal prediction. IEEE Trans. Netw. Sci. Eng. **11**(2), 1999–2011 (2024)
32. Manca, G., Kunze, F.C., Fay, A.: Addressing uncertainty in online alarm flood classification using conformal prediction. IEEE Access **12**, 165626–165652 (2024)
33. Meister, J.A.: Confident COVID-19 detection with Conformal Prediction. Ph.D. thesis, University of Brighton (2024)
34. Melluish, T., Saunders, C., Nouretdinov, I., Vovk, V.: Comparing the Bayes and typicalness frameworks. In: De Raedt, L., Flach, P. (eds.) Machine Learning: ECML 2001, pp. 360–371. Springer, Berlin, Heidelberg (2001)
35. Meng, F., Song, T.: Uncertainty forecasting system for tropical cyclone tracks based on conformal prediction. Expert Syst. Appl. **249**, 123743 (2024)
36. Miao, J.: Research on an artificial olfactory system based on information mining methods. Ph.D. thesis, Zhejiang University, China (2016)
37. Molchanov, P., Mallya, A., Tyree, S., Frosio, I., Kautz, J.: Importance estimation for neural network pruning. In: 2019 IEEE/CVF Conference on Computer Vision and Pattern Recognition (CVPR), pp. 11256–11264 (Jun 2019)

38. Muandet, K., Fukumizu, K., Dinuzzo, F., Schölkopf, B.: Learning from distributions via support measure machines. In: Proceedings of the 26th International Conference on Neural Information Processing Systems - Volume 1. NIPS'12, vol. 1, pp. 10–18. Curran Associates Inc., Red Hook, NY, USA (Dec 2012)
39. Muandet, K., Fukumizu, K., Sriperumbudur, B., Schölkopf, B.: Kernel mean embedding of distributions: A review and beyond. Foundations and Trends® in Machine Learning **10**(1-2), 1–141 (2017)
40. Nagaraja Hegde, D., R M, K., Chatterjee, U.: Reliable and efficient beam selection using conformal prediction in 6G systems. In: 2025 National Conference on Communications (NCC), pp. 1–6 (Mar 2025)
41. Nouretdinov, I., Li, G., Gammerman, A., Luo, Z.: Application of conformal predictors to tea classification based on electronic nose. In: Papadopoulos, H., Andreou, A.S., Bramer, M. (eds.) Artificial Intelligence Applications and Innovations, pp. 303–310. Springer, Berlin Heidelberg, Berlin, Heidelberg (2010)
42. Papadopoulos, H.: Inductive conformal prediction: Theory and application to neural networks. In: Fritzsche, P. (ed.) Tools in Artificial Intelligence. InTech (Aug 2008)
43. Papadopoulos, H.: Cross-conformal prediction with ridge regression. In: Gammerman, A., Vovk, V., Papadopoulos, H. (eds.) Statistical Learning and Data Sciences, pp. 260–270. Springer International Publishing, Cham (2015)
44. Papadopoulos, H.: Guaranteed coverage prediction intervals with gaussian process regression. IEEE Trans. Pattern Anal. Mach. Intell. **46**(12), 9072–9083 (2024)
45. Papadopoulos, H., Georgiou, N., Eliades, C., Konstantinidis, A.: Android malware detection with unbiased confidence guarantees. Neurocomputing **280**, 3–12 (2018)
46. Papadopoulos, H., Vovk, V., Gammerman, A.: Regression conformal prediction with nearest neighbours. J. Artif. Int. Res. **40**(1), 815–840 (2011)
47. Poczos, B., Singh, A., Rinaldo, A., Wasserman, L.: Distribution-free distribution regression. In: Proceedings of the Sixteenth International Conference on Artificial Intelligence and Statistics, pp. 507–515. PMLR (Apr 2013)
48. Qian, X., Wu, J., Wei, L., Lin, Y.: Random projection ensemble conformal prediction for high-dimensional classification. Chemom. Intell. Lab. Syst. **253**, 105225 (2024)
49. Rahimi, A., Recht, B.: Random features for large-scale kernel machines. In: Advances in Neural Information Processing Systems, vol. 20. Curran Associates, Inc. (2007)
50. Ren, Y., et al.: System log detection model based on conformal prediction. Electronics **9**(2) (2020)
51. Ruta, D., Gabrys, B.: Classifier selection for majority voting. Information Fusion **6**(1), 63–81 (2005)
52. Shafer, G., Vovk, V.: A tutorial on conformal prediction. J. Mach. Learn. Res. **9**, 371–421 (2008)
53. Shen, Z., Yao, S., Chen, Y., Ding, X.: Uncertainty quantification of cuffless blood pressure with deep evidential regression model. In: 2024 46th Annual International Conference of the IEEE Engineering in Medicine and Biology Society (EMBC), pp. 1–4 (2024)
54. Singh, G., Moncrieff, G., Venter, Z., Cawse-Nicholson, K., Slingsby, J., Robinson, T.B.: Uncertainty quantification for probabilistic machine learning in earth observation using conformal prediction. Sci. Reports **14**(16166) (2024)
55. Stankeviciute, K., M. Alaa, A., van der Schaar, M.: Conformal time-series forecasting. In: Advances in Neural Information Processing Systems, vol. 34, pp. 6216–6228. Curran Associates, Inc. (2021)

56. Steinberger, L., Leeb, H.: Leave-one-out prediction intervals in linear regression models with many variables (Feb 2016)
57. Szabo, Z., Gretton, A., Poczos, B., Sriperumbudur, B.: Two-stage sampled learning theory on distributions. In: Proceedings of the Eighteenth International Conference on Artificial Intelligence and Statistics, pp. 948–957. PMLR (Feb 2015)
58. Szabó, Z., Sriperumbudur, B.K., Póczos, B., Gretton, A.: Learning theory for distribution regression. J. Mach. Learn. Res. **17**(152), 1–40 (2016)
59. Tang, L., Luo, R., Zhou, Z., Colombo, N.: Enhanced route planning with calibrated uncertainty set. Mach. Learn. **114**(5) (Mar 2025)
60. Tibshirani, R., Hastie, T., Narasimhan, B., Chu, G.: Diagnosis of multiple cancer types by shrunken centroids of gene expression. Proc. Natl. Acad. Sci. **99**(10), 6567–6572 (2002)
61. Toccaceli, P.: Introduction to conformal predictors. Pattern Recogn. **124**, 108507 (2022)
62. University of Electronic Science and Technology of China: A blood pressure measurement system based on conformal prediction. Patent (2024)
63. Vazquez, J., Facelli, J.C.: Conformal prediction in clinical medical sciences. J. Healthcare Inform. Res. **6**(14), 241–252 (09 2022)
64. Vovk, V.: Cross-conformal predictors. Ann. Math. Artif. Intell. **74**(1), 9–28 (2015)
65. Vovk, V., Gammerman, A., Shafer, G.: Algorithmic Learning in a Random World. Springer International Publishing, Cham (2022)
66. Wang, D., Wang, P., Ji, Z., Yang, X., Li, H.: Conformal loss-controlling prediction. IEEE Trans. Neural Netw. Learn. Syst. **36**(2), 2973–2983 (2025)
67. Wang, D., Wang, P., Shi, J.: A fast and efficient conformal regressor with regularized extreme learning machine. Neurocomputing **304**, 1–11 (2018)
68. Wang, D., Wang, P., Wang, C., Zhuang, S., Shi, J.: A conformal prediction inspired approach for distribution regression with random fourier features. Appl. Soft Comput. **97**, 106807 (2020)
69. Wang, D., Wang, P., Zhuang, S., Wang, C., Shi, J.: Asymptotic analysis of locally weighted jackknife prediction. Neurocomputing **417**, 10–22 (2020)
70. Wang, H., et al.: Unsupervised cross-user adaptation in taste sensation recognition based on surface electromyography. IEEE Trans. Instrum. Meas. **71**, 1–11 (2022)
71. Wang, P., Wang, P., Wang, D., Xue, B.: A conformal regressor with random forests for tropical cyclone intensity estimation. IEEE Trans. Geosci. Remote Sens. **60**, 1–14 (2022)
72. Wang, P., et al.: A conformal regressor for predicting negative conversion time of omicron patients. Medical & Biological Engineering & Computing (Feb 2024)
73. Wang, W., Li, K., Zhang, X.: Bitcoin prediction based on OSL-SCN and conformal prediction method. In: 2024 6th International Conference on Frontier Technologies of Information and Computer (ICFTIC), pp. 882–885 (2024)
74. Wang, Y., Yang, M., Luo, Z., Wang, Y., Li, G., Hu, R.: Rapid coal classification based on confidence machine and near infrared spectroscopy. Spectroscopy Spectral Anal. **36**(06), 1685–1689 (2016)
75. Wang, Y., Wang, Z., Diao, J., Sun, X., Luo, Z., Li, G.: Discrimination of different species of dendrobium with an electronic nose using aggregated conformal predictor. Sensors **19**(4) (2019)
76. Wang, Y., Zhang, M., Wu, R., Wang, H., Luo, Z., Li, G.: Speech neuromuscular decoding based on spectrogram images using conformal predictors with bi-lstm. Neurocomputing **451**, 25–34 (2021)

77. Wang, Z., Sun, X., Miao, J., Wang, Y., Luo, Z., Li, G.: Conformal prediction based on k-nearest neighbors for discrimination of ginsengs by a home-made electronic nose. Sensors **17**(8) (2017)
78. Wang, Z., Yang, F., Li, H., Yang, C., Liu, X.: Threat information utilization and propagation method based on statistical learning (2018), patent, CN109462578B, 11 Jan., 2022
79. Wang, Z., Yu, P., Sun, X., Wei, R., Qiu, K.: Multi-model cross-detection method for malicious code based on statistical learning (2018), patent, CN109033836B, 20 July, 2021
80. Xi, Z.: Distributions-free martingale test distributions-shift for swarm behavior prediction. In: 2022 International Joint Conference on Neural Networks (IJCNN), pp. 1–8 (2022)
81. Xi, Z., Chen, H., Chen, X., Yao, W.: Mondrian conformal prediction of boosting for swarm behavior recognition. In: 2022 IEEE International Conference on Unmanned Systems (ICUS), pp. 1–6 (2022)
82. Xi, Z., Chen, H., Chen, X., Yao, W.: Conformal prediction enhanced svms for swarm behavior classification. In: 2023 International Joint Conference on Neural Networks (IJCNN), pp. 1–8 (2023)
83. Xi, Z., Zhuang, X., Chen, H.: Conformal prediction for hypersonic flight vehicle classification. In: Johansson, U., Boström, H., An Nguyen, K., Luo, Z., Carlsson, L. (eds.) Proceedings of the Eleventh Symposium on Conformal and Probabilistic Prediction with Applications. Proceedings of Machine Learning Research, vol. 179, pp. 118–206. PMLR (24–26 Aug 2022)
84. Xie, X., Jin, Z., Wang, J., Yang, L., Lu, Y., Li, T.: Confidence guided anomaly detection model for anti-concept drift in dynamic logs. J. Netw. Comput. Appl. **162**, 102659 (2020)
85. Xu, C., Xie, Y., Vazquez, D.A.Z., Yao, R., Qiu, F.: Spatio-temporal wildfire prediction using multi-modal data. IEEE J. Select. Areas Inform. Theory **4**, 302–313 (2023)
86. Xu, C., et al.: A tri-light warning system for hospitalized covid-19 patients: credibility-based risk stratification for future pandemic preparedness. Europ. J. Radiol. Open **13**, 100603 (2024)
87. Yang, K., Ma, H.: Multi-mode network flow prediction method and device based on meta-learning framework (2022), patent, CN115225520B, 26 Sept., 2023
88. Yang, M., et al.: Development and validation of an interpretable conformal predictor to predict sepsis mortality risk: Retrospective cohort study. J. Med. Internet Res. **26** (2024)
89. Yang, M., et al.: Enhancing patient selection in sepsis clinical trials design through an ai enrichment strategy: Algorithm development and validation. J. Med. Internet Res. **26** (2024)
90. Zhan, X., Guan, X., Wu, R., Wang, Z., Wang, Y., Li, G.: Discrimination between alternative herbal medicines from different categories with the electronic nose. Sensors **18**(9) (2018)
91. Zhan, X., et al.: Online conformal prediction for classifying different types of herbal medicines with electronic nose. In: IET Doctoral Forum on Biomedical Engineering, Healthcare, Robotics and Artificial Intelligence 2018 (BRAIN 2018), pp. 1–8 (2018)
92. Zhan, X., Wang, Z., Yang, M., Luo, Z., Wang, Y., Li, G.: An electronic nose-based assistive diagnostic prototype for lung cancer detection with conformal prediction. Measurement **158**, 107588 (2020)

93. Zhang, J., Li, G., Hu, M., Li, J., Luo, Z.: Recognition of hypoxia EEG with a preset confidence level based on EEG analysis. In: 2008 IEEE International Joint Conference on Neural Networks (IEEE World Congress on Computational Intelligence), pp. 3005–3008 (2008)
94. Zhang, M., Wang, Y., Zhang, W., Yang, M., Luo, Z., Li, G.: Inductive conformal prediction for silent speech recognition. J. Neural Eng. (Mar 2020)
95. Zhang, R., Zhou, P.: Uncertainty quantification based on conformal prediction for industrial time series with distribution shift. IEEE Trans. Indust. Inform. 1–10 (2025)
96. Zhang, R., Zhou, P., Chai, T.: Improved copula-based conformal prediction for uncertainty quantification of multi-output regression. J. Process Control **129**, 103036 (2023)
97. Zhang, Y., Wang, Z., Wu, Y., Du, Z.: Cyber threat intelligence propagation based on conformal prediction. Netinfo Security **20**(06), 90–95 (2020)
98. Zhang, Z., Dong, Y., Hong, W.C.: Long short-term memory-based twin support vector regression for probabilistic load forecasting. IEEE Trans. Neural Netw. Learn. Syst. **36**(1), 1764–1778 (2025)
99. Zhao, X., Bellotti, A.: Pruning neural networks for inductive conformal prediction. In: Proceedings of the Eleventh Symposium on Conformal and Probabilistic Prediction with Applications, pp. 273–293. PMLR (Aug 2022)
100. Zhao, X., Farjudian, A., Bellotti, A.: Pruning convolutional neural networks for inductive conformal prediction. Neurocomputing **611**, 128704 (2025)
101. Zhou, P., Wen, C., Zhao, P., Li, M.: Ironmaking process modeling uncertainty quantification via conformal prediction based on random vector functional link networks. Comput. Electr. Eng. **123**, 110247 (2025)
102. Zhou, P., Zhang, Y., Lu, S.: Semi-supervised concept drift detection and adaptation method based on conformal martingale framework (2024), patent, CN118885983B, 18 Feb., 2025
103. Zhu, M., Zecchin, M., Park, S., Guo, C., Feng, C., Simeone, O.: Federated inference with reliable uncertainty quantification over wireless channels via conformal prediction. IEEE Trans. Signal Process. **72**, 1235–1250 (2024)
104. Ziyue, B., Huan, L., Jingtao, Z., Shu, Z., Xiaoyan, Z.: Distribution network topology identification based on cnn and improved conformal prediction. In: 2024 China International Conference on Electricity Distribution (CICED), pp. 587–592 (2024)

Application of Confidence and Probabilistic Models to Practical Problems

Lars Carlsson[1,2(✉)] [iD], Johan Hallberg Szabadváry[1,4] [iD], Ernst Ahlberg[2,3] [iD], and James Gammerman[5] [iD]

[1] Jönköping University, Jönköping, Sweden
{lars.carlsson,johan.hallberg.szabadvary}@ju.se
[2] Centre for Reliable Machine Learning, Royal Holloway, University of London, Egham, Surrey, UK
[3] Mölnlycke Health Care, Gothenburg, Sweden
ernst.ahlberg@molnlycke.com
[4] Stockholm University, Stockholm, Sweden
[5] Vitality UK, London, UK
http://www.ju.se

Abstract. This chapter surveys the practical applications of conformal prediction and related methods across diverse domains. We examine how these techniques, which provide rigorous uncertainty quantification with minimal distributional assumptions, have been adapted to address real-world challenges in fields such as drug discovery, medical diagnostics, financial risk management, autonomous systems, natural language processing, and industrial engineering. Key developments include the use of conformal methods to enhance safety in autonomous driving, mitigate hallucinations in large language models, detect anomalies in data streams, and optimize maintenance schedules. We also review the growing ecosystem of open-source software that has facilitated adoption of these methods in industry. By providing guaranteed error control and calibrated probabilities, conformal prediction is enabling more reliable and trustworthy machine learning systems in high-stakes applications. This survey demonstrates how the theoretical foundations laid by Alexander Gammerman and colleagues have evolved into practical tools for building safer and more robust AI systems.

1 Introduction

When deploying machine learning predictions to high-stakes decisions, it is crucial to quantify uncertainty in a principled manner. Unfortunately, most traditional models provide only point estimates or heuristic confidence scores with no rigorous uncertainty guarantees. Classical statistical approaches (for example, Gaussian error assumptions and asymptotic confidence intervals) and even Bayesian methods often rely on strong distributional assumptions that may not hold in practice.

K. An Nguyen and Z. Luo (Eds.): Alexander Gammerman Festschrift, LNCS 16290, pp. 227–268, 2026.
https://doi.org/10.1007/978-3-032-15120-9_12

In complex, high-dimensional problems, it is "very complicated" to understand the prediction uncertainty without robust assumptions. Both easy and hard cases may yield a single-point prediction, and the model itself cannot distinguish between them. This gap motivates the need for distribution-free methods that can wrap around any predictor to deliver honest measures of uncertainty without trusting any particular data distribution or model.

1.1 The Need for Distribution-Free Uncertainty Estimation

Traditional machine learning and statistical frameworks struggle when we cannot assume anything such as normality, independent errors, or a well-specified prior. For example, Bayesian credible intervals provide conditional uncertainty under a model but do not guarantee frequentist coverage; they can be miscalibrated if the model or prior is wrong. By contrast, recent techniques based on algorithmic learning theory offer distribution-free guarantees. Conformal prediction and Venn prediction are two such frameworks developed to handle the scenario of "no strong assumptions" on data.

These methods only assume that data points are generated exchangeably (i.e. as if drawn i.i.d. in no particular order) [63]. Exchangeability means that the joint distribution is invariant under permutations of examples – essentially the minimal assumption of randomness.

Under this mild condition, conformal and Venn predictors can sit on top of any machine learning model [63] and produce valid uncertainty estimates that hold for any underlying distribution. In other words, "the conformal framework allows us to use any model, while ensuring validity with respect to any distribution" (as long as examples are i.i.d.). This is a remarkable departure from traditional methods, which typically break their guarantees if the data deviate from the assumed distributions (Table 1). Conformal predictors guarantee coverage in finite samples without distributional assumptions, whereas Bayesian intervals are not guaranteed to cover the true value unless the model assumptions (prior, likelihood) happen to be correct.

The strength of these new methods lies in offering calibrated, guaranteed uncertainty, even in the absence of a trustworthy probabilistic model. Conformal prediction constructs prediction sets or intervals that provably contain the true outcome with a user-specified probability (e.g. 90% coverage) regardless of the data distribution. Crucially, this coverage guarantee holds in finite samples without relying on asymptotic theory or parametric assumptions. Formally, given training data $(x_1, y_1), \ldots, (x_{n-1}, y_{n-1})$ drawn from an unknown distribution P that is exchangeable, a conformal predictor outputs a set $\Gamma_n^\varepsilon(x_n)$ of possible labels for a new object x_n such that

$$P\{Y_n \in \Gamma_n^\varepsilon(x_n)\} \geq 1 - \varepsilon,$$

for the chosen error rate, ε.

Table 1. Comparison of Conformal Prediction and Bayesian Credible Intervals

Method	Coverage	Optimality
Conformal Prediction	Yes (finite-sample guarantee)	Distribution-free near-optimality proven
Bayesian Credible Intervals	Conditional, not guaranteed	Optimality proven only under correct prior/model specification

Remarks: Conformal methods provide the strongest general guarantees. Bayesian methods are optimal *if and only if* prior and model assumptions hold.

In other words, with (for example) $\varepsilon = 0.1$, the method guarantees that no more than 10% of the future test points will fall outside the predicted set. This guarantee was rigorously proven under the sole assumption of exchangeability.

This means we can have statistical confidence in the uncertainty intervals produced – a new data point's prediction interval will cover the true value at least 90% of the time by construction. Such finite-sample validity is something that standard confidence intervals usually achieve only asymptotically or under exact distributional models.

1.2 Conformal Prediction: Confidence Sets with Coverage Guarantees

Conformal prediction (CP) is a flexible wrapper that can be applied to virtually any prediction algorithm to yield valid uncertainty sets. The key idea is to use the algorithm's fitted model to define a nonconformity score that measures how unusual a candidate prediction is compared to the training examples.

For a regression model, for instance, a natural nonconformity score is the absolute residual $|y - \hat{f}(x)|$; for classification, it could be related to the model's output probability or margin for a class.

To predict a new input x_n, the conformal method tentatively assigns a label and computes its nonconformity score relative to the training scores. By comparing this new score with the distribution of scores from the training data, we can derive a p-value for the candidate label. The conformal prediction set $\Gamma_n^\varepsilon(x_n)$ is a collection of all labels for which the nonconformity score is not extreme at the α level.

Intuitively, if a potential label for the new point is too inconsistent with what has been seen before (falling in the worst α fraction of conformity), it is excluded. Otherwise, it remains in the prediction set. Owing to the symmetry under exchangeability, the method guarantees that the true label will be excluded at most ε fraction of the time.

Importantly, conformal predictors adapt to the difficulty of each example. If the new input x_n is in a region well covered by the training data and the underlying model is confident, the conformal prediction set will often be small, perhaps a single label (in classification) or a narrow interval (in regression). If, instead, x_n lies in an ambiguous or novel region (a "hard" prediction), the conformal algorithm will output a larger set, reflecting higher uncertainty.

In extreme cases, the prediction set could even be the entire label space (no information) or empty (no label meets the confidence criterion), although such outcomes are rare with well-chosen scoring functions. This adaptiveness addresses the earlier issue that "an easy prediction can't often be distinguished from a hard prediction" by a traditional model. In conformal prediction, an easy case yields a tight prediction, while a hard case yields a broad one, making uncertainty explicit.

The user sets the desired error rate (e.g. 5% or 10%), and the method guarantees that level of confidence by sometimes outputting multi-valued predictions as the price to pay. Indeed, unlike a single point estimate, a conformal predictor might output a set of possible classes (e.g. both "0" and "1" as plausible labels) or an interval $[y_{min}, y_{max}]$ for a numeric outcome. This willingness to hedge with a set is precisely how strict coverage guarantees are achieved.

Conformal prediction thus provides a distribution-free confidence set, one that is valid for any data distribution (under exchangeability) and that automatically broadens or narrows based on how much uncertainty is present.

1.3   Venn Prediction: Calibrated Probabilistic Estimates

While conformal predictors output confidence sets or intervals, Venn predictors focus on producing probability estimates with guaranteed calibration. In many applications, practitioners prefer a single probability or score for each class rather than a prediction set. However, an impossibility result in online learning states that no algorithm can produce perfectly calibrated single-case probabilities in a distribution-free manner.

Venn prediction circumvents this by allowing multiple probability outputs per test example. In practice, a Venn predictor assigns a new instance to a category (or taxonomy) based on the training data and a tentative label. For each category, it computes the frequency of each class label among similar training examples. By temporarily including the new example with a hypothetical label in that category, the method derives a set of probable label frequencies. The result is that, for each test object, the Venn predictor outputs two or more probability values – essentially, a small interval or range of probabilities for each class – with the guarantee that the true label's probability is calibrated.

In other words, one of the probability values in the output is the "valid" probability, in the sense that over many such predictions, the long-run frequency of the true label equals the predicted probability. For example, a Venn predictor might say a new example belongs to class 1 with a probability between 0.68 and 0.70 (or simply 0.69, if it outputs a midpoint). This means that across all examples for which it predicts a 0.69 probability for class 1, approximately 69% of them will indeed belong to class 1 – a perfect calibration by construction.

The range $[0.68, 0.70]$ indicates minimal uncertainty around that probability. In most cases, the range is narrow, yielding essentially a single probability value, and only if the model is very unsure do we obtain a wider probability interval. Thus, Venn predictors provide credible probabilistic predictions with guaranteed calibration, complementing the guaranteed confidence sets of conformal prediction.

1.4   Conformal Predictive Systems (CPSs): Distribution-Free Probabilistic Forecasts

While conformal classification and conformal regression produce confidence sets/intervals with guaranteed finite-sample coverage, many applications need a full predictive distribution—a calibrated probability distribution over outcomes for each new x.

Conformal predictive systems (CPSs) meet this requirement. Under the same minimal assumption of exchangeability, they wrap around any base predictor to produce a distribution-free, finite-sample-calibrated probabilistic forecast. A CPS outputs conformal predictive distributions (CPDs), which are p-values arranged into a distribution function.

Under mild assumptions on the conformity measure, a CPD is strongly calibrated in probability [64], in the sense of satisfying the probability integral transform (PIT). In other words, a CPD behaves like the distribution function. Venn predictors output multi-probabilities (a tiny set or narrow range of probabilities) with perfect calibration guarantees for classification. CPSs play a similar role in regression or continuous outcomes, offering a full CDF with PIT-style calibration

CPDs can be used to calibrate any base predictor (quantile regression, Gaussian processes, random forests, and deep nets). If a domain model is trustworthy, CPDs can refine it; if not, CPDs repair its miscalibration distribution-freely. Because CPDs are full distributions, they can be evaluated with proper scores (for example, CRPS and log score, with caution regarding discreteness) and used in decision-theoretic pipelines (risk curves and expected utilities).

1.5   Computationally Efficient Versions of CP Methods

To handle the often intractable computational cost of full or transductive conformal predictors, the computationally efficient inductive conformal predictors (ICPs) were introduced in [52] for the regression problem, and [53] for classification. The computational efficiency of ICPs arises from splitting the training set into a proper training set, used to construct a fixed prediction rule (the inductive step), and a calibration set, used to calibrate the predictions. This key adaptation has enabled conformal methods to be applied to large-scale real world problems, using essentially any base machine learning method. The efficiency, however, does not come for free, as the validity of ICPs is of the probably approximately correct (PAC) type, with a parameter δ, in addition to the significance level ε, which depends on the calibration set size [61].

1.6 From Assumption-Free Guarantees to Research Frontiers

By providing distribution-free uncertainty quantification, conformal and Venn prediction methods address scenarios where traditional ML inference falls short. They enable machine learning practitioners to make reliable decisions in the absence of trustable model assumptions, which is a critical capability for building trustworthy AI systems.

These approaches are grounded in strong theoretical guarantees, yet are practical to implement with modern machine learning models. In fact, one can even combine them with classical methods. If domain knowledge suggests a valid parametric model or prior, one may apply Bayesian inference for efficiency and still layer a conformal predictor on top to retain coverage guarantees.

In the next section, we transition from this conceptual introduction to a literature review, surveying some applications in which conformal prediction and Venn prediction have been used.

2 A Survey of Real-World Applications

Conformal prediction (CP) has transitioned from a theoretical concept into a versatile tool applied across a wide array of high-stakes domains. Its ability to provide model-agnostic, distribution-free uncertainty guarantees is uniquely suited to fields where the cost of an incorrect prediction is substantial. This section provides a high-level summary of the key application areas detailed in this chapter, showcasing how conformal methods are being used to build more reliable, trustworthy, and transparent AI systems. An overview of these applications can be found in Table 2.

In the **life sciences, toxicology, and medicine**, conformal prediction has become a valuable tool for Quantitative Structure-Activity Relationship (QSAR) modelling. Instead of relying on the coarse concept of an "applicability domain," CP provides a statistically sound confidence level for each individual prediction. This helps researchers prioritize compounds for screening or flag potential toxicity with a known error rate. In clinical medicine, the framework adds reliability to diagnostic models, ensuring that predictions for disease risk come with well-calibrated probabilities that clinicians can trust for making critical decisions.

The **energy and environmental forecasting** sector uses conformal methods to quantify the uncertainty in volatile processes like electricity prices and renewable energy output from wind and solar sources. Because time-series data often violates the core assumption of exchangeability, researchers have developed adaptive versions of CP to handle these dynamic, non-stationary environments, providing the robust prediction intervals necessary for grid stability and economic planning.

In **finance and risk management**, conformal methods facilitate a shift from making simple point forecasts to implementing principled risk management. Applications include constructing portfolios with improved worst-case performance, creating reliable credit scoring models that can flag uncertain applicants for manual review, and developing fraud detection systems that guarantee a controlled rate of false alarms.

For safety-critical systems like **autonomous systems and computer vision**, standard coverage guarantees are often not enough. In areas like autonomous driving and critical infrastructure monitoring, advanced methods like Conformal Risk Control are used to manage specific, asymmetric risks, such as minimizing the chance of a false negative when detecting a railway signal. This allows engineers to build perception systems that are not just accurate but demonstrably safe.

Conformal prediction is also emerging as a powerful tool in **Natural Language Processing (NLP)** to address the inherent unreliability of Large Language Models (LLMs). By producing a set of plausible outputs instead of a single, overconfident one, conformal methods can effectively signal model uncertainty. This provides a mechanism to mitigate model "hallucinations" and is used in Automatic Speech Recognition (ASR) to identify and flag potentially mistranscribed words in high-risk settings.

The "industrialization" of conformal prediction is evident in **industrial engineering**, with applications in manufacturing and supply chain management for tasks like quality control and demand forecasting. By generating full predictive distributions, these methods allow for more nuanced decision-making. In predictive maintenance, they provide probabilistic estimates of an asset's Remaining Useful Life (RUL), enabling operators to move beyond simple point forecasts to make cost-effective, risk-informed maintenance decisions.

Finally, in **online monitoring and anomaly detection**, Conformal Test Martingales provide a direct way to test the underlying exchangeability assumption of a model in real time. This has significant practical value for online change-point detection, offering a principled answer to the question of when a model needs to be retrained due to dataset shift. The same mechanism is widely used for anomaly detection in data streams, providing a robust, non-parametric method for identifying novel or abnormal events.

Table 2. Summary of Real-World Application Areas for Conformal Methods

Application Domain	Specific Task(s)	Key Challenge Addressed
Life Sciences & Medicine	– Drug discovery, QSAR modelling – Clinical diagnostics & risk prediction	Replacing the coarse "applicability domain" with instance-specific confidence; providing trustworthy, calibrated risk scores for clinical decisions.
Toxicology & Environmental Science	– Chemical safety assessment – Skin sensitization prediction	Handling highly imbalanced datasets; providing the transparent error guarantees required for regulatory acceptance.
Energy & Environmental Forecasting	– Electricity price forecasting – Renewable energy output prediction	Adapting to non-exchangeable time-series data; producing reliable prediction intervals to manage grid stability and market volatility.
Finance & Risk Management	– Portfolio selection – Credit scoring – Anomaly and fraud detection	Moving from simple point forecasts to principled risk management; controlling error rates for lending decisions and false alarms in fraud detection.
Autonomous Systems & Computer Vision	– Perception for autonomous driving – Railway signal detection	Controlling specific, asymmetric risks (e.g., false negatives) beyond simple coverage guarantees in safety-critical applications.
Natural Language Processing (NLP)	– Mitigating LLM "hallucinations" – Reliable Automatic Speech Recognition	Quantifying model uncertainty to combat overconfidence; flagging specific words or outputs that are likely to be incorrect.
Industrial Engineering	– Predictive maintenance (RUL) – Manufacturing quality control	Providing full predictive distributions (CPS, Venn-Abers) for nuanced, cost-effective operational decisions.
Online Monitoring & Anomaly Detection	– Real-time dataset shift detection – Anomaly detection in data streams	Providing a principled, non-parametric method to determine when a model needs retraining; detecting novel events with guaranteed error control.

3 Life Sciences and Medicine

3.1 Drug Discovery and QSAR Models

One of the first forays of CP into life sciences was ligand-based drug discovery, specifically quantitative structure–activity relationship (QSAR) modelling. QSAR models predict properties of molecules (e.g. biological activity, solubility, toxicity) from chemical structure descriptors, and are widely used in pharma for tasks such as virtual screening (prioritising compounds for testing) and toxicity alerts.

In [23], conformal prediction was introduced to QSAR modelling, demonstrating how CP could generate prediction intervals for QSAR regression models.

In their study, CP was applied to multiple drug discovery datasets using support vector machine (SVM) regression. Notably, the resulting prediction intervals closely matched the actual distribution of prediction errors, indicating that the intervals were efficient and useful for assessing model reliability. This was a critical innovation, as traditional QSAR models often lack a rigorous measure of confidence for individual predictions.

3.2 Clinical Diagnostics and Decision Support

In [19], conformal prediction (specifically Mondrian conformal predictors, a label-conditional variant) was applied in an ovarian cancer screening trial. Using proteomic blood test data from the UKCTOCS trial, they built a model to distinguish between cancer and healthy cases and provided each prediction with a measure of confidence. The main advantage of the conformal predictor was its proven validity: the team could guarantee the error rate of the risk predictions, making the risk scores far more trustworthy for patient screening. In fact, by analysing how predictions changed over time for each patient, this method helped identify cancer cases up to 11 months earlier than standard screening in some instances. Such earlier detection was statistically significant, illustrating how valid probabilistic predictions can directly improve clinical outcomes. The ability to output rigorous confidence measures means that when the model flags a high-risk patient, clinicians can be confident in risk stratification. This early work suggested that conformal predictors could be valuable in healthcare, where knowing the certainty of a prediction (e.g. for a disease diagnosis) is as important as the prediction itself.

Subsequent studies have applied CP to diverse bioinformatics problems, from analysing proteomic patterns for cancer diagnostics to assessing stroke risk from medical images, consistently finding that CP methods can provide important insights into the accuracy of individual predictions in clinical applications.

3.3 Broader Adoption in Pharma

Following these initial demonstrations, CP gained traction in the pharmaceutical industry in the mid-2010s. A landmark publication [46], advocating CP as a "transparent and flexible alternative" to the traditional concept of a QSAR applicability domain. In QSAR modelling, the applicability domain (AD) is typically a coarse characterisation of the chemical space beyond which a model's predictions are considered unreliable.

In [46], it was argued that the rigorous mathematical framework of conformal prediction provides a more nuanced way to estimate prediction reliability: instead of a binary "inside/outside AD" label, CP gives a confidence level tied to each prediction. Not only is the mathematics statistically sound, but the interpretation is intuitive; for instance, a CP at 80% confidence will make errors on at most 20% of new cases by design.

Importantly, the confidence level is adjustable by the user, allowing flexibility depending on the application's risk tolerance (e.g. using a higher confidence for critical decisions). This approach aligns well with regulatory guidelines for QSAR [49], which require transparency and confidence in their predictions. By demonstrating CP on 10 public QSAR datasets, Norinder's team showed that CP could satisfy these requirements while inherently solving the AD problem through its calibrated prediction regions.

3.4   Key Studies and Evolution

After 2014, the use of CP in drug discovery expanded rapidly. A series of studies have integrated CP into real-world pharmaceutical modelling pipelines:

- **Introduction and Review (2021):** In [3] an accessible introduction and review of conformal prediction in the drug discovery context was provided. In particular, conformal prediction (and its variants) is applied to QSAR-driven drug discovery tasks to attach valid, object-specific confidence to model outputs. Concretely, it's used to generate prediction intervals for regression (e.g., LogD at pH 7.4) and label sets for classification (e.g., active vs non-active), inherently handling applicability domain by widening intervals/sets for "strange" compounds. The authors illustrate use cases such as off-target binding risk profiling and a full case study on classifying inhibitors of ABC transporters (P-gp/ABCB1, BCRP/ABCG2, BSEP/ABCB11), reporting validity/efficiency across significance levels. They explained that CP is a model-agnostic layer that produces valid confidence estimates for each prediction – in regression, a tailored prediction interval for the compound's activity - in classification, a prediction set of possible classes (e.g. active/inactive) that might even be empty or contain multiple labels, depending on uncertainty. The authors emphasised that CP offers mathematically proven error rate guarantees and integrates the model uncertainty and applicability domain in a unified manner. They also discussed practical implications for drug discovery, showing through an example (predicting inhibitors of ABC transporter proteins) how CP-derived confidence intervals guide better decision-making in hit selection. This work underlined the maturation of CP from a novel idea to a tool ready for everyday use by medicinal chemists.
- **Prospective Large-Scale Comparison (2019):** In [11] a large-scale study comparing conformal prediction models to traditional QSAR models across 550 protein targets from the ChEMBL bioactivity database was conducted. Each target had both a QSAR model and an analogous CP model (Mondrian inductive CP) trained to predict ligand binding outcomes. The study found that, in terms of traditional metrics, the CP models had a predictive performance comparable to that of standard QSAR models. However, CP additionally provided information on prediction certainty, aiding decision-making by highlighting which predictions were reliable. An important observation was that using CP's extra information is not always straightforward; the study noted that it is crucial to determine how to best use the confidence information in practice. For instance, one might decide to act only on predictions made at 90% confidence or above, depending on the context. The authors

reported that CP and non-CP methods have similar behaviours in many cases, but there are notable differences in how out-of-domain or uncertain cases are handled, which practitioners must understand. This kind of head-to-head evaluation, performed in what amounted to a prospective prediction scenario (models were tested on new data that became available after model training), helped convince industry scientists of CP's value at scale. The take-away was that CP could be integrated into large workflows without loss of accuracy while providing a layer of risk control for each prediction.

– **High-Throughput Screening and Iterative Testing:** Conformal prediction has also been applied to high-throughput screening (HTS) and other iterative discovery processes. For example, [56] explored using CP to maximise the yield of actives in an HTS campaign. By combining CP with chemogenomic models, they developed strategies to prioritise compounds such that screening focuses on those most likely to be active (at a given confidence level), thereby improving screening efficiency. The use of CP in this context ensures that the selection of compounds for follow-up is performed with a known error rate – effectively, fewer false positives or negatives in the hit list, as controlled by the specified confidence. This illustrates a broader trend of CP adoption in active learning or iterative experimentation, where each prediction's confidence can indicate whether to trust the prediction or gather more data.

3.5 Impact on Decision-Making

Across most applications, a common theme is that conformal predictors enable better-informed decisions in life sciences by providing actionable confidence metrics. For instance, in drug discovery projects, teams often have to decide which compound series to pursue or which predicted hits to synthesise. Using CP, a model might predict 10 compounds as active, but perhaps only five of them with high confidence (tight prediction intervals or single-label prediction sets). Project leaders can prioritise high-confidence hits, thus allocating resources more efficiently.

As noted [11], CP "provides information on the certainty of a prediction, and thus helps in decision-making" in practical drug discovery settings. Additionally, CP's ability to flag when it is unsure – for example, outputting a prediction set containing both "active" and "inactive" or a very wide activity range – is effectively an automatic applicability domain indicator. A compound that is very structurally different from the training chemistry may yield an undecided prediction, signaling chemists that "this molecule is out of scope; proceed with caution."

This intrinsic treatment of the applicability domain was explicitly highlighted by Norinder et al., who showed that CP's validity guarantees hold even as the confidence level varies, making the trade-off between coverage and error transparent. In regulated environments (e.g. submitting QSAR results to a regulatory agency), this is crucial – it means that a model can be tuned to the required confidence and will reliably not exceed the allowed error rate.

Overall, the evolution in this domain can be summarised as follows: initial small-scale tests lead to broader adoption in pharma R&D (with software such as [9]) operational in multiple organisations, which in turn leads to consideration in regulatory modelling and validation guidelines. With increasing data and computational power, CP has moved from a novel academic concept to a practical component of drug discovery toolkits, complementing traditional machine-learning methods by adding a layer of trust. This breadth of use underscores that CP is becoming a general approach for predictive modelling with a built-in error control, benefiting any life science application where decisions hinge on the reliability of predictions.

4   Toxicology and Environmental Science

The toxicology domain, including chemical safety assessment and environmental risk modelling, has also embraced conformal prediction techniques. Predictive toxicology faces many of the same challenges as drug discovery (e.g. limited and imbalanced data, need for an applicability domain) and often operates under an additional regulatory mandate to justify model predictions. CP's guarantee of valid confidence levels and its robustness to distributional uncertainties are highly attractive in this setting.

4.1   Chemical Toxicity Prediction

A notable application of CP in environmental toxicology is the modelling of large screening programs, such as ToxCast and Tox21. These programs, led by the U.S. EPA and collaborators, have tested thousands of environmental chemicals and drugs across hundreds of in vitro assays (covering endpoints such as endocrine receptor activity, cytotoxicity, etc.). The application of conformal prediction was made to classify compounds based on their activity in ToxCast/Tox21 oestrogen receptor (ER) assays [44].

This study used a massive dataset of environmental chemicals tested for oestrogenic activity and built CP-based classifiers to predict whether a new chemical would be active or inactive in the assay. The use of CP allowed the model to output, for each chemical, a prediction set (e.g. "active" or "inactive" or "active, inactive") at a chosen confidence level, rather than a single hard call. This enhanced the utilisation of the data by identifying predictions that were made with high certainty.

One practical outcome was that many compounds could be confidently predicted as inactive (with narrow prediction sets), which is important for screening non-hazardous chemicals, while compounds in ambiguous zones were flagged for further testing rather than being misclassified. By calibrating on a dedicated subset of data, the conformal classifier ensured that, say, at 90% confidence, no more than 10% of the active/inactive calls would be wrong. In a domain such as toxicology, where false negatives (missing a toxic chemical) can be critical, this error control is invaluable.

4.2   Safety Pharmacology and Imbalanced Data

Toxicological datasets are frequently imbalanced. For example, only a small fraction of chemicals might be highly toxic (the "positives") while the majority are relatively safe (the "negatives"). Traditional machine learning can struggle in this scenario or produce models that are overconfident in predicting the majority class.

Conformal prediction addresses this by calibrating predictions per class (using approaches such as Mondrian CP, which maintains error rates separately for each class). One study [45] specifically investigated binary classification on imbalanced toxicity datasets using conformal prediction. They demonstrated that CP, especially with class-conditional normalisation, could maintain valid error rates for both minority (toxic) and majority (non-toxic) class predictions, whereas conventional models might provide unreliable probability estimates in the low-frequency class. Techniques such as Mondrian CP effectively ensure that, for example, a prediction of "toxic" at 80% confidence truly reflects about 20% false-positive rate in that toxic subset, independent of how many non-toxics dominate the data.

This is a more principled alternative to traditional applicability domain heuristics. Instead of declaring an "AD" beyond which all predictions are invalid (an approach which often fails for minority classes), CP simply outputs wider intervals or multi-label prediction sets for out-of-distribution instances, naturally indicating higher uncertainty. In practice, toxicologists can use this feature to identify chemicals for which the model is unsure – those that might be candidates for additional laboratory testing or analogue searching.

4.3   Applications in Specific Toxicity Endpoints

Conformal predictors have been developed for various toxicological endpoints. In [57], for instance, CP was applied to compound cytotoxicity data from Pub-Chem high-throughput screens to quantify the uncertainty in predicting the cell viability effects of compounds.

Another prominent example is skin sensitisation prediction, an important endpoint for chemical safety (it measures if a substance can trigger allergic skin reactions). A study [27] incorporated conformal prediction into a machine-learning assay called Genomic Allergen Rapid Detection (GARD), which uses genomic readouts to classify skin sensitisers. The goal was to determine the applicability domain of GARD via CP.

By calibrating the GARD predictive model with CP, Forreryd and colleagues could identify which test substances fell into areas of model uncertainty. In their results, the CP framework enabled "predicting skin sensitizers with confidence" by providing a clear criterion for when a GARD prediction is reliable. If a new chemical yielded a conformal prediction that was inconclusive (i.e., a prediction set containing both "sensitizer" and "non-sensitizer"), that was effectively a signal that the chemical was outside the model's domain of competency. This approach provided a quantitative handle on the applicability domain, which is especially important for regulatory acceptance of such non-animal tests. Regulators could be shown that the model will only make a high-confidence call when similar chemistry has been seen before (as reflected in narrow CP prediction sets); otherwise, it flags the result as indeterminate, prompting additional investigation.

4.4   Environmental Risk Assessment Models

In addition to in vitro assay predictions, CP has been integrated into broader risk-assessment pipelines. A good illustration is the KnowTox project [42], which developed a pipeline for predicting the various toxic effects of compounds during the early development stages. The pipeline combines QSAR models for 88 toxicity endpoints with structural alerts and read-across tools. Notably, the KnowTox pipeline employs conformal prediction to ensure confidence in its QSAR model outputs.

As the authors explain, when applying machine learning to predict toxic effects, "applicability and reliability of predictions for new chemicals are of utmost importance". They therefore added an extra CP calibration step to their models, which "per definition creates internally valid predictors at a given significance level". In other words, each toxicity prediction comes with a built-in guarantee of error rate (e.g. a 5% significance level means the model will misclassify at most 5% of truly toxic/non-toxic compounds if we interpret it in a binary context). This gave risk assessors greater confidence in using the model's output for regulatory decisions.

In a case study on androgen receptor antagonism (one of the endpoints in KnowTox), they even introduced additional techniques, such as kNN-based normalisation within the CP framework, to further improve the validity of predictions. The result was a 23% increase in the validity of the predictions on an external test set, illustrating that CP can be augmented and tuned for maximum reliability. The KnowTox example underscores how CP is adopted in practice to meet stringent regulatory requirements. By guaranteeing a controlled error rate and clearly identifying which predictions are trustworthy, CP-enabled models become more acceptable in risk assessment workflows (where false assurances of safety could have serious consequences).

4.5   Impact on Toxicology Decision-Making

The integration of conformal prediction in toxicology and environmental science has led to more robust decision-making in several ways. First, it provides transparent risk metrics. Instead of a black-box model that simply says "chemical X is predicted toxic", a CP model might say that "chemical X is predicted toxic with 90% confidence" or even "undecided at 90% confidence", information that is crucial for regulators and risk assessors. Such transparency aligns with the precautionary principle often used in environmental decisions. If a model is unsure, uncertainty is explicitly communicated, preventing over-reliance on possibly erroneous predictions.

Second, CP facilitates the handling of data drift and model calibration issues. A recent study noted that CP can be used to diagnose data drift in toxicological in vitro models by checking calibration plots; if the empirical error rates deviate from the nominal, it signals potential changes in data distribution or model performance [43]. This is particularly relevant when models are applied to a new chemical space over time.

Third, CP's ability to manage imbalanced datasets means that models remain useful even when the incidence of toxicity is low. In practice, this could help prioritise chemicals that are flagged for further testing. For example, an imbalanced model might otherwise mark nearly everything as "non-toxic" to achieve accuracy, but a CP model will indicate low confidence for those few chemicals where it cannot be sure, thereby rescuing potentially toxic cases from being overlooked.

Lastly, conformal prediction contributes to the paradigm shift away from animal testing by making in silico predictions more reliable and interpretable. Regulatory frameworks such as REACH increasingly allow (and even prefer) computational predictions to screen chemicals for hazards, provided the models come with defined applicability domains and error estimates. By inherently fulfilling these requirements (with sound statistical underpinning), CP models ease the regulatory acceptance of computational toxicology. Indeed, by controlling the error rate, a conformal predictor can be set to the level of confidence required for a given decision threshold.

As observed in one CP study for risk assessment, combining multiple modelling approaches with CP-based confidence produced a higher impact in guiding toxicity testing and de-selecting likely harmful compounds early compared to using a single predictive method [42]. This means that safer compounds move forward in development with quantifiable assurance, and hazardous ones are more reliably caught – a win-win for both industry efficiency and public/environmental safety.

In summary, the practical applications of conformal prediction in toxicology and environmental science mirror those in drug discovery. Initial proofs of concepts have grown into robust frameworks adopted by the industry and recommended by regulators. CP handles the field's unique challenges (imbalanced data, diverse chemical space, and the need for clear applicability domains) by providing mathematically guaranteed confidence for each prediction. The result is improved decision-making – whether it is an environmental agency evaluating a chemical's risk or a pharmaceutical company ensuring a new drug candidate does not have hidden toxicities – all achieved by predicting with confidence and a known error rate [3].

This and other peer-reviewed studies highlight that conformal prediction, through both conformal classification and conformal regression approaches, has moved from theory into practice. Its adoption in life sciences, medicine, and toxicology is growing, driven by the need for predictive models that can not only predict but also quantify their own uncertainty. With CP, practitioners in these domains can make decisions (e.g. selecting a drug hit, diagnosing a patient, or flagging a toxic chemical) with a predefined confidence level, thus carrying out predictions "with confidence" in the most literal sense of the phrase.

5   Energy Forecasting and Environmental Prediction

Conformal methods have proven their value in energy forecasting and environmental prediction, which include problems such as electricity price forecasting, renewable energy output prediction, and weather modelling. In these areas, quantifying uncertainty is crucial for planning and risk management. However, the assumption of exchangeability is often violated in time-series data, which has led researchers to develop several methods inspired by CP with the aim of retaining some of its desirable properties.

Adaptive Conformal Inference (ACI) was introduced by [28] to retain a form of finite-sample error rate guarantee for non-exchangeable data. Instead of using the same significance level for all predictions, ACI uses an online update of the significance level, treating it as a control parameter. The basic procedure provides a finite-sample error rate guarantee simply by decreasing the significance level if an error is committed in the step before and increasing it otherwise.

It should be noted that ACI is a control mechanism applied on top of the CP procedure to steer the average error rate to a desired level. As noted by [31], there is nothing conformal about this procedure, which works with exactly the same guarantee for any set predictor that predicts with confidence. Several extensions of ACI have been suggested, including more advanced control mechanisms [5] and adaptive step sizes in control updates [29].

In the energy forecasting domain, [68] used Online Expert Aggregation on ACI (AgACI) to forecast French electricity spot prices with predictive intervals.

Another method for handling non-exchangeable data was suggested by [10]. The idea is to introduce weights to the examples that indicate the contribution of each example to the p-value. Intuitively, larger weights should be assigned to examples originating from (nearly) the same distribution as the test point. The method was evaluated using synthetic and real-world data, including electricity usage and prices in Australia.

In contrast to ACI and related methods, which are limited to set prediction, weights can also be applied to conformal predictive systems (CPSs) [36], thereby allowing users to construct predictive distributions as well as prediction sets. The energy sector is a particularly compelling field for the application of conformal methods. The increasing reliance on intermittent renewable energy sources, such as wind and solar, coupled with the volatile nature of electricity markets, makes reliable uncertainty quantification not only beneficial but also essential for grid stability, economic efficiency, and risk management.

Traditional point forecasts are insufficient because they fail to provide a measure of confidence. CP addresses this gap by offering a model-agnostic framework to produce prediction intervals (PIs) or full predictive distributions that provide a formal guarantee of valid coverage.

Research in this area has explored several key applications, including forecasting electricity prices, solar photovoltaic (PV) power generation, and wind power output. A common theme across these applications is the need to adapt CP methodologies to handle the unique challenges of energy time-series data, most notably, the violation of the exchangeability assumption due to factors such as seasonality, weather dependence, and market dynamics, which lead to heteroscedasticity (non-constant variance).

CP applications are diverse and geographically widespread. For instance, [17] demonstrated its use in the Mexican electricity market by employing various machine learning models to forecast locational marginal prices and using ensemble-based CP to quantify uncertainty. Recent studies have pushed boundaries by applying more advanced CP variants to various energy sources.

[54] focused on solar PV power forecasting, investigating a wide array of methods, including Mondrian binning (which creates adaptive intervals based on data categories), k-nearest neighbours (KNN) to estimate prediction difficulty, and Conformal Predictive Systems (CPS) to generate full conformal predictive distributions (CPDs). Similarly, [35] tackled large-scale regional wind power forecasting by combining deep Convolutional Neural Networks (CNNs) with Conformalized Quantile Regression (CQR). They were the first to propose the use of the Split Conformal Predictive System (SCPS) for this task, even introducing a novel adaptive distribution method called Split Conformal Distribution Regression Forests (SCDRF).

As an example of probabilistic classification in the environmental forecasting domain, [59] suggested a model for forecasting Large-Scale Travelling Ionospheric Disturbances (LSTIDs) over Europe up to three hours in advance. To make the model's predictions reliable for decision-making in a high-risk setting, the authors employed Venn-ABERS predictors to calibrate the output scores. This approach yields not only better-calibrated probability estimates of an LSTID event but also a valid measure of confidence for each forecast, which is critical for end-user risk management.

5.1   Pioneering Work in Power Markets

In one of the first comprehensive investigations in this area, [37] explored the application of CP to short-term electricity price forecasting across three European power markets (Nord Pool, EPEX intraday, and GEFCom). Their work serves as an excellent case study on the practical utility and inner workings of CP in a real-world high-stake environment.

A key aspect of their research was to demonstrate the model-agnostic nature of CP. They coupled it with a range of underlying point predictors, from a simple naive benchmark to more advanced models, such as lasso, k-nearest neighbours (KNN), and Support Vector Machines (SVM), and compared its performance to state-of-the-art Quantile Regression Averaging (QRA).

Perhaps their most significant contribution was a "path-dependent evaluation" designed to isolate the specific components of CP that drive its performance. They systematically analysed the impact of the three core characteristics.

- **Symmetry:** The practice of creating symmetric PIs by calculating absolute errors and adding/subtracting a single quantile from the point forecast.

- **Sampling:** The splitting of data into a proper training set and a calibration set to prevent overfitting and ensure the non-conformity scores are calculated on out-of-sample data.
- **Normalization:** The key feature of Normalized Conformal Prediction (NCP), where the non-conformity score (absolute error) is normalized by a separate prediction of the error magnitude. This makes the intervals adaptive to the local volatility of the data, which is a critical feature of heteroscedastic electricity prices.

Their findings showed that CP, particularly the normalised variant (NCP), produces sharp and reliable prediction intervals that are competitive with or superior to other methods. The study concluded that no single model configuration was universally superior. The optimal blend of point predictor and CP variant depended heavily on the specific market characteristics, advising practitioners to test different setups to find the most effective combination for their particular use case.

5.2   Tackling Heteroscedasticity in Wind Forecasting

Although standard ICP methods provide training-conditional valid coverage [61], their tendency to produce fixed-width intervals is a major limitation for forecasting highly variable sources, such as wind power, where the uncertainty is far from constant. [33] addressed this challenge directly by proposing a novel hybrid model named Conformalized Temporal Convolutional Quantile Regression Networks (CTCQRN). This model integrates two powerful technologies.

- **Temporal Convolutional Network (TCN):** As the underlying regression algorithm, a TCN is a deep learning architecture adept at capturing long-range temporal dependencies in sequence data. It offers advantages over recurrent architectures like LSTMs, such as parallel processing, and stable gradients, making it well suited for time-series forecasting.
- **Conformalized Quantile Regression (CQR):** Instead of starting with a point forecast and adding an error margin, CQR begins with an initial quantile forecast (e.g., the 5th and 95th percentiles) generated by a base model (here, the TCN). The calibration set was then used to calculate a conformity score based on how often the true values fell outside the initial interval. This score was used to produce a single correction factor that uniformly adjusted the initial interval to guarantee the desired coverage level.

The key advantage of the CTCQRN framework is that it explicitly generates intervals of varying widths that are adaptive to the data's local volatility (heteroscedasticity) while simultaneously ensuring that the final calibrated intervals meet the rigorous theoretical guarantee of valid coverage. In case studies of two wind-power datasets, Hu et al. demonstrated that their proposed model had a "distinct edge" over benchmarks, including other neural network architectures combined with CQR, standard conformal regression, and classical quantile regression. It successfully produced narrower and more precise intervals without sacrificing the guaranteed coverage rate, proving its value for reliable wind power forecasting.

6 Probabilistic Applications

The conformal framework includes probabilistic prediction as well as set prediction. In the case of classification, this is achieved by Venn predictors, while conformal predictive systems provide probabilistic predictions for regression problems.

Venn predictors are a family of probabilistic predictors that wrap around any underlying machine-learning model to produce valid probability outputs. The core idea is to partition the space of examples into "Venn categories" using a taxonomy (often induced by the base model's scores or features). For a new object, the method temporarily assigns each possible label in turn, places the object into a category under that assumption, and recomputes category frequencies from the training data. This yields a multiprobability prediction instead of a single probability estimate (which is impossible without making stronger assumptions [64]). In the case of binary problems, the output is the pair (p^0, p^1), where intuitively, the probability that the label is 1 is either p^0 or p^1. The multiprobability prediction is typically interpreted as an interval $[p_0, p_1]$ for binary problems. The intervals are guaranteed to be well-calibrated in the large (frequentist validity); on average over many predictions, the reported probabilities match empirical frequencies.

In practice, using Venn-Abers predictors, which combine a base classifier's scores with isotonic regression to get tight probability intervals, and sometimes a single probability via a suitable aggregation (e.g., the midpoint) when a point estimate is preferred. Compared with plain calibration, Venn predictors provide a built-in notion of uncertainty; wide intervals flag regions with unreliable evidence, while narrow intervals indicate confident predictions. They are closely related to conformal prediction. Both types ensure distribution-free validity, but Venn predictors focus specifically on probability calibration rather than set-valued predictions, making them handy for decision-theoretic tasks (thresholding, ranking, expected-utility optimization) where trustworthy probabilities matter.

Conformal predictive systems (CPSs) are the conformal probabilistic predictors in the case of regression. The output of a CPS is a conformal predictive distribution (CPD), which consists of p-values arranged into a distribution function. Importantly, CPDs are valid under unconstrained randomness, as a consequence of p-values being independent and uniformly distributed on $[0, 1]$ (strong calibration in probability). When constructing CPSs, it is convenient to work with conformity measures rather than nonconformity measures. However, not all conformity measures determine CPSs, because the p-values cannot be arranged into distribution functions. A simple, but restrictive condition is that the conformity measure is monotonic (monotonically increasing in y_n and monotonically decreasing in y_i, $i = 1, \ldots, n - 1$). Other sufficient conditions are discussed in [64].

An important advantage of CPDs over conformal prediction intervals is that CPDs contain more information. From a CPD, it is possible to extract prediction intervals at any significance level ε. The main advantage, however, is that CPDs can be used in combination with utility functions to make optimal decisions. [62] introduced the method of conformal predictive decision-making (CPDM), and illustrated its use on a small dataset.

Below, we explore how these techniques have been applied in practice – focusing on medical diagnosis/risk and chemistry/drug discovery – and how their adoption has evolved over time.

6.1   Medical and Clinical Applications

In medicine, reliable probability estimates can be as important as the prediction itself. For instance, a model that predicts cancer with 0.9 probability versus 0.6 probability can lead to very different clinical decisions. Venn and conformal predictors have been used to ensure that diagnostic probabilities are trustworthy for clinicians. As noted earlier, adding valid confidence measures is especially beneficial in clinical diagnosis [19], where it informs the user "how much he can rely on each prediction". Over the past decade, researchers have applied Venn methods to various medical problems, such as early disease detection, diagnosis, and risk stratification, with promising results. The first notable application in this area was the work by [51]. This work addressed the common problem that many medical diagnostic models do not indicate the reliability of each individual prediction.

The proposed solution was to use Venn Prediction (VP), a variation of conformal prediction that produces valid probabilistic predictions, in conjunction with neural networks. For test cases with medical datasets, the VP approach produced well-calibrated confidence measures, showing empirically that it could flag uncertainty in predictions and outperform standard neural networks in terms of reliability. This is a clear example of Venn probabilities that enable more trustworthy risk assessments in practice.

In each case, the method provides doctors with confidence-informed probabilistic predictions – e.g., "This patient has a $0.7 - 0.85$ probability of complication X" instead of just a binary result. Such outputs allow medical professionals to incorporate algorithmic predictions into their decision-making processes with appropriate caution.

Overall, the use of Venn predictors in medicine leads to reliable probability outputs for diagnoses and risk estimates, which is invaluable for patient counselling and making high-stakes decisions (e.g. whether to perform further tests or treatments). By quantifying uncertainty, these methods help avoid false certainty, thereby improving patient safety and trust in AI-assisted diagnosis.

6.2   Chemistry and Drug Design Applications

Drug discovery and chemistry have also embraced Venn prediction methods to better handle uncertainty in predictive modelling. In these fields, datasets are often highly imbalanced (e.g. very few "active" compounds among a vast library of inactives), and making decisions under uncertainty is routine (selecting which compounds to synthesise or test next, estimating safety risks, etc.).

Traditional quantitative structure-activity relationship (QSAR) models might give a single probability that a molecule will be active or toxic, but if that probability is not well-calibrated, it could mislead scientists. Venn predictors and conformal predictive distributions help by producing calibrated probabilities even on imbalanced data, ensuring that, say, a 5% predicted chance of activity truly means about one in 20 compounds like this turn out active in the long run.

Moreover, conformal frameworks can be incorporated into iterative decision-making loops that are common in drug discovery. For example, an algorithm can predict compound activities with confidence, and those confidence-informed predictions guide which compounds to test in the next experimental round, creating a feedback loop that systematically reduces uncertainty. Some practical applications in cheminformatics and drug design include the following:

- **Metabolism Site Prediction:** [8] used Cross
  Venn-ABERS Predictors (CVAP) to predict sites of metabolic transformation on drug molecules. In this context, the model must estimate the probability that a particular site (atom) on a molecule is metabolised by enzymes. The CVAP approach produces well-calibrated probability predictions for each potential metabolic site across thousands of biotransformations. The results showed strong predictive power and proper calibration on all tested datasets. This means that chemists can trust the model's probability; for instance, if the model says a certain bond has an 80% chance of being metabolised, that probability has a solid empirical basis. Well-calibrated metabolism predictions help in designing drug molecules (for example, avoiding metabolic "hotspots" that cause rapid clearance) because scientists can gauge confidence that a given structural change will alter metabolism risk. The study concluded that CVAP is an "interesting method for further exploration in drug discovery applications" owing to its reliable probabilistic outputs.
- **Iterative Compound Screening:** In early-stage drug discovery, it's common to perform iterative screening – test a batch of compounds, use the results to update the model, then select another batch, and so on. [13] introduced a novel strategy using Venn-Abers predictors to improve iterative screening decisions. A key question in such campaigns is how many compounds should be screened in the next round to achieve the desired number of hits (active compounds). Traditionally, this was chosen somewhat arbitrarily because one lacked a reliable estimate of how many hits a given screening size would yield. The Venn-Abers method tackled this by providing accurate probabilistic estimates of the hit rates. Essentially, for any proposed batch size, the model (with Venn-Abers calibration) could predict how many actives are likely in that batch, with confidence bounds. The authors found that these predictors enabled a "reliable and flexible decision" on the portion of the library to screen each iteration. Using this approach, they could dynamically adjust batch sizes

to meet hit goals rather than guessing. Moreover, Venn Abers-guided selection yielded enriched compound subsets, meaning that the screened sets had higher hit fractions and chemically diverse hits compared to uncalibrated approaches. In practice, this helps R&D teams allocate resources efficiently: for example, "Screen the top 5% most promising compounds to retrieve 10 hits with 90% confidence" – a level of quantitative planning made possible by conformal calibration. This application highlights how multiprobability predictions plug into decision loops, actively steering the experimental design with the measured risk. It directly leverages probability estimates (and their confidence) to make better choices in an otherwise imbalanced, uncertain search space.

- **Cardiovascular Risk Prediction (Drug Safety):** Another important area is predicting drug toxicity risk (such as cardiac toxicity) from early experimental data [1]. The work demonstrated the use of Venn-Abers predictors to assess cardio-toxic risk of compounds based on in vitro assays. They trained support vector machine models on high-throughput ion channel assay data (related to heart safety, e.g. hERG channel inhibition which can cause arrhythmia) and then calibrated the predictions with Venn-Abers. The output of this system is the probability that a compound is high-risk, accompanied by a valid confidence range. Importantly, this study introduced a "more intuitive way to reflect cardiac risk" using these probabilistic predictors. Instead of a black-box score or a yes/no flag for toxicity, chemists obtain a quantified risk with an uncertainty measure. For instance, a compound might come with a prediction "Risk of cardiac toxicity 5% (with a [0–10%] confidence interval)" which is far more informative for decision-making than a raw score. By accounting for uncertainty, the team could prioritise compounds more rationally, perhaps advancing a slightly risky compound if the confidence interval is wide (indicating uncertainty) or rejecting a compound if even the lower bound of risk is unacceptably high. In summary, conformal calibration in this context ensures that safety risk predictions are dependable, reducing false reassurance or undue alarms when selecting drug candidates.

In all scenarios, the common thread is that calibration on imbalanced, noisy chemical data leads to more reliable models. Data scientists in drug discovery often deal with skewed datasets (few actives vs. many inactives, few toxic vs. many non-toxic), and Venn/CP methods help maintain valid error rates in such conditions. By using techniques such as Mondrian labelling (calibrating per class) or cross-conformal ensembles, one can trust that the model's 0.1 probability for toxicity truly corresponds to a 10% chance in practice, even if only 1% of compounds are toxic in the dataset. This reliability is critical when models are used to make costly decisions, such as which compound to synthesise next or which lead to drop due to risk.

6.3 Uptake and Outlook

Adoption over time: The conformal prediction framework was first formalised in the mid-2000s [63] as a theoretical guarantee for algorithmic learning. In the late 2000s and the early 2010s, applications began to appear in medicine and bioinformatics. By 2012, interest had grown enough to merit the first dedicated workshop (COPA 2012) on conformal prediction, and researchers such as Nouretdinov, Gammerman, and Vovk were demonstrating the practicality of these methods on real data.

The term "Venn predictor" emerged around that time to denote multi-probability extensions, and by 2012, Vovk had also proposed the Venn-Abers method (combining Venn predictors with isotonic regression for probabilistic calibration [65]).

Throughout the mid-2010s, uptake in drug discovery accelerated. Groups in pharma and academia (e.g. at AstraZeneca and Uppsala University) started publishing conformal prediction studies, as we saw with metabolism (2017) and screening (2018) examples. The advantage of guaranteed calibration resonates with chemoinformatics researchers dealing with uncertainties in predictive toxicology and hit-finding.

By the late 2010s, conformal prediction had moved from academic curiosity to a tool considered in production settings. The method's performance in enriching hits and saving resources demonstrated tangible value to R&D organisations.

In recent years, this uptake has broadened. A 2021 review in the Journal of Pharmaceutical Sciences [3] detailed how to "predict with confidence" using conformal prediction in drug discovery, reflecting growing mainstream awareness in that community.

Specialised software tools have been developed to bring these methods to data scientists. One notable example is CPSign, an open-source package for chem- and bioinformatics that fully implements conformal prediction for classification, regression, and Venn-Abers probability estimation. Published in 2024, CPSign allows users to seamlessly train models that output prediction intervals or probability distributions, using familiar machine learning algorithms (SVMs, deep learning models, etc.) under the hood. The availability of such tools has lowered the barrier to adoption: CPSign, [7], has already been utilized in several studies and even deployed in production by multiple organizations in pharma. This indicates that conformal and Venn predictors are no longer purely academic but are entering real-world data science workflows. Likewise, Python libraries (for example, Nonconformist and MAPIE) provide conformal prediction utilities for general machine learning, further expanding access.

Looking forward, the influence of these methods is expected to grow in any domain where decision-making under uncertainty is critical. Research on integrating conformal predictive distributions with deep learning and large-scale data is ongoing. For instance, recent work applied conformal prediction to pathology image analysis, quantifying diagnostic uncertainty for AI models that assist pathologists. [50]. In machine learning safety, Venn and conformal approaches are being explored to provide transparent risk metrics for black-box models (so that regulators and practitioners can understand the model confidence).

In conclusion, Venn predictors, multi-probability predictions, and conformal predictive distributions have proven their value in delivering reliable probability estimates across fields. Their applications range from guiding clinicians with trustworthy diagnosis probabilities to helping chemists pick molecules in a smarter way using calibrated risk, all the way to improving model transparency in emerging AI applications.

The core principle of valid confidence is to transform the manner in which predictions are consumed. Instead of taking a model's output at face value, decision makers are empowered with an understanding of uncertainty. For data scientists, this means models that are not just accurate but also honest about their uncertainty, which is essential for responsible AI.

The steady uptake—from pioneering medical studies in 2009–2013, through extensive chemistry/pharmaceutical usage in 2015–2020, to today's tooling and cross-disciplinary adoption—suggests that probabilistic predictors with confidence guarantees are becoming a practical staple in data science toolkits. As more practitioners realise the benefits of calibrated predictions on imbalanced or high-risk data, these methods are poised for even wider application in the coming years, promoting safer and more informed decision-making in AI-driven systems.

7   Financial Applications

The financial sector, characterized by high volatility, inherent uncertainty, and significant consequences for error, represents fertile ground for the application of conformal methods [15,55]. The adoption of these techniques signals a crucial evolution in the use of machine learning in finance. This marks a transition from using models solely for generating point forecasts to leveraging them as tools for principled and quantifiable risk management. A point prediction, such as an expected stock return or the probability of loan default, provides no information about the model's confidence. In contrast, a conformal prediction interval or set explicitly delineates a range of outcomes that is guaranteed to contain the true value with a pre-specified probability. This allows financial institutions to make decisions based not on an opaque best guess but on a well-defined worst-case scenario, thereby aligning the statistical output of a model directly with the institution's risk appetite.

7.1   Calibrated Uncertainty in Portfolio Selection and Volatility Forecasting

Forecasting asset returns and market volatility is a cornerstone of quantitative finance, but the dynamic and nonstationary nature of financial markets makes this task notoriously difficult [15]. Conformal prediction provides a robust, model-agnostic framework for constructing prediction intervals around these forecasts, enabling more sophisticated risk management strategies. A prime example is the **Conformal Predictive Portfolio Selection (CPPS)** framework, which directly integrates conformal prediction into the portfolio construction process [38]. The standard approach to portfolio management involves forecasting the returns of various assets and optimising portfolio weights. The CPPS framework enhances this by first forecasting the returns for a set of candidate portfolios and then using conformal prediction to compute a valid prediction interval for each portfolio's future return.

The final portfolio was then selected based on the objective function defined over these intervals. For instance, an investor might choose the portfolio with the highest predicted lower bound of its return interval at the 95% confidence level. This strategy explicitly aims to improve worst-case portfolio performance. Empirical studies on US and Japanese stock data have shown that portfolios constructed using CPPS, with underlying models ranging from autoregressive models to neural networks, deliver superior and more stable cumulative returns while effectively reducing the impact of extreme losses compared to strategies based on point forecasts alone [38].

Beyond asset returns, market volatility is a fundamental input for pricing derivatives and managing risk. In a study presented at COPA 2023, [15] proposed an algorithm to estimate the market-implied uncertainty of the future realised volatility. This method treats a market-based volatility forecast, such as the VIX index, as an underlying point prediction. It then applies Adaptive Conformal Inference (ACI) to construct a valid prediction interval around this forecast. By analysing over 15 years of daily data for the S&P 500 index, this study demonstrated that this approach can successfully infer market-implied prediction intervals for realised volatility. The use of ACI is particularly relevant here because it is designed to handle the distribution shifts common in financial time series, thereby providing more reliable uncertainty estimates in a dynamic market environment [15].

7.2   Reliable Credit Scoring and Lending Decisions

Credit scoring is a critical function in the lending industry in which the primary goal is to assess the risk of a borrower defaulting on a loan. While machine learning models can achieve high accuracy in predicting default, they are often poorly calibrated, meaning that their output probabilities do not accurately reflect the true likelihood of default. An overconfident model that assigns a low probability of default to high-risk applicants can lead to significant financial losses.

Conformal prediction is a powerful solution for this problem [26]. Rather than attempting to recalibrate the model's internal probabilities, conformal classification provides a post-hoc "wrapper" that produces a prediction *set* for each applicant. For a given significance level ε, the prediction set may be `{Default}`, `{No Default}`, or crucially, `{Default, No Default}` or `{}`, the empty set. The first two outcomes represent confident predictions, whereas the last two, containing both possible labels or none, is an unambiguous signal of the model's uncertainty.

Applied in this mode, a conformal classifier is transformed to a classifier with reject option [32]. The output is directly actionable for financial institutions. An applicant receiving the set `{Default, No Default}` or `{}` can be flagged for an additional manual review by a loan officer, preventing an automated decision based on unreliable prediction. This transforms the model from a black-box decision maker into a more transparent decision support tool.

This provides a formal guarantee that, in the long run, the proportion of true labels falling outside their respective prediction sets will not exceed ε. By analysing the error probability, conditioned on set size, [32] demonstrated that it is also possible to estimate the error probability of the single-label predictions. This allows institutions to manage their risk exposure with statistical rigour. However, practitioners must carefully consider the use case and the interpretation of the guarantee. The long-run coverage property does not guarantee a specific outcome for any single finite batch of loan applications, which is a crucial distinction in a regulated industry.

7.3 Anomaly and Fraud Detection in Financial Transactions

The detection of fraudulent financial transactions is an ideal application of conformal methods, framed as an unsupervised anomaly detection problem. Fraudulent activities are, by definition, rare and deviate from normal patterns of behaviour, making them "non-conformal" with the vast majority of legitimate transactions [40].

The core mechanism of conformal prediction is the calculation of a non-conformity score for each new example, which measures how "strange" or "atypical" the example is compared to a set of previous examples [55, 58].

In the context of anomaly detection, a model is trained exclusively on normal, non-fraudulent data. For a new transaction, a non-conformity score was calculated. This score is then used to compute a p-value that represents the fraction of normal examples in the calibration set that are at least as strange as the new transaction [63]. A legitimate transaction, similar to the training data, will yield a high p-value, whereas a fraudulent transaction will yield a very low p-value.

This provides a statistically principled method for flagging suspicious activities. By setting a significance level ϵ (for example, 0.01), the system can automatically flag any transaction with a p-value less than ϵ as anomalous. The key advantage of this approach is the validity guarantee: under the assumption of exchangeability, the rate of false alarms (i.e. flagging a normal transaction as fraudulent) is guaranteed to not exceed ϵ in the long run [63]. This provides a formal, tunable mechanism for balancing security with customer convenience, representing a significant improvement over ad hoc methods that rely on arbitrary thresholds for anomaly scores [40].

8 Enhancing Reliability in Autonomous Systems and Computer Vision

The deployment of artificial intelligence in safety-critical domains, such as autonomous driving and critical infrastructure monitoring, necessitates a higher standard of reliability than in many other fields. A simple guarantee of marginal coverage, although powerful, is often insufficient. The primary concern may not be whether a prediction set contains the true label but whether the system avoids specific, catastrophic errors, such as failing to detect a pedestrian or a stop signal. The applications in this section demonstrate a significant maturation of the conformal framework, moving beyond foundational coverage guarantees towards the control of more specific, application-relevant risks. This evolution is driven by the pragmatic engineering requirements of building AI systems that are not only accurate but also demonstrably safe and trustworthy.

8.1 Safe Perception for Autonomous Driving

Conformal prediction is emerging as a key technology for providing the formal safety guarantees required for autonomous vehicle perception systems [21,22]. Research has progressed toward the development of sophisticated frameworks that tailor the general principles of conformal prediction to the specific challenges of navigating complex, dynamic environments. One such advancement is the **Knowledge-Refined Prediction Sets (KRPS)** framework, which addresses the problem of semantic inconsistency in multitask perception. An autonomous vehicle must simultaneously perform multiple classification tasks, such as identifying an agent (e.g. vehicle, pedestrian), its location (e.g. road, sidewalk), and its actions (e.g. moving, stopped). Applying standard conformal prediction independently to each task can lead to logically inconsistent results, for example, classifying an object as a 'vehicle' but its location as being on the 'sidewalk'. KRPS solves this by integrating a knowledge graph that encodes semantic relationships between classes across different tasks. The framework first generates a conformal prediction set for a primary task (e.g. agent classification) and then uses the knowledge graph to prune the possible labels for subsequent tasks before applying another conformal step. This sequential refinement ensures that the final prediction sets are not only statistically valid but also semantically coherent, resulting in smaller and more contextually appropriate outputs for the vehicle's decision-making module [20].

Another innovative application is the **Augmented Reality Conformal Prediction (AR-CP)** framework, which focuses on the human-in-the-loop aspect of assisted driving [21]. This system is designed to help human drivers perceive their surroundings in adverse conditions, such as bad weather or poor lighting. When the underlying object detection model is uncertain about an object, it produces a large conformal prediction set (e.g. {`Car`, `Truck`, `Van`}). Instead of displaying this ambiguous set, which could increase the driver's cognitive load, the AR-CP system uses a class taxonomy to find the nearest common parent class (for example, "Vehicle") and visualises this higher-level concept on an augmented reality windshield display. This provides the driver with a simplified yet reliable piece of information. User studies have shown that this approach significantly improves object identification, reduces mental load, and increases overall situational awareness compared with showing no information or showing raw bounding boxes [21].

More recent work has even begun to integrate conformal prediction with Large Language Models (LLMs) for navigation. The **SafePath** framework uses an LLM to generate candidate paths by reasoning about traffic scenarios and then employs conformal prediction to refine the path selection and filter out risky options [22]. This approach aims to provide formal safety guarantees for the final motion plan, with initial results showing a significant reduction in collision rates compared to state-of-the-art planning methods [22].

8.2   Conformal Risk Control for Critical Infrastructure Monitoring

Railway signal detection provides a compelling case study for the application of advanced conformal methods to a critical infrastructure. Deploying a deep-learning-based object detector in a certified railway system requires confidence estimates that are not just heuristic but formally guaranteed [4]. A standard conformal guarantee of coverage is not sufficient, as the operational risk is asymmetric; A false negative (failing to detect a signal) is far more dangerous than a false positive.

To address this, [4] applied the framework of Conformal Risk Control (CRC) [6] to this problem in a study presented at COPA 2023. CRC generalises standard conformal prediction by allowing for the control of arbitrary bounded loss functions, not just the 0-1 loss corresponding to the coverage error. This allows engineers to define a loss function that accurately reflects the real-world operational risks of a system, such as a higher penalty for false negatives. The CRC procedure then constructs prediction sets that provide a finite-sample guarantee that the expected value of this specific loss function is below the user-defined risk level α [6].

Using a novel dataset of images collected from the perspective of a training operator, the authors demonstrated how CRC, combined with state-of-the-art object detectors, can produce more reliable and trustworthy predictions. Their work provides a practical methodology for evaluating model performance and achieving formally guaranteed uncertainty bounds that are directly tied to the safety requirements of the application. This represents a crucial step in moving machine learning models from research labs into real-world certified systems where safety and reliability are paramount [4].

9   Quantifying Uncertainty in Natural Language Processing (NLP)

The rapid proliferation of Large Language Models (LLMs) and other sophisticated NLP systems has been accompanied by growing concerns about their reliability. These models are known to be poorly calibrated and prone to "hallucinations"—generating fluent, confident-sounding text that is factually incorrect or nonsensical [14]. This fundamental weakness stems from an inability to "know when they do not know."

Conformal prediction is emerging as a theoretically sound and practically useful framework for addressing this challenge [14]. By its very nature, it provides an antidote to overconfidence. Instead of being forced to output a single, potentially erroneous answer, a conformalized model produces a *set* of plausible answers. A large prediction set serves as a direct and quantifiable signal of high model uncertainty, thereby providing a crucial mechanism for building more trustworthy NLP systems.

9.1 From Calibrated Classification to Trustworthy Generation

The model-agnostic nature of conformal prediction allows it to be applied across a wide spectrum of NLP tasks, from foundational classification problems to the cutting edge of text generation. A comprehensive survey by Campos et al. [14] outlined the extensive applications and opportunities for conformal methods in NLP.

At the foundational level, Inductive Conformal Prediction (ICP) has been successfully applied to core tasks such as part-of-speech (POS) tagging and text infilling (also known as masked language modelling) [25]. In these applications, a base model such as BERT or a BiLSTM is used to generate initial predictions (e.g. softmax probabilities for each possible POS tag). ICP is then used as a wrapper to convert these scores into prediction sets with guaranteed coverage. Experiments have shown that this approach is highly practical, producing usefully small prediction sets—for instance, averaging only one to two POS tags at a 99% confidence level—while rigorously maintaining the nominal coverage rate. This provides a reliable way to quantify the uncertainty for each word in a sentence [25].

The most pressing challenge in modern NLP is the unreliability of LLMs. Conformal prediction offers a promising path for mitigating the risk of hallucinations. Because model uncertainty is strongly correlated with the tendency to hallucinate, the size of a conformal prediction set can serve as a warning signal. At the token level, instead of greedily selecting the single most likely next token, a conformalized LLM can produce a set of plausible next tokens. If this set is large, it indicates that the model is uncertain how to proceed. At the sentence level, this approach could be extended to generate a set of possible continuations. In a human-in-the-loop system, a large prediction set can trigger a request for clarification or indicate that the model's output should not be trusted, thereby forming a critical safety layer for the deployed generative models [14].

9.2 Conformal Methods for Automatic Speech Recognition (ASR)

Automatic Speech Recognition (ASR) is a critical technology in many high-risk environments, such as communication with air traffic control or in-vehicle command systems, where a mistranscribed word can have severe consequences. A paper by Ernez et al. [24] from the COPA conference demonstrated two distinct and powerful applications of conformal methods to improve the reliability of an end-to-end ASR model based on the Wav2vec 2.0 architecture.

The first application targets the overall quality of a transcript by controlling the **Word Error Rate (WER)**, a standard metric in ASR. The authors applied the **Conformal Risk Control (CRC)** framework, similar to its use in railway signalling. Here, the goal is not just to produce one transcript, but also a *set of candidate sentences*. The CRC procedure provides a formal guarantee that this set contains at least one sentence whose WER is below user-specified risk level α. In their experiments on the LibriSpeech corpus, they were able to generate prediction sets that contained a transcript with a WER below 2 % with 80 % confidence. While the average set size was 29 sentences, for a critical application, this provides a valuable set of high-quality options for a human operator to review [24].

The second application focused on identifying specific points of failure within a single transcript. Using Inductive Conformal Prediction (ICP), the authors treated the problem as a word-level binary classification task: for each word in the ASR output, is it correct or incorrect? By training a conformal classifier, they produced a prediction set for each word. If the set for a given word is {incorrect} or {correct, incorrect}, it is flagged as uncertain. This method proved highly effective and successfully detected 90% of incorrectly transcribed words in their test set. This capability allows a system to automatically highlight potentially erroneous words in a transcript, directing user attention to the most critical parts of the text and enabling more efficient verification and correction [24].

10   Conformal Prediction in Industrial and Engineering Contexts

The adoption of conformal methods in an industrial setting marks a significant phase of maturation in the framework. This signals a transition from a primarily academic pursuit to a practical deployable tool for industrial data scientists. This "industrialization" is facilitated by the development of accessible open-source libraries such as MAPIE and Crepes, which lower the barrier to entry and allow companies to apply sophisticated uncertainty quantification techniques to their own proprietary, real-world data [58]. Case studies from manufacturing, supply chain management, and predictive maintenance demonstrate that the rigorous guarantees provided by conformal prediction are not just of theoretical interest but offer tangible value in operational decision-making, where managing uncertainty is key to efficiency and cost reduction [58].

10.1   Conformal Predictive Systems for Manufacturing and Demand Forecasting

While many publications on conformal prediction focus on popular open-source datasets, a paper from COPA 2023, [58] details the application of conformal regression to two real-world industrial use cases at the Husqvarna Group, a major manufacturer of outdoor power products. This work is significant because it demonstrates the direct application of these methods to solve concrete business problems using proprietary industrial data. The use cases spanned manufacturing analytics, such as quality control for injection moulding processes and supply chain management, specifically demand prediction [58].

A key element highlighted in these industrial applications is the utility of Conformal Predictive Systems (CPS). While standard conformal regression produces a single prediction interval, CPS outputs a full, valid predictive distribution (in the form of a cumulative distribution function or CDF) for each new instance [58]. This richer output enables far more nuanced decision-making. For example, in a manufacturing setting, instead of merely knowing the 95% interval for a quality control parameter, an engineer can use a conformal predictive distribution to calculate the probability that the parameter will exceed a critical failure threshold.

Similarly, in demand forecasting, a supply chain manager can estimate the probability of demand falling below a certain level, allowing for more precise inventory management and risk assessment [58]. The successful implementation of these systems in a major industrial company, facilitated by Python libraries such as MAPIE and Crepes, underscores the growing practicality and business relevance of the conformal framework [58].

10.2   Venn-Abers Predictive Distributions for Predictive Maintenance

Predictive maintenance (PdM) is a critical area in modern industry that aims to predict equipment failures before they occur in order to minimise downtime and maintenance costs [39]. A central task in PdM is estimating the Remaining Useful Life (RUL) of an asset. However, a simple point forecast of the RUL is often insufficient for making optimal, cost-effective maintenance decisions. What an operator truly needs is a probabilistic assessment of the failure risk over time.

This is a problem uniquely suited for probabilistic prediction methods, such as **Venn Prediction**. Recent studies have applied **Inductive Venn-Abers Predictive Distributions** (IVAPD) in PdM contexts (e.g., degradation measures) and to energy consumption forecasting [47]. First proposed by Nouretdinov et al. (Nouretdinov, Volkhonskiy, et al., 2018), this technique extends the Venn-Abers predictor, originally designed for classification, to the regression setting. It does so by mapping the regression problem onto a series of binary classification problems at different value thresholds t, yielding a well-calibrated predictive distribution for the quantity of interest (e.g., RUL or a degradation measure) [48].

This allows an operator to ask and answer sophisticated probabilistic questions, such as, "What is the probability that this machine will fail within the next 100 operating hours?" or "What is the warranty period we can offer such that the probability of failure is less than 1%?"

Further research presented at COPA 2024 explored the application of IVAPD to real-time predictive maintenance and energy consumption forecasting tasks and investigated appropriate evaluation metrics such as the Continuous Ranked Probability Score (CRPS) and interval width to assess the accuracy and sharpness of the predictive distributions [47].

Complementing this, Kharazian et al. [39] introduced an algorithm called **Conformal Prediction for Active Learning (CoPAL)**. CoPAL uses the uncertainty quantified by conformal prediction, specifically the width of the prediction intervals, to guide an active learning process. In the PdM context, it intelligently selects the most uncertain assets for inspection and labelling, thereby improving the performance of the RUL model in the most data-efficient manner possible [39].

11   Testing Randomness: Conformal Test Martingales in Practice

The assumption of exchangeability, that the ordering of data points does not affect their joint probability distribution, is a cornerstone of most statistical and machine-learning theories [63]. However, in practice, this assumption is frequently violated, especially in online settings, where data distributions can drift over time. Conformal Test Martingales (CTMs), a concept deeply rooted in the algorithmic theory of randomness, provide a powerful, non-parametric framework for testing this foundational assumption online [60,66]. This capability bridges the gap between abstract statistical theory and the concrete operational challenges of deploying and maintaining machine learning models in a dynamic world. It provides a principled, data-driven answer to the critical practical question: "Is my model still valid, or does it need to be retrained?"

11.1   Online Change-Point Detection for Model Maintenance

A predictive model, when trained on historical data, will exhibit a decline in performance if there is a change in the underlying data-generating process. The crucial operational challenge is to detect this "dataset shift" or "change-point" in real time to trigger a model retraining event. Conformal Test Martingales offer a general and theoretically sound solution to this problem [66].

The design of this methodology is elegant. First, a conformal predictor is used to transform the incoming stream of data $z_1, z_2, \ldots$ into a sequence of p-values $p_1, p_2, \ldots$. If the data remain exchangeable, these p-values will be uniformly distributed on $[0, 1]$. A test martingale is then constructed as a betting strategy that starts with a capital of one and gambles against this uniformity. If the p-values begin to deviate from uniformity—for instance, if they become systematically small because the model is encountering data it deems "strange"—the martingale's capital will grow. Thus, a large martingale value serves as strong evidence against the exchangeability assumption [63,66].

[66] proposed concrete schemes for using CTMs to decide when to retrain a prediction algorithm. These schemes involve running a statistical procedure in addition to the martingale process. For example, the **Ville procedure** raises an alarm when the martingale S_n exceeds a threshold c, which guarantees that the probability of a false alarm is at most $1/c$. To improve the detection efficiency over time, this can be supplemented with conformal versions of the popular **CUSUM** and **Shiryaev–Roberts** procedures, which are more sensitive to changes that occur after the initial deployment. These methods have been demonstrated on real-world data, showing that they can effectively detect change points and provide a valid, automated signal for model maintenance [66].

11.2 Anomaly Detection in Industrial and Geospatial Data Streams

The same mechanism used to detect dataset shifts for model retraining can be applied directly to online anomaly detection. To date, this is arguably one of the most widespread and impactful industrial applications of conformal methods. The approach, often using **Inductive Conformal Martingales (ICMs)** for computational efficiency, is non-parametric and requires very few assumptions about the data, making it highly robust and applicable to diverse data streams.

In [60], ICMs were compared to classical change-point detection methods (such as CUSUM) and specialised "oracles" that have partial knowledge of data distributions. The study found that ICMs were comparably efficient to oracles and only slightly less efficient than optimal classical methods, but with the major advantage of being far more general and not requiring the strong, often unrealistic, distributional assumptions of classical techniques [60].

The versatility of this approach is further demonstrated by its recent application in detecting change points in high-dimensional data from geospatial object detectors [67]. In scenarios where an autonomous satellite monitors a region, the environment can change owing to seasonal shifts, new construction, or other events. A CTM-based system can monitor the output of an object detector and automatically detect when the domain has shifted significantly, indicating that the detector must be adapted or retrained. This provides a critical capability for maintaining the performance of automated systems deployed in dynamic real-world environments without constant human supervision [67].

12 A Survey of Open-Source Software for Conformal Prediction

Conformal Prediction (CP) is a model-agnostic framework for creating statistically rigorous prediction intervals and sets with guaranteed coverage. Its practical adoption has been accelerated by a growing ecosystem of open-source software that translates complex statistical theory into deployable tools. This survey covered the most prominent libraries for CP and related methods, such as Venn Prediction and Conformal Test Martingales, across Python, R, Julia, and Java.

The Python Ecosystem

The Python ecosystem is the most mature and diverse, with libraries for general-purpose use, deep learning, and specialised methodologies.

General-Purpose and Scikit-Learn-Centric Libraries. These libraries integrate with `scikit-learn` to "conformalize" nearly any compatible model.

- **MAPIE** [16]: A flagship library in the `scikit-learn-contrib` ecosystem, offering a wide range of methods like Jackknife+, CV+, and Conformalized Quantile Regression (CQR) for regression, and modern set-prediction algorithms like RAPS for classification.
- **PUNCC** [41]: Features a dual API for both non-experts (`scikit-learn` compatible) and researchers, with interoperability for PyTorch and TensorFlow.
- **crepes** [12]: A model-agnostic library whose unique contribution is the implementation of Conformal Predictive Systems (CPS). Instead of just intervals, CPS outputs full predictive cumulative distribution functions (CDFs) for each test instance, enabling richer probabilistic forecasts.
- **nonconformist:** A foundational but now unmaintained library that pioneered CP integration with `scikit-learn`.

Libraries for Deep Learning. Specialised libraries offer GPU acceleration and tight integration with deep-learning frameworks.

- **TorchCP** [34]: A PyTorch-native library with a modular design (`Trainer`, `Score`, `Predictor`) that supports GPU acceleration and batch processing. It uniquely enables "training-integrated" CP, where models are optimised during training to produce more efficient prediction sets and supports advanced applications such as Graph Neural Networks (GNNs) and Large Language Models (LLMs).
- **Fortuna** [18]: A library for uncertainty quantification built on the JAX framework, which includes functionalities for conformal prediction alongside other methods.

Specialized Implementations. These packages are dedicated to specific branches of the CP family.

- **online-cp** [30]: Focuses on the online setting where data arrives sequentially. Its key feature is the implementation of conformal test martingales to test the exchangeability assumption on the fly, allowing for the detection of distribution shifts.

– **Venn-ABERS Libraries:** Dedicated libraries also exist for Venn Prediction, which produces multi-probabilistic outputs.

Key implementations include `venn-abers` by `Ivan Petej` and `VennABERS` by `Paolo Toccaceli`.

The R Ecosystem

The R ecosystem features high-quality packages that are often canonical implementations of seminal research methods.

– **conformalInference:** The reference implementation for influential papers on CP for regression and covariate shift from leading researchers like Tibshirani, Candès, and Wasserman.
– **AdaptiveConformal:** Implements state-of-the-art Adaptive Conformal Inference (ACI) algorithms for non-exchangeable data like time series, which adjust prediction intervals in response to ongoing model performance.

Emerging Ecosystems: Julia and Java

CP is also expanding to other languages for both general and domain-specific uses.

– **ConformalPrediction.jl (Julia) [2]:** The primary CP package for the Julia language, designed for seamless integration with its main machine learning framework, `MLJ.jl`.
– **CPSign (Java) [9]:** A specialized, end-to-end application for cheminformatics (QSAR modelling), which uses CP and the Venn-ABERS methodology to provide reliable predictions for chemical compounds.

Summary and Comparative Analysis

The open-source landscape for conformal prediction is diverse, and Python offers the broadest selection of tools for general use (`MAPIE`), deep learning (`TorchCP`), and specialised methods (`online-cp`). The R ecosystem provides canonical implementations of cutting-edge research (`conformalInference`, `AdaptiveConformal`), whereas libraries in Julia (`ConformalPrediction.jl`) and Java (`CPSign`), [7], demonstrate the framework's expanding reach. The following table (Table 3) provides a comparative summary.

Table 3. Comparative Summary of Conformal Prediction Libraries

Library	Lang.	Primary Focus	Backend/ Framework	GPU Accel.	Key Strengths	CPS?	Venn?	Test Martingales?
MAPIE	Python	General-Purpose	NumPy/Scikit-learn	No	Comprehensive methods, scikit-learn integration, risk control	No	No	No
PUNCC	Python	General-Purpose	Scikit-learn, PyTorch, TF	No	Ease of use, interoperability, low-level API for research	No	No	No
TorchCP	Python	Deep Learning	PyTorch	Yes	GPU-native, modular design, training-integrated methods, GNN/LLM support	No	No	No
Fortuna	Python	Deep Learning	JAX	Yes	JAX-native, part of a broader UQ library	No	No	No
nonconformist	Python	General-Purpose	NumPy/Scikit-learn	No	Foundational library (note: no longer actively maintained)	No	No	No
crepes	Python	General-Purpose/CPS	NumPy/Scikit-learn	No	Canonical implementation for Conformal Predictive Systems (CPS)	Yes	No	No
online-cp	Python	Online Learning	NumPy	No	Online setting, exchangeability testing with conformal test martingales	Yes	No	Yes
conformalInference	R	Regression	R-native	No	Reference implementation for seminal research papers	No	No	No
AdaptiveConformal	R	Time Series	R-native	No	State-of-the-art for non-exchangeable data (ACI methods)	No	No	No
ConformalPrediction.jl	Julia	General-Purpose	Julia/MLJ.jl	No	Idiomatic CP for the Julia ecosystem, MLJ integration	No	No	No
CPSign	Java	Chem-informatics	JVM	No	End-to-end QSAR application, includes Venn-ABERS	No	Yes	No
venn-abers	Python	Venn Prediction	NumPy/Scikit-learn	No	scikit-learn-compatible Venn-ABERS calibration	No	Yes	No
VennABERS	Python	Venn Prediction	NumPy/Python	No	Fast, pure-Python implementation of inductive Venn-ABERS	No	Yes	No

13 Concluding Remarks

The journey of conformal methods, from their theoretical origins in the work of Alexander Gammerman and his colleagues on algorithmic randomness to their current deployment in a vast landscape of real-world applications, is a testament to the power of principled statistical thinking. The diverse examples explored in this chapter—spanning finance, autonomous systems, natural language processing, and industrial engineering—are unified by a single compelling theme: the critical need for **trustworthy, reliable, and rigorously quantified uncertainty** in modern machine learning.

As this survey has shown, the conformal framework has proven to be remarkably versatile. In finance, it provides a direct bridge from statistical prediction to principled risk management. The safety-critical domains of autonomous driving and infrastructure monitoring have evolved beyond simple coverage guarantees to enable the control of specific application-relevant risks. The formidable challenge of unreliable Large Language Models offers a theoretically sound antidote to overconfidence and hallucination. In industry, the development of accessible software libraries has facilitated its "industrialization", turning it into a practical tool for improving manufacturing processes and supply chains. Finally, the application of conformal test martingales to online change-points and anomaly detection demonstrates a profound connection between the deepest concepts of the randomness theory and the most practical challenges of maintaining live AI systems.

The continued expansion of these methods is certain. Future research will undoubtedly focus on further relaxing the exchangeability assumption, developing more powerful methods for achieving granular conditional guarantees, and creating more computationally efficient algorithms for high-dimensional and streaming data. The foundational work of Alexander Gammerman has not only enriched the fields of statistics and machine learning but has also provided an indispensable toolkit for building the next generation of safe, reliable, and ultimately more intelligent artificial systems.

Acknowledgement. It is with profound pleasure and deep respect that we dedicate this chapter to the commemorative volume celebrating Professor Alexander Gammerman on the occasion of his 80th birthday. The methodologies we examine herein, spanning domains such as drug discovery, finance, and autonomous systems, are all grounded in the theoretical foundations he was instrumental in establishing. Professor Gammerman's pioneering contributions in transforming the abstract principles of algorithmic randomness into a robust, practical framework of conformal prediction have fundamentally reshaped the landscape of trustworthy machine learning. This survey of real-world applications stands as a testament to the profound and enduring impact of his intellectual legacy.

References

1. Ahlberg, E., Buendia, R., Carlsson, L.: Using Venn-Abers predictors to assess cardio-vascular risk. In: Gammerman, A., Vovk, V., Luo, Z., Smirnov, E., Peeters, R. (eds.) Proceedings of the Seventh Workshop on Conformal and Probabilistic Prediction and Applications. Proceedings of Machine Learning Research, vol. 91, pp. 132–146. PMLR (2018)
2. Altmeyer, P.: Conformalprediction.jl (2022)
3. Alvarsson, J., McShane, S.A., Norinder, U., Spjuth, O.: Predicting with confidence: using conformal prediction in drug discovery. J. Pharm. Sci. **110**(1), 42–49 (2021)
4. Andéol, L., Fel, T., De Grancey, F., Mossina, L.: Confident object detection via conformal prediction and conformal risk control: an application to railway signaling. In: Conformal and Probabilistic Prediction with Applications, pp. 36–55. PMLR (2023)
5. Angelopoulos, A., Candes, E., Tibshirani, R.J.: Conformal PID control for time series prediction. In: Advances in Neural Information Processing Systems, vol. 36, pp. 23047–23074 (2023)
6. Angelopoulos, A.N., Bates, S., Fisch, A., Lei, L., Schuster, T.: Conformal risk control. arXiv preprint arXiv:2208.02814 (2022)
7. Aros Bio AB. Cpsign (2024). https://github.com/arosbio/cpsign. Version 2.0.0 (commit 96b183e). Accessed 06 Oct 2025
8. Arvidsson, S., Spjuth, O., Carlsson, L., Toccaceli, P.: Prediction of metabolic transformations using cross Venn-ABERS predictors. In: Gammerman, A., Vovk, V., Luo, Z., Papadopoulos, H. (eds.) Proceedings of the Sixth Workshop on Conformal and Probabilistic Prediction and Applications. Proceedings of Machine Learning Research, vol. 60, pp. 118–131. PMLR (2017)
9. Arvidsson McShane, S., Norinder, U., Alvarsson, J., Ahlberg, E., Carlsson, L., Spjuth, O.: CPSign: conformal prediction for cheminformatics modelling. J. Cheminform. **16**(1), 75 (2024)
10. Barber, R.F., Candes, E.J., Ramdas, A., Tibshirani, R.J.: Conformal prediction beyond exchangeability. Ann. Stat. **51**(2), 816–845 (2023)
11. Bosc, N., Atkinson, F., Felix, E., Gaulton, A., Hersey, A., Leach, A.R.: Large scale comparison of QSAR and conformal prediction methods and their applications in drug discovery. J. Chem. **11**(1), 4 (2019)
12. Boström, H.: Conformal prediction in python with crepes. In: Proceedings of the 13th Symposium on Conformal and Probabilistic Prediction with Applications, pp. 236–249. PMLR (2024)
13. Buendia, R., Engkvist, O., Carlsson, L., Kogej, T., Ahlberg, E.: Venn-Abers predictors for improved compound iterative screening in drug discovery. In: Conformal and Probabilistic Prediction and Applications, pp. 201–219 (2018)
14. Campos, M., Farinhas, A., Zerva, C., Figueiredo, M.A.T., Martins, A.F.T.: Conformal prediction for natural language processing: a survey. Trans. Assoc. Comput. Linguist. **12**, 1497–1516 (2024)
15. Canete, A.: Market implied conformal volatility intervals. In: Conformal and Probabilistic Prediction with Applications, pp. 89–99. PMLR (2023)
16. Cordier, T., Blot, V., Lacombe, L., Morzadec, T., Capitaine, A., Brunel, N.: Flexible and systematic uncertainty estimation with conformal prediction via the MAPIE library. In: Conformal and Probabilistic Prediction with Applications (2023)

17. De la Torre, J., Rodriguez, L.R., Monteagudo, F.E., Arredondo, L.R., Enriquez, J.B.: Electricity price forecast in wholesale markets using conformal prediction: case study in Mexico. Energy Sci. Eng. **12**(3), 524–540 (2024)
18. Detommaso, G., Gasparin, A., Donini, M., Seeger, M., Wilson, A.G., Archambeau, C.: Fortuna: a library for uncertainty quantification in deep learning. arXiv preprint arXiv:2302.04019 (2023)
19. Devetyarov, D., et al.: Conformal predictors in early diagnostics of ovarian and breast cancers. Prog. Artif. Intell. **1**(3), 245–257 (2012)
20. Doula, A., Güdelhöfer, T., Mühlhäuser, M., Guinea, A.S.: Conformal prediction for semantically-aware autonomous perception in urban environments. In: Agrawal, P., Kroemer, O., Burgard, W. (eds.) Proceedings of the 8th Conference on Robot Learning. Proceedings of Machine Learning Research, vol. 270, pp. 2629–2654. PMLR (2025)
21. Doula, A., Mühlhäuser, M., Guinea, A.S.: AR-CP: uncertainty-aware perception in adverse conditions with conformal prediction and augmented reality for assisted driving. In: Proceedings of the IEEE/CVF Conference on Computer Vision and Pattern Recognition, pp. 216–226 (2024)
22. Doula, A., Mühlhäuser, M., Guinea, A.S.: Safepath: conformal prediction for safe LLM-based autonomous navigation. arXiv preprint arXiv:2505.09427 (2025)
23. Eklund, M., Norinder, U., Boyer, S., Carlsson, L.: Application of conformal prediction in QSAR. In: Iliadis, L., Maglogiannis, I., Papadopoulos, H., Karatzas, K., Sioutas, S. (eds.) AIAI 2012. IAICT, vol. 382, pp. 166–175. Springer, Heidelberg (2012). https://doi.org/10.1007/978-3-642-33412-2_17
24. Ernez, F., Arnold, A., Galametz, A., Kobus, C., Ould-Amer, N.: Applying the conformal prediction paradigm for the uncertainty quantification of an end-to-end automatic speech recognition model (wav2vec 2.0). In: Conformal and Probabilistic Prediction with Applications, pp. 16–35. PMLR (2023)
25. Fisch, A., Schuster, T., Jaakkola, T., Barzilay, R.: Conformal prediction sets with limited false positives. In: Chaudhuri, K., Jegelka, S., Song, L., Szepesvari, C., Niu, G., Sabato, S. (eds.) Proceedings of the 39th International Conference on Machine Learning. Proceedings of Machine Learning Research, vol. 162, pp. 6514–6532. PMLR (2022)
26. Fontana, M., Zeni, G., Vantini, S.: Conformal prediction: a unified review of theory and new challenges. Bernoulli **29**(1), 1–23 (2023)
27. Forreryd, A., Norinder, U., Lindberg, T., Lindstedt, M.: Predicting skin sensitizers with confidence: using conformal prediction to determine applicability domain of GARD. Toxicol. Vitro **48**, 179–187 (2018)
28. Gibbs, I., Candes, E.: Adaptive conformal inference under distribution shift. Adv. Neural. Inf. Process. Syst. **34**, 1660–1672 (2021)
29. Gibbs, I., Candès, E.J.: Conformal inference for online prediction with arbitrary distribution shifts. J. Mach. Learn. Res. **25**(162), 1–36 (2024)
30. Szabadváry, J.H., Löfström, T., Matela, R.: online-cp: a python package for online conformal prediction, conformal predictive systems and conformal test martingales. In: Nguyen, K.A., Luo, Z., Papadopoulos, H., Löfström, T., Carlsson, L., Boström, H. (eds.) Proceedings of the Fourteenth Symposium on Conformal and Probabilistic Prediction with Applications. Proceedings of Machine Learning Research, vol. 266, pp. 595–614. PMLR (2025)
31. Szabadváry, J.H., Löfström, T.: Beyond conformal predictors: adaptive conformal inference with confidence predictors. Pattern Recogn. **170**, 111999 (2026)

32. Szabadváry, J.H., Löfström, T., Johansson, U., Sönströd, C., Ahlberg, E., Carlsson, L.: Classification with reject option: distribution-free error guarantees via conformal prediction. Mach. Learn. Appl. **20**, 100664 (2025)
33. Jianming, H., Luo, Q., Tang, J., Heng, J., Deng, Y.: Conformalized temporal convolutional quantile regression networks for wind power interval forecasting. Energy **248**, 123497 (2022)
34. Huang, J., Song, J., Zhou, X., Jing, B., Wei, H.: Torchcp: a python library for conformal prediction (2024)
35. Jonkers, J., Avendano, D.N., Van Wallendael, G., Van Hoecke, S.: A novel day-ahead regional and probabilistic wind power forecasting framework using deep CNNs and conformalized regression forests. Appl. Energy **361**, 122900 (2024)
36. Jonkers, J., Van Wallendael, G., Duchateau, L., Van Hoecke, S.: Conformal predictive systems under covariate shift. In: Vantini, S., Fontana, M., Solari, A., Boström, H., Carlsson, L. (eds.) Proceedings of the Thirteenth Symposium on Conformal and Probabilistic Prediction with Applications. Proceedings of Machine Learning Research, vol. 230, pp. 406–423. PMLR (2024)
37. Kath, C., Ziel, F.: Conformal prediction interval estimation and applications to day-ahead and intraday power markets. Int. J. Forecast. **37**(2), 777–799 (2021)
38. Kato, M.: Conformal predictive portfolio selection. arXiv preprint arXiv:2410.16333 (2024)
39. Kharazian, Z., Lindgren, T., Magnusson, S., Boström, H.: Copal: conformal prediction in active learning an algorithm for enhancing remaining useful life estimation in predictive maintenance. In: The 13th Symposium on Conformal and Probabilistic Prediction with Applications, pp. 195–217. PMLR (2024)
40. Laxhammar, R.: Conformal anomaly detection: detecting abnormal trajectories in surveillance applications. Ph.D. thesis, University of Skövde (2014)
41. Mendil, M., Mossina, L., Vigouroux, D.: PUNCC: a python library for predictive uncertainty calibration and conformalization. In: Conformal and Probabilistic Prediction with Applications, pp. 582–601. PMLR (2023)
42. Morger, A., et al.: Knowtox: pipeline and case study for confident prediction of potential toxic effects of compounds. J. Chem. **12**(1), 24 (2020)
43. Morger, A., et al.: Assessing the calibration in toxicological in vitro models with conformal prediction. J. Chem. **13**(1), 35 (2021)
44. Norinder, U., Boyer, S.: Conformal prediction classification of a large data set of environmental chemicals from toxcast and tox21 estrogen receptor assays. Chem. Res. Toxicol. **29**(6), 1003–1010 (2016)
45. Norinder, U., Boyer, S.: Binary classification of imbalanced datasets using conformal prediction. J. Mol. Graph. Model. **72**, 256–265 (2017)
46. Norinder, U., Carlsson, L., Boyer, S., Eklund, M.: Introducing conformal prediction in predictive modelling: a transparent and flexible alternative to applicability domain determination. J. Chem. Inf. Model. **54**(6), 1596–1603 (2014)
47. Nouretdinov, I., Gammerman, J.: Inductive Venn-Abers predictive distributions: new applications & evaluation. Proc. Mach. Learn. Res. **230**, 1–18 (2024)
48. Nouretdinov, I., Volkhonskiy, D., Lim, P., Toccaceli, P., Gammerman, A.: Inductive Venn-Abers predictive distribution. In: Conformal and Probabilistic Prediction and Applications, pp. 15–36. PMLR (2018)
49. OECD. (Q)SAR Assessment Framework: Guidance for the regulatory assessment of (Quantitative) Structure Activity Relationship models and predictions. OECD Series on Testing and Assessment 386, OECD Publishing, Paris (2023)
50. Olsson, H., et al.: Estimating diagnostic uncertainty in artificial intelligence assisted pathology using conformal prediction. Nat. Commun. **13**(1), 7761 (2022)

51. Papadopoulos, H.: Reliable probabilistic prediction for medical decision support. In: Iliadis, L., Maglogiannis, I., Papadopoulos, H. (eds.) AIAI/EANN -2011. IAICT, vol. 364, pp. 265–274. Springer, Heidelberg (2011). https://doi.org/10.1007/978-3-642-23960-1_32
52. Papadopoulos, H., Proedrou, K., Vovk, V., Gammerman, A.: Inductive confidence machines for regression. In: Elomaa, T., Mannila, H., Toivonen, H. (eds.) ECML 2002. LNCS (LNAI), vol. 2430, pp. 345–356. Springer, Heidelberg (2002). https://doi.org/10.1007/3-540-36755-1_29
53. Papadopoulos, H., Vovk, V., Gammerman, A.: Qualified predictions for large data sets in the case of pattern recognition. In: Proceedings of the International Conference on Machine Learning and Applications, CSREA Press, Las Vegas, NV, pp. 159–163. CSREA Press (2002)
54. Renkema, Y., Visser, L., AlSkaif, T.: Enhancing the reliability of probabilistic PV power forecasts using conformal prediction. Solar Energy Adv. **4**, 100059 (2024)
55. Shafer, G., Vovk, V.: A tutorial on conformal prediction. J. Mach. Learn. Res. **9**(3) (2008)
56. Svensson, F., et al.: Conformal regression for QSAR modelling: quantifying prediction uncertainty. J. Chem. Inf. Model. **58**(5), 1132–1140 (2018)
57. Svensson, F., Norinder, U., Bender, A.: Modelling compound cytotoxicity using conformal prediction and pubchem HTS data. Toxicol. Res. **6**(1), 73–80 (2017)
58. Uddin, N., Lofstrom, T.: Applications of conformal regression on real-world industrial use cases using crepes and MAPIE. In: Conformal and Probabilistic Prediction with Applications, pp. 147–165. PMLR (2023)
59. Ventriglia, V., et al.: An explainable machine learning model for large-scale travelling ionospheric disturbances forecasting. J. Space Weather Space Clim. **15**, 25 (2025)
60. Volkhonskiy, D., Burnaev, E., Nouretdinov, I., Gammerman, A., Vovk, V.: Inductive conformal martingales for change-point detection. In: Conformal and Probabilistic Prediction and Applications, pp. 132–153. PMLR (2017)
61. Vovk, V.: Conditional validity of inductive conformal predictors. In: Hoi, S.C.H., Buntine, W. (eds.) Proceedings of the Asian Conference on Machine Learning. Proceedings of Machine Learning Research, vol. 25, pp. 475–490, Singapore Management University, Singapore. PMLR (2012)
62. Vovk, V., Bendtsen, C.: Conformal predictive decision making. In: Gammerman, A., Vovk, V., Luo, Z., Smirnov, E., Peeters, R. (eds.) Proceedings of the Seventh Workshop on Conformal and Probabilistic Prediction and Applications. Proceedings of Machine Learning Research, vol. 91, pp. 52–62. PMLR (2018)
63. Vovk, V., Gammerman, A., Shafer, G.: Algorithmic Learning in a Random World. Springer (2005)
64. Vovk, V., Gammerman, A., Shafer, G.: Algorithmic Learning in a Random World, 2nd edn. Springer (2022)
65. Vovk, V., Petej, I., Fedorova, V., Nouretdinov, I., Gammerman, A.: Venn-Abers predictors. Mach. Learn. **92**(2–3), 349–376 (2013)
66. Vovk, V., Petej, I., Nouretdinov, I., Ahlberg, E., Carlsson, L., Gammerman, A.: Retrain or not retrain: conformal test martingales for change-point detection. In: Conformal and Probabilistic Prediction and Applications, pp. 191–210. PMLR (2021)
67. Wang, G., Zhiying, L., Wang, P., Zhuang, S., Wang, D.: Conformal test martingale-based change-point detection for geospatial object detectors. Appl. Sci. **13**(15), 8647 (2023)

68. Zaffran, M., Féron, O., Goude, Y., Josse, J., Dieuleveut, A.: Adaptive conformal predictions for time series. In: International Conference on Machine Learning, pp. 25834–25866. PMLR (2022)

Conformal Prediction in the Age of Multimodal Foundation Models: A Survey

Christian Stavan[2], Kunal Tilaganji[1], Sai Mathura Krishnan[1], Sai Srinivas Kancheti[1], and Vineeth N. Balasubramanian[1,3(✉)]

[1] Indian Institute of Technology, Hyderabad, Hyderabad, India
`kunal.tilaganji@prjt.cse.iith.ac.in,`
`{saimathura.krishnan,vineethnb}@cse.iith.ac.in, cs21resch01004@iith.ac.in`
[2] New York University, New York, USA
`snc8114@nyu.edu`
[3] Microsoft Research India, Bengaluru, India
`vineeth.nb@microsoft.com`

Abstract. Foundation Models and their multimodal extensions have transformed AI, enabling flexible reasoning and generation across diverse modalities such as text, vision, audio, and video. However, their black-box nature, large output spaces, and susceptibility to hallucinations raise concerns about reliability, particularly in high-stakes applications. Conformal Prediction offers a principled, distribution-free framework for uncertainty quantification, but its traditional formulations are challenged by the complexity of FM pipelines, multimodal interactions, and streaming or partially supervised scenarios. This chapter provides a detailed survey of recent efforts to adapt CP to deep neural networks and multimodal foundation models. We review innovations spanning generative modeling, predictive pipelines, reasoning and control, and retrieval-augmented generation, highlighting techniques such as sampling-aware calibration, trust-score conditioning and input-adaptive error control. The chapter also synthesizes design principles for robust conformal prediction in multimodal contexts, including per-modality normalization, dependency-aware calibration, and representative evaluation strategies. We conclude by outlining open research directions that can shape reliable and interpretable uncertainty quantification for the next generation of foundation models.

1 Introduction

Foundation models (FMs) have emerged as a central paradigm in modern machine learning. These are typically large-scale transformer [16, 100] architectures pretrained on vast, uncurated web-scale corpora through unsupervised or

---

C. Stavan, K. Tilaganji, S. M. Krishnan and S. S. Kancheti—These authors contributed equally to this work.

K. An Nguyen and Z. Luo (Eds.): Alexander Gammerman Festschrift, LNCS 16290, pp. 269–305, 2026.
https://doi.org/10.1007/978-3-032-15120-9_13

self-supervised objectives [10,33]. The resulting models exhibit strong generalization capabilities across diverse tasks in fields such as natural language processing [69,79], computer vision [70,77], speech [78], reinforcement learning [104] as well as in multimodal domains [1,60]. Their ability to perform general tasks in a zero-shot manner i.e. without task-specific finetuning, has led to widespread practical adoption in domains such as healthcare [95], law [92], scientific discovery [9,35] and as conversational systems [69]. Their remarkable zero-shot and few-shot performance stems from their parameter count and the breadth and diversity of their training data, but this same generality obscures their internal working, and FMs are often treated as black-boxes whose predictions are difficult to interpret. Furthermore, FMs are susceptible to hallucination [36] and may fail unpredictably for certain inputs [21,57]. As a consequence, assessing the reliability of FM predictions especially in open-ended or high-stakes settings remains a central concern.

Why Revisit Conformal Prediction for FMs. Conformal prediction (CP) provides a distribution-free framework for quantifying uncertainty, offering formal coverage guarantees under the assumption of exchangeable data [90,101]. Traditionally applied to supervised learning problems with finite label spaces, CP constructs calibrated prediction sets that contain the true label with a user-specified probability $1-\alpha$. However, when applied to foundation models, classical CP techniques encounter several structural mismatches:

- *Open-ended output spaces*: Unlike standard classification or regression, FM outputs may be free-form text, structured plans, or high-dimensional action sequences, complicating set construction.
- *Non-exchangeable generation processes*: Many FMs operate autoregressively, generating outputs token-by-token where the conditional dependencies violate exchangeability.
- *Partial or delayed supervision*: In settings like program synthesis or multi-step reasoning, supervision may only be available in binary form (e.g., whether a final answer is correct), limiting feedback granularity.
- *Restricted model access*: Foundation models are often deployed as black boxes, exposing only interface-level outputs such as logits or samples, which constrains how nonconformity scores are computed.

Despite these challenges, the deployment of FMs in high-stakes domains – ranging from medicine and law to autonomous systems – calls for robust uncertainty quantification that does not rely on strong distributional assumptions. These requirements have spurred a growing body of work that revisits and extends CP principles to accommodate the unique statistical and algorithmic structure of FM pipelines.

Conformal Prediction for Multimodal FMs. Multimodal FMs (MMFMs) have emerged as important classes of large-scale, pre-trained architectures that can ingest and understand multiple modalities – such as vision, text, audio, and video, while responding to user queries. The adoption of CP in MMFMs presents additional challenges beyond those discussed above:

- *Heterogeneity of Modalities*: Different modalities exhibit distinct statistical properties that pose significant challenges for CP methods. The dimensionality and structure of data vary dramatically across modalities, ranging from high-dimensional pixel arrays in images to discrete token sequences in text to continuous waveforms in audio. Each modality also has characteristic noise models, including motion blur in images, background noise in audio, and sensor drift in continuous measurements.
- *Temporal and Quality Issues*: Temporal characteristics in some modalities further complicate the picture, as some modalities like images represent instantaneous snapshots while others like video or audio streams exhibit temporal dependencies. Additionally, real-world deployments often encounter missing or degraded modalities, where certain types of input may be partially available or of reduced quality due to sensor limitations or transmission issues.
- *Cross-Modal Alignment Uncertainty*: Alignment uncertainty represents another fundamental challenge when modality representations are not reliably semantically matched. This issue is particularly prevalent in web-scale training corpora, which often contain weakly paired data where the relationship between modalities may be loose or imprecise. For example, image captions may describe general scenes while the corresponding images contain many specific objects not mentioned in the text, creating semantic gaps that can mislead alignment processes.

These challenges have implications for adoption of conformal prediction. Standard nonconformity scores may over-expand prediction sets in noisy regions while under-representing uncertainty in other areas, leading to miscalibrated confidence estimates. Moreover, temporal data violates the independent and identically distributed assumptions underlying basic conformal methods, necessitating dependency-aware calibration approaches that can account for correlations across time.

Model limitations also contribute to alignment uncertainty, as global similarity measures may successfully capture broad correspondences between modalities while failing to establish fine-grained alignment at the detail level. This is particularly problematic for conformal prediction because incorrect alignment can create false confidence, leading to overly small prediction sets that fail to achieve proper conditional coverage. Furthermore, alignment quality varies significantly across different domains and data types, potentially causing calibration drift when models encounter data that differs from their training distribution.

In this work, we review existing efforts on adaptation of CP to MMFMs; however, in this process, we also discuss how CP has been adapted to deep neural networks (DNNs) in general before discussing specific efforts for MMFMs. We also share our gleanings from this review in the form of emerging trends and directions as well as open challenges in the field. The remainder of this work is organized as follows: Sect. 2 briefly discusses the background required in CP and MMFMs; Sect. 3 reviews existing efforts that adapt CP to DNNs;

Sect. 4 describes the various efforts that attempt to bring CP to MMFMs; Sect. 5 discusses emerging trends and open challenges; and Sect. 6 makes conclusive remarks.

2 Background

2.1 Conformal Prediction Fundamentals

Conformal prediction (CP) provides a distribution-free, model-agnostic framework to equip any machine learning model with rigorous uncertainty quantification [4,90,101]. In the most widely used form of CP, one reserves a calibration set and computes a nonconformity score $s(x,y)$ for each example – for instance $1 - \hat{p}(y \mid x)$ from a learning model – then selects a threshold τ at the $(1 - \alpha)$-quantile of these scores. The conformal prediction set is then defined as $\Gamma_{1-\alpha}(x) = \{y : s(x,y) \leq \tau\}$ which satisfies the marginal coverage guarantee $\Pr\left[Y \in \Gamma_{1-\alpha}(X)\right] \geq 1 - \alpha$ [3]. Thus, CP can turn a point predictor (e.g., a classifier) into a set predictor that covers the true label with probability at least $1 - \alpha$ under exchangeability. We now briefly discuss the background and variants of CP that are most commonly used.

2.1.1 Exchangeability and Coverage Guarantees

Conformal prediction (CP) relies on the exchangeability assumption, i.e., data points are drawn from the same distribution regardless of order. This enables the *validity guarantee* of marginal coverage: for significance level α, a conformal predictor outputs a prediction region $\Gamma(X_{n+1})$ such that:

$$\mathbb{P}(Y_{n+1} \in \Gamma(X_{n+1})) \geq 1 - \alpha \tag{1}$$

This guarantee holds distribution-free under exchangeability as explained by Shafer et al. [90]. However, CP's marginal coverage does not guarantee *conditional coverage* (per-class reliability), an active research area.

2.1.2 Inductive vs. Transductive Variants

Conformal prediction methods can be broadly categorized into two key variants based on their computational approach and theoretical guarantees. *Transductive CP (TCP)* evaluates conformity scores for all possible labels, providing tighter guarantees but requiring intensive computation for each prediction. In contrast, *Inductive CP (ICP)* splits data into training and calibration sets, training once and computing conformity scores over calibration data for efficiency. The original transductive procedure requires retraining the model for every test input. ICP instead relies on a calibration split: non-conformity scores are computed once on held-out data and reused for test-time inference, reducing the additional cost of conformalization to a single forward pass.

Papadopoulos et al. [73] introduced *normalized nonconformity functions* that adapt to heteroskedasticity, improving prediction intervals by accounting for input-dependent noise while maintaining inductive efficiency. This advancement allows conformal methods to handle varying levels of uncertainty across different regions of the input space more effectively.

2.1.3   Split Conformal Prediction

Split Conformal Prediction (SCP) [29,72] provides a distribution-free framework for constructing prediction intervals with finite-sample marginal coverage guarantees. The method begins by splitting the dataset into a proper training set I_1 and a calibration set I_2. A regression algorithm A is applied to I_1 to estimate conditional quantiles of the response given the covariates, yielding lower and upper bounds $\{\hat{q}_\alpha^{lo}, \hat{q}_\alpha^{hi}\}$. Using the calibration set, nonconformity scores are computed as deviations of observed responses from these bounds:

$$E_i := \max\{\hat{q}_{\alpha_{lo}}(X_i) - Y_i,\ Y_i - \hat{q}_{\alpha_{hi}}(X_i)\}, \quad i \in I_2,$$

This quantifies the extent to which the initial quantile estimates fail to capture the data. The $(1-\alpha)(1+1/|I_2|)$-th empirical quantile of these scores, denoted $Q_{1-\alpha}(E, I_2)$, serves as a correction factor to ensure valid coverage. The final conformal prediction interval for a new instance X_{n+1} is thus obtained as:

$$C(X_{n+1}) = [\hat{q}_{\alpha_{lo}}(X_{n+1}) - Q_{1-\alpha}(E, I_2),\ \hat{q}_{\alpha_{hi}}(X_{n+1}) + Q_{1-\alpha}(E, I_2)],$$

which guarantees $\Pr\{Y_{n+1} \in C(X_{n+1})\} \geq 1-\alpha$. This approach balances model-based estimation with empirical calibration, providing robust predictive intervals that do not rely on strong distributional assumptions.

2.1.4   Mondrian Conformal Prediction

Mondrian conformal prediction addresses this by enforcing validity within defined groups or categories. The term 'Mondrian' (after the Piet Mondrian's grid patterns [20]) was introduced by Vovk et al. (2003) for conformal predictors that operate separately on data partitions. Given a categorical function $G(z)$ that assigns each example $z = (x, y)$ to a group $g \in \mathcal{G}$, Mondrian CP runs separate conformal inference within each category. That is, nonconformity scores and calibration are performed using only points that share the same group g. The result is a stronger guarantee.

$$P\{y_{n+1} \in C_n(x_{n+1}) \mid G(x_{n+1}, y_{n+1}) = g\} \geq 1 - \alpha, \quad \text{for each group } g.$$

In practice, one may split the calibration set by $\kappa(z) = g$ and compute the p-value using only scores from the group g. Mondrian CP is useful in heterogeneous data (e.g., different classes or demographics) because it guarantees nominal coverage separately in each subgroup, at the cost of using fewer points per group (which can make intervals wider for small groups).

2.1.5   Venn–Abers Predictor

Venn–Abers predictors are a recently introduced class of Venn predictors for binary classification that was introduced by Vovk et al. [102], which uses isotonic regression to calibrate the output of any scoring classifier (a classifier that produces a real value score). In this method, a scoring function s is trained on the data and, for a new test instance x, two isotonic 'calibrators' are constructed by temporarily augmenting the training set with $(x, 0)$ and with $(x, 1)$, respectively. In practice, scoring functions s_0, s_1 (on the extended data labeled 0 or 1) are obtained, and increasing functions g_0, g_1 are fitted via isotonic regression to the pairs of scores and labels. The Venn–Abers predictor then outputs the pair of probabilities (p_0, p_1) defined by $p_0 = g_0(s_0(x))$ and $p_1 = g_1(s_1(x))$. Thus, the prediction is multiprobabilistic, i.e., a set or interval of probabilities (here given by the two values p_0, p_1) rather than a single point estimate. (If a single probability is needed, one can combine these, for example, by $p = p_1/(1 - p_0 + p_1)$, which is the minimax-optimal choice under log loss.) A key theoretical property of Venn–Abers predictors is that they inherit the validity or calibration guarantee of Venn predictors: under the standard exchangeability assumption, the long-run frequency of the event will match the predicted probability range. In summary, Venn–Abers predictors provide a nonparametric, isotonic-regression-based calibration of scoring classifiers that yield multivalued probability outputs with provable calibration.

2.2   Multimodal Foundation Models (MMFMs)

Multimodal foundation models (MMFMs) represent a transformative class of large-scale, pre-trained architectures capable of processing and reasoning across multiple modalities – such as vision, text, audio, and video – without requiring specialized architectural modifications. These models are typically trained on heterogeneous datasets using techniques like contrastive learning or cross-attention, enabling robust cross-modal alignment and exceptional zero-shot and few-shot generalization through natural language grounding. As a result, MM-FMs excel across a broad range of downstream tasks.

Early examples in this domain include: CLIP, which is contrastively trained on image-text pairs for zero-shot classification [77]; Flamingo, which incorporates visual context into language models for few-shot visual question answering (QA) and captioning [1]; GPT-4V, a multimodal extension of GPT-4 that supports multimodal instructions and reasoning [116]; and Kosmos-2, which grounds language in perception and action across modalities [74]. Recent advancements mark a significant evolution in MM-FMs. OpenAI introduced GPT-4o ("omni"), capable of processing text, image, and audio inputs and generating corresponding outputs—bringing full multimodal interaction into mainstream use through ChatGPT interfaces. Meanwhile, Google recently unveiled Gemma 3, an open-source, multimodal, multilingual, long-context model, which can process text and image inputs with context windows up to 128K tokens, alongside variants optimized for device constraints (e.g., Gemma 3n). Globally, Baidu recently launched ERNIE 4.5 Turbo and X1 Turbo multimodal models, featuring capabilities in text, audio, image, and video processing designed for efficiency and enterprise integration. Alibaba released Qwen2.5-Omni, a model capable of processing text, images, audio, and video while generating both speech and textual responses in a streaming fashion through a dual-module Thinker–Talker architecture [113].

These advancements open fertile ground for conformal prediction (CP), enabling rich uncertainty quantification across multiple modalities. In visual-language tasks, conformal prediction sets can be calibrated not only over textual logits but also across multimodal representations that involve vision and speech. Architectures like Qwen2.5-Omni, with synchronous audio-text streams, provide new opportunities for conformal calibration in streaming scenarios, potentially adapting CP methods through sequential or control-theoretic extensions. In summary, the rapid emergence of models such as LLama, GPT-4o, GPT-5, Gemma 3, and Qwen2.5-Omni offers a compelling foundation for advancing conformal prediction methods tailored to the multimodal era. These methods will support reliable, interpretable, and efficient uncertainty estimation in increasingly complex data modalities across a variety of deployment scenarios.

3   Conformal Prediction in Deep Learning

Deep neural networks (DNNs) have become the backbone of modern vision, language, and decision-making systems, yet their predictions often lack a reliable measure of confidence. Conformal prediction (CP) provides a principled mechanism to address this gap by attaching rigorously calibrated uncertainty estimates to model outputs. Unlike heuristic confidence scores, CP guarantees that the error rate of the prediction set remains bounded by a user-specified level, even for finite samples and without requiring distributional assumptions. Mapping this framework into the landscape of deep learning has yielded several methodological innovations that strike a balance between computational efficiency, coverage reliability, and interpretability.

3.1   Conformal Prediction in Neural Networks

As stated earlier, the CP framework requires a nonconformity score $s(x, y)$ for each example to equip a neural network with uncertainty quantification [4,11]. This, for instance, can come from a network's softmax output probabilities, an absolute residual or distances in the latent representation space. While this approach has minimal assumptions when providing exact finite-sample coverage [3], it faces challenges such as the need for a sufficiently large calibration set and handling complex outputs (e.g., multilabel or structured outputs) within the prediction set framework. We now briefly discuss how the CP framework has been adapted to different neural network (NN) architectures, before discussing their use in foundation models in Sect. 4.

Angelopoulos et al. [6] propose Regularized Adaptive Prediction Sets (RAPS), an algorithm that calibrates any pretrained CNN (e.g. ResNet) to output valid conformal sets. RAPS first applies a Platt-scaling [75] step to the CNN's softmax logits, then regularizes the small class scores to produce smaller, stable sets. The resulting method is fast (like Platt scaling) but yields finite-sample coverage guarantees: on ImageNet it produces prediction sets that are often 5–10× smaller than naïve baselines, while always containing the true class with the desired probability. Similarly, a conformal RNN method (CF-RNN) [96] measures nonconformity as the vector of absolute residuals over a forecasting horizon and applies a Bonferroni correction to obtain valid prediction intervals for each time step. The work also work guarantees that the resulting intervals satisfy $\Pr\left(\forall h \in \{1,\ldots,H\} : y_{t+h} \in [\hat{y}_{t+h} - \hat{\varepsilon}_h,\ \hat{y}_{t+h} + \hat{\varepsilon}_h]\right) \geq 1 - \alpha$, the joint coverage over all horizons. This yields distribution-free confidence bands for time-series without retraining the model (though it assumes conditional exchangeability across time steps).

In NLP tasks with transformers (BERT, GPT, etc.), CP has likewise been used to calibrate classification and generation models. For instance, Mondrian and Venn–Abers conformal schemes on top of BERT or RoBERTa have been shown to give calibrated label sets for paraphrase detection, sentiment analysis and even sequence tagging [11]. Empirically, conformal sets from transformers remain tight; e.g. a CP-calibrated BERT POS tagger at 99% confidence produced almost singleton label sets (fewer than 4% of sets contained multiple tags). In general, CP offers a unified way to get valid uncertainty sets from CNNs, RNNs or transformers with no model-specific assumptions, even though each architecture poses specific issues – for example, high output dimensionality or temporal/spatial dependence must be addressed via tailored conformal score functions.

3.2 Beyond Calibration: Other Related Efforts Toward Conformal Prediction on DNNs

Beyond vanilla calibration, recent years have seen different directions of efforts toward conformal prediction on Deep Neural Network (DNN) models. We discuss salient directions briefly below.

3.2.1 Confidence-Conditional Coverage with Trust Scores

Standard conformal prediction ensures marginal validity but can yield unbalanced coverage: models may be highly confident yet wrong on hard examples, causing those regions to be undercovered. Kaur et al. [47] observe that miscoverage in classification is concentrated where the model's confidence is high but the prediction deviates from the Bayes optimum. To target these failures, they propose a conditional CP method using a two-dimensional statistic (Conf, Trust). Here, Conf is the classifier's predicted confidence (top softmax score) and Trust is a nonparametric "trust score" [41] that estimates agreement with the Bayes-optimal classifier. The trust score for a test point x is defined as (after excluding low-density regions):

$$\mathrm{Trust}(x) = \frac{\text{distance}\,(x,\ \text{nearest training point of the predicted class})}{\text{distance}\,(x,\ \text{nearest training point of a different class})} \qquad (2)$$

Intuitively, a low trust score indicates the classifier may be mis-ranking the true label (disagreeing with Bayes).

Using these two variables, the method learns an adaptive threshold for the conformity score by conditional quantile regression. Concretely, given $S(x,y)$

is a conformity score (e.g. a one-vs-rest score from the network), let $\phi(x) =$ $(\mathrm{Conf}(x), \mathrm{Trust}(x))$. On the calibration set $\{(S_i, \phi_i)\}$, one fits a regression $\hat{q}(\phi)$ by minimizing a pinball loss so that $\hat{q}(\phi)$ estimates the $(1 - \alpha)$-quantile of S conditional on ϕ. At test time, the prediction set is:

$$C(x) = \{y : S(x, y) \leq \hat{q}(\phi(x))\} \tag{3}$$

which provides coverage on subsets of instances with similar $(\mathrm{Conf}, \mathrm{Trust})$ values. This yields an approximate conditional guarantee: the sets adapt to both model confidence and estimated Bayes error, giving more coverage in regions where the model is overconfident.

In experiments on large image datasets such as ImageNet [87], Places365 [124], long-tailed variants, and Fitzpatrick-17k [26] skin data, Kaur et al. [47] evaluate the conditional performance via "coverage gaps" over bins of confidence/trust. They find that their method significantly reduces the coverage gap compared to standard CP. For example, on ImageNet and other datasets, conditional CP achieves near-target coverage across most confidence/trust bins, whereas standard CP undercovers some bins. Notably, on Fitzpatrick-17k, the trust-based method boosts coverage for underrepresented skin types without needing their labels. In summary, by stratifying calibration on (confidence, trust), this method provides empirically tighter guarantees on hard subsets, yielding more balanced conditional coverage (and often smaller set sizes) than marginal CP.

3.2.2   Class-Clustering for Label-Conditional CP

A natural goal when using CP in classification problems is class-conditional coverage, i.e. each class c should have probability $(1 - \alpha)$ coverage when $Y = c$. Achieving this via per-class CP (splitting calibration by class) fails when many classes have few examples (few-shot or long-tailed settings). Ding et al. [17] address this by clustering classes with similar score distributions and pooling their calibration data. The idea is to find groups of classes whose conformal scores behave alike, so that one can compute a shared quantile per cluster. Formally, let q_c be the vector of quantiles of the base classifier's conformity score (or softmax score) for class c at several confidence levels. These quantile embeddings are clustered (e.g., by k-means) to assign classes to clusters k. During calibration, one pools all calibration examples from classes in cluster k and computes a single $(1 - \alpha)$-quantile ϵ_k. At test time, for any input x, its predicted class y lies in some cluster $k = \kappa(y)$, and the prediction set includes class y if $S(x, y) \leq \epsilon_k$. Intuitively, such a clustered CP approach balances the extremes of "standard" (one quantile for all) and "classwise" (one per class) by grouping similar classes. The method guarantees cluster-conditional coverage: each cluster's pooled procedure attains coverage for points in that cluster. In the ideal (oracle) case where each class in a cluster truly shares an identical score distribution, one recovers class-conditional coverage.

Empirically, Ding et al. [17] test clustered CP on four image benchmarks (ImageNet, CIFAR-100, Places365, iNaturalist) with up to 1000 classes. They compare standard CP, classwise CP, and clustered CP under varying calibration set sizes. The results show that clustered CP substantially improves class-conditional coverage gaps relative to baselines, especially in the limited-data regime. For modest per-class sample sizes, clustered CP achieves the smallest average coverage gap (CovGap) across classes, whereas standard CP undercovers rare classes and classwise CP is too conservative (very large sets). In many cases, clustered CP gives nearly the best of both: good coverage with only slightly larger prediction sets. Across all datasets, the "clustered" method typically outperforms existing approaches on class-conditional metrics and yields competitive set sizes. Thus, by sharing calibration data among related classes, clustered conformal prediction delivers tighter class-wise guarantees in many-class problems where pure per-class calibration would fail.

3.2.3   Self-supervision to Enhance Adaptive Coverage

Locally adaptive conformal prediction (e.g., Conformalized Residual Fitting, CRF) augments standard conformal intervals by normalizing the non-conformity score with a learned "σ" model, yielding per-instance intervals $[f(x) \pm \epsilon \cdot \sigma(x)]$ where $\gamma(x) = |y - f(x)|/\sigma(x)$. Although CRF attains valid marginal coverage, it may be inefficient because it does not take advantage of additional information to predict which samples are hard. Seedat et al. propose Self-Supervised Conformal Prediction (SSCP) [89] that addresses this by leveraging a self-supervised pretext task: after fitting the base regressor f, one trains a self-supervised model f_{ss} (sharing the encoder) to predict some pretext target (e.g., masked feature reconstruction). At calibration time, a self-supervised "error" $L_{ss}(x)$ is calculated on each instance. The conformal normalizer σ is then trained to predict the residuals of f using characteristics $(x, L_{ss}(x))$. Finally, one computes the usual quantile ϵ of normalized residuals on the calibration set and forms adaptive intervals.

$$[\ell(x), r(x)] = [f(x) - \epsilon \, \hat{\sigma}(\hat{x}), \ f(x) + \epsilon \, \hat{\sigma}(\hat{x})] \tag{4}$$

where $\hat{x} = \{x, L_{ss}(x)\}$. Alternatively, SSCP can be seen as using a residual model $\hat{\sigma}(x) = \sigma(x, f_{ss}(x))$ such that $\gamma(x) = |y - f(x)|/\hat{\sigma}(x)$. SSCP is performed in three phases: (i) *Train base and self-supervised models:* Fit the predictor $f = g \circ e$ on labeled data, then train an auxiliary predictor $f_{ss} = h \circ e$ on a self-supervised task using the shared encoder e; (ii) *Train conformal normalizer:* Using a separate residual set, train σ (for example, a regressor) to predict $|y - f(x)|$ from inputs $(x, L_{ss}(x))$; and (iii) *Calibrate and predict:* On the calibration set, compute normalized scores $\gamma(x) = |y - f(x)|/\sigma(\hat{x})$ and find the $(1 - \alpha)$-quantile ϵ. For a test point x, output $C(x) = [f(x) \pm \epsilon \cdot \sigma(\hat{x})]$.SSCP retains the finite-sample marginal coverage guarantees of CRF under exchangeability. Empirical results (on synthetic and real regression benchmarks) show that including $L_{ss}(x)$ narrows the interval width while preserving coverage. Overall, SSCP leverages self-supervised error signals to improve the adaptivity of conformal intervals, achieving narrower (more efficient) prediction sets, especially in challenging regions, while maintaining the $(1 - \alpha)$ coverage guarantee.

3.2.4 Conformal Prediction for Sequential Decision Making in DNNs

Einbinder et al. (2022) [18] introduce a novel technique for training uncertainty-aware classifiers with conformalized deep learning that embeds conformal-style constraints directly into the training loop. This approach produces deep models that inherently output well-calibrated uncertainties without requiring post hoc adjustment. The training objective combines the standard classification loss, such as cross-entropy, with a conformal calibration loss that aligns empirical p-value distributions from a held-out validation set with uniformity. This end-to-end approach ensures that the model allocates expressive capacity to regions where uncertainty is most decision-critical while maintaining predictive accuracy.

Building upon this foundation, PlanCP [97] extends the conformalized training concept to trajectory prediction models using diffusion-based learned dynamics. During training, a quantile-enhanced loss encourages the model's outputs to produce p-values that match target coverage across diverse trajectories. At inference time, classical conformal calibration is applied on a held-out set of expert trajectories to form prediction regions in trajectory space, providing end-to-end uncertainty-aware planning guarantees. Empirically, PlanCP demonstrates significant improvements in trajectory uncertainty estimation and planning performance on offline reinforcement learning benchmarks, substantially outperforming baselines in safety-critical robotic tasks. Together, these approaches represent a fundamental advancement in conformal prediction, evolving it from a post-processing wrapper to a training-aware methodology that enables neural networks to internalize coverage goals and produce uncertainty-native predictions.

While not yet fully formulated with complete CP integration, these methods already illustrate the potential for conformal uncertainty to guide adaptive decision loops effectively. The emerging paradigm represents a significant shift toward decision-aware conformal systems where uncertainty estimates serve a dual purpose: they are not only calibrated to provide reliable coverage guarantees but also actively inform and regulate policy or planning choices. This integration marks an important evolution in the field, moving beyond passive uncertainty quantification toward active uncertainty-informed decision making in complex, dynamic environments.

4   Conformal Prediction for Multimodal Foundation Models

This section surveys recent progress in adapting CP to multimodal foundation models. For purposes for clarity and relevance, we also include efforts that broadly pertain to foundation models, even if not multimodality directly. We organize the literature around four core application areas, each highlighting distinctive constraints and calibration strategies:

- *Generative Pipelines*, where sampling-based generation is paired with conformal filtering for output admissibility;

- *Predictive Pipelines and (V)QA*, which explore how CP can be applied to ranking, scoring, or partial-feedback settings in large candidate spaces;
- *Adaptive Reasoning and Control*, where CP principles are used to guide intermediate decision-making steps or abstention behavior in FM responses;
- *Retrieval*, where CP is integrated into document ranking, re-ranking, or retrieval-augmented generation (RAG) to provide guarantees on relevance.

We conclude with a final subsection on *Challenges and Limitations*, discussing how classical assumptions break down in FM contexts, and what new theory, algorithms, and empirical studies are needed to advance CP's role in trustworthy FM deployment. We discuss how the tools offered by CP can be reframed to handle weak supervision, black-box inference, multimodal uncertainty and other such desiderata.

4.1   Generative Pipelines

4.1.1   Redefining Sampling: From Label Sets to Candidate Sets

In generative tasks such as open-domain QA, document summarization, or code generation, the output space is combinatorially large and cannot be enumerated, making it infeasible to select candidate responses from the entire output space for CP. Instead, existing works focus on calibrating the sampling process to draw a finite candidate set $C(x)$ for input x, by controlled sampling, filtering of low-quality or redundant outputs, and targeted stopping. This 'sample-filter-stop' paradigm provides:

$$\Pr\big[\exists\, y \in C(x)\colon A(y) = 1\big] \geq 1 - \alpha$$

for an admission function [19] A which evaluates to 1 for acceptable outputs.

Conformal Language Modeling (CLM) [76] exemplifies this approach for free-form text generation. It formalizes prediction sets over sequences by defining three components: (i) a *quality* score $Q(x, y)$, e.g. length-normalized log-likelihood; (ii) a *diversity* metric $S(y, y')$, e.g. n-gram overlap; and (iii) a *set-confidence* function $F(C)$, which aggregates evidence that the candidate set C contains at least one acceptable output (e.g. $F(C) = \max_{y \in C} Q(x, y)$). Given thresholds $\lambda = (\lambda_1, \lambda_2, \lambda_3)$ and a sampling budget $k_{\max}$, CLM samples $y_k \sim \hat{\pi}(\cdot \mid x)$, skips any y_k with $Q(x, y_k) < \lambda_2$ or $\max_{y \in C} S(y_k, y) > \lambda_1$, adds accepted samples to C, and stops once $F(C) \geq \lambda_3$. Calibration is performed via Learn-Then-Test [5], which adapts thresholds $\hat{\lambda}$ so that

$$\Pr\big[\exists\, y \in C_{\hat{\lambda}}(x)\colon A(y) = 1\big] \geq 1 - \alpha.$$

CLM achieves valid coverage on tasks like TriviaQA [45], CNN/DM [32] summarization, and MIMIC-CXR (chest X-ray) [44] report generation, demonstrating its ability to isolate acceptable outputs in unstructured spaces.

4.1.2   Regression-Based Reduction via Generative Prediction Sets

Shahrokhi et al. [91] (GPS) reframes prediction-set construction for deep generative models as a conformal *regression* problem: rather than attempting to enumerate correct outputs, GPS predicts the number of samples needed to obtain an admissible output and then draws that many samples to form the prediction set. This explicitly accounts for finite sampling budgets and provides a principled abstention mechanism when admissible outputs are unlikely within the available budget.

Concretely, for each calibration input X_i GPS draws up to a budget M samples and records the minimum index

$$K_i = \min\{j : A(X_i, Y_{ij}) = 1\} \quad \text{(or } M + 1 \text{ if no admissible sample).}$$

Treating (X_i, K_i) as augmented calibration data, GPS fits a conditional $1 - \alpha$ quantile predictor $\widehat{q}_{1-\alpha}(x)$ (via a success-probability estimator $\widehat{f}$ and the geometric quantile formula). Because this point estimate is not finite-sample valid by itself, GPS applies a conformalized quantile regression correction: with the one-sided CQR score

$$S(x, k) = \max\{0, k - \widehat{q}_{1-\alpha}(x)\},$$

the conformal adjustment is the empirical $1 - \alpha$ quantile of $\{S(X_i, K_i)\}_{i=1}^n$, and the calibrated sample budget is

$$\widehat{K}(x) = \widehat{q}_{1-\alpha}(x) + Q_{1-\alpha}\big(\{S(X_i, K_i)\}_{i=1}^n\big).$$

If $\widehat{K}(x) \leq M$ GPS draws $\widehat{K}(x)$ samples from the generative model and returns the distinct sampled outputs as the prediction set; if $\widehat{K}(x) > M$ the protocol may abstain. This procedure yields the conformal guarantee that the returned set contains an admissible output with probability at least $1 - \alpha$ while explicitly managing sampling cost and abstention.

Empirically, GPS is evaluated on code generation (MBPP [7], DS-1000 [52]), math problem solving (GSM8K [14], MATH [31]), and open-domain QA (TriviaQA [45]), primarily using GPT-4o Mini, with supplement results for Phi-2, LLaMA-3-8B [99], and Gemma-2-27B. Across tasks, GPS consistently outperforms the CLM [76] baseline in low-α regimes ($\alpha \leq 0.3$) and maintains valid prediction sets where CLM fails, for example DS-1000, where CLM produces no valid sets across $\alpha \in [0.1, 0.5]$.

4.1.3   Task-Aware Scoring and Enhanced Calibration

While sampling and stopping mechanisms determine the number and timing of candidate generations, the reliability of those candidates ultimately depends on the nonconformity score. In many generative applications such as fact verification in summarization, hallucination mitigation in medical QA, or semantic alignment in Visual Question Answering (VQA), raw model likelihood often fails to reflect task-relevant correctness. Furthermore, prompt difficulty can vary substantially across inputs, rendering global thresholds either too stringent for easy cases or too lax for challenging ones. Achieving robust coverage under such conditions requires enhancing nonconformity scores with task-specific signals and adapting calibration procedures to input heterogeneity. Two recent studies that exemplify such refinements are discussed below.

Validity via Enhanced CP. Cherian et al. [13] define a general monotone loss

$$L\big(\hat{F}(C;\tau), W\big)$$

where $\hat{F}(C;\tau) = \{c \in C : p(c) \geq \tau\}$ filters the candidate set C by retaining only those claims whose quality score $p(c)$ exceeds threshold τ, and W denotes the ground-truth annotations. They then introduce the conformity score

$$S(C_i, W_i) = \inf\big\{\tau : L\big(\hat{F}(C_i;\tau), W_i\big) \leq \lambda\big\},$$

i.e. the smallest cutoff τ that drives the loss below the user-specified tolerance λ. To predict S for new inputs, they fit a quantile-regression model $g_S(X)$ over features X extracted from the prompt and its candidate claims, by minimizing the pinball loss on observed $\{S_i\}$. At test time, the conformal threshold is set to

$$\hat{\tau}(X) = \sup\{\tau \mid \tau \leq g_S(X)\},$$

so that any claim scoring above $\hat{\tau}$ is kept. A randomized variant of $\hat{\tau}$ ensures exact marginal validity:

$$P\big[L\big(\hat{F}(C_{n+1};\hat{\tau}(X_{n+1})), W_{n+1}\big) \leq \lambda\big] \geq 1 - \alpha.$$

Building on this, they propose two refinements:

- *Conditional Boosting:* This refinement parameterizes the scorer $p_\theta(c)$ and optimizes:

$$\theta^* = \arg\max_\theta \sum_{(C_i, W_i) \in \text{hold-out}} \big|\hat{F}(C_i; \hat{\tau}_i(\theta)) \cap W_i\big|,$$

  i.e. it adjusts θ to maximize the number of true claims retained under the conformal filter, effectively "boosting" the score function to align better with the loss criterion.
- *Level-Adaptive CP:* Rather than a global α, this refinement learns an input-dependent error level $\alpha(x)$ by replacing the pinball loss ℓ_α with $\ell_{\alpha(x)}$ in the regression, so that "easy" prompts (where g_S is accurate) allow a higher $\alpha(x)$ (smaller sets) and "hard" prompts a lower $\alpha(x)$ (larger sets). This yields the stronger guarantee:

$$P\big[L(\hat{F}(C_{n+1};\hat{\tau}(X_{n+1})), W_{n+1}) \leq \lambda \mid X_{n+1}\big] \geq 1 - \alpha(X_{n+1}).$$

On the MedLFQA medical-QA benchmark [39] and a Wikipedia biography dataset analyzed in [67], these enhancements both increase retention of valid claims and reduce coverage variance across prompts of differing difficulty, while preserving the formal $1 - \alpha$ validity guarantee for the chosen loss bound λ.

Conformal Linguistic Calibration (CLC). Jiang et al. [42] propose a novel approach to risk-controlling factuality in natural language generation by leveraging *imprecision* as a tunable output attribute. Rather than predicting confidence scores or abstaining from uncertain outputs, CLC treats hedging through linguistic generalization as a mechanism for calibration. Given a model-generated claim y, a rewrite operator $R_\gamma(y)$ is used to produce less specific variants, where γ denotes the degree of imprecision. The goal is to select a threshold γ^* such that the rewritten outputs satisfy:

$$\Pr\big[f_{\text{fact}}(R_{\gamma^*}(y)) = 1\big] \geq 1 - \alpha,$$

where f_{fact} is a factuality classifier. The rewriting process is implemented using the Learn-Then-Test (LTT) [5] pipeline on decomposed atomic claims sourced from QA datasets. A pool of candidate rewrites is generated by clustering sampled answers from an LLM, constructing nested semantic subsets based on confidence and proximity. Conformal selection over a labeled calibration set is then used to choose the minimal γ^* ensuring the desired risk control.

To amortize this process, CLC fine-tunes a LLaMA-3-8B model on synthetic rewrites from SimpleQA [109] and NaturalQuestions [51]. When evaluated on Core-filtered subclaims from the FActScore dataset [66], the calibrated model improves factuality while maintaining appropriate imprecision. Unlike methods that discard uncertain claims or output confidence scores, CLC enables models to generate verifiably calibrated text via adaptive control over specificity.

The above efforts demonstrate how CP can be enhanced through: (i) differentiable quantile-based training, (ii) input-conditional calibration, and (iii) reinterpretation of generative uncertainty as set prediction, while rigorously preserving marginal validity at level $1 - \alpha$.

Language Models with Conformal Factuality Guarantees. Mohri et al. [67] introduce the framework of conformal factuality as a way to give high-probability correctness guarantees for open-ended FM outputs. A claim y is said to be factual if it is entailed [2] by the ground-truth y^*. They define an entailment set $E(y) = \{y' : y' \Rightarrow y\}$, so that $y^* \in E(y)$ is a correctness guarantee for y. The LLM output $y = L(x)$ is split into a set of sub-claims, and a 'back-off' function $F_t(x, L(x))$ is designed that evaluates each sub-claim of $L(x)$ and keeps only those which as deemed 'correct' with a threshold t. As t increases, $F_t(x, L(x))$ gets less specific, with $F_{sup\ t}(x) = \phi$. Under the assumption that the entailment E of a set of sub-claims is the intersection of individual entailments, the set $E(F_t(x, L(x)))$ satisfies the nested property, and the split-CP approach for nested sets [27] can be leverage to give prediction sets. A calibration threshold $\hat{q}_\alpha$ is computed using a validation set, yielding the guarantee

$$P(Y^* \in E(F_{\hat{q}_\alpha}(X))) \geq 1 - \alpha$$

Thus the modified LLM generation $F_{\hat{q}_\alpha}(x)$ for test input x is shorter and factually correct with a user specified probability. The back-off function is implemented via GPT-4 [68], and they design prompts for generating the sub-claims and for scoring their correctness. Experiments are performed using GPT-4 on FActScore [66] (biographical fact checking), NaturalQuestions [51] (open-domain QA) & MATH [31] (mathematical problem solving). Their results show that conformal factuality consistently achieves the target correctness guarantees while preserving the majority of content, for ex. on FActScore, correctness improves from roughly 30% to over 80%.

4.2 Predictive Pipelines and (V)QA

Vision–language and visual question answering (VQA) systems increasingly operate in settings where the model must select one or more correct answers from a structured but potentially large candidate set. This includes tasks such as multiple-choice VQA, caption ranking, image-text retrieval, and open-ended question answering. Traditional conformal prediction (CP) frameworks, particularly split conformal prediction, offer a theoretical guarantee of the form:

$$\Pr[y^* \in C(x)] \geq 1 - \alpha,$$

where $C(x)$ denotes the predicted set and y^* the true label. However, applying CP in these multimodal settings introduces challenges, including the size of the output space, partial or weak supervision (e.g., in retrieval), and uncertainty due to the interaction of visual and linguistic modalities. These issues motivate the development of adapted calibration strategies that remain valid under such complexities.

4.2.1   CP for Multiple-Choice Question Answering

In multiple choice QA, the outputs space is the discrete list of choices and evaluation reduces to classification. Two representative approaches are Kumar et al. [50] and Ye & Wei [118], who adapt split-CP to large foundation models in textual and multimodal settings respectively.

Kumar et al. [50] frame MCQ answering with LLMs as a classification task. Each option y's probability is extracted from the model using carefully designed prompts which include GPT-4–generated exemplar questions and answers in-context. Let y be a multi-token string, then the answer probability is obtained by teacher-forcing $p(y|x) = \prod_t p(y_t|x, y_{<t})$. Final probabilities are obtained by normalizing across the options $\mathcal{Y}$, and averaging across ten prompt variants. The nonconformity score is defined as the Least Ambiguous Classifier (LAC) [88] score $S(x, y) = 1 - P(y|x)$, where $P(y|x)$ is the probability of predicting option y. . Given a calibration set, the threshold τ is set according to standard split-CP, and the prediction set is $C(x) = \{y \in \mathcal{Y} : S(x, y) \leq \tau\}$. Experiments are performed using LLaMA-13B [99] on 16-subject MMLU [30] dataset. Split CP with a 50/50 calibration–evaluation split delivers empirical coverage matching the nominal level, and prediction set size adapts to task difficulty, with larger

sets for hard subjects such as formal logic and smaller ones for easier domains like marketing. Experiments also show that set size is strongly correlated with accuracy, enabling selective classification, while naive top-k sets fail to guarantee coverage.

Ye & Wei [118] extend the same recipe to VLMs but in a black-box setting, where the logits are inaccessible. Each question is sampled $P = 36$ times at a temperature of 1.0 to encourage diverse, free-form generations. A smaller LM then maps these generations into the fixed set of options, producing a distribution

$$P(y|x) = \frac{1}{P} \sum_{i=1}^{P} 1\{\hat{y}^i \mapsto y\}$$

where $\hat{y}^i$ is the i-th free-form response. The LAC score is used as the non-conformity score $S(x, y)$, and a validation set is used to compute a threshold τ and consequently a prediction set, using split-CP. Experiments are performed on ScienceQA [63] and MMMU [120] with eight VLMs such as LLaVA-1.5 [60], Qwen2-VL [105], InternVL2 [12]. They run two protocols: (i) Fixed validation split ratio (0.5) while varying confidence α, where they observe that empirical error is strictly below α for both datasets indicating marginal coverage; and (ii) fixed $alpha = 0.2$ while varying split ratios, where empirical errors remain below α. Overall the method achieves valid marginal coverage and shows the expected inverse relationship between set size and confidence level.

CP for Short-Answer Tasks. CP can be used to quantify uncertainty in the free-form QA task, where the answer is a short phrase that is not restricted to fixed choices. We explore ConU [107] & SConU [108] proposed by Wang et al. to this end.

ConU adapts split-CP to free-form short-answer QA with black-box LLMs to generate prediction sets with guarantees. For input prompt x, M answers $\{\hat{y}_m\}$ are generated and semantically clustered into K groups using Sentence-BERT [83] embeddings. They denote a representative generation in the k-th cluster as $\hat{r}_k$, and the size of the cluster by V_k. For any generation $\hat{y}_m$ the uncertainty is

$$U(\hat{y}_m) = 1 - \lambda \frac{V_l}{M} - (1 - \lambda) \cdot \frac{1}{K} \sum_{i=1}^{K} S(\hat{y}_m, \hat{r}_k) \cdot \frac{V_k}{M}$$

where $\hat{y}_m$ belongs to cluster l. This measure uses self-consistency so that generation is more uncertain if it belongs to a smaller cluster and it is dissimilar to other generations. They propose a nonconformity score based on this uncertainty measure and sampling. For any prompt x and any completion y, they first sample $\{\hat{y}_m\}$ generations and pick a representative $\hat{y}_{rep}$ that is closest to y i.e. the one with highest $S(\hat{y}_m, y)$. The score function is thus $S(x, y) = U(\hat{y}_{rep})$. Following split-CP, a threshold τ is found using a calibration set and a prediction set is generated for each test point as $C(x) = \{y \in \{\hat{y}_m\} : S(x, y) \leq \tau\}$. To provide the coverage guarantee, they assume that at least one sample among M is acceptable. Experiments are performed on seven LLMs such as GPT-3.5-turbo, LLaMA-3, Mistral-7B [40], Vicuna-13B [123] across four free-form QA datasets – CoQA [82] for conversational QA, TriviaQA [45] for reading comprehension, MedQA [43] & MedMCQA [71] for medical problems. The proposed uncertainty score U is evaluated using AUROC metric which measures if U can effectively distinguish between correct and incorrect generations. In addition, conformal prediction sets generated with ConU achieve targeted coverage for all datasets while maintaining small average set sizes.

SConU [108] extends ConU by proposing a test for exchangeability of test samples. For each test point, SConU computes a conformal p-value using an uncertainty score u_{N+1} (e.g., predictive entropy), under the null hypothesis that it is exchangeable with the calibration scores $\{u_i\}$:

$$p_{N+1} = \frac{1 + \sum_{i=1}^{N} \mathbb{1}\{u_i \geq u_{N+1}\}}{N+1}.$$

This measures how unusual the test point's uncertainty score is compared to the calibration set $\{x_i, y_i^\star\}_{i=1}^{N}$. They also propose a refined variant that only compares against calibration points that themselves achieve coverage at level α. Since a p-value is obtained for each test point, SConU applies the Benjamini–Hochberg [8] procedure to decide which test points to abstain on, while the remaining points are passed to ConU, which generates prediction sets using nonconformity scores. SConU also formalizes the minimum manageable risk level α_l, the smallest error rate that can be guaranteed for a given finite calibration set. This is estimated by checking whether each calibration example's true answer is included in the M sampled generations from the LLM. If the fraction of such uncovered calibration samples is $L_N(1)$, then

$$\alpha_l = \frac{N \cdot L_N(1)}{N+1}.$$

Coverage guarantees are only valid when $\alpha \geq \alpha_l$, since attempting smaller values would be unsupported by the calibration set. They follow ConU for their experimental setup, and achieve better cross-domain calibration and improved approximate conditional coverage compared to ConU.

4.2.2 Online Semi-bandit Calibration

In streaming or interactive VQA settings, full feedback is often unavailable; systems observe only whether their prediction set contains the correct answer. Ge et al. [22] address this challenge by developing an online conformal prediction algorithm under *semi-bandit feedback*, where only hit/miss indicators are observed for the prediction set. At each round t, the learner selects a threshold τ_t and constructs the prediction set

$$C_t = \{\, y : f(x_t, y) \geq \tau_t \,\},$$

receiving feedback only on whether $y_t^* \in C_t$. To guarantee marginal coverage, the algorithm maintains a high-probability upper confidence bound G_t on the

CDF of the scoring function $f(x_t, y_t^*)$, estimated using the Dvoretzky–Kiefer–Wolfowitz inequality. The threshold is then updated as

$$\tau_{t+1} = \max\left\{\tau_t, \sup\{\tau : G_t(\tau) \leq 1 - \alpha\}\right\},$$

ensuring $\tau_t \leq \tau^*$ with probability at least $1 - 2/T$, and achieving $\tilde{O}(\sqrt{T})$ regret relative to the optimal static threshold τ^*. In experiments on ImageNet classification, SQuAD QA dataset [80] retrieval, and auction pricing with $T = 10{,}000$ and $\alpha = 0.9$, the proposed Semi-bandit Prediction Set (SPS) algorithm attains valid coverage, adapts quickly under sparse feedback, and minimizes cumulative regret. Notably, SPS avoids undercoverage and rapidly converges to efficient prediction sets without access to full labels.

4.2.3   CP as a Tool for Uncertainty-Aware Evaluation of FMs

To comprehensively evaluate FMs, benchmarking efforts should focus on uncertainty quantification (UQ) along with predictive accuracy. UQ allows for models to provide a confidence score on their outputs, helping address concerns of hallucination and reliability. CP can serve as a unifying framework for uncertainty quantification (UQ) of FMs, with Ye et al. [117] focusing on LLMs and Kostumov et al. [48] extending the approach to VLMs.

Ye et al. [117] propose to use CP to evaluate the uncertainty of FM outputs. For input x, the set-size of the calibrated prediction sets $|C(x)|$ can act as an interpretable proxy for model uncertainty. Ye et al. recast diverse downstream tasks such as question answering (MMLU [30]), reading comprehension (CosmosQA [37]), commonsense inference (HellaSwag [121]), dialogue response selection & document summarization (subsets of HaluEval [53]) into multiple choice questions. They include two additional options 'I don't know' and 'None of the above' to standardize difficulty and to allow the LLM to abstain from choosing from the given choices. They use two standard nonconformity scores, LAC score $S(x, y) = 1 - P(y|x)$ and APS score $S(x, y) = \sum_{\{y' \in \mathcal{Y} : P(y'|x) \geq P(y|x)\}} P(y'|x)$. The paper studies three prompting strategies to generate answers. For each prompt the option probability is computed by applying a softmax function on logits corresponding to each option letter, i.e. A to F, generated by the language modeling head.

They use CP to report coverage (CR), accuracy (Acc) and average set size (SS) for 9 open-source LLMs such as Qwen-14B, Gemma-7B, Mistral-7B, LLaMA-14B and 2 closed-source LLMs GPT-4 & GPT-3.5 on the above tasks. For calibration a 50/50 calibration-test split is taken, $\alpha = 0.1$, and results are averaged over score functions and prompting strategies. All models achieve coverage guarantee, however higher Acc does not necessarily imply lower SS, and instruction tuning increases SS. Their findings establish CP as a practical and model-agnostic tool for uncertainty-aware benchmarking of LLMs, outperforming entropy/perplexity as an uncertainty measure.

Kostumov et al. [48] extend CP for evaluation of VLMs. Following Ye et al. [117] they recast VQA tasks such as image captioning (COCO [59]), visual question answering (ScienceQA [63], TextVQA [94]), and visual grounding (Ref-COCO [119]) into a standardized MCQ format. The LAC and APS scores are used to construct prediction sets $C(x)$ and set-size $|C(x)|$ serves as a measure of uncertainty. They report average set size (SS), coverage rate ($\geq$90 %), accuracy, and standard calibration errors (ECE, MCE) across a suite of eleven VLMs such as LLaVA [60], CogVLM [106], Qwen-VL [105]. Results reveal that while all models achieve the target coverage, SS rankings often diverge markedly from accuracy alone e.g., Monkey-Chat [58] has high accuracy but also a large SS. Furthermore ECE/MCE poorly correlate with conformal uncertainty, underscoring the need for distribution-free set-based evaluation in multimodal settings.

4.2.4 Conformal Prediction Under Zero-Shot Domain Shift

Large vision-language foundation models (e.g., CLIP) introduce a new challenge: they are pretrained once on broad source data, then deployed zero-shot to target domains that may differ substantially from calibration sets. Since split CP presumes exchangeability between calibration and test inputs, naive adaptation on target data breaks the underlying validity guarantees.

Silva-Rodríguez et al. [93] address this gap with *Conf-OT*, a transductive transfer method that aligns calibration and test distributions via optimal transport (OT). Given a similarity matrix S between calibration and test points, they solve for a transport plan P under marginal constraints, with entropic regularization for tractability. The resulting soft reweighting adjusts calibration scores to reflect the target distribution, after which standard split-CP quantiles are applied.

Experiments across fifteen vision benchmarks (ImageNet [87] variants, fine-grained, and specialized datasets) show that Conf-OT preserves marginal coverage while reducing average prediction set sizes by up to ~20% relative to vanilla split CP, and remains substantially faster than alternative transductive baselines. Conceptually, Conf-OT functions as a lightweight post-hoc adapter: it requires no additional labeled data, yet improves set efficiency under domain drift while maintaining distribution-free validity guarantees.

4.3 Adaptive Reasoning and Control via Conformal Prediction

Chain-of-Thought [110] reasoning enables LLMs and VLMs to generate intermediate rationales before generating the answer, and significantly improves their problem solving capabilities [112]. As FMs transition from single-shot generation to multi-step reasoning and decision-making pipelines, it is necessary to ensure that the intermediate steps are not redundant and are factually grounded. CP offers tools to manage inference cost and avoid dangerous hallucinations during reasoning, as we describe below.

4.3.1 Controlling Chain-of-Thought Quality and Cost

When LLMs generate chains-of-thought, two risks arise: dropping low-confidence steps can break logical support, and over-long reasoning wastes resources without proportional gains. Both can be addressed by applying CP to structured fragments of the reasoning trace, pruning or halting under controlled risk.

Rubin-Toles et al. [86] introduce coherent factuality, a notion of correctness of multi-step reasoning chains where the factuality of a reasoning step is evaluated in context of the previous steps. They build a deducibility graph over a full reasoning trace, where nodes are individual claims and edges encode inferential dependencies. They assign each connected subgraph G a nonconformity score $s(G)$ (e.g. aggregate self-consistency loss), then calibrate the split-CP quantile τ on held-out problems so that

$$\Pr\big[s(G) \leq \tau\big] \geq 1 - \alpha.$$

At runtime, they retain the largest ancestor-connected subgraph with $s(G) \leq \tau$, guaranteeing that at least one coherent, factual sub-proof remains with risk $\leq\alpha$, and preserving logical support.

Wu et al. [111] address reasoning cost by training lightweight probes on hidden states to estimate the probability r_t that further steps will alter the final answer. Treating each potential cutoff t as a CP hypothesis "$\text{risk}_t > \delta$?", they compute p-values via fixed-sequence CP on calibration chains and select the earliest step t^* where risk $\leq \delta$. This "thought calibration" stops reasoning early under formal risk control, cutting inference tokens by up to 60% with negligible accuracy loss.

4.3.2   Risk-Controlled Abstention

Even well-pruned reasoning can produce low-confidence answers. A principled abstention policy can bound hallucination risk while maintaining usability. Abbasi-Yadkori et al. [115] wrap any LLM in a peer-agreement scheme: sample k candidate answers, compute a consensus score g by averaging pairwise similarities, and define nonconformity $r = 1 - g$. On a small calibration set, they set the split-CP threshold λ as the $\lceil (n+1)(1-\alpha) \rceil / n$ quantile of $\{r_i\}$. At inference, if the top answer's agreement $g < 1 - \lambda$, the system replies "I don't know"; otherwise it answers. This guarantees hallucination probability $\leq\alpha$, while abstaining significantly less often than naive log-probability baselines. By grouping claim filtering and early stopping under chain-of-thought control and handling low-confidence outputs via risk-controlled abstention, the authors gain end-to-end CP-governed pipelines that manage both the quality and safety of LLM reasoning.

4.4   Retrieval

Retrieval tasks—ranging from ad hoc document ranking to Retrieval-Augmented Generation (RAG)—involve identifying a small subset of relevant items from large and dynamic corpora. Considering the widespread use of RAG, this forms an important component of modern-day foundation models. In these settings, practitioners often seek coverage guarantees that at least one item in the returned set is relevant or supports a correct downstream prediction. Adapting conformal prediction (CP) to retrieval entails managing large candidate spaces, uncertain or partial feedback, cross-modal signals, and end-to-end grounding across pipeline stages.

4.4.1 Score-Level Calibration and Multimodal Alignment

A central approach to conformal retrieval treats model-generated similarity scores as nonconformity measures, enabling retrieval sets with statistical guarantees via split conformal prediction. However, raw scores often vary unpredictably across queries, inflating the size of retrieval sets and undermining efficiency.

Streamlining Conformal IR: Intrator et al. [38] proposes a score refinement strategy that preserves coverage while reducing retrieval set sizes. Rather than altering the underlying retrieval model, their method introduces a monotonic transformation applied to the similarity scores before calibration. Given a query q and its candidate scores $S = \{s_1, s_2, \ldots, s_N\}$, the scores are first normalized by the maximum value $s_{\max} = \max_i s_i$. Then, a position-dependent penalty is applied to downweight lower-ranked items:

$$T(s_{(r)}, r) = \frac{s_{(r)}}{s_{\max}} \cdot \frac{1}{\log(1 + r^\lambda)}$$

where $s_{(r)}$ denotes the rth largest score and λ is a tunable hyperparameter. This transformation maintains score order while controlling calibration thresholds across queries of differing difficulty. The authors apply this technique on BEIR benchmarks (FEVER [98], SCIFACT [103], FIQA [65]) using dense retrieval models (BGE-large-1.5 and E5-Mistral-7B), and demonstrate that their refined scoring consistently yields smaller retrieval sets when compared to vanilla CP and baselines like TopK (which utilizes Vanilla CP but calibrates to a fixed set size K for all queries), APS [84], and RAPS [6]. While the empirical coverage remains comparable—albeit slightly lower in some settings—the reduced prediction set size significantly improves the practical deployability of CP-based IR systems.

Leveraging LLM-Based Entailment: Zhao et al. [85] introduces a framework for conformal prediction in RAG pipelines by leveraging LLM-based entailment or factuality scoring to guide retrieval decisions. The goal is to ensure, with high confidence, that the retrieved context contains sufficient information to answer a user query. The method calibrates a nonconformity threshold on a development set and then uses it to guide inclusion of retrieved chunks at inference time.

Concretely, each retrieved passage is assigned a confidence score $\text{score}_i \in [0, 1]$, typically based on entailment probability or semantic similarity between the chunk and the query. The nonconformity score is defined as:

$$r_i = 1 - \text{score}_i$$

Calibration proceeds by collecting these nonconformity scores over a representative set of queries known to be answerable given the retrieved documents. Given a user-specified error rate α, the conformal threshold is set as the $\lceil (n+1)(1-\alpha) \rceil$-th smallest nonconformity value in the calibration set:

$$\tau = \text{Quantile}_{1-\alpha}(r_1, \ldots, r_n)$$

At inference, the system retrieves all document chunks for which the confidence score exceeds $1 - \tau$. This guarantees, under the standard exchangeability assumption, that the retrieved set includes at least one chunk sufficient to answer the question with probability at least $1 - \alpha$. Unlike traditional RAG, which retrieves a fixed number of top-k chunks, CONFLARE adaptively expands the retrieval set based on the learned threshold, offering a coverage guarantee at the cost of potentially longer contexts. The method is agnostic to the downstream LLM and focuses entirely on the retrieval stage. Notably, it assumes access to a dataset or synthetic pool of answerable calibration queries and corresponding gold document chunks to compute meaningful retrieval confidence scores.

Bridging Modalities: Li et al. [54] addresses retrieval under incomplete modality scenarios, where either the query or reference instances may lack one or more modality streams (e.g., missing audio in a video). The key challenge in such settings lies in computing and comparing cross-modal similarity scores when no modality is consistently shared across all pairs. Any2Any proposes a two-stage calibration pipeline that transforms heterogeneous similarity scores into conformal probabilities, enabling well-calibrated retrieval decisions despite asynchronous and incomplete observations.

In the first stage, Any2Any applies split conformal prediction to each cross-modal similarity space independently. Given a similarity score (e.g., from a CLIP or LiDAR-text encoder) and a binary label indicating whether the retrieval was correct, it calibrates the nonconformity scores to produce a probability that reflects the likelihood of a correct match. Since each modality pair (e.g., image-audio, text-LiDAR) exhibits distinct distributions, this per-space calibration yields a set of conformal matrices that normalize similarity values to a shared probabilistic scale. The second stage aggregates these calibrated scores into a unified retrieval score. A scalar summary—typically the mean of all valid entries in the conformal matrix—is computed and itself calibrated using a second round of conformal prediction. This produces a final conformal probability that grounds retrieval decisions across missing and mismatched modalities. The returned set then contains the top-ranked reference instances with the highest calibrated retrieval probabilities, ensuring coverage guarantees even in the absence of modality synchronization.

Experiments on datasets such as KITTI (LiDAR, video, text) [23], MSR-VTT (video, audio, text) [114], and Monash Bitcoin (text-to-time-series) [25] show that Any2Any achieves retrieval performance competitive with oracle baselines using full modality access. Notably, it remains effective when up to 50% of modalities are missing, highlighting the robustness of its calibration strategy. Furthermore, ablations reveal that the two-stage calibration process outperforms naive heuristics and unimodal retrieval pipelines, particularly in cross-domain scenarios with high modality variance.

4.4.2 Interactive and End-to-End Certification

In retrieval-augmented generation (RAG) systems, both the retrieval and the subsequent generative components contribute to the final output. However, traditional conformal prediction pipelines focus on certifying a single model stage. Recent methods aim to extend CP to cover entire end-to-end pipelines and interactive settings, offering formal guarantees even when supervision is sparse or delayed.

Merging Prediction Sets for Retrieval and LLM Outputs: Li et al. [55] introduces TRAQ, a conformal prediction framework for retrieval-augmented generation (RAG), constructing prediction sets for both retriever and language model (LLM) outputs that guarantee inclusion of semantically correct answers with high probability. It forms retriever sets by thresholding calibrated similarity scores to control the error rate of relevant passage retrieval. For each passage, multiple LLM responses are generated and clustered by semantic similarity to build LLM prediction sets under a separate error budget. These are aggregated using a Bonferroni correction to ensure overall coverage guarantees. TRAQ employs Bayesian optimization to select error budgets that optimize the trade-off between coverage and prediction set size, resulting in more compact prediction sets without sacrificing statistical guarantees. Extensive experiments on multiple QA datasets with fine-tuned retrievers and LLMs (GPT-3.5, Llama-2) demonstrate TRAQ consistently meets coverage targets and outperforms baselines in set size and robustness across clustering strategies and prompt designs. While it relies on i.i.d. data and sufficient model accuracy, TRAQ proposes mitigations for assumption violations and acknowledges computational overhead from repeated retrievals and response sampling as future work.

Bounding Generation Risk: Kang et al. [46] introduces C-RAG, a conformal prediction framework for retrieval-augmented generation (RAG) models by bounding the generation risk (e.g., 1 - ROUGE) via a constrained generation protocol. This protocol controls retrieval size, number of generations, and diversity to produce sets of candidate outputs. C-RAG provides two guarantees: a high-probability upper bound on risk for any fixed configuration, and a valid set of configurations that keep risk below a desired level α. It further extends these guarantees to settings with distribution shifts between calibration and test data. Theoretically, C-RAG proves that RAG models achieve lower conformal generation risk than standalone LLMs, with benefits scaling with retrieval quality and model calibration. Empirical results across multiple datasets confirm the soundness and tightness of these risk bounds under both in-distribution and shifted conditions. Overall, C-RAG offers a principled, distribution-free method to certify and control uncertainty in modern retrieval-augmented language generation systems. The abovementioned efforts illustrate the growing importance of uncertainty estimation and calibration when using modern day large-scale foundation models. Below, we discuss a few takeaways from the above review.

4.5   Takeaways

For robust conformal prediction on multimodal foundation models, we identified a few central issues that need to be addressed:

- Per-modality normalization represents a crucial first step, requiring normalization of nonconformity scores by modality-specific variance estimates to handle heteroskedasticity effectively across different data types.

- Uncertainty-weighted fusion provides another important strategy, forming composite scores through the formula:

$$s(x, y) = \sum_m w_m(x)\, s_m(x_m, y) \tag{5}$$

  where weights reflect the reliability of each modality for a given input. This approach allows the system to dynamically adjust the influence of different modalities based on their expected accuracy and relevance.
- Representative calibration ensures that calibration sets comprehensively cover expected modality combinations, noise conditions, and temporal correlations that the system will encounter in deployment. This requires careful curation of calibration data to reflect the full range of operational conditions rather than just clean, idealized examples.
- For temporally correlated data such as video or sensor streams, dependency-aware methods become essential, utilizing block or sliding-window calibration techniques that respect the underlying temporal structure. When alignment quality varies across different contexts, conditional approaches such as trust-score conditioning or group-aware calibration help maintain reliability across diverse scenarios.
- Finally, a comprehensive evaluation extends beyond simple marginal coverage to monitor performance across modality presence patterns, different device types, multiple languages, and various noise conditions. This holistic evaluation approach ensures that conformal prediction methods maintain their reliability guarantees across the full spectrum of multimodal deployment scenarios, enabling trustworthy uncertainty quantification in complex multimodal systems.

In the next section, we discuss open challenges and emerging trends relevant for this direction of research.

5   Challenges and Emerging Directions

Extending conformal prediction to MMFM-based pipelines presents several challenges that go beyond the classical CP theory. Expanding on the discussion in the previous section, we summarize key limitations practitioners must navigate:

Breakdown of Exchangeability. CP's marginal coverage guarantee relies on calibration and test examples being exchangeable. In multimodal settings, however, data often come from non-stationary or heterogeneous sources, thus resulting in: (i) *cross-modal drift*, i.e. different modalities (e.g. images vs. text) may evolve independently, violating joint exchangeability; and/or (ii) *temporal shifts*, i.e. streaming inputs may change distribution over time (e.g. new imaging protocols or visual scenes).

Task-Specific Admission Predicates. Constructing a binary admission function $A(y)$—which flags whether a candidate output is acceptable—often requires bespoke, domain-expert heuristics, thus resulting in: (i) *domain dependence*, i.e. an admission rule tuned for medical reports may not transfer to open-domain QA or VQA without redesign; and/or (ii) *complex outputs*, i.e. defining $A(y)$ on structured or compositional outputs (e.g. action sequences, multimodal answers) can be laborious.

Budget–Utility Trade-Off. Achieving CP's theoretical bounds often demands large calibration sets and, in generative pipelines, many model samples. This may suffer when there is: (i) *insufficient sampling*, i.e. low sampling budgets can force CP to return empty sets or trivial guarantees; and/or (ii) *conservative sets*, i.e. tight error targets (small α) push prediction sets to grow unwieldy, undermining their practical utility.

Sensitivity to Scoring and Aggregation Choices. Multimodal CP pipelines hinge on two design choices—nonconformity score $r(y)$ and set-aggregation functions $F(C)$; however, optimal forms for these choices is non-trivial. This results in issues around: (i) *nonconformity design*, i.e. deciding between model likelihoods, embedding distances, or learned reward scores can dramatically affect coverage and set size; and/or (ii) *aggregation heuristics*, i.e. whether to use maximum, sum, entropy, or more elaborate set-confidence metrics which may in turn influence stopping behavior and efficiency.

The abovementioned challenges reiterate that applying CP in FM settings requires careful validation of exchangeability, deliberate design of admission criteria, prudent allocation of sampling budgets, and empirical tuning of scoring and aggregation functions to ensure both statistical validity and practical value.

5.1   Emerging Trends and Opportunities

As CP methods become more integrated with MMFM pipelines, newer research is pushing beyond single-task adaptations toward more robust, adaptive, and hybrid frameworks. Our discussion on existing efforts in the previous section point to specific directions of focus, such as: relaxing exchangeability via conditional guarantees, developing fine-grained calibration strategies to shrink prediction sets, and/or CP with complementary uncertainty quantification tools. In the sections that follow, we discuss our perspectives on these emerging directions.

5.1.1   Addressing Data Exchangeability and Conditional Guarantees

As stated earlier, standard conformal prediction provides only marginal coverage, i.e.

$$\Pr_{(X,Y)\sim\mathcal{D}}\left[Y \in C(X)\right] \geq 1 - \alpha,$$

which can hide undercoverage on subgroups or in data drift scenarios. Gibbs et al. [24] introduced *conditional* CP, wherein by choosing a finite family of groups or features $\mathcal{G}$, they compute separate thresholds per group to guarantee:

$$\Pr\left[Y \in C(X) \mid G(X) = g\right] \geq 1 - \alpha, \quad \forall g \in \mathcal{G},$$

and, more generally, provide uniform finite-sample bounds over richer shift classes via hyperparameterized correction schemes. Cherian et al. [13] advance this idea by learning thresholds that vary smoothly with input features, rather than relying on coarse group partitions. Instead of one cutoff per group, their method maps each input to its own confidence threshold and error tolerance, unifying subgroup validity with per-instance calibration. This approach reduces undercoverage in heterogeneous or drifting settings, but depends on informative difficulty features and sufficient calibration data.

Practically, these developments imply a tiered deployment strategy. Start with simple group-conditional checks (entropy, length, coarse domain tags) to detect glaring undercoverage; where groups are well populated, apply group-wise split-CP to recover subgroup guarantees. If calibration resources and predictive features permit, transition to input-adaptive quantile models to obtain tighter sets on "easy" instances and wider, safer sets on "hard" ones. Finally, when unlabeled test pools are available, lightweight transductive reweighting (e.g., OT-style alignment) can increase set efficiency while preserving the marginal validity that practitioners rely on.

5.1.2 Score Functions Calibrations

Native confidence signals in foundation models, be it generation likelihood or similarity scores, are often imperfect surrogates for task correctness. The practical research agenda therefore treats these signals as explicit calibration targets: refine them post-hoc so the nonconformity score better aligns with the admission predicate used by downstream validators. Representative instantiations include supervised regressors and monotone score transforms: Shahrokhi et al. [91] learn sample-budget regressors to predict how many draws are needed to find an admissible output, Intrator et al. [38] apply monotonic refinements to retrieval similarities before calibration, and Liu et al. [61] train binary correctness scorers on human-validated outputs to produce cleaner residuals for conformal filtering.

Across these approaches, the recurring engineering tradeoffs are clear: (i) supervised scorers typically yield the strongest alignment to task correctness but require labeled calibration data; (ii) monotone transforms and unsupervised refinements are cheaper to deploy but produce coarser gains; and (iii) any learned score adjustment must be validated under likely distributional shifts to avoid overfitting calibration idiosyncrasies.

5.1.3 Decoding-Strategy Calibrations

Beyond post-hoc score adjustments, CP can be woven into the decoding process itself, dynamically setting sampling or beam parameters to meet coverage targets without extraneous computation.

Conformal Nucleus Sampling: Ravfogel et al. [81] formally evaluate whether nucleus (top-p) decoding admits the intended probabilistic interpretation, viz. that the true next token lies in the top-p set with probability p. Their calibration procedure applies split CP to next-token softmax outputs. Concretely, for each validation timestep t they compute the APS-style score:

$$s_t = \sum_{i:\, p_t(i) \geq p_t(y_t)} p_t(i),$$

where $p_t(\cdot)$ is the distribution of the next token of the model and y_t the observed token. The calibrated nucleus mass $\hat{q}$ is chosen as the $(1-\alpha)$-quantile of $\{s_t\}$; by standard split-CP theory, this yields marginal coverage up to the usual finite-sample slack:

$$1-\alpha \leq \Pr\big[y \in C_{\hat{q}}(x)\big] \leq 1-\alpha + \tfrac{1}{n+1}.$$

To accommodate heteroskedastic uncertainty, the authors estimate $\hat{q}$ conditionally on entropy percentiles so that the calibrated p varies with the entropy of the next-token distribution. Empirically, the OPT family models exhibit systematic overconfidence (the calibrated $\hat{q}$ exceeds the nominal $1-\alpha$), with the strongest miscalibration occurring in low-entropy contexts; calibration degrades modestly with model scale. Downstream generation quality (MAUVE, BERTScore) remains comparable under conformal p-sampling, and the paper documents practical mitigations for token-sequence dependence (e.g., per-sentence token sampling for calibration).

Beam Search with Coverage Guarantees: Deutschmann et al. [15] adapt split-CP to greedy decoding via two practical beam-search variants. The first, a *conformal beam-subset* method, runs a standard beam search, then post-hoc calibrates and prunes the beam using split-CP thresholds to produce a sequence-level marginal guarantee $\Pr[S^{\star} \in C_{\alpha|\beta}(X)] \geq 1-\alpha$ (up to finite-sample slack). The second, *dynamic conformal beams*, integrates CP into decoding by expanding/pruning the beam at each timestep until per-step conformal criteria are met; this can attain chosen coverage levels but may produce very large beams or require a maximum length to control accumulated risk. Empirical results on NLP and chemistry tasks show both methods meet their coverage targets; the authors recommend length-conditional/group-conditional calibration to tighten guarantees in heterogeneous or long-tailed settings. These decoding-level procedures complement supervised score refinements: the former allocates compute where uncertainty is highest, while the latter sharpens the nonconformity signal used by sampling or search.

5.1.4   Calibration in Retrieval Techniques

Recent work in conformal retrieval is shifting from calibrating individual similarity scores to certifying retrieval as a full pipeline. One direction focuses on stabilizing similarity signals: for example, monotonic score adjustments can reduce query-specific variability, producing more reliable inputs for conformal calibration [38]. Another direction derives retrieval confidence from semantic signals such as entailment or factuality checks from large language models, ensuring that at least one retrieved passage meets a calibrated threshold for downstream use

[85]. The handling of incomplete modalities further highlights the need to reconcile heterogeneous similarity spaces on a shared probabilistic scale; [54] achieves this by first calibrating per-modality scores and then aggregating them, leading to robust retrieval decisions. Finally, end-to-end certifications for retrieval-augmented generation extend guarantees beyond the retriever, either by merging calibrated retriever sets with clustered LLM outputs under a combined error budget [55] or by bounding generation risk through constrained configuration sets with distribution-shift robustness [46]. Together, these directions emphasize a calibration that is query-adaptive, modality-aware, and pipeline-level, aligning retrieval outputs with downstream generative use while maintaining conformal guarantees.

5.1.5 Evaluation Frameworks for Confidence Measures

Robust evaluation of confidence and uncertainty estimators for foundation models is an emerging concern. MCQA-Eval [61] tackles this by repurposing multiple-choice QA items: each option is treated as a candidate response, white-box (logit/likelihood) and black-box (consistency) measures are applied, and discriminative power is scored against the dataset's correct choice (AUROC [49] as the primary metric). This removes reliance on noisy correctness heuristics such as LLM judges or ad-hoc similarity thresholds. Though not conformal, MCQA-Eval complements CP evaluation in two ways. Gold-label benchmarks provide low-noise ground truth that can validate conformal wrappers, and option-based testbeds reveal how well estimators separate correct from incorrect outputs before or alongside distribution-free CP guarantees.

More broadly, evaluation in VQA and diffusion tasks could benefit from integrating gold-label testbeds with CP-based guarantees. Existing efforts—such as uncertainty-aware VQA benchmarks [48], inductive conformal calibration for VLMs [118], and generative prediction sets for deep generative models [91]—do not employ gold-label discrimination directly, but their calibration and certification pipelines would be strengthened by combining such benchmarks with conformal coverage.

5.1.6 Applications in Newer Domains

While most existing research efforts on CP for MMFMs have concentrated on language and vision–language modeling, we posit that the application of adapted conformal techniques to newer domains where uncertainty bounds have immediate operational impact would be another important direction of future work. For example, in *crowd counting* [28,62], conformal intervals on counting predictions yield statistically valid upper and lower bounds on population estimates, supporting safety-critical planning in surveillance, transportation, and environmental monitoring. In *data poisoning detection* [122], CP wraps anomaly scores from embedding-based detectors to achieve finite-sample false alarm control, ensuring that flagged samples correspond to statistically significant deviations from clean-data distributions.

In *molecular property prediction* and *drug discovery* [34], conformal wrappers around graph neural networks or large chemical language models can provide rigorous uncertainty limits on predicted bioactivity or toxicity, allowing safer candidate selection in early screening stages. In *medical imaging* [64], CP calibrated segmentation masks and classification outputs can support decision-making by pairing model predictions with per case coverage guarantees, a critical factor in high-stakes diagnostic settings. In *time-series forecasting* [56], particularly for the modeling of energy demand and climate, CP-derived predictive bands can allow operators to plan within validated risk tolerances, accommodating non-stationary and seasonal data.

Across these domains, adaptations often hinge on crafting task-specific non-conformity scores (e.g., density residuals for crowd counting, embedding space margins for detection, chemical similarity measures for molecular tasks) while preserving the distribution-free nature of CP guarantees. By aligning CP with domain semantics, these approaches extend its utility beyond canonical classification/regression setups into operational pipelines where interpretable finite-sample coverage is mission-critical.

6   Conclusions

Conformal prediction offers a powerful, theoretically grounded approach to providing rigorous uncertainty guarantees for modern AI systems, yet its direct application to foundation models and multimodal settings is far from straightforward. As our survey shows, recent research has made significant strides in reimagining CP for these systems, ranging from sample-efficient calibration in generative models and multi-step reasoning pipelines to retrieval-augmented generation and multimodal alignment. Common trends emerge: the need for modality-aware scoring, domain-adaptive calibration, and risk-controlled abstention; integration of CP with generative and reasoning processes; and growing emphasis on evaluation that reflects real-world deployment conditions. However, challenges remain, including addressing non-exchangeability in temporal and interactive settings, scalability to web-scale data, and achieving conditional guarantees beyond marginal coverage. We hypothesize that future progress hinges on unifying CP theory with advances in large-scale modeling, exploring tighter bounds for structured outputs, and developing practical calibration protocols for dynamic, streaming environments. By bridging these gaps, CP can become a cornerstone for building trustworthy, interpretable, and robust multimodal AI systems.

References

1. Alayrac, J.B., et al.: Flamingo: a visual language model for few-shot learning. In: Advances in Neural Information Processing Systems (NeurIPS), vol. 35 (2022)
2. Angeli, G., Manning, C.D.: NaturalLI: natural logic inference for common sense reasoning. In: Moschitti, A., Pang, B., Daelemans, W. (eds.) Proceedings of the 2014 Conference on Empirical Methods in Natural Language Processing (EMNLP), pp. 534–545. Association for Computational Linguistics, Doha, Qatar (2014)

3. Angelopoulos, A.N., Barber, R.F., Bates, S.: Theoretical foundations of conformal prediction (2025). https://doi.org/10.48550/arXiv.2411.11824

4. Angelopoulos, A.N., Bates, S.: Conformal prediction: a gentle introduction. Found. Trends Mach. Learn. **16**(4), 494–591 (2023)

5. Angelopoulos, A.N., Bates, S., Candès, E.J., Jordan, M.I., Lei, L.: Learn then test: calibrating predictive algorithms to achieve risk control. Ann. Appl. Stat. **19**(2), 1641–1662 (2025)

6. Angelopoulos, A.N., Bates, S., Malik, J., Jordan, M.I.: Uncertainty sets for image classifiers using conformal prediction. In: International Conference on Learning Representations (2021)

7. Austin, J., Odena, A., Nye, M., Bosma, M., Michalewski, H., Dohan, D., et al.: Program synthesis with large language models (2021). https://doi.org/10.48550/arXiv.2108.07732

8. Benjamini, Y., Hochberg, Y.: Controlling the false discovery rate: a practical and powerful approach to multiple testing. J. Royal Stat. Soc. Ser. B-Methodol. **57**, 289–300 (1995)

9. Bodnar, C., Bruinsma, W.P., Lucic, A., et al.: A foundation model for the earth system. Nature **641**, 1180–1187 (2025)

10. Bommasani, R., et al.: On the opportunities and risks of foundation models (2021). https://doi.org/10.48550/arXiv.2108.07258

11. Campos, M., Farinhas, A., Zerva, C., Figueiredo, M.A.T., Martins, A.F.T.: Conformal prediction for natural language processing: a survey. Trans. Assoc. Comput. Linguist. **12**, 1497–1516 (2024)

12. Chen, Z., et al.: How far are we to GPT-4V? closing the gap to commercial multimodal models with open-source suites (2024). https://doi.org/10.48550/arXiv.2404.16821

13. Cherian, J., Gibbs, I., Candès, E.: Large language model validity via enhanced conformal prediction methods. In: Advances in Neural Information Processing Systems (2024)

14. Cobbe, K., et al.: Training verifiers to solve math word problems (2021). https://doi.org/10.48550/arXiv.2110.14168

15. Deutschmann, N., Alberts, M., Rodríguez Martínez, M.: Conformal autoregressive generation: beam search with coverage guarantees. In: Proceedings of the Thirty-Eighth AAAI Conference on Artificial Intelligence (AAAI), pp. 1–9. AAAI Press (2024)

16. Devlin, J., Chang, M.W., Lee, K., Toutanova, K.: Bert: pre-training of deep bidirectional transformers for language understanding. In: North American Chapter of the Association for Computational Linguistics (2019)

17. Ding, T., Angelopoulos, A.N., Bates, S., Jordan, M.I., Tibshirani, R.J.: Class-conditional conformal prediction with many classes (2023). https://doi.org/10.48550/arXiv.2306.09335

18. Einbinder, B.S., Romano, Y., Sesia, M., Zhou, Y.: Training uncertainty-aware classifiers with conformalized deep learning (2022). https://doi.org/10.48550/arXiv.2205.05878

19. Fisch, A., Schuster, T., Jaakkola, T., Barzilay, R.: Efficient conformal prediction via cascaded inference with expanded admission. In: International Conference on Learning Representations (2021)

20. Fontana, M., Zeni, G., Vantini, S.: Conformal prediction: a unified review of theory and new challenges. Bernoulli **29**(1), 1–23 (2023)

21. Ganguli, D., et al.: Red teaming language models to reduce harms: methods, scaling behaviors, and lessons learned (2022). https://doi.org/10.48550/arXiv.2209.07858
22. Ge, H., Bastani, H., Bastani, O.: Stochastic online conformal prediction with semi-bandit feedback. In: Forty-Second International Conference on Machine Learning (2025)
23. Geiger, A.: Are we ready for autonomous driving? The kitti vision benchmark suite. In: Proceedings of the IEEE Conference on Computer Vision and Pattern Recognition (CVPR), pp. 3354–3361. IEEE Computer Society (2012)
24. Gibbs, I., Cherian, J.J., Candès, E.J.: Conformal prediction with conditional guarantees. J. Roy. Stat. Soc. Ser. B Stat. Methodol. qkaf008 (2025)
25. Godahewa, R.W., Bergmeir, C., Webb, G.I., Hyndman, R., Montero-Manso, P.: Monash time series forecasting archive. In: NeurIPS Datasets and Benchmarks Track (Round 2) (2021)
26. Groh, M., et al.: Evaluating deep neural networks trained on clinical images in dermatology with the fitzpatrick 17k dataset. In: Proceedings of the IEEE/CVF Conference on Computer Vision and Pattern Recognition (CVPR) Workshops, pp. 1820–1828 (2021)
27. Gupta, C., Kuchibhotla, A.K., Ramdas, A.: Nested conformal prediction and quantile out-of-bag ensemble methods. Pattern Recogn. **127**, 108496 (2022)
28. Han, T., Bai, L., Gao, J., Wang, Q., Ouyang, W.: Dr.vic: decomposition and reasoning for video individual counting (2022). https://doi.org/10.48550/arXiv.2203.12335
29. Han, X., Tang, Z., Ghosh, J., Liu, Q.: Split localized conformal prediction (2023). https://doi.org/10.48550/arXiv.2206.13092
30. Hendrycks, D., et al.: Measuring massive multitask language understanding. In: International Conference on Learning Representations (2021)
31. Hendrycks, D., et al.: Measuring mathematical problem solving with the math dataset. In: NeurIPS 2021 Datasets and Benchmarks Track (2021)
32. Hermann, K.M., et al.: Teaching machines to read and comprehend. In: Advances in Neural Information Processing Systems, vol. 28 (2015)
33. Hoffmann, J., et al.: Training compute-optimal large language models. In: Proceedings of the 36th International Conference on Neural Information Processing Systems. NIPS 2022. Curran Associates Inc., Red Hook, NY, USA (2022)
34. Hu, W., et al.: Deep learning methods for small molecule drug discovery: a survey. IEEE Trans. Artif. Intell. **5**(02), 459–479 (2024)
35. Huang, K., Chandak, P., Wang, Q., et al.: A foundation model for clinician-centered drug repurposing. Nat. Med. **30**, 3601–3613 (2024)
36. Huang, L., et al.: A survey on hallucination in large language models: principles, taxonomy, challenges, and open questions. ACM Trans. Inf. Syst. **43**, 1–55 (2023)
37. Huang, L., Le Bras, R., Bhagavatula, C., Choi, Y.: Cosmos QA: machine reading comprehension with contextual commonsense reasoning. In: Proceedings of the 2019 Conference on EMNLP-IJCNLP, pp. 2391–2401. Association for Computational Linguistics, Hong Kong, China (2019)
38. Intrator, Y., Cohen, R., Kelner, O., Goldenberg, R., Rivlin, E., Freedman, D.: Streamlining conformal information retrieval via score refinement. In: Proceedings of the Seventh Fact Extraction and VERification Workshop (FEVER), pp. 186–191. Association for Computational Linguistics, Miami, Florida, USA (2024)
39. Jeong, M., Hwang, H., Yoon, C., Lee, T., Kang, J.: Olaph: improving factuality in biomedical long-form question answering (2024). https://doi.org/10.48550/arXiv.2405.12701

40. Jiang, A.Q., et al.: Mistral 7b (2023). https://doi.org/10.48550/arXiv.2310.06825
41. Jiang, H., Kim, B., Guan, M.Y., Gupta, M.: To trust or not to trust a classifier (2018). https://doi.org/10.48550/arXiv.1805.11783
42. Jiang, Z., Liu, A., Van Durme, B.: Conformal linguistic calibration: trading-off between factuality and specificity (2025). https://doi.org/10.48550/arXiv.2502.19110
43. Jin, D., Pan, E., Oufattole, N., Weng, W.H., Fang, H., Szolovits, P.: What disease does this patient have? A large-scale open domain question answering dataset from medical exams. arXiv abs/2009.13081 (2020)
44. Johnson, A.E., et al.: MIMIC-CXR-JPG, a large publicly available database of labeled chest radiographs. Sci. Data **6**(1), 317 (2019)
45. Joshi, M., Choi, E., Weld, D.S., Zettlemoyer, L.: Triviaqa: a large scale distantly supervised challenge dataset for reading comprehension. In: Proceedings of the 55th Annual Meeting of the Association for Computational Linguistics (Volume 1: Long Papers), pp. 1601–1611 (2017)
46. Kang, M., Gürel, N.M., Yu, N., Song, D., Li, B.: C-RAG: certified generation risks for retrieval-augmented language models. In: Proceedings of the 41st International Conference on Machine Learning (ICML), pp. 1–38. JMLR.org (2024)
47. Kaur, J.N., Jordan, M.I., Alaa, A.: Conformal prediction sets with improved conditional coverage using trust scores (2025). https://doi.org/10.48550/arXiv.2501.10139
48. Kostumov, V., Nutfullin, B., Pilipenko, O., Ilyushin, E.: Uncertainty-aware evaluation for vision-language models (2024). https://doi.org/10.48550/arXiv.2402.14418
49. Kuhn, L., Gal, Y., Farquhar, S.: Semantic uncertainty: Linguistic invariances for uncertainty estimation in natural language generation. In: Proceedings of the Eleventh International Conference on Learning Representations (2023)
50. Kumar, B., et al.: Conformal prediction with large language models for multi-choice question answering (2023). https://doi.org/10.48550/arXiv.2305.18404
51. Kwiatkowski, T., et al.: Natural questions: a benchmark for question answering research. Trans. Assoc. Comput. Linguist. **7**, 452–466 (2019)
52. Lai, Y., et al.: DS-1000: a natural and reliable benchmark for data science code generation. In: Proceedings of the 40th International Conference on Machine Learning, vol. 202, pp. 18319–18345. PMLR (2023)
53. Li, J., Cheng, X., Zhao, X., Nie, J.Y., Wen, J.R.: HaluEval: a large-scale hallucination evaluation benchmark for large language models. In: Bouamor, H., Pino, J., Bali, K. (eds.) Proceedings of the 2023 Conference on Empirical Methods in Natural Language Processing, pp. 6449–6464. Association for Computational Linguistics, Singapore (2023)
54. Li, P.H., Yang, Y., Omama, M., Chinchali, S., Topcu, U.: Any2any: incomplete multimodal retrieval with conformal prediction (2024). https://doi.org/10.48550/arXiv.2411.10513
55. Li, S., Park, S., Lee, I., Bastani, O.: Traq: trustworthy retrieval augmented question answering via conformal prediction. In: Proceedings of the 2024 Conference of the North American Chapter of the Association for Computational Linguistics: Human Language Technologies (NAACL-HLT), pp. 3799–3821. Association for Computational Linguistics (2024)
56. Li, W., Law, K.L.E.: Deep learning models for time series forecasting: a review. IEEE Access **12**, 92306–92327 (2024)

57. Li, X., Wang, W., Li, M., Guo, J., Zhang, Y., Feng, F.: Evaluating mathematical reasoning of large language models: a focus on error identification and correction. In: Annual Meeting of the Association for Computational Linguistics (2024)
58. Li, Z., et al.: Monkey: image resolution and text label are important things for large multi-modal models. In: Proceedings of the IEEE/CVF Conference on Computer Vision and Pattern Recognition (2024)
59. Lin, T.Y., et al.: Microsoft coco: common objects in context. In: European Conference on Computer Vision (2014)
60. Liu, H., Li, C., Wu, Q., Lee, Y.J.: Visual instruction tuning. In: Advances in Neural Information Processing Systems (NeurIPS) (2023)
61. Liu, X., Lin, Z., Da, L., Chen, C., Trivedi, S., Wei, H.: MCQA-Eval: efficient confidence evaluation in NLG with gold-standard correctness labels (2025). https://doi.org/10.48550/arXiv.2502.14268
62. Liu, X., et al.: Weakly supervised video individual counting. In: Proceedings of the IEEE/CVF Conference on Computer Vision and Pattern Recognition (CVPR), pp. 19228–19237 (2024)
63. Lu, P., et al.: Learn to explain: multimodal reasoning via thought chains for science question answering. In: Advances in Neural Information Processing Systems (2022)
64. Ma, S., et al.: Deep learning approaches for medical imaging under varying degrees of label availability: a comprehensive survey (2025). https://doi.org/10.48550/arXiv.2504.11588
65. Maia, M., et al.: WWW'18 open challenge: financial opinion mining and question answering. In: Companion Proceedings of the Web Conference 2018, pp. 1941–1942. ACM (2018)
66. Min, S., et al.: Factscore: fine-grained atomic evaluation of factual precision in long form text generation. In: Advances in Neural Information Processing Systems, vol. 36 (2023)
67. Mohri, C., Hashimoto, T.: Language models with conformal factuality guarantees. In: Proceedings of the Forty-First International Conference on Machine Learning (2024)
68. OpenAI: GPT-4 technical report (2023). https://doi.org/10.48550/arXiv.2303.08774
69. Achiam, J., et al.: GPT-4 technical report (2024). https://doi.org/10.48550/arXiv.2303.08774
70. Oquab, M., et al.: Dinov2: learning robust visual features without supervision (2023). https://doi.org/10.48550/arXiv.2304.07193
71. Pal, A., Umapathi, L.K., Sankarasubbu, M.: Medmcqa: a large-scale multi-subject multi-choice dataset for medical domain question answering. In: ACM Conference on Health, Inference, and Learning (2022)
72. Papadopoulos, H.: Inductive conformal prediction: theory and application to neural networks. In: Fritzsche, P. (ed.) Tools in Artificial Intelligence, chap. 18. IntechOpen, Rijeka (2008)
73. Papadopoulos, H., Gammerman, A., Vovk, V.: Normalized nonconformity measures for regression conformal prediction. In: Proceedings of the IASTED International Conference on Artificial Intelligence and Applications (AIA), pp. 64–69 (2008)
74. Peng, Z., et al.: Kosmos-2: grounding multimodal large language models to the world (2023). https://doi.org/10.48550/arXiv.2306.14824

75. Platt, J.C.: Probabilistic outputs for support vector machines and comparisons to regularized likelihood methods. In: Advances in Large Margin Classifiers, pp. 61–74. MIT Press (1999)

76. Quach, V., et al.: Conformal language modeling. In: The Twelfth International Conference on Learning Representations (2024)

77. Radford, A., et al.: Learning transferable visual models from natural language supervision. In: Proceedings of the 38th International Conference on Machine Learning (ICML), pp. 8748–8763 (2021)

78. Radford, A., Kim, J.W., Xu, T., Brockman, G., McLeavey, C., Sutskever, I.: Robust speech recognition via large-scale weak supervision. In: International Conference on Machine Learning (2022)

79. Raffel, C., et al.: Exploring the limits of transfer learning with a unified text-to-text transformer. J. Mach. Learn. Res. 21(140), 1–67 (2020)

80. Rajpurkar, P., Zhang, J., Lopyrev, K., Liang, P.: SQuAD: 100,000+ questions for machine comprehension of text. In: Su, J., Duh, K., Carreras, X. (eds.) Proceedings of the 2016 Conference on Empirical Methods in Natural Language Processing, pp. 2383–2392. Association for Computational Linguistics, Austin, Texas (2016)

81. Ravfogel, S., Goldberg, Y., Goldberger, J.: Conformal nucleus sampling. In: Findings of the Association for Computational Linguistics: ACL, pp. 27–34. Association for Computational Linguistics (2023)

82. Reddy, S., Chen, D., Manning, C.D.: CoQA: a conversational question answering challenge. Trans. Assoc. Comput. Linguist. 7, 249–266 (2018)

83. Reimers, N., Gurevych, I.: Sentence-BERT: sentence embeddings using Siamese BERT-networks. In: Inui, K., Jiang, J., Ng, V., Wan, X. (eds.) Proceedings of the 2019 Conference on Empirical Methods in Natural Language Processing and the 9th International Joint Conference on Natural Language Processing (EMNLP-IJCNLP), pp. 3982–3992. Association for Computational Linguistics, Hong Kong, China (2019)

84. Romano, Y., Sesia, M., Candès, E.: Classification with valid and adaptive coverage. In: Advances in Neural Information Processing Systems, vol. 33, pp. 3581–3591. Curran Associates, Inc. (2020)

85. Rouzrokh, P., Faghani, S., Gamble, C.U., Shariatnia, M., Erickson, B.J.: Conflare: conformal large language model retrieval (2024). https://doi.org/10.48550/arXiv.2404.04287

86. Rubin-Toles, M., Gambhir, M., Ramji, K., Roth, A., Goel, S.: Conformal language model reasoning with coherent factuality. In: The Thirteenth International Conference on Learning Representations (2025)

87. Russakovsky, O., et al.: ImageNet large scale visual recognition challenge. Int. J. Comput. Vision (IJCV) 115(3), 211–252 (2015)

88. Sadinle, M., Lei, J., Wasserman, L.A.: Least ambiguous set-valued classifiers with bounded error levels. J. Am. Stat. Assoc. 114, 223–234 (2016)

89. Seedat, N., Jeffares, A., Imrie, F., van der Schaar, M.: Improving adaptive conformal prediction using self-supervised learning. In: International Conference on Artificial Intelligence and Statistics. PMLR (2023)

90. Shafer, G., Vovk, V.: A tutorial on conformal prediction (2007). https://doi.org/10.48550/arXiv.0706.3188

91. Shahrokhi, H., Roy, D.R., Yan, Y., Arnaoudova, V., Doppa, J.R.: Conformal prediction sets for deep generative models via reduction to conformal regression (2025). https://doi.org/10.48550/arXiv.2503.10512

92. Shu, D., Zhao, H., Liu, X., Demeter, D., Du, M., Zhang, Y.: Lawllm: law large language model for the us legal system. In: Proceedings of the 33rd ACM International Conference on Information and Knowledge Management (2024)

93. Silva-Rodríguez, J., Ben Ayed, I., Dolz, J.: Conformal prediction for zero-shot models (2025). https://doi.org/10.48550/arXiv.2505.24693

94. Singh, A., et al.: Towards VQA models that can read. In: Proceedings of the IEEE/CVF Conference on Computer Vision and Pattern Recognition (CVPR) (2019)

95. Singhal, K., et al.: Large language models encode clinical knowledge. Nature **620**, 172–180 (2022)

96. Stankeviciute, K., Alaa, A.M., van der Schaar, M.: Conformal time-series forecasting. In: Ranzato, M., Beygelzimer, A., Dauphin, Y., Liang, P., Vaughan, J.W. (eds.) Advances in Neural Information Processing Systems, vol. 34, pp. 6216–6228. Curran Associates, Inc. (2021)

97. Sun, J., Jiang, Y., Qiu, J., Nobel, P., Kochenderfer, M.J., Schwager, M.: Conformal prediction for uncertainty-aware planning with diffusion dynamics model. In: Advances in Neural Information Processing Systems (NeurIPS), vol. 36 (2023)

98. Thorne, J., Vlachos, A., Christodoulopoulos, C., Mittal, A.: Fever: a large-scale dataset for fact extraction and verification. In: Proceedings of the 2018 Conference of the North American Chapter of the Association for Computational Linguistics: Human Language Technologies (NAACL-HLT), pp. 809–819. Association for Computational Linguistics, New Orleans, Louisiana (2018)

99. Touvron, H., et al.: Llama: open and efficient foundation language models (2023). https://doi.org/10.48550/arXiv.2302.13971

100. Vaswani, A., et al.: Attention is all you need. In: Neural Information Processing Systems (2017)

101. Vovk, V., Gammerman, A., Shafer, G.: Algorithmic Learning in a Random World. Springer, Heidelberg (2005)

102. Vovk, V., Petej, I.: Venn-Abers predictors. In: Proceedings of the Thirtieth Conference on Uncertainty in Artificial Intelligence, UAI 2014, pp. 829–838. AUAI Press, Arlington, Virginia, USA (2014)

103. Wadden, D., et al.: Fact or fiction: verifying scientific claims. In: Proceedings of the 2020 Conference on Empirical Methods in Natural Language Processing (EMNLP), pp. 7534–7550. Association for Computational Linguistics (2020)

104. Wang, G., et al.: Voyager: an open-ended embodied agent with large language models. Trans. Mach. Learn. Res. (2024)

105. Wang, P., et al.: Qwen2-VL: enhancing vision-language model's perception of the world at any resolution (2024). https://doi.org/10.48550/arXiv.2409.12191

106. Wang, W., et al.: Cogvlm: visual expert for pretrained language models. In: Advances in Neural Information Processing Systems (NeurIPS) (2024)

107. Wang, Z., et al.: ConU: conformal uncertainty in large language models with correctness coverage guarantees. In: Al-Onaizan, Y., Bansal, M., Chen, Y.N. (eds.) Findings of the Association for Computational Linguistics: EMNLP 2024, pp. 6886–6898. Association for Computational Linguistics, Miami, Florida, USA (2024)

108. Wang, Z., et al.: SConU: selective conformal uncertainty in large language models. In: Che, W., Nabende, J., Shutova, E., Pilehvar, M.T. (eds.) Proceedings of the 63rd Annual Meeting of the Association for Computational Linguistics (Volume 1: Long Papers), pp. 19052–19075. Association for Computational Linguistics, Vienna, Austria (2025)

109. Wei, J., et al.: Measuring short-form factuality in large language models (2024). https://doi.org/10.48550/arXiv.2411.04368
110. Wei, J., et al.: Chain of thought prompting elicits reasoning in large language models. In: Oh, A.H., Agarwal, A., Belgrave, D., Cho, K. (eds.) Advances in Neural Information Processing Systems (2022)
111. Wu, M., Zhou, C., Bates, S., Jaakkola, T.: Thought calibration: efficient and confident test-time scaling (2025). https://doi.org/10.48550/arXiv.2505.18404
112. Xu, F., et al.: Towards large reasoning models: a survey of reinforced reasoning with large language models (2025). https://doi.org/10.48550/arXiv.2501.09686
113. Xu, J., et al.: Qwen2.5-omni technical report (2025). https://doi.org/10.48550/arXiv.2503.20215
114. Xu, J., Mei, T., Yao, T., Rui, Y.: MSR-VTT: a large video description dataset for bridging video and language. In: Proceedings of the IEEE Conference on Computer Vision and Pattern Recognition (CVPR) (2016)
115. Yadkori, Y.A., et al.: Mitigating LLM hallucinations via conformal abstention (2024). https://doi.org/10.48550/arXiv.2405.01563
116. Yang, Z., et al.: The dawn of LMMs: preliminary explorations with GPT-4V(ision) (2023). https://doi.org/10.48550/arXiv.2309.17421
117. Ye, F., et al.: Benchmarking LLMs via uncertainty quantification. In: The Thirty-eight Conference on Neural Information Processing Systems Datasets and Benchmarks Track (2024)
118. Ye, Y., Wen, W.: Data-driven calibration of prediction sets in large vision-language models based on inductive conformal prediction (2025). https://doi.org/10.48550/arXiv.2504.17671
119. Yu, L., Poirson, P., Yang, S., Berg, A.C., Berg, T.L.: Modeling context in referring expressions. In: Leibe, B., Matas, J., Sebe, N., Welling, M. (eds.) Computer Vision - ECCV 2016, pp. 69–85. Springer, Cham (2016)
120. Yue, X., et al.: MMMU: a massive multi-discipline multimodal understanding and reasoning benchmark for expert AGI. In: Proceedings of the IEEE/CVF Conference on Computer Vision and Pattern Recognition, pp. 9556–9567 (2024)
121. Zellers, R., Holtzman, A., Bisk, Y., Farhadi, A., Choi, Y.: HellaSwag: can a machine really finish your sentence? In: Korhonen, A., Traum, D., Màrquez, L. (eds.) Proceedings of the 57th Annual Meeting of the Association for Computational Linguistics, pp. 4791–4800. Association for Computational Linguistics, Florence, Italy (2019)
122. Zhao, P., Zhu, W., Jiao, P., Gao, D., Wu, O.: Data poisoning in deep learning: a survey (2025). https://doi.org/10.48550/arXiv.2503.22759
123. Zheng, L., et al.: Judging LLM-as-a-judge with MT-bench and chatbot arena (2023). https://doi.org/10.48550/arXiv.2306.05685. NeurIPS 2023 Datasets & Benchmarks track poster
124. Zhou, B., Lapedriza, A., Khosla, A., Oliva, A., Torralba, A.: Places: a 10 million image database for scene recognition. IEEE Trans. Pattern Anal. Mach. Intell. **40**(6), 1452–1464 (2018)

Conformal Prediction for Offensive Security

Giovanni Cherubin[(✉)]

Microsoft, Cambridge, UK
gcherubin@microsoft.com

Abstract. Despite its introduction more than a quarter century ago, Conformal Prediction (CP) has seen surprisingly few applications to the cyber security world thus far. In particular, we observe that, while CP has been employed as a defensive measure in many recent works, its use for carrying out attacks (i.e., for *offensive security*) is hard to trace in the literature. We explore this gap, by presenting initial findings in two key areas of offensive security: Privacy-Preserving Machine Learning, and network traffic analysis.

Keywords: Conformal Prediction · Cyber Security · Privacy-preserving Machine Learning · Traffic analysis · Side Channels

> To Alex: PhD supervisor, mentor,
> and friend, on the occasion of his
> 80th birthday, in recognition of
> his guiding influence and
> pioneering work in Machine
> Learning and Conformal
> Prediction.

1 Introduction

Cyber Security offers multiple and varied opportunities for problem solving and scientific discovery. Importantly, security has a direct impact on society; this can lead to success stories (e.g., it is nowadays possible for two people to communicate in full privacy over the Internet thanks to end-to-end encryption protocols [1]), as well as unsuccessful ones (e.g., the use of poor anonymization techniques can lead to user identification [40]). Differently from many fields in Computer Science, security deals with adversarial parties; this forces researchers to continuously look at the security of a system from two sides, the defender's and the attacker's, and it encourages a continuous improvement from both.

Offensive research, which aims to improve attacks against computer systems, is an essential part of cyber security. First, it helps grounding work on defenses, because of an asymmetry between attacks and defenses: a defense needs to

K. An Nguyen and Z. Luo (Eds.): Alexander Gammerman Festschrift, LNCS 16290, pp. 306–324, 2026.
https://doi.org/10.1007/978-3-032-15120-9_14

protect against most (or all) attacks in a threat scenario, whereas an attack just needs to bypass one defense to be considered successful. Secondly, offensive research informs the public and product vendors on what attack techniques can be used: it would be a fallacy to think that reducing research on attacks is a good way of preventing cyber crime.

Machine Learning (ML), and more recently Conformal Prediction (CP), have demonstrated a plethora of successful applications to *defensive* security; here, "defensive" refers to using ML or CP for attack prevention or detection. In particular, numerous defenses have been developed in recent years where CP is the main basis for protecting a system. They range from network intrusion [18,20,53], and malware detection [34], to a large body of works evaluating CP's ability to prevent ML evasion attacks (so-called "adversarial examples") and poisoning attacks [6,14,25,28,36,38,39]; more recently, CP has also been investigated as a defense mechanism for privacy-preserving ML [3,4].

On *offensive* security, whilst traditional ML has seen a similarly successful adoption [9,11,12,15,27,29,30,33,35,41,46,47,49], CP has been largely unexploited; in fact, to the best of our knowledge, no attack to date employs CP in an offensive manner. This is made even more surprising by the fact that security researchers have historically been early adopters of new ML techniques for improving attacks; this was quite evident, for example, in the Website Fingerprinting community, which rapidly adopted increasingly advanced neural network techniques for improving their attacks [41,47,48].

We attribute the lack of CP applications to offensive security research to a simple fact: CP is not a plug & play alternative to traditional ML. Simply, and this is part of its beauty, conformal inference requires a different way of framing the problem. Far from being an obstacle, we argue that this switch of perspective opens up many new possibilities for offensive security research. In this work, we support this claim by introducing new CP-based attacks in two areas of cybersecurity. Along the way, we identify features that make these attacks particularly well suited for CP, such as the ability to generate arbitrarily large datasets and the value of producing prediction sets rather than point predictions. These aspects highlight how CP provides unique advantages over conventional ML approaches in offensive security settings.

2   Preliminaries

We describe how traditional ML models (classifiers or regressors) are typically used for carrying out attacks, and then introduce how CP can be adapted for the same purpose.

Traditional ML Applications to Offensive Security

Many security attacks can be modeled as instances of a statistical inference game [15]: an adversary tries to predict some secret information $s \in \mathbb{S}$ based on side information (or "observation") $o \in \mathbb{O}$ they have available; in attacks of this form, we model the relation between o and s via their joint distribution $P(s,o) = P(s)\,P(o \mid s)$. For example, consider a password guessing attack: the secret s is the user's password, and the observation o could be partial knowledge of the password, system responses to attempted logins (e.g., timing hints from login failures), or any other signals that help narrow down the correct guess. Typically, the attacker can further observe (and learn from) examples of the form $(o_i, s_i) \in \mathbb{O} \times \mathbb{S}$, which are sampled from the joint $P(s,o)$. The goal of the attacker is to guess the password s based on this knowledge o.

In this formulation, it is perhaps not surprising that traditional ML has been an extremely successful tool in carrying out attacks: this setup has the same form of classical statistical learning problems, (classification or regression), where a learner aims to train a predictor $f : \mathbb{O} \mapsto \mathbb{S}$ that makes guesses for secret information s based on the side information o.

More concretely, traditional ML models can be applied to security attacks of this form as follows. An attacker obtains a training set $T \in (\mathbb{O} \times \mathbb{S})^n$ by sampling from $P(s,o)$.[1] For example, in the password guessing scenario, the attacker could collect a variety of passwords s_i alongside timing responses or other feedback signals from login attempts (o_i). On this basis, the attacker trains an ML predictor (classifier or regressor) by minimizing some loss; e.g., for classification problems (i.e., the secret information s is a categorical value), the attacker trains:

$$f = \arg\min_f \sum_{s,o \in T} I(f(o) \neq s)$$

where I is the indicator function. The attacker can then use f for carrying out the attack, to make a prediction $s' = f(o)$ for the true secret s.

The Conformal Prediction Way to Offensive Security

Conformal Prediction (CP) can be readily applied to security attacks of the form described above, although it often requires reframing the inference problem. As we argue in this chapter, this adjustment is not a drawback but rather an opportunity to devise novel and often more informative attacks.

In what follows, we adopt a simplified notation for CP and refer to the literature for a more formal introduction [51]. We denote by $C_T^\alpha : \mathbb{O} \mapsto \mathcal{P}(\mathbb{S})$ a conformal predictor, where $T \in (\mathbb{O} \times \mathbb{S})^n$ is a training dataset and $\mathcal{P}(\cdot)$ is the power set. Informally, $C_T^\alpha(o)$ produces a subset of $\mathbb{S}$ that contains all secrets deemed plausible for the new observation o; for brevity, we omit the underlying nonconformity function, which is typically based on a traditional ML model (e.g., a regressor or probabilistic classifier). A conformal predictor is defined for a significance level $\alpha \in [0, 1]$ and it assumes that T and each new example (o, s)

---

[1] In practice, this sampling usually happens in two steps, by first sampling the prior $P(s)$ and then the conditional $P(o \mid s)$. In general, we assume the attacker can sample from this distribution (oracle access), but they may not know the distribution itself.

come from an exchangeable distribution. Under these conditions, CP provides the following *validity* guarantee:[2]

$$P(s \notin C_{\mathcal{T}}^{\alpha}(o)) \leq \alpha.$$

Integrating conformal prediction (CP) into an attack that relies on a traditional ML model f is often straightforward: one can simply "wrap" CP around f [2,42,51]. However, there are scenarios where such a direct approach brings limited benefits, while CP's emphasis on rethinking the problem can lead to new interesting and impactful attacks (Subsect. 3.2).

For an attacker using CP, the most immediate consequence is that, rather than a point prediction, CP returns a set of labels (the prediction region). This, as shown in various examples throughout this chapter (Sect. 3–4), can be very informative to an attacker:

> **Remark**
>
> When used for offensive security, CP changes the attacker's objective from making a single best guess to producing a set of plausible secrets, with a guaranteed error rate (α). This can make attacks more efficient, as it gives the attacker a principled way to focus on a smaller set of candidates. For instance, in a password-guessing scenario, CP might return a relatively small set of possible passwords, which an attacker can then systematically test; by focusing on this smaller subset, the attacker significantly reduces the time and computational cost of a full brute-force search.

On this aspect, one may wonder if applying CP to security is equivalent to using multi-label ML methods (i.e., returning the top-k candidate labels from a traditional ML predictor, or returning all candidates whose confidence exceeds a certain threshold). A crucial difference is that CP offers a principled solution to the problem – namely it does not require setting arbitrary thresholds on the number of candidates to return, as in CP the size of the prediction set is determined dynamically depending on the underlying probability distribution.

We also observe that CP's validity guarantee comes handy to an attacker: from the attacker's perspective, there is a risk in reducing the set of candidate values to explore when trying to guess a secret, since it may lead to missing the correct secret entirely. CP's validity guarantee provides a formal way to quantify and control this risk directly via α.

In the next sections, we study attacks from two different fields of Cyber Security and how they benefit from using CP. We split this discussion between case studies (🔍), which contain initial experiments that showcase and study the attacks, and research directions (⅄), which set out the basis for ideas to explore.

---

[2] Here, the probability space is the realization of the exchangeable distribution, and the guarantee is marginal: it holds true on average for new samples $\mathcal{T} \cup \{(o, s)\}$, but not conditionally on each (o, s).

3   Privacy-Preserving ML

The field of Privacy-preserving ML (PPML) is concerned with the privacy of the data that is used for training an ML model [11,27,46]. The main question being asked is as follows. Suppose an ML model was trained on private data, and its predictions (or the trained parameters of the model itself) were released to the public: what information could be inferred about the private training data by a member of the public, such that this information could have not been inferred prior to the model's release? In this sense, PPML has to do with measuring and exploiting the information leakage of an ML model about its training data.

Numerous PPML attacks and defenses have been proposed over years. For a broader look, we refer the reader to Salem et al. [45], who offers a comprehensive overview and formalization. Here, we focus on one: reconstruction attacks.

3.1   🔍 Case Study: Data Record Reconstruction

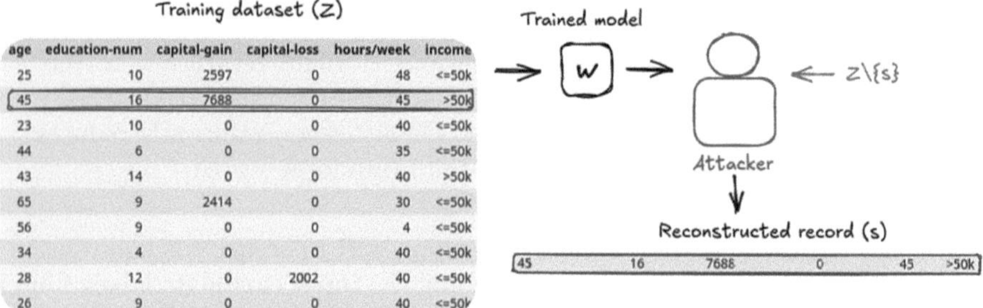

Fig. 1. A reconstruction attack. The attacker tries to infer the features (and label) of one of the data records that were used for training an ML model. The attacker has access to the model, and (in this chapter) the entire training dataset minus the data record.

In reconstruction attacks, an attacker gets access to a trained ML model, the *target* model, and they try to infer (*reconstruct*) the values of one of the training data points that were used for training the model. Among PPML attacks, they provide the highest leakage but can also be among the hardest to carry out [5,11]. We explore the potentials of CP when used to carry out a reconstruction attack.

Problem Setup and Threat Model. In reconstruction attacks, an ML model w is trained on a private training dataset $Z \in (\mathbb{X} \times \mathbb{Y})^N$; at this stage, it is not important whether the model's task is classification or regression. It is useful to think of this model in terms of its trained parameters: these are, for example, the intercept and coefficients if the model is a linear model, or the trained neuron weights in the case of a neural network. We shall henceforth interchangeably use $w \in \mathbb{W}$ for referring to the target model and its trained parameters.

An attacker's goal is to infer the value of a data record, $s \in Z$, using the following information:[3] the model's parameters w, and the entire training dataset minus the data record $Z \setminus \{s\}$; we also assume that the attacker knows the hyperparameters of the model, and any randomness source: that is, they should be able to train a model that is identical to w if they somehow had access to the full dataset Z. These strong assumptions, while idealised, allow us to explore the theoretical limits of reconstruction attacks and are consistent with prior work [5].

According to the notation defined in Sect. 2, a reconstruction attacker needs to infer the value of a data record (and its label), $s \in (\mathbb{X} \times \mathbb{Y}) = \mathbb{S}$, based on the information they have available: $w \in \mathbb{W} = \mathbb{O}$.[4] To carry out the attack, the attacker trains an ML regressor model $r : \mathbb{S} \mapsto \mathbb{O}$, which enables them making a guess $s' = r(w, Z \setminus \{s\})$ for the true data record s; we measure the goodness of an attack as the distance between s' and s (e.g., L_2 distance). To train the regressor, the attacker generates examples by training ML models $\{\tilde{w}_i\}_{i=1}^{k}$ based on datasets $\{Z \setminus \{z\} \cup \{\tilde{s}_i\}\}_{i=1}^{k}$, where $\tilde{s}_i \in (\mathbb{X} \times \mathbb{Y})$ are random samples from some distribution[5]. We observe that, in principle, this allows generating *arbitrarily many* training examples for training the attacker's regressor.

To the uninitiated, it may be confusing that there are two ML models at play here: the target model w, which was trained by a data curator, and the attacker's model r, which is used for carrying out the attack. This fact should not surprise: ML models are very efficient for performing security attacks [15], and it is a mere coincidence that the target of the attack in this case involves an ML model.

CP for Reconstruction. Reconstruction attacks apply a traditional ML regressor r for inferring a data record s from the original training set Z based on the target model's parameters w. Reformulating the attack for an adversary using CP is straightforward: they can perform CP regression, which outputs a prediction region, rather than a point prediction. In turn, CP gives a reconstruction attacker prediction bands, which are arguably more informative than a point prediction when trying to infer a person's private information: the attacker might use the confidence bands as a signal, to be combined with information from other sources (e.g., OSINT [8]), to cause a more impactful attack.

Experimental Setup

Target Model Training. In this experiment, the target model is a Logistic Regression classifier trained on the Adult dataset [7] for the typical classification

---

[3] Here we use the "informed attacker" threat model by Balle et al. [5].

[4] Technically, the information available to our attacker is $(w, Z \setminus \{s\}) \in \mathbb{O} = \mathbb{W} \times (\mathbb{X} \times \mathbb{Y})^{N-1}$, but we omit $Z \setminus \{s\}$ for compact notation, and we implicitly assume that the attack is conditioned on a specific $Z \setminus \{s\}$ and s throughout.

[5] In our experiments, we use the empirical distribution of the set $Z \setminus \{z\}$.

task: to predict whether the income of a person in the dataset is above or below 50k (`income`); the prediction is based on these features: `age`, `education-num`, `capital-gain`, `capital-loss`, `hours-per-week`. We train 5,000 target models; each model is trained on the dataset $Z = Z' \cup \{s_i\}$, where Z' is the Adult dataset's training set, whereas s_i are points taken from the Adult dataset's test split. The goal of the reconstruction attack is to reconstruct each s_i based on the respectively trained models. We also repeat the experiment with Logistic Regression target models that are trained with Differential Privacy (DP) guarantees [23], implemented via `diffprivlib` [32]; for an adequate DP parameter ε, this defense mechanism ensures that models are robust against reconstruction attacks, thanks to DP's formal guarantees; we select $\varepsilon = 5$, which in this case is a sufficient level of protection to counter reconstruction attacks [5].

Attacker's Model Training. We employ Conformalized Quantile Regression (CQR) by Romano et al. [44], which enables tailoring coverage bands to individual data points, and overall improves CP's performance in regression settings. We use gradient boosting (LightGBM [37]) to build the underlying nonconformity measure; the implementation for this experiment is based on the MAPIE library [19]. For simplicity, we run a separate attack for each feature (and for the label) of the data record to reconstruct; the attack's output is obtained by running each of the regressors independently, and then joining their outputs. For training the regressors, we generate a dataset of 100k data points; each of them is a target model $\tilde{w}_i$ trained on $Z' \cup \{\tilde{s}_i\}$, where Z' is the Adult dataset's training split, and $\tilde{s}_i$ is a random point sampled from the empirical distribution of Z'. We sample $\tilde{s}_i$ from the empirical distribution of Z' to simulate the process of drawing a random person's data and then training the corresponding model $\tilde{w}_i$, thus approximating the joint distribution of (s, w).

Performance Metrics. We evaluate the CP-based attack in terms of its coverage and the size of its prediction regions. For the point predictor (LightGBM) used to define the nonconformity measure, we measure the success of the attack by computing the L_2 distance between the guessed record s' and the true record s. We also include a naive baseline attack that simply outputs the median value for each feature in $Z \setminus \{s_i\}$, allowing us to see how much an informed attack improves over one that ignores the target model entirely.

Results. Table 1 shows the performance of the CP-based attack; the reconstruction success metrics are presented individually for each feature (and for the label) of the data records. In the no-privacy setting (i.e., the target model is a Logistic Regression classifier trained without any privacy protection), CP achieves small prediction sets for each of the features and labels; furthermore, the point predictions of its underlying classifier indicate a remarkable advantage over the naive baseline, showing that the attack is effective. When a DP privacy-preserving mechanism is applied (with DP $\varepsilon = 5$), the attack is unsuccessful as expected. In both cases, the CP coverage of $\alpha = 0.1$ is attained.

Table 1. In reconstruction attacks, an entire data record is reconstructed (including its label), based on having access to an ML model trained on it. We look at two cases: one where the adversary attacks a standard Logistic Regression model, without any privacy mechanism in place; one where a Differential Privacy (DP) mechanism is applied to the Logistic Regression model. This table shows the results over a test set of 5k examples. CP's coverage guarantees (for $\alpha = 0.1$) are met, and we observe that the prediction sets are small relatively to the features' range.

	Age	Education	Gain	Loss	Hours	Label
No privacy						
Coverage	0.91	0.91	0.98	0.99	0.91	1.00
Average Prediction Size	15.32	2.67	828.86	24.23	10.91	0.12
Prediction L_2 norm	4.69	1.18	3146.16	60.90	8.73	0.05
Baseline improvement	0.67	0.53	0.62	0.85	0.30	0.89
DP $(\varepsilon = 5)$						
Coverage	0.90	0.92	0.95	0.95	0.92	1.00
Average Prediction Size	44.58	8.81	4899.16	257.55	43.43	1.00
Prediction L_2 norm	14.12	2.55	8364.72	411.70	12.39	0.49
Baseline improvement	0.00	0.00	0.00	0.00	0.00	0.00

Figure 2 shows three reconstruction examples, cherry-picked to illustrate the variety of outcomes we observed. The first two examples show valid predictions for each feature. In the third example, the coverage interval for "Capital Gain" falls out of the prediction region, illustrating a case where CP's validity fails. This highlights a disadvantage of CP: because CP's validity guarantee does not hold conditionally for an example, but on average, there will be individual examples (for $\alpha > 0$) for which validity is not met.

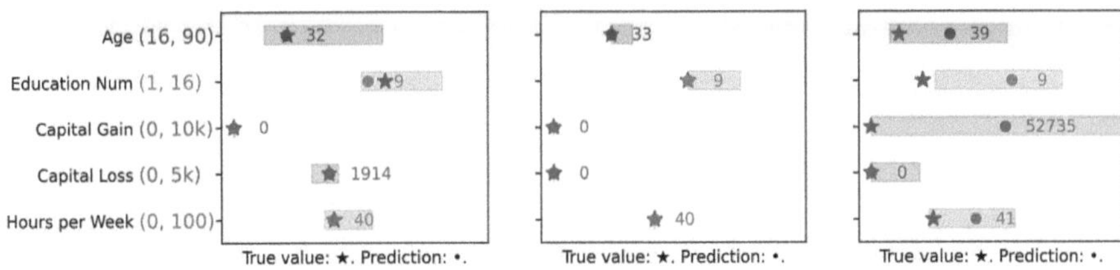

Fig. 2. Cherry-picked reconstruction examples: the first one from the left shows perfect validity; in the second one, some of the reconstructed values are on the bands' edges; the third one shows the largest deviation from validity that we observed in our experiments for the feature "Capital Gain".

Discussion. Results confirm that CP is suitable for tackling reconstruction attacks, and it is a promising avenue for PPML attacks in general. Its main advantage is clear: an attacker can use CP to get an informative coverage band for their prediction, which they can combine with further information from other attacks. This encourages more security research to take into consideration CP as part of the threat model. We highlight two characteristics that make reconstruction (as well as several other attacks) especially suitable for CP:

> **Learnings**
>
> **Virtually unlimited training examples.** In some attacks, the attacker can generate as many training examples as they like For example, in the case of reconstruction attacks, the attacker can generate arbitrarily many $(\tilde{s}_i, \tilde{w}_i)$ training pairs, by sampling $\tilde{s}_i$ from a known distribution and fitting the respective model $\tilde{w}_i$. This makes it possible to use methods such as split CP and CQR, which are typically less data-efficient but much more computationally efficient than full CP.
>
> **Data is IID.** In many security attacks, the data available to an attacker for training and in evaluation can be realistically assumed to come as independent samples from the same distribution. In our reconstruction attack, we similarly assume that the target model w is drawn from the same distribution as the training examples we generate, allowing conformal prediction to retain its validity guarantees.

3.2 ⅄ Research Direction: Membership Inference Attacks

The field of PPML has a lot of varied attacks, and we foresee CP being significant for carrying out many of them. Some attacks are trivial to adapt to the CP framework; for example, in attribute inference, where an attacker reconstructs part of a data record, one can readily apply our learnings from the reconstruction attack described above. On the other hand, we observe that CP may induce new research directions for other attacks, such as membership inference. We propose two new attacks that leverage CP for membership inference.

In traditional membership inference attacks, an attacker gets access to a model trained on private dataset Z, and to a data record z, and their goal is to figure out if $z \in Z$ (i.e., if the record was a *member* of the training data). For an attacker, this resolves in a binary classification problem, whose goal is to predict the indicator $I(z \in Z)$. This makes the direct application of CP classification to this setting straightforward, but possibly uninsightful. There are, however, two ways of adapting membership inference to exploit CP to its fullest.

Membership Inference from First Principles. As Carlini et al. [10] observed, membership inference is an asymmetric attack: the attacker is more interested in making a reliable prediction for true positives (i.e., predicting "member" when $z \in Z$), rather than true negatives; this is because non-members are much more prevalent (they are represented by all the data in the world that is not in Z). A way to exploit this is through *shadow models*: the attacker trains multiple models on different "mock" datasets, labeling their training points as members and held-out points as non-members [46].

On the other hand, observe that CP is based, at its core, on a randomness test that allows determining if a sample comes from some distribution or not. By applying CP to the outputs of shadow models, the attacker can determine whether a queried point's behaviour is typical of the training data, which gives rise to an effective membership inference attack. Furthermore, CP's coverage guarantees can lead to more reliable true-positive detections: if the shadow models data is generated properly, in such a way that it can be considered to be IID (or exchangeable) with the true distribution, then CP guarantees $\geq 1 - \alpha$ true positives, for any desired significance level α.

Subset Membership Inference. We propose another variant of membership inference, which is directly inspired from CP: the attacker gets access to a model trained on Z, and to a set of data records $A = \{z_i\}_{i=1}^{k}$; their goal is to find the set $S = Z \cap A$ of elements from the challenge set A that are member of the model's training data Z; the attack's success can be measured more finely, by having the attacker output a set $S' \subseteq A$, and measuring its precision and recall with respect to the true set S.

There are at least two ways of carrying out this attack. The first is to apply the shadow-based attack, as sketched in the previous paragraph, to each individual record $z \in A$: a data record is included if, for a certain α, CP judges it to come from the same distribution as other members. The second way is slightly more involved: the attacker could set up a CP classification problem, where each data record $z \in A$ is treated as a label, and the CP's prediction set (which can contain any number of labels from the set A) is returned as the attacker's output S'. Further work is likely needed to understand which of the two approaches is more effective.

While this may seem like a minor variation on membership inference attacks, we argue it opens up an opportunity for real-world attackers, who are often interested in identifying possible multiple users from large datasets. For example, an attacker might want to sell information about what data was used for training a model. Through this membership inference attack, which we refer to as *subset membership inference*[6], the adversary gets to reduce the space with error guarantees. We leave these research directions as future work.

---

[6] This should not be confused with set membership inference, where the attacker is given a set of records, all of which are either members or non-members [31].

4   Traffic Analysis Attacks

Traffic analysis refers to attacks against communication networks aiming to infer information from the messages in transit. Typically, one assumes that the network channel is appropriately protected by encryption, which forces the attacker to look for patterns in the network traffic that leak information about the communication's content.

For example, consider an instant messaging app protected via end-to-end encryption. An attacker looking at the encrypted traffic might see the volume and timing of a user's messages, even though they are unreadable. If the user consistently sends (or receives) large bursts of messages late at night, the attacker can guess they are chatting with specific contacts before bed. Over time, the attacker can build a profile of the user's sleep schedule, social activity patterns, and maybe even their closest friends, all without reading the actual messages.

The field of traffic analysis is very broad, and new interesting attack vectors are continuously discovered to this day [54]. In this section, we explore the use of CP for carrying out so-called Website Fingerprinting attacks against Tor, an anonymity network, and then discuss other opportunities that CP offers for traffic analysis attacks in general.

4.1   🔍 Case Study: Website Fingerprinting

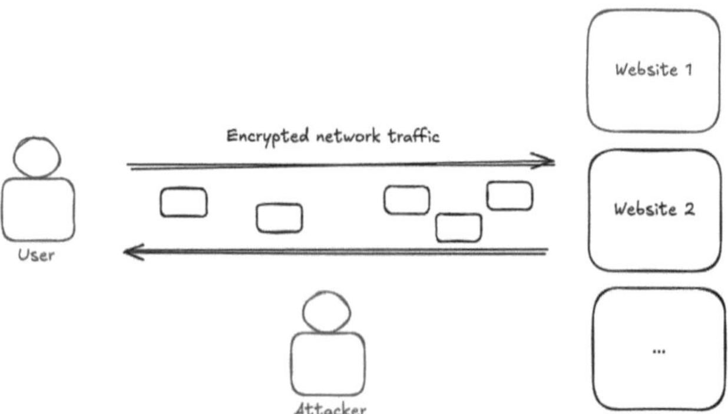

Fig. 3. An overview of Website Fingerprinting attacks. An attacker monitors the encrypted traffic between a user and the web, and tries to infer which webpages the user is visiting.

Anonymity networks, such as Tor [22], enable users to communicate over the Internet in such a way that an attacker controlling a single point on the network between them cannot figure out who is talking to whom. Unfortunately, some of these networks are susceptible to traffic analysis attacks, whereby an attacker who monitors the (encrypted) network traffic between two points infers information about the communication thanks to patterns in the traffic.

Website Fingerprinting (WF) is possibly one of the most studied traffic analysis attacks against the Tor network [9,12,16,24,29,33,35,41,47]. In WF, an attacker observes network traffic from an encrypted tunnel (e.g., Tor, a VPN), between a user and the web, and their goal is to infer which webpages the user is visiting. In traditional WF, the attacker employs an ML classifier to make this inference. Over the past 15 years, WF attacks have improved substantially, reaching almost 100% accuracy in lab conditions. On the other hand, it is unclear whether WF can be effective in real-world settings [35], and empirical measurements suggest WF might only be successful in special conditions, such as when targeting a restricted number of individuals or webpages [16].

Problem Setup and Threat Model. Our setup is deeply tied to a well-known dataset for WF attacks [47]. This dataset contains Tor network traces, each corresponding to a user loading one of 95 websites, 1,000 traces per website. This simulates the case where an attacker is interested in determining whether a user visits one of these websites by looking at their Tor network traffic; we henceforth refer to this set of these websites as the *monitored* websites. The dataset also includes an open world (OW) set, containing 40,716 traces from websites that are not in the monitored set. This reflects the fact that, in practice, a user rarely visits a predictable set of websites; therefore, while the attacker is interested in predicting whether they visit a monitored website, the attacker also needs to discern them from visits to unmonitored (OW) websites [35].

We emulate a WF attack via a standard ML classification pipeline: the classifier is trained on 90% of the dataset – the split is done in a stratified manner, and we evaluate its performance on the rest of the traces. We employ the attack by Sirinam et al. [47], Deep Fingerprinting (DF), which uses a convolutional neural network (CNN) to make the prediction. For the implementation, we adapt the `wflib` library by Deng et al. [21].

CP for WF. We extend CP for carrying out WF attacks as follows. We use split (or "inductive") CP [42,43], because it can scale to a larger and more complex nonconformity measure while retaining validity [50]. We use 10% of the training set as the calibration set, and use the remainder to train the nonconformity measure (proper training set); the nonconformity measure is defined as the complement of the softmax probability scores (for a desired label) of the CNN trained as per the DF attack on the proper training set.

Remark: CP with OW Traces. In traditional WF attacks, one typically needs to train the ML model both on monitored and unmonitored (i.e., OW) websites; this is to capture the "unmonitored" label. Interestingly, CP lends itself to handling unmonitored (OW) websites by design. All one needs is to train the nonconformity measure on all monitored traces, and *ignore the unmonitored websites* during training; this reflects real-world scenarios, where an attacker only has reliable information on what are the monitored websites (since they

318   G. Cherubin

are chosen by the attacker). Then, when evaluating the attack, CP captures the OW class automatically: if its prediction set contains no label, this means the trace likely corresponds to an unmonitored website. Importantly, CP's ability to return an empty prediction set when a trace does not match any monitored website provides a principled way to handle open-world scenarios, without the need to explicitly model the unmonitored class.

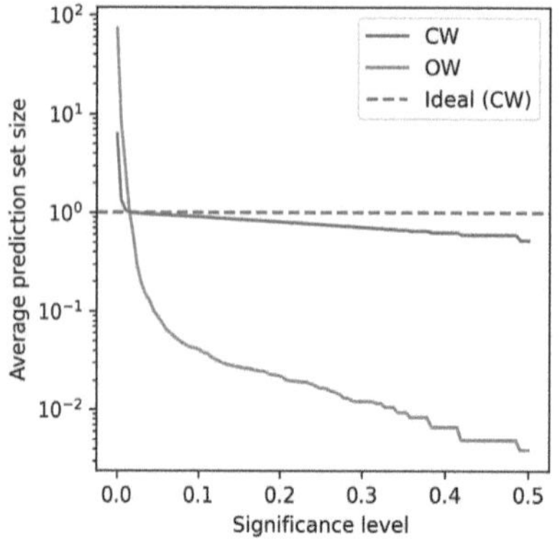

(a) Prediction set sizes for different significance levels.

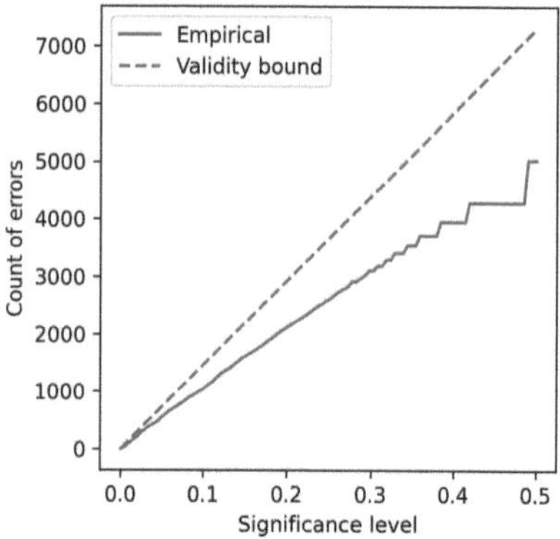

(b) Miscoverage (error) rates vs. significance levels.

Fig. 4. WF attack via CP, using DF [47] as the basis for its nonconformity measure. Results in (a) are shown separately for monitored (CW) and unmonitored (OW) websites.

Results. Figure 4a reports both the prediction set size and the empirical error (i.e., number of times the true website is not in the prediction set), as functions of the significance level α. We measure the prediction set size separately for monitored (CW) and unmonitored (OW) websites. Because in this dataset there are no overlaps between websites (e.g., websites that look very similar), the ideal prediction set size is exactly 1 (i.e., exactly one label per network trace). For monitored traces, we observe that the prediction set size is as expected for reasonably small α, whereas the predictor becomes more conservative as the significance level increases. This is desirable, and it confirms the effectiveness of CP for carrying out WF attacks. The prediction set size for unmonitored traces has an interesting behavior: Whereas negligible values of α force the CP attack to output many websites, for more common values ($\alpha > 0.015$) the prediction set size becomes smaller than 1; this is highly informative, as it communicates to the attacker that empty prediction sets likely correspond to unmonitored traces.

We learn two characteristics, that likely extend beyond WF, and that support CP's suitability to offensive research:

> **Learnings**
>
> **Open world settings.** In some security attacks we are only interested in a subset of labels, and for other labels we may not even have any data to train on. For example, in WF the attacker is only interested in inferring which websites a user is visiting from a set of *monitored* websites, which is only a small subset of all websites. CP offers a great opportunity for attacks with this characteristic: rather than returning a wrong label, it will tend to return an empty set; this informs the attacker that the true label is not one they are necessarily interested in.
>
> **Multiple choices are correct.** Similarly, there are attacks where multiple labels are acceptable. In the WF case, it can be that many websites are indistinguishable in terms of network traffic (e.g., their pages' content is very similar), so CP will naturally return many labels in the prediction set. For the attacker, this result conveys crucial information: if the traffic signature is ambiguous, the set of possible websites remains large, and CP captures this uncertainty explicitly. This can be valuable in practice, prompting the attacker to collect more data or apply additional analyses to further refine which monitored website is relevant.

Limitations, and How CP Can Help. WF research has received multiple critiques over the years, suggesting that successful attacks in lab conditions (e.g., network data collected from a single machine, with the same browser, during the same period of time) may not succeed in the real world [16,35]. We posit that tools from the conformal inference literature could help with deploying more practical attacks. For instance, consider the following problem: websites change over time, and therefore a WF attack that works today may not work tomorrow because of concept drift. There are two tools that an attacker can use. First, they can keep monitoring the size of the prediction set in a CP-based WF attack: this is highly correlated with the attack's performance, and it can quickly inform the attacker that their attack needs improving or retraining. Secondly, they can use methods such as exchangeability martingales to monitor websites and their network traffic for any distribution changes [14,26,52].

4.2 ⅄ Research Direction: Other Attacks to Explore

Traffic analysis is a very broad field, and we foresee CP being useful for many attacks beyond WF. We provide two further examples below.

CP for Keystroke Analysis. A well-known traffic analysis attack against SSH was demonstrated by Song et al. [49]. Since SSH transmits one keystroke per packet, an attacker who measures the time between packets can infer what the user is typing, including passwords. Indeed, the time it takes a user to type two keys depends on factors such as their distance on the keyboard, and this can be exploited to predict the corresponding characters based on timing information.

In their original approach, the authors estimated parameters of a hidden Markov model (HMM) from empirical keystroke-timing data. This allowed them to narrow down a list of candidate character sequences. We suggest that, by applying CP to this problem, the attacker can explicitly control the error rate (and, indirectly, the size of the candidate sequences) in a principled manner. For example, CP-HMM [17] would output a set of likely character sequences, with a controlled error rate, allowing the attacker to focus on a smaller set of candidates. In practice, this ensures that the probability of excluding the correct sequence from the final list is at most α, thereby giving the adversary a more direct control on the attack's precision.

Topic Prediction from AI Chatbot. AI chatbots typically stream LLM predicted tokens progressively to the user's browser: this improves the user experience when interacting with slower LLMs. On the flip side, this means that each token in an LLM response is sent individually, which implies that an eavesdropping attacker gets to see the *size* of each token, even if the content of the actual communication is protected by encryption. It was recently shown [54] that a sequence of tokens' sizes, corresponding to an LLM response, might expose information about the text, such as the topic of the conversation that a user is having with an LLM. We foresee CP can be beneficial for this attack. For example, rather than having the attack predict a single topic, a CP-based attack might make a judgement for each topic independently; this accounts for cases where multiple topics are discussed in the same conversation with the chatbot.

5   Discussion

In this paper, we observed the surprisingly scarce application of CP to offensive security. In fact, despite a brief suggestion that CP could be used to this end [13], we were unable to find any evidence of CP being employed in the field. Based on our findings, this is remarkable, as CP is particularly suitable to characteristics that many security threat models offer, such as: i) some attacks allow generating virtually unlimited training examples ii) that can be safely considered to be IID; furthermore, CP's set predictions (with validity guarantees) are very informative in cases where iii) the attacker is only concerned with (and can only train on) a subset of all labels such as open-world scenarios, and iv) where multiple labels are valid predictions when carrying out an attack.

More in general, we argued that CP matches a quintessential aspect of real world security, where a set prediction is more effective than a point prediction: in many cases, attackers are more interested in reducing the set of possible candidates (e.g., when guessing a password), which they can then explore in subsequent and more costly attacks (e.g., brute-forcing). Furthermore, we argued that CP's validity guarantee on the errors is a useful tool, even for attackers, to calculate the risks involved when reducing the set of viable candidates.

Overall, we believe that CP has significant potential in the context of offensive security. While this chapter provided somewhat diverse examples, we suspect that new interesting CP-based attacks will be devised, and we hope this work can help towards that.

Acknowledgments. I thank Andrew Paverd for useful discussion.

References

1. The new textsecure: Privacy beyond SMS. https://signal.org/blog/the-new-textsecure/
2. Angelopoulos, A.N., Bates, S.: A gentle introduction to conformal prediction and distribution-free uncertainty quantification. arXiv preprint arXiv:2107.07511 (2021)
3. Angelopoulos, A.N., Bates, S., Zrnic, T., Jordan, M.I.: Private prediction sets. Harvard Data Sci. Rev. **4**(2) (2022)
4. Balinsky, A.D., Krzeminski, D., Balinsky, A.: Conformal prediction for privacy-preserving machine learning. arXiv preprint arXiv:2507.09678 (2025)
5. Balle, B., Cherubin, G., Hayes, J.: Reconstructing training data with informed adversaries. In: 2022 IEEE Symposium on Security and Privacy (SP), pp. 1138–1156. IEEE (2022)
6. Bao, J., Dang, C., Luo, R., Zhang, H., Zhou, Z.: Enhancing adversarial robustness with conformal prediction: a framework for guaranteed model reliability. arXiv preprint arXiv:2506.07804 (2025)
7. Becker, B., Kohavi, R.: Adult. UCI Machine Learning Repository (1996). https://doi.org/10.24432/C5XW20
8. Block, L.: The long history of OSINT. J. Intell. Hist. **23**(2), 95–109 (2024)
9. Cai, X., Zhang, X.C., Joshi, B., Johnson, R.: Touching from a distance: website fingerprinting attacks and defenses. In: Proceedings of the 2012 ACM Conference on Computer and Communications Security, pp. 605–616 (2012)
10. Carlini, N., Chien, S., Nasr, M., Song, S., Terzis, A., Tramer, F.: Membership inference attacks from first principles. In: 2022 IEEE Symposium on Security and Privacy (SP), pp. 1897–1914. IEEE (2022)
11. Carlini, N., et al.: Extracting training data from large language models. In: 30th USENIX Security Symposium (USENIX Security 21), pp. 2633–2650 (2021)
12. Cherubin, G.: Bayes, not Näive: security bounds on website fingerprinting defenses. arXiv preprint arXiv:1702.07707 (2017)
13. Cherubin, G.: Black-box security: measuring black-box information leakage via machine learning. Ph.D. thesis, Royal Holloway, University of London (2019)
14. Cherubin, G., Baldwin, A., Griffin, J.: Exchangeability martingales for selecting features in anomaly detection. In: Conformal and Probabilistic Prediction and Applications, pp. 157–170. PMLR (2018)
15. Cherubin, G., Chatzikokolakis, K., Palamidessi, C.: F-BLEAU: fast black-box leakage estimation. In: 2019 IEEE Symposium on Security and Privacy (SP), pp. 835–852. IEEE (2019)
16. Cherubin, G., Jansen, R., Troncoso, C.: Online website fingerprinting: evaluating website fingerprinting attacks on tor in the real world. In: 31st USENIX Security Symposium (USENIX Security 22), pp. 753–770 (2022)

17. Cherubin, G., Nouretdinov, I.: Hidden Markov models with confidence. In: Symposium on Conformal and Probabilistic Prediction with Applications, pp. 128–144. Springer (2016)
18. Cherubin, G., et al.: Conformal clustering and its application to botnet traffic. In: International Symposium on Statistical Learning and Data Sciences, pp. 313–322. Springer (2015)
19. Cordier, T., Blot, V., Lacombe, L., Morzadec, T., Capitaine, A., Brunel, N.: Flexible and systematic uncertainty estimation with conformal prediction via the MAPIE library. In: Conformal and Probabilistic Prediction with Applications (2023)
20. Dang, Q.V., Pham, T.H.: Kernel methods for conformal prediction to detect botnets. In: International Conference on Artificial Intelligence on Textile and Apparel, pp. 29–41. Springer (2023)
21. Deng, X., Li, Q., Xu, K.: Robust and reliable early-stage website fingerprinting attacks via spatial-temporal distribution analysis. In: Proceedings of the 2024 ACM SIGSAC Conference on Computer and Communications Security (2024)
22. Dingledine, R., Mathewson, N., Syverson, P.: Tor: the second-generation onion router. In: 13th USENIX Security Symposium (USENIX Security 04). USENIX Association, San Diego, CA, August 2004. https://www.usenix.org/conference/13th-usenix-security-symposium/tor-second-generation-onion-router
23. Dwork, C.: Differential privacy. In: International Colloquium on Automata, Languages, and Programming, pp. 1–12. Springer (2006)
24. Dyer, K.P., Coull, S.E., Ristenpart, T., Shrimpton, T.: Peek-a-boo, I still see you: why efficient traffic analysis countermeasures fail. In: 2012 IEEE Symposium on Security and Privacy, pp. 332–346. IEEE (2012)
25. Ennadir, S., Alkhatib, A., Bostrom, H., Vazirgiannis, M.: Conformalized adversarial attack detection for graph neural networks. In: Conformal and Probabilistic Prediction with Applications, pp. 311–323. PMLR (2023)
26. Fedorova, V., Gammerman, A., Nouretdinov, I., Vovk, V.: Plug-in martingales for testing exchangeability on-line. arXiv preprint arXiv:1204.3251 (2012)
27. Fredrikson, M., Jha, S., Ristenpart, T.: Model inversion attacks that exploit confidence information and basic countermeasures. In: Proceedings of the 22nd ACM SIGSAC Conference on Computer and Communications Security, pp. 1322–1333 (2015)
28. Gendler, A., Weng, T.W., Daniel, L., Romano, Y.: Adversarially robust conformal prediction. In: International Conference on Learning Representations (2021)
29. Hayes, J., Danezis, G.: k-fingerprinting: a robust scalable website fingerprinting technique. In: 25th USENIX Security Symposium (USENIX Security 16), pp. 1187–1203 (2016)
30. Hettwer, B., Gehrer, S., Güneysu, T.: Applications of machine learning techniques in side-channel attacks: a survey. J. Cryptogr. Eng. **10**(2), 135–162 (2020)
31. Hilprecht, B., Härterich, M., Bernau, D.: Monte Carlo and reconstruction membership inference attacks against generative models. In: Proceedings on Privacy Enhancing Technologies (2019)
32. Holohan, N., Braghin, S., Mac Aonghusa, P., Levacher, K.: Diffprivlib: the IBM differential privacy library. ArXiv e-prints 1907.02444 [cs.CR], July 2019
33. Johnson, A., Wacek, C., Jansen, R., Sherr, M., Syverson, P.: Users get routed: traffic correlation on tor by realistic adversaries. In: Proceedings of the 2013 ACM SIGSAC Conference on Computer & Communications Security, pp. 337–348 (2013)

34. Jordaney, R., et al.: Transcend: detecting concept drift in malware classification models. In: 26th USENIX Security Symposium (USENIX Security 17), pp. 625–642 (2017)
35. Juarez, M., Afroz, S., Acar, G., Diaz, C., Greenstadt, R.: A critical evaluation of website fingerprinting attacks. In: Proceedings of the 2014 ACM SIGSAC Conference on Computer and Communications Security, pp. 263–274 (2014)
36. Kang, M., Gürel, N.M., Li, L., Li, B.: COLEP: certifiably robust learning-reasoning conformal prediction via probabilistic circuits. arXiv preprint arXiv:2403.11348 (2024)
37. Ke, G., et al.: LightGBM: a highly efficient gradient boosting decision tree. In: Advances in Neural Information Processing Systems, vol. 30 (2017)
38. Luo, R., Bao, J., Zhou, Z., Dang, C.: Game-theoretic defenses for robust conformal prediction against adversarial attacks in medical imaging. arXiv preprint arXiv:2411.04376 (2024)
39. Messoudi, S., Rousseau, S., Destercke, S.: Deep conformal prediction for robust models. In: International Conference on Information Processing and Management of Uncertainty in Knowledge-Based Systems, pp. 528–540. Springer (2020)
40. Narayanan, A., Shmatikov, V.: How to break anonymity of the Netflix prize dataset. arXiv preprint cs/0610105 (2006)
41. Panchenko, A., et al.: Website fingerprinting at internet scale. In: NDSS, vol. 1, p. 23477 (2016)
42. Papadopoulos, H.: Inductive Conformal Prediction: Theory and Application to Neural Networks. INTECH Open Access Publisher Rijeka (2008)
43. Papadopoulos, H., Proedrou, K., Vovk, V., Gammerman, A.: Inductive confidence machines for regression. In: Elomaa, T., Mannila, H., Toivonen, H. (eds.) ECML 2002. LNCS (LNAI), vol. 2430, pp. 345–356. Springer, Heidelberg (2002). https:// doi.org/10.1007/3-540-36755-1_29
44. Romano, Y., Patterson, E., Candes, E.: Conformalized quantile regression. In: Advances in Neural Information Processing Systems, vol. 32 (2019)
45. Salem, A., et al.: SoK: let the privacy games begin! a unified treatment of data inference privacy in machine learning. In: 2023 IEEE Symposium on Security and Privacy (SP), pp. 327–345. IEEE (2023)
46. Shokri, R., Stronati, M., Song, C., Shmatikov, V.: Membership inference attacks against machine learning models. In: 2017 IEEE Symposium on Security and Privacy (SP), pp. 3–18. IEEE (2017)
47. Sirinam, P., Imani, M., Juarez, M., Wright, M.: Deep fingerprinting: undermining website fingerprinting defenses with deep learning. In: Proceedings of the 2018 ACM SIGSAC Conference on Computer and Communications Security, pp. 1928–1943 (2018)
48. Sirinam, P., Mathews, N., Rahman, M.S., Wright, M.: Triplet fingerprinting: more practical and portable website fingerprinting with N-shot learning. In: Proceedings of the 2019 ACM SIGSAC Conference on Computer and Communications Security, pp. 1131–1148 (2019)
49. Song, D.X., Wagner, D., Tian, X.: Timing analysis of keystrokes and timing attacks on SSH. In: 10th USENIX Security Symposium (USENIX Security 01) (2001)
50. Vovk, V.: Conditional validity of inductive conformal predictors. In: Asian Conference on Machine Learning, pp. 475–490. PMLR (2012)
51. Vovk, V., Gammerman, A., Shafer, G.: Algorithmic Learning in a Random World. Springer (2005)

52. Vovk, V., Nouretdinov, I., Gammerman, A.: Testing exchangeability on-line. In: Proceedings of the 20th International Conference on Machine Learning (ICML-03), pp. 768–775 (2003)
53. Wechsler, H.: Cyberspace security using adversarial learning and conformal prediction. Intell. Inf. Manag. **7**(4), 195–222 (2015)
54. Weiss, R., Ayzenshteyn, D., Mirsky, Y.: What was your prompt? A remote keylogging attack on AI assistants. In: 33rd USENIX Security Symposium (USENIX Security 24), pp. 3367–3384 (2024)

Temporal Distribution of Clusters of Investors and Their Application in Prediction with Expert Advice

Wojciech Wisniewski[1], Yuri Kalnishkan[1(✉)], David Lindsay[2], and Siân Lindsay[2]

[1] Department of Computer Science, Royal Holloway, University of London, London, UK
`wojciech.wisniewski.2019@live.rhul.ac.uk`, `yuri.kalnishkan@rhul.ac.uk`
[2] AlgoLabs, Bracknell, UK
`{david,sian}@algolabs.com`

Abstract. Financial organisations such as brokers face a significant challenge in servicing the investment needs of thousands of their traders worldwide. This task is further compounded since individual traders will have their own risk appetite and investment goals. Traders may look to capture short-term trends in the market which last only seconds to minutes, or they may have longer-term views which last several days to months. To reduce the complexity of this task, client trades can be clustered. By examining such clusters, we would likely observe many traders following common patterns of investment, but how do these patterns vary through time? Knowledge regarding the temporal distributions of such clusters may help financial institutions manage the overall portfolio of risk that accumulates from underlying trader positions. This study contributes to the field by demonstrating that the distribution of clusters derived from the real-world trades of 20k Foreign Exchange (FX) traders (from 2015 to 2017) is described in accordance with Ewens Sampling Distribution. Further, we show that the Aggregating Algorithm (AA), an on-line prediction with expert advice algorithm, can be applied to the aforementioned real-world data in order to improve the returns of portfolios of trader risk. However we found that the AA "struggles" when presented with too many trader "experts", especially when there are many trades with similar overall patterns. To help overcome this challenge, we have applied and compared the use of Statistically Validated Networks (SVN) with a hierarchical clustering approach on a subset of the data, demonstrating that both approaches can be used to significantly improve results of the AA in terms of profitability and smoothness of returns.

Keywords: statistically validated networks · Ewens sampling distribution · foreign exchange · behavioural finance · clusters of investors · aggregating algorithm

K. An Nguyen and Z. Luo (Eds.): Alexander Gammerman Festschrift, LNCS 16290, pp. 325–344, 2026.
https://doi.org/10.1007/978-3-032-15120-9_15

1   Introduction

In recent years, published research has highlighted a growing interest in methods to cluster together traders based on their strategic and behavioural features, as well as studying how they influence each other. Since 2012, various studies have investigated clustering of investors in financial markets. In particular, [21] used Statistically Validated Networks (SVN) and the Infomap method to detect clusters of investors with similar trading decisions. Similarly, [7] identified clusters by studying the relationship between portfolio and trading decisions of Swedish investors. In [17], hierarchical clustering and SVN are used to detect clusters of investors with similar trading decisions. Paper [8] extended the SVN methodology and inferred a lead-lag network of clusters of investors trading in FX, which showed that most of the trading activity has a market endogenous origin. In [20] traders are detected and clustered with respect to limit-order and market-order strategies, and trading strategies are classified based on their response pattern to historical price changes. In [12], mutual information and transfer entropy are used to identify a network of synchronization and anticipation relationships between financial traders. Paper [4] constructed multilink networks covering two years after IPOs and obtained clusters of investors characterized by synchronization in the timing of trading decisions. In [9] a method is proposed to detect lead-lag networks between the states of traders determined at different timescales and it is observed that institutional and retail traders have different causality structures of lead-lag networks. Paper [6] proposed a deep learning architecture, ExNet, for both investor clustering and modeling. Finally, Viet et al. [5] analyzed the structure of investor networks during a financial crisis and showed changes in investor trading behavior and mutual interactions in the stock market.

In his analysis of the topic entitled "Clusters of Traders in Financial Markets" [16], Mantegna describes a working hypothesis for analysing the dynamics of clusters of investors trading in financial markets, by remarking that the empirical results of [18] are fully consistent with Aoki's modeling hypothesis in [3]. Aoki proposed to use the framework of the Ewens sampling formula [11] for the characterisation of clusters of economic agents having reached a dynamical equilibrium in a specific market.

Using a proprietary dataset derived from traders investing in the Foreign Exchange (FX) market, we set out to contribute to the body of research concerning the clustering of financial market traders by focusing on the clusters' temporal distributions. For this purpose we construct sliding window investor networks (Statistically Validated Networks) based on statistically significant trade time synchronisation and show that the Ewens Sampling Distribution is a good fit. Having greater insights into the variations of trader clusters throughout time will likely assist financial institutions as they manage the overall portfolio of risk that builds up from underlying trader positions.

Whilst the literature concerning clustering in portfolio selection is very broad, there is no analysis of how clusters may be leveraged by prediction with expert advice techniques such as the Aggregating Algorithm (AA). This may be because the theoretical dependency of the AA on the number of experts is mild (under uniform initial distribution, the dependency is logarithmic). This suggests that with a larger number of experts, over time the algorithm will manage to work out if they are needed. In the domain of AA in several financial contexts, several researchers have contributed significantly. In [23,25] the AA is developed in the general context of prediction with expert advice. In [24] a portfolio selection method based on prediction with expert advice is proposed, thus bringing in realistic trading scenarios. In [26] a universal trading strategy based on well-calibrated forecasts is constructed using prediction with expert advice methods, namely, AdaHedge-type algorithms. In [28] constant rebalanced portfolios are considered as experts and a universal portfolio strategy using the weak aggregating algorithm is constructed. Recent papers such as [2] have demonstrated how the AA can be applied to FX data for improving returns of portfolios of trader risk. However, in this practical situation the number of experts present a problem, they overwhelm the advantages of the best expert. Thus we hypothesise that clustering of the trading experts in our dataset should make prediction more reliable and robust by reducing noise.

The organisation of the chapter is as follows. The chapter is split into two major parts. First, we demonstrate how the cluster distributions of trading activity can be described according to Ewens sampling distribution. We also investigate whether these distributions are stationary and depend on the way the clusters are detected. Secondly, we show how clusters of traders over time can be used to improve the profitability of the AA.

2   Clustering of Retail Traders by Their Synchronicity

We largely follow the methods developed in [22], i.e., introduce a behavioural synchronicity measure between traders and then construct a statistically validated network on which an unsupervised clustering method is used.

2.1   Synchronicity Between Traders

A simple way to infer if two traders have similar trading behaviours is to compare their scaled trading volume (referred to as the imbalance ratio) in a specific time frame. In order to do so, we partition the time line into disjoint intervals $\cup[t, t+\delta t[$ and let $r(i, t)$ be the imbalance ratio of trader i in interval $[t, t + \delta t[$:

$$r(i, t) = \frac{b(i, t) - s(i, t)}{b(i, t) + s(i, t)}. \tag{1}$$

where $b(.,.)$ and $s(.,.)$ denote the total volumes bought and sold by the trader in the given time frame.

For a given threshold a we define a trader state as follows:

$$\text{state}(i, t) = \begin{cases} \text{buying state,} & \text{if } r(i, t) > a \\ \text{selling state,} & \text{if } r(i, t) < -a \\ \text{neutral state,} & \text{if } -a \leq r(i, t) \leq a \\ \text{inactive state,} & \text{if } b(i, t) + s(i, t) = 0 \end{cases} \tag{2}$$

The synchronicity between a pair of traders is measured by counting the co-occurrences in the time series of their states, and attributing a p-value that reflects the statistical significance of this synchronicity assuming pure randomness. The hypergeometric distribution is used to calculate the p-value. The p-value is the probability that in a series of n trades, where one trader was n_p times in a state p and the other was n_q times in a state q, these occurrences overlapped $n_{p,q}$ times or more:

$$p(n_{p,q}) = 1 - \sum_{i=0}^{n_{p,q}-1} H(i|n, n_p, n_q)$$

where

$$H(i|n, n_p, n_q) = \frac{\binom{n_p}{i}\binom{n - n_p}{n_q - i}}{\binom{n}{n_q}} .$$

This method is often used to measure similarity in genetic sequences. In the literature, there exist many other alternative methods; however, the hyper-geometric test can be used with sparse data and it is not sensitive to outliers. Therefore it suits our needs.

To deal with the testing of all pairs of traders and all types of co-occurrences a multiple hypothesis testing correction is needed. For this purpose, we use the Bonferroni correction, which is the statistical significance (0.05) divided by the number of tests. It is worth pointing out that alternatively one could use the false discovery rate for multiple test correction.

2.2 Statistically Validated Network

A statistically validated network is a network built by validating links between pairs of traders if the p-value of their synchronisation is smaller than the corrected threshold. Traders without any links are dropped. In the resulting network we exclude links between opposite actions (buy-sell), links between neutral states and links between inactive states. The reason is that we are mostly interested in active traders with the same kind of behaviour (buy-buy and sell-sell) and which manifest high trading activity.

3 Clustering Distribution

This section will define Ewens sampling formula, which is the backbone of Aoki's modeling hypothesis concerning the dynamics of trading behaviour.

3.1 Partition Vector

We will now introduce a partition vector, which will be useful for modelling.

Let c_i be the number of clusters with exactly i traders. Then $K_n = \sum_{i=1}^{n} c_i$ is the total number of clusters formed by n traders and $\sum_{i=1}^{n} i c_i = n$. We will call the vector $c = (c_1, c_2, \ldots, c_n)$ a partition vector.

3.2   The Ewens Sampling Distribution

Ewens sampling formula describes a specific probability for the partition of the positive integer into parts. It was discovered by Ewens [11] as providing the probability of the partition of a sample of n selectively equivalent genes into a number of different gene types (alleles). For positive integers $c_1, c_2, ..., c_n$ satisfying $\sum_j j c_k = n$ and being a realisation of a random partition vector $(C_1(n), C_2(n), \ldots, C_n(n))$, we have:

$$\mathbb{P}_\theta(C_1(n) = c_1, \ldots, C_n(n) = c_n) = \frac{n!}{\theta_{(n)}} \prod_{j=1}^{n} \left(\frac{\theta}{j}\right)^{c_j} \frac{1}{c_j!}, \tag{3}$$

for $\theta \in (0, \infty)$, $\theta_{(n)} := \theta(\theta + 1) \cdots (\theta + n - 1) = \Gamma(n + \theta)/\Gamma(\theta), n \geq 1$ and $\theta_{(0)} = 1$.[1]

We are interested in groups of traders with strategies manifesting similar synchronisation, and since the SVN clustering process builds a network with strong links between pairs of traders we conjecture there should be no (or only a small number of) mono-communities. Thus it is natural to fit a distribution conditional on the event $c_1 = 0$.

Let us define the conditional distribution $(\tilde{C}_2(n), \ldots, \tilde{C}_n(n))$ (see [10]) in the following way:

$$\mathcal{L}(\tilde{C}_2(n), \ldots, \tilde{C}_n(n)) = \mathcal{L}(C_2(n), \ldots, C_n(n)|C_1(n) = 0) \tag{4}$$

The probability of the condition is given by:

$$\lambda_n(\theta) := \mathbb{P}(C_1(n) = 0) = \frac{1}{\theta_{(n)}} \sum_{k=1}^{n} \theta^k D(n, k), \tag{5}$$

(with $\lambda_0(\theta) = 1$ and $\lambda_1(\theta) = 0$.), where $D(n, k)$ is the number of derangements of size n having k cycles:

$$D(n, k) := \sum_{l=0}^{k} (-1)^l \binom{n}{l} \begin{bmatrix} n - l \\ k - l \end{bmatrix} ; \tag{6}$$

here $\begin{bmatrix} n \\ k \end{bmatrix}$ is the unsigned Stirling number of the first kind.

Accurate computation of alternating series, present in $D(n, k)$, is a well-known hard problem therefore it is useful to give the recursive relation of $\lambda_n(\theta)$ which verifies:

$$\lambda_{n+1}(\theta) := \frac{n}{n + \theta} \left[\lambda_n(\theta) + \frac{\theta}{n + \theta - 1} \lambda_{n-1}(\theta) \right]. \tag{7}$$

Table 1 summarises relevant characteristics of the Ewens conditional and non conditional distribution.

---

[1] We define $\theta_{(-k)} = 0$, for $k \in \mathbb{N}$.

4 Experiments

In this section we describe the proprietary dataset and the experiments.

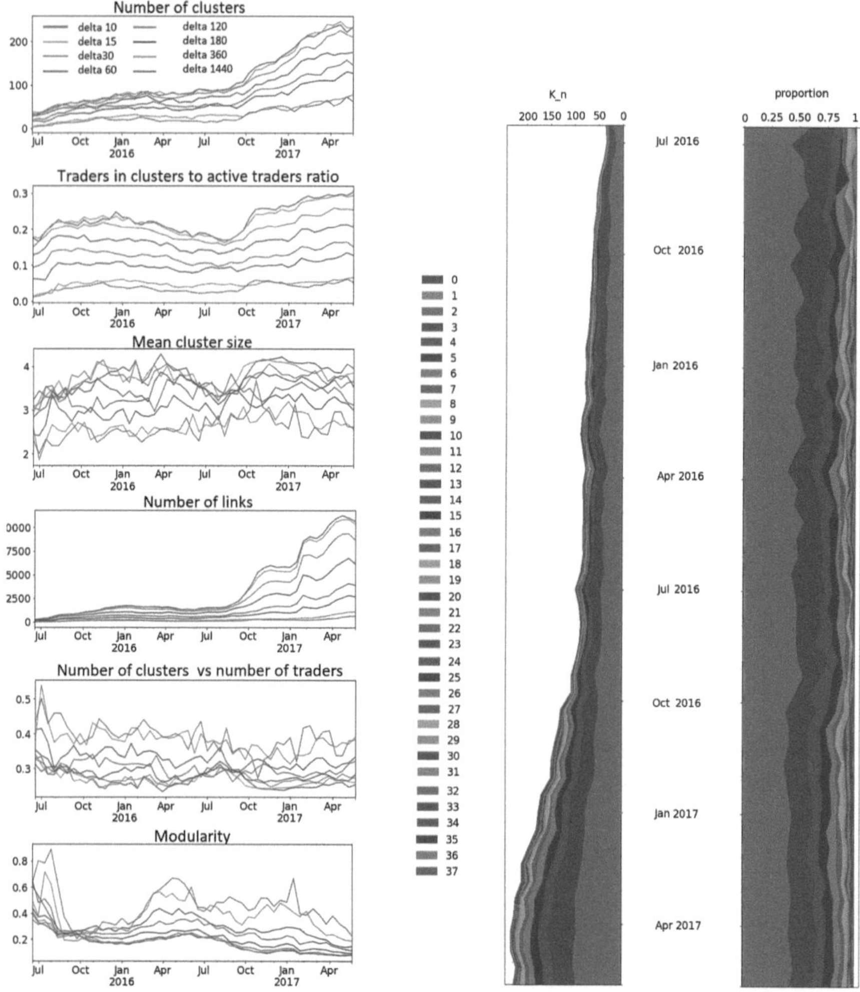

Fig. 1. Evolution of some statistics (number of clusters, number of links, number of traders in clusters vs active traders ratio, number of clusters vs number of traders, mean cluster size, modularity) over time for a network of traders at deltas (10, 15, 30, 60, 120, 180, 360, and 1440 min) for EUR/USD currency pair.

Fig. 2. Proportion vector and normalised proportion vector of temporal evolution for clustering on EUR/USD for 10min delta and cutoff 100. The Infomap algorithm was used to identify clusters after the SVN network was constructed.

Table 1. Comparison of Ewens conditional and non-conditional distributions

Feature	*Ewens distribution*	*Conditional Ewens distribution*
Probability	$\frac{n!}{\theta_{(n)}}\prod_{j=1}^{n}\left(\frac{\theta}{j}\right)^{c_j}\frac{1}{a_j!}$,	$\frac{n!}{\theta_{(n)}\lambda_n(\theta)}\prod_{j=2}^{n}\left(\frac{\theta}{j}\right)^{c_j}\frac{1}{a_j!}$.
Expected cycle counts	$\mathbb{E}C_j(n)=\frac{n!}{\theta_{(n)}}\frac{\theta_{(n-j)}}{(n-j)!}\frac{\theta}{j}$,	$\frac{\lambda_{(n-j)}(\theta)}{\lambda_{(n)}(\theta)}\mathbb{E}C_j(n)$.
Expected number of cycles	$\mathbb{E}K_n=\sum_{i=0}^{n-1}\frac{\theta}{\theta+i}$,	$\sum_{j=2}^{n}\frac{\lambda_{(n-j)}(\theta)}{\lambda_{(n)}(\theta)}\mathbb{E}(C_j(n))$.

4.1 The Dataset Description

We consider financial data gathered from the client trades of a retail Foreign Exchange (FX) broker. A typical retail broker will provide their clients with an online trading platform software such as MetaTrader 4 (MT4) where they can place trades, monitor positions, track both historic and live movements in prices, and access the latest world economic news. Online trading platforms often operate under the stipulation that once an order is placed (opened), it must be closed in its entirety. Source data is essentially stored in a temporal table with each row representing a client order that provides the opening and closing time, as well as the currency traded (symbol), amount traded and side (buy or sell) of the order.

The proprietary dataset comprises the trades made by over 20 thousand clients during 2015–2017. Each client was allowed to buy or sell any of available currency pairs and they could place trades as many times as they wanted, at any time of day provided they stayed within the confines of their leveraged funds. The dataset contains only necessary features for further investigation namely an investor's anonymised ID, opening and closing trade times, amount of lots traded, sign (long or short position), and the traded symbol.

4.2 Experimental Protocol

It is convenient to use sliding windows in order to track the temporal evolution of clustering. For each in-sample time window, we filtered out traders with less than 100, 500 or 1000 trades (referred to as the cut-off). We observe that the number of traders grows in an approximately linear fashion throughout time which is related directly with the business growth. We focus our investigation on trading activity that occurs during standard business days within the most active hours (6am–6pm). Investigations are conducted solely considering the EUR/USD currency pair. We construct a sliding window of size 6 months and shift it every 2 weeks. Then we build a SVN network at every step using the imbalance ratio time series for δt ranging from 10, 15, 30, 60, 120, 180, 360, and 1440 min (referred to as deltas).

4.3 SVN Clustering and Its Descriptive Statistics over Time

To categorize traders into distinct groups, we used Infomap clustering algorithm [19] since its popularity can be attributed to its information-theoretic approach, scalability, high quality clusters, flexibility, and statistical significance. According to the study [13] the Infomap clustering algorithm empirically gave the best results in [13] when applied to different benchmarks on Community Detection methods. Our empirical findings indicate that evolution of the proportion vector (with respect to its normalised version) allure satisfies our conjecture of a sparse number of mono-communities (see Fig. 2). From the figures, we notice a smooth evolution of proportions, and also the appearance of new and larger clusters – this is to be expected since the number of traders is growing over time. Moreover we observe a pattern of having fewer clusters of significant cardinality. The existence of a very big cluster (and many very small ones) would negate the heterogeneity of trading strategies. We observe that Infomap is consistent with the resolution scale and number of trades cut-off. We calculated several pertinent statistics to evaluate how the SVN's are affected by different time resolutions sampled throughout the lifespan of the entire dataset (i.e., from 2015–2017), as illustrated in Fig. 1. As previously stated, the number of traders in the dataset increases over time, however, we notice a sudden increase in the number of links and clusters from July 2016.

This results in an increase in the number of clusters and links in the sliding networks. We remark stability over time in the ratio of numbers of traders against the number of clusters. At each slide an SVN is built and some traders are never taken into consideration and the ratio of existent traders is increasing slightly with increase of the resolution delta. The modularity is slowly decreasing and is low besides deltas of 360 and 1440 min, which testifies about rather weak connections between clusters.

4.4 Goodness of Fit

In order to assess the goodness of fit to the data we refer to what is conventionally used: a classical χ^2 test. The parameter θ was estimated for every sliding window and since the formula is not explicit for $\mathbb{E}K_n$ (see Table 1) we approximate it to the closest integer. It is worth noting that for a non-conditional Ewens distribution one can readily find an explicit formula for θ using $\mathbb{E}K_n$.

Taking the example for δ equal to 10 mins and cut-off of minimum 100 trades we apply the χ^2 test for 50 sliding windows at significance of 0.05. We find a 95% pass rate which confirms that most of the time the conditional Ewens distribution is a good fit.

Figure 5 shows that for all studied scenarios in most cases we have a high pass rate. In general for a cut-off of 100 the pass rate is above 85%, for others it seems to increase with delta. Figure 4 illustrates a typical comparison between empirical and theoretical fit on a given sliding window which is satisfying. Figure 3 shows the evolution of the Ewens distribution fitted parameter. It is more or less stationary for bigger deltas and increasing for smaller ones. Larger estimated parameter $\hat{\theta}$ indicates a higher so-called mutation rate, therefore the existence of more clusters.

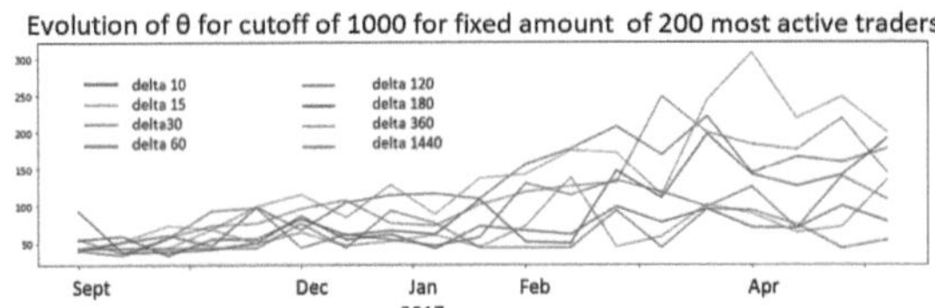

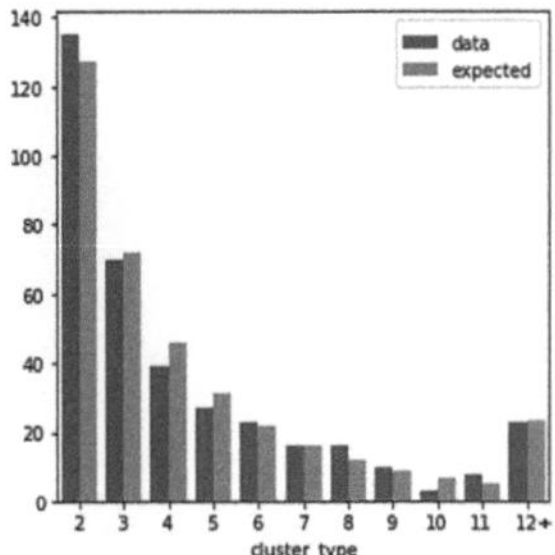

Fig. 3. Evolution of θ parameter for all δ time slices and cutoff of 1000 in the fixed amount of 200 most active traders. Other scenarios bear similarities in the shape of the curves, i.e., for deltas 360 and 1440 the parameter θ stays more or less stationary and others increase suddenly at some point.

Fig. 4. A comparison of empirical and theoretical fit on last sliding window. The plots were obtained using EUR/USD data for 10min scale and cutoff 100.

4.5 Temporal Cluster Evolution and Consistent Grouping Identification Issue

In some cases we require consistent grouping identification and the main difficulty comes from the lack of consistent naming of clusters for subsequent time frames. The latter allows us to, amongst other things, produce meaningful visualisations. The technique used relies on a total consistency measure which is in close relation to the Jaccard index (for more details see [14]).

In Fig. 6 we see a so-called alluvial plot where at a given time, traders belonging to the same group are stacked together to form a continuous flow. The stability of group composition is shown when the same colouring persists between two time steps. However, a group can split, merge, die out, appear suddenly or persist throughout time. These changes in groups are to be expected as traders' investment strategies evolve over time, and existing traders leave and new traders join. Overall we remark some stability, however as expected eventually there are die outs, merges, splits and new appearances. When we considered different deltas (results not shown), we found that larger groups were more prevalent for smaller time frames.

5   Clusterised Aggregating Algorithm

We wish to study the temporal evolution of clusters of trading activity and investigate how they can be used for practical purposes. Clustering evolution could be used in prediction problems since grouping has the advantage of simplifying the description of the system state by reducing the dimensionality of the prediction problem. In the literature there are numerous examples of the latter set in a financial context. For example, in [8] the authors used SVN's to demonstrate improvement in predicting both the sign of the order flow and the direction

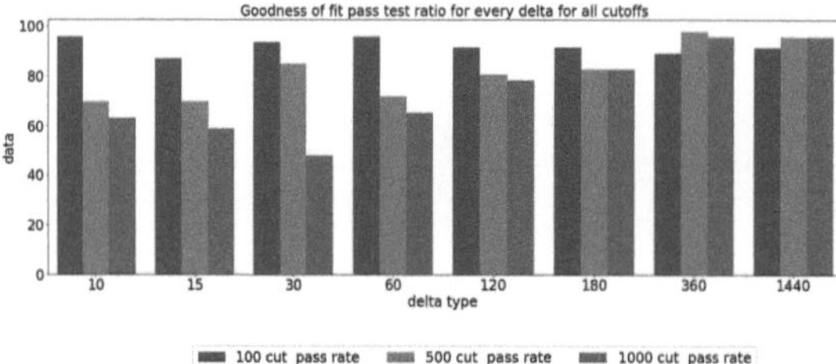

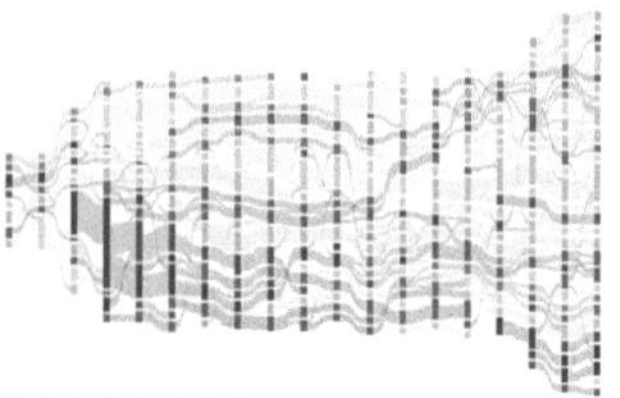

Fig. 5. Pass rate in percent for all δ time slices and 100, 500, and 1000 cutoffs. This rate represents the ratio of non-rejected null χ^2 hypotheses for all sliding windows.

Fig. 6. Alluvial plot with 1-step history (see [14] for more details) for 200 most active traders with cutoff 1000 and delta of one day.

of the average transaction price for a retail trader dataset. In this study we have applied the clustering evolution to prediction with an online expert advice model, namely the Aggregating Algorithm (AA) from [23,25]. The AA is given a series of online predictions from a pool of experts (in our case the traders). At each time epoch, the loss of each experts' prediction (in our case a trader's investment decision) is fed back into the AA and over time adjusts its trust in each expert to make future predictions. In the next subsections we introduce the framework of the AA and the games of investment with expert advice.

5.1 Aggregating Algorithm

Suppose that the learner L is tasked with predicting elements of a sequence $\omega_1, \omega_2, \ldots$ called outcomes. The outcomes occur in discrete time. Before seeing outcome ω_t, the learner is outputting a prediction γ_t. The quality of the prediction is measured by a loss function $\lambda(.,.)$. The expert aims to suffer low cumulative loss $\mathrm{Loss}_T(L) = \sum_{t=1}^{T} \lambda(\omega_t, \gamma_t)$.

We assume that the set of all possible outcomes (outcome space) Ω is known to us in advance and we are allowed to draw predictions from a known prediction space Γ, which may or may not be the same as Ω. The function λ is also known and maps $\Gamma \times \Omega$ to a subset of the extended real line, typically $[0, +\infty]$. The choice of a triple $G = \langle \Omega, \Gamma, \lambda \rangle$, is referred to as a game.

Suppose that the learner gets help from experts. The experts predict the same sequence and their predictions are made available to the learner before it commits to its own predictions. We are not concerned with their internal mechanics, which may well be inaccessible to us (e.g., the experts may rely on some sources of information unavailable or even unknown to us). The interaction with experts may be described by the following protocol. Here we assume that experts are parameterised by $\theta \in \Theta$.

Algorithm 1 Aggregating Algorithm

Require: η, C, q, N
1: Initialize weights $\omega_0^i \sim q_i$ for $i = 1, \ldots, N$
2: Choose loss function $\lambda(.,.)$
3: **for** $t = 1, 2, \ldots$ **do**
4:     Read experts' predictions γ_t^i
5:     Normalize the weights: $p_t^i = \dfrac{\omega_{t-1}^i}{\sum_j \omega_{t-1}^j}$
6:     Output $\gamma_t \in \Gamma$ satisfying:

$$\forall \omega \in \Omega, \quad \lambda(\gamma, \omega) \le \frac{C}{\eta} \ln \sum_{i=1}^{N} p_i e^{-\eta \lambda(\gamma, \omega)}$$

7:     Observe outcome ω_t
8:     Update the weights: $\omega_t^i = \omega_{t-1}^i \cdot e^{-\eta \cdot \lambda(\gamma_t^i, \omega_t)}$

Expert E_θ suffers loss $\mathrm{Loss}_T(E_\theta) = \sum_{t=1}^{T} \lambda(\gamma_t^\theta, \omega_t)$. The goal of the learner is to merge experts' predictions γ_t^θ into its own prediction γ_t in such a way that the learner's loss $\mathrm{Loss}_T(L)$ is low as compared to retrospectively best experts. It may use information about past outcomes and predictions. Formally, we are seeking a merging strategy $S : (\Gamma^\Theta \times \Omega)^* \times \Gamma^\Theta \to \Gamma$.

We typically want S to guarantee an upper bound on $\mathrm{Loss}_T(L)$ in terms of $\inf_{\theta \in \Theta} \mathrm{Loss}_T(E_\theta)$; we want $\mathrm{Loss}_T(L)$ to be low whenever $\mathrm{Loss}_T(E_\theta)$ is low for some θ. We assume that the pool of experts is finite, i.e., $|\Theta| = N < +\infty$.

Consider a game $G = \langle \Omega, \Gamma, \lambda \rangle$ a constant $C > 0$ is admissible for a learning rate $\eta > 0$ if for every $N = 1, 2, \ldots$, every set of predictions $\gamma_1, \ldots, \gamma_N \in \Gamma$, and every distribution $(p_1, p_2, \ldots, p_N) \in \Delta_{N-1}$, there is $\gamma \in \Gamma$ ensuring for all outcomes $\omega \in \Omega$ the inequality: $\lambda(\gamma, \omega) \le \frac{C}{\eta} \ln \sum_{i=1}^{N} p_i e^{-\eta \lambda(\gamma, \omega)}$. The mixability constant C_η is the infimum of all $C > 0$ admissible for η. This infimum is usually achieved. The admissibility is required to ensure the learner's predictions exist and belong to Γ since for example the learner's prediction of the form $\gamma_t = \sum_{i=1}^{N} p_i \gamma_t^i$ is a linear combination and Γ may not be convex. The AA takes as parameters a set of prior experts' weights $(q_1, \ldots, q_N) \in \Delta_{N-1}$, a learning rate $\eta > 0$ and an admissible $C > 0$. The algorithm works as shown in Algorithm 1.

The validity of the AA holds under some mild regularity assumptions on the game and assuming the uniform initial distribution, it can be shown (as in Eq. 8) that the constants in the following inequality are optimal:

$$\mathrm{Loss}_T(L) \le C\mathrm{Loss}_T(E_i) + \frac{C}{\eta} \ln N \tag{8}$$

5.2 Long Short Game

The problem of portfolio selection is a natural special case of a prediction with expert advice problem where in [24] considered realistic trading scenarios, i.e., the Long Short game.

The Long-Short game aims to represent a realistic trading scenario. A trader is allowed to open positions, both long and short, within certain limits based on their deposit and money they had earned previously. The limits aim to minimise the chance of bankruptcy. Given the wealth W_{t-1} at time $t-1$ trader i opens a position of size $W_{t-1}\gamma_t^i$ when the return ω_t is known, the trader's wealth changes accordingly:

$$W_t = W_{t-1} \cdot \lambda(\gamma_t^i, \omega_t) = W_{t-1} \cdot (1 + \gamma_t \cdot \omega_t)$$

In this framework one can apply the AA with $\eta = 1, C = 1$ and the substitution rule given by $\gamma_t = \sum_{i=1}^{N} p_i \gamma_t^i$ to the general long-short game. If $1 + \gamma_t \cdot \omega_t > 0$ for $t = 1, \ldots, T$, i.e., the learner does not get bankrupt along the way, the bound (8) will hold.

5.3 AA with Modifications

In [2], an evaluation of the performance of the AA was made using a real-life trading dataset. Some modifications of the AA were proposed in order to improve the practical performance of the resulting portfolio. In particular, a downside loss and weighted average between the latter and the long short loss were introduced. Downside loss, in contrast to long short loss (originally used in [24]), penalises financial losses but does not reward gains since a strategy not to lose money may be more important than the ability to earn money.

$$\begin{aligned}
\lambda_{\text{Long Short Loss}}(\rho, \gamma, r) &= -\log[\max(1 + \rho \cdot \gamma \cdot r, 0)] \\
\lambda_{\text{Downside Loss}}(\rho, \gamma, r) &= -\log\{\max[1 + \rho \cdot \min(\gamma \cdot r, 0), 0]\}
\end{aligned} \tag{9}$$

where ρ is the scaling factor, γ is the investment decision from $[-1, 1]$, and r is the return.

In our research we faced one particular challenge with our dataset: the pool of traders constantly changes through time. For example, traders may choose to cease trading with the broker at any time, they may take breaks from trading, new ones may join, or traders may close their account entirely. The AA requires such experts to continually provide predictions through time – a natural way to encode such activities is to use the so-called "sleeping" experts extension shown in Algorithm 2.

5.4 Clusterised Aggregating Algorithm (CAA) and Decision Rules

The classical AA learner prediction is:

$$\gamma_t = \sum_k p_t^k \gamma_{t-1}^k \,, \tag{10}$$

Algorithm 2 Aggregating Algorithm With Sleeping Experts

Require: η, ρ, N
1: Initialize weights: $\omega_0^i = 1$ for $i = 1, \ldots, N$
2: Choose loss function $\lambda(\gamma, r)$
3: **for** $t = 1, 2, \ldots$ **do**
4:     Get sets of awake experts A_t and sleeping experts S_t
5:     Read investments of awake experts γ_t^i for $i \in A_t$
6:     Normalize weights of awake experts: $p_t^i = \dfrac{\omega_{t-1}^i}{\sum_{j \in A_t} \omega_{t-1}^j}$
7:     Compute investment prediction: $\gamma_t = \sum_{j \in A_t} p_t^j \cdot \gamma_{t-1}^j$
8:     Observe return r_t
9:     **for** $i \in A_t$ **do**
10:         Update weights: $\omega_t^i = \omega_{t-1}^i \cdot \exp[-\eta \cdot \lambda(\gamma_t^i, r_t)]$
11:     **for** $i \in S_t$ **do**
12:         Update weights: $\omega_t^i = \omega_{t-1}^i \cdot \exp[-\eta \cdot \lambda(\gamma_t, r_t)]$

---

which is a weighted average of experts' predictions. For clusterised aggregating algorithm (CAA) we introduced two decision rules:

$$\gamma_t^{\text{MEAN}} = \sum_i \sum_j^{n_i} p_t^{i,j} \cdot \sum_k \frac{\gamma_{t-1}^{i,k}}{n_i} \qquad \text{take the mean of experts' predictions in a cluster}$$

$$\gamma_t^{\text{PEN}} = \sum_i \sum_j^{n_i} p_t^{i,j} \frac{\gamma_{t-1}^{i,j}}{n_i} \qquad \text{penalise by dividing by the cardinality of a cluster}$$

where n_i is the cardinality of i-th cluster and p^i is the sum of probabilities of i-th cluster.

The decision rule of γ_t^{MEAN} is interesting in a trivial case scenario, i.e., having the same duplicated experts in every cluster. Let's suppose that we have m identical experts in the pool. It appears desirable to collate them into one. However, this is done by the AA automatically. The behaviour of the AA would be the same as if one expert with the combined weight is present in the pool. Assuming the uniform distribution on the initial experts, the weight of the combined expert will be m/N and the loss bound for the duplicated experts E_i (again assuming the mixable case $C = 1$) turns into:

$$\text{Loss}_T(L) \leq \text{Loss}_T(E_i) + \frac{1}{\eta} \ln \frac{N}{m}$$

However, if duplicate experts are bad, this creates a problem: needlessly increasing n worsens the bound for good experts. For example, if there were two clusters, with each having different duplicated experts and the bigger cluster had better-performing experts then the AA bound would be improved.

The second decision rule, i.e., γ_t^{PEN} has an interpretation of partially awake experts if the penalising factor is normalised, i.e., $\dfrac{\frac{1}{n_i}}{\sum_{k \in \text{Clusters}} \frac{1}{n_k}}$. This idea was generalised in [27]. Apart from a prediction γ_t such an expert produces a confidence value $c_t \in [0, 1]$, which quantifies its confidence (a fully sleeping expert would output confidence of 0 and a fully awake expert would output a confidence of 1). Here the confidence would be inverse proportional to the cardinality of the cluster. This is similar to inverse-variance weighting in portfolio selection problems in particular the equal risk contributions portfolio [15].

5.5   Experts as Clusters Approach to AA (ECAA)

Up until now we only clusterised via the decision rules, and the experts were identified as the traders. It seems natural to consider treating clusters of traders as meta-experts. We averaged experts' investement decisions per cluster in order to obtain the meta-experts' predictions. In Appendix 11, we derive a condition to which these extensions to the AA would outperform the original set up of the AA with duplicated experts. In practice we identified the flow of meta-experts according to the alluvial plot (see Fig. 6). There are several things to consider in this scenario especially the splitting and merging of clusters on every epoch. We suggested the following approach:

- If the cluster is split then the children would inherit the parents weight divided by number of splits.
- If clusters are merged then the resultant weight is the sum of the parents weights.

5.6   Experiments

First we applied a data staging technique known as DAPRA (see [1]) which, when applied to data streams pertaining to trades and prices, allows one to sample the data at regular time intervals (required for this study). We then compared the performance of the AA with its clusterised counterparts (CAA and ECAA) with the expectation that these extensions would improve scalability and reduce noise. The CAA extension simply takes the mean of investments of awake experts γ in a given cluster (MEAN), or divides their decision by the cardinality of the cluster (PEN). As a benchmark we used the equally weighted portfolio strategy. We compared the CAA and the ECAA using the SVN-infomap approach with hierarchical clustering based on correlations of the traders' net positions (i.e., difference between total open long (buy) and open short (sell) positions in USD dollars) with a chosen distance metric: $1 - |\text{correlation}|$. The latter approach has a possibility of adjusting the construction of clusters by changing the dissimilarity threshold. The rationale behind clustering based on net position correlation is that it is a desirable feature for the broker since it is a measure of risk. The SVN approach is focused on trading synchronicity therefore we have less control on the quality of clustering in regards to the net position. Ideally all traders would trade all the time or have a high trading intersection period but since it is not the case one can end up with "noisy" clusters.

Table 2. Summary of experimental results for CAA.

Strategy	Type	Scaling factor	Return	Sharpe Ratio	Max Drawdown	Calmar Ratio
EW	Benchmark	–	1.4%	0.6	1.2%	1.2
AA	Sleeping Experts	70	**2.8%**	1.1	1.8%	1.5
CAA	MEAN/ SVN	70	3%	1.2	1.85%	1.8
CAA	MEAN/Hierarchical	70	**4.8%**	**2**	1.15%	**4**
CAA	PEN/SVN	70	**2.5%**	1.35	0.9%	2.5
CAA	PEN/Hierarchical	70	2.5%	**1.4**	**0.9%**	**2.6**
ECAA	Hierarchical 80	200	1%	**1.65**	**0.3%**	**3.5**
ECAA	SVN	1	0.5%	0.4	**0.8%**	0.6

We obtained optimistic results – especially for the downside loss (9), which is more appropriate in this framework. We evaluated the performance using four well established portfolio risk measures: the return of the portfolio, sharpe ratio is the amount of return an investor receives per unit of risk, the maximum drawdown is the maximum observed loss from a peak to a trough of a portfolio, before a new peak is attained and Calmar ratio measures the risk-adjusted performance of a portfolio by comparing the return to the maximum drawdown.

The distribution of traders' returns is close to symmetric and the mean is approximately zero. Performances of CAA are on the whole comparable with those of the MEAN clustering decision rule for the clusters constructed with the SVN-infomap method. However the results using the hierarchical clustering are significantly better across all risk measures. The best performing cutoff for the distance metric is around 70%. On the other hand, the results for the PEN clustering decision rule are comparable for the return on investment but for other metrics we noticed significantly better results for both clustering techniques. Figures 7 and 8 show the comparison among all results for a return scaling factor up to 400.

For ECAA we consider the scenario of treating clusters as meta-experts. Using the alluvial chart we can readily identify the flow of clusters over time since without it we could not identify clusters at different time epochs since they are unlabeled. Overall performance of the ECAA using SVN-infomap clusters is poor, manifesting lowest return, Sharpe Ratio and Calmar Ratio. However for hierarchical clustering all other risk measures are significantly better than the standard AA besides the return (see Fig. 9). Moreover, ECAA has smoother net profit as seen by much smaller drawdown than CAA, AA and the benchmark.

Table 2 summarises the experimental results for near optimal variations of all algorithms. Figures 10 and 11 show the evolution of returns and drawdowns throughout time, using the behaviour of a very basic strategy averaging traders' positions with constant equal weights as the benchmark. It is worth mentioning that when the scaling factor gets bigger (larger than 100) more and more traders go bankrupt because of the nature of the loss (9). Moreover, the algorithm could

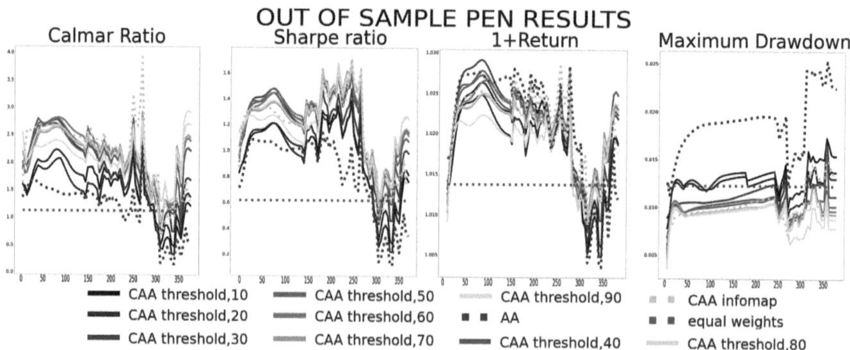

Fig. 7. Comparison of results among all four considered measures of risk in the out-of-sample scenario, where the CAA learner prediction is the expert's predictions divided by the cardinality of each cluster. The return-to-maximum-drawdown ratio, Sharpe ratio, 1 + return (= the proportional change of wealth), and maximum drawdown are shown for different return scaling factors. The green, blue, and pink dotted lines denote the equal-weight portfolio, AA, and CAA for SVN-Infomap performances. Other curves represent CAA using clusters created with hierarchical clustering at different thresholds.

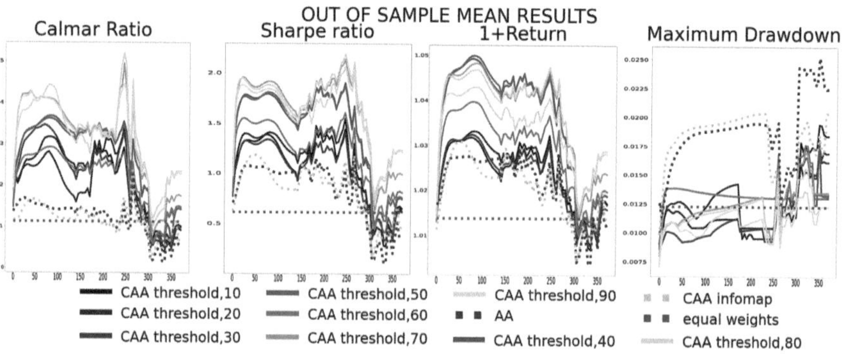

Fig. 8. Comparison of results among all four considered measures of risk in the out-of-sample scenario, where the CAA learner prediction is the mean of the expert's predictions for each cluster. The return-to-maximum-drawdown ratio, Sharpe ratio, 1 + return, and maximum drawdown are shown for different return scaling factors. The green, blue, and pink dotted lines denote the equal-weight portfolio, AA, and CAA for SVN-Infomap performances. Other curves represent CAA using clusters created with hierarchical clustering at different thresholds.

suddenly stop investing when the scaling factor gets too big therefore one must be cautious when interpreting the results.

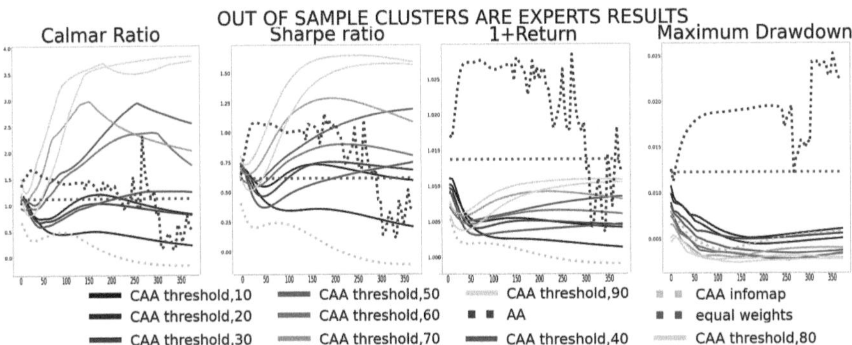

Fig. 9. Comparison of results among all four considered measures of risk in the out-of-sample scenario, where the ECAA learner prediction is the mean of the expert's predictions for each cluster. The return-to-maximum-drawdown ratio, Sharpe ratio, 1 + return, and maximum drawdown are shown for different return scaling factors. The green, blue, and pink dotted lines denote the equal-weight portfolio, AA, and ECAA for SVN-Infomap performances. Other curves represent ECAA using clusters created with hierarchical clustering at different thresholds.

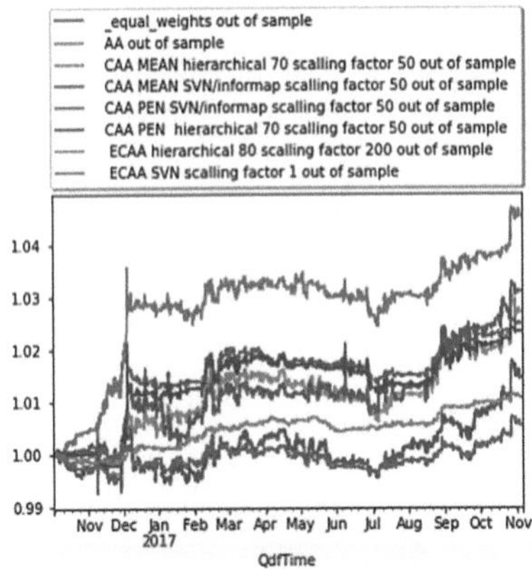
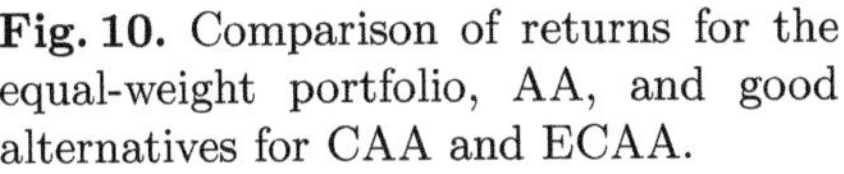

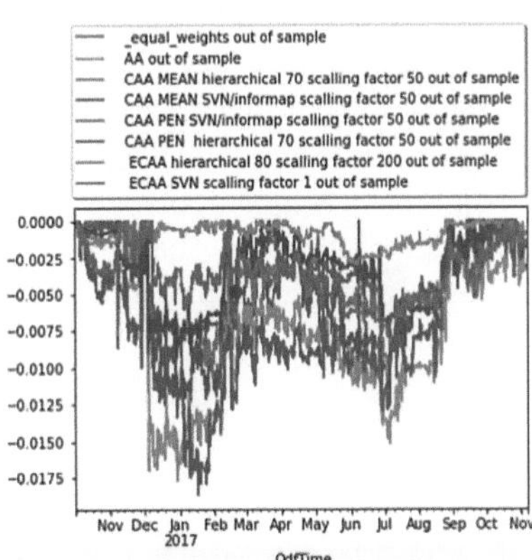

Fig. 10. Comparison of returns for the equal-weight portfolio, AA, and good alternatives for CAA and ECAA.

Fig. 11. Comparison of relative drawdowns for the equal-weight portfolio, AA, and good alternatives for CAA and ECAA.

6   Conclusion

In this chapter our findings confirm that clustering of traders' investments can be described by Ewens distribution. The temporal clustering distribution depends on many parameters and market conditions however its clustering could be leveraged to make better investment decisions. We adjusted the aggregating algorithm with sleeping experts to test the latter hypothesis using two clustering techniques, namely SVN-infomap and hierarchical clustering. In this framework the latter approach gives better results and gives more meaningful clusters since it is based on correlations of the investors' net positions and not on their trading synchronicity. In particular we compared CAA (used aggregated traders' decisions per cluster to calculate the investment prediction) and ECAA (clusters played the role of experts) with AA and the equally weighted portfolio strategy. Our introduced modifications to the AA indicate clear performance benefits in our experimental results in terms of four well established portfolio risk measures: return, Sharpe ratio, maximal drawdown and Calmar ratio.

Acknowledgments. The authors acknowledge the support of AlgoLabs Ltd in establishing and developing this research. Special thanks go to Xudong Li, Tzyy Tong and Samuel Manoharan for setting up the servers necessary to run our experiments. Further thanks go to Simon Tavaré for useful insights.

Appendix: Clusterised AA bound

We will discuss when it is beneficial to run AA on (equally weighted) cluster experts rather than the original experts and connect this with our intuition about the performance of traders. The analysis will be done on an artificial example but the conclusion is instructive.

Suppose that we have m identical experts in a pool of N. One may want to collate them into one; there is no need though as this is done by the AA automatically. The behaviour of the AA would be the same as if one expert with the combined weight is present in the pool. Assuming the uniform distribution on N original experts, the weight of the combined expert will be m/N and the loss bound for the duplicated experts E_i (assuming the mixable case $C = 1$) turns into

$$\mathrm{Loss}_T(L) \leq \mathrm{Loss}_T(E_i) + \frac{1}{\eta} \ln \frac{N}{m}.$$

This is a stronger bound and if the performance of the expert is actually good, it leads to lower $\mathrm{Loss}_T(L)$. However, if duplicate experts perform badly, they create a problem: increasing N worsens the bound for good experts.

Suppose that we have M clusters of experts of cardinalities $c_1, .., c_M$. Let all experts in each cluster be identical and suffer the same cumulative loss. Applying AA to cluster meta experts (with equal initial weights) will give us the loss bound U_- and applying AA to the original experts will give us the loss bound U_*:

$$U_- = \min_{i=1,2,\ldots,M} \left\{ \mathrm{Loss}_T(E_{C_i}) + \frac{1}{\eta} \ln M \right\} = \mathrm{Loss}_T(E_*) + \frac{1}{\eta} \ln M,$$

$$U_* = \min_{i=1,2,\ldots,M} \left\{ \mathrm{Loss}_T(E_{C_i}) + \frac{1}{\eta} \ln \frac{N}{c_i} \right\} = \mathrm{Loss}_T(E_{C_{i_0}}) + \frac{1}{\eta} \ln \frac{N}{c_{i_0}},$$

where E_{C_i} is an expert from cluster i, E_* is the best expert overall, and i_0 is the number of the cluster where the minimum in U_* is achieved.

We get that

$$U_- \leq U_* \iff c_{i_0} \leq \frac{N}{M} e^{\eta[\mathrm{Loss}_T(E_{C_{i_0}}) - \mathrm{Loss}_T(E_*)]}, \tag{11}$$

where $\mathrm{Loss}_T(E_{C_{i_0}}) - \mathrm{Loss}_T(E_*) \geq 0$. This means that the bound with cluster meta experts is better when there are no good experts in large clusters.

As the practice of trading shows, good trades are usually few and make a minority, which is one of the justification for the cluster AA. Cluster AA gives an advantage to smaller clusters.

References

1. Al-baghdadi, N., Wisniewski, W., Lindsay, D., Lindsay, S., Kalnishkan, Y., Watkins, C.: Structuring time series data to gain insight into agent behaviour. In: 2019 IEEE International Conference on Big Data (Big Data), pp. 5480–5490. IEEE (2019)
2. Al-baghdadi, N., Lindsay, D., Kalnishkan, Y., Lindsay, S.: Practical investment with the long-short game. In: Proceedings of the Ninth Symposium on Conformal and Probabilistic Prediction and Applications, volume 128 of Proceedings of Machine Learning Research, pp. 209–228, Verona, Italy, 09–11 Sep (2020). PMLR
3. Aoki, M.: Cluster size distributions of economic agents of many types in a market. J. Math. Anal. Appl. **249**, 32–52 (2000)
4. Baltakiene, M., Baltakys, K., Kanniainen, J., Pedreschi, D., Lillo, F.: Clusters of investors around initial public offering. Palgrave Communications (2019)
5. Baltakys, K., Le Viet, H., Kanniainen, J.: Structure of investor networks and financial crises. Entropy **23**(4) (2021). ISSN 1099-4300
6. Barreau, B., Carlier, L., Challet, D.: Deep prediction of investor interest: A supervised clustering approach. Algorithmic Finance 1–13 (2020)
7. Bohlin, L., Rosvall, M.: Stock portfolio structure of individual investors infers future trading behavior. PLoS ONE **9**(7) (2014). ISSN 1932-6203
8. Challet, D., Chicheportiche, R., Lallouache, M., Kassibrakis, S.: Statistically validated lead-lag networks and inventory prediction in the foreign exchange market. Adv. Complex Syst. (2018)
9. Cordi, M., Challet, D., Kassibrakis, S.: The market nanostructure origin of asset price time reversal asymmetry (2020)
10. da Silva, P.H., Jamshidpey, A., Tavaré, S.: Random derangements and the Ewens sampling formula (2020)
11. Ewens, W.J.: The sampling theory of selectively neutral alleles. Theor. Population Biol. **3**(1), 87–112 (1972). ISSN 0040-5809
12. Gutiérrez-Roig, M., Borge-Holthoefer, J., Arenas, A., Perelló, J.: Mapping individual behavior in financial markets: Synchronization and anticipation. EPJ Data Sci. **8** (2019)
13. Lancichinetti, A., Fortunato, S.: Community detection algorithms: a comparative analysis. Phys. Rev. E **80**, 056117 (2009)
14. Liechti, J.I., Bonhoeffer, S.: A time resolved clustering method revealing long-term structures and their short-term internal dynamics (2020)
15. Maillard, S., Roncalli, T., Teïletche, J.: The properties of equally weighted risk contribution portfolios. J. Portfolio Manag. **36**(4), 60–70 (2010). ISSN 0095-4918
16. Mantegna, R.N.: Clusters of Traders in Financial Markets, pp. 203–212. Springer Singapore (2020). ISBN 978-981-15-4806-2
17. Musciotto, F., Marotta, L., Micciche, S., Piilo, J., Mantegna, R.N.: Patterns of trading profiles at the Nordic stock exchange: a correlation-based approach. Chaos, Solitons & Fractals **88**, 267–278 (2016). ISSN 0960-0779

18. Musciotto, F., Marotta, L., Piilo, J., Mantegna, R.: Long-term ecology of investors in a financial market. Palgrave Commun. **4**, 92 (2018)
19. Rosvall, M., Bergstrom, C.T.: Maps of random walks on complex networks reveal community structure. Proc. Nat. Acad. Sci. **105**(4), 1118–1123 (2008). ISSN 1091-6490
20. Sueshige, T., Kanazawa, K., Takayasu, H., Takayasu, M.: Ecology of trading strategies in a forex market for limit and market orders. PLoS ONE **13**(12), 1–14 (2018)
21. Tumminello, M., Lillo, F., Piilo, J., Mantegna, R.: Identification of clusters of investors from their real trading activity in a financial market. New J. Phys. **14** (2011)
22. Tumminello, M., Micciche, S., Lillo, F., Piilo, J., Mantegna, R.: Statistically validated networks in bipartite complex systems. PLoS ONE **6**, e17994 (2011)
23. Vovk, V.: A game of prediction with expert advice. J. Comput. Syst. Sci. **56**(2), 153–173 (1998). ISSN 0022-0000
24. Vovk, V., Watkins, C.: Universal portfolio selection. In: Proceedings of the Eleventh Annual Conference on Computational Learning Theory, COLT' 98, pp. 12–23, New York, NY, USA (1998). Association for Computing Machinery. ISBN 1581130570
25. Vovk, V.G.: Aggregating strategies. Proceedings of Computational Learning Theory 1990 (1990)
26. V'yugin, V.: Universal algorithm for trading in stock market based on the method of calibration. In: Sanjay Jain, Rémi Munos, Frank Stephan, and Thomas Zeugmann, editors, Algorithmic Learning Theory, pp. 53–67, Berlin, Heidelberg (2013). Springer Berlin Heidelberg. ISBN 978-3-642-40935-6
27. V'yugin, V., Trunov, V.: Online aggregation of probability forecasts with confidence. Pattern Recogn. **121**(C) (2022). ISSN 0031-3203
28. Zhang, Y., Yang, X.: Online portfolio selection strategy based on combining experts' advice. Comput. Econ. **50**(1), 141–159 (2017). ISSN 0927-7099

Known Unknowns: Trading Scheduled Surprises with Conformal Prediction

David Lindsay[1,2(✉)] [ID] and Siân Lindsay[1,2] [ID]

[1] AlgoLabs Ltd, Bracknell, UK
{david,sian}@algolabs.com
[2] QuantBee Ltd, Bracknell, UK

Abstract. Scheduled macroeconomic announcements are among the few market events known in advance, and they are well documented to move financial markets in systematic ways [2,10,12,18,21]. In this chapter we present a case study that implements Conformal Prediction (CP) for financial markets trading (specifically, Foreign Exchange). Our case study uses more than 17 years' worth of publicly available 1-minute EUR/USD quotes [16], and a historical economic release calendar [26] to forecast *pre-release* and *post-release* price-movement bounds of EUR/USD around scheduled economic events. We will show how it is possible to compress millions of EUR/USD price movements using an online retracement segmentation that yields alternating up/down "intrinsic-time"' akin to a directional-change framework and the "zig-zag" indicator used by traders [9,13,25]. Resulting price segments preserve localised highs and lows in price history, enabling us to align segments around economic event releases. For each of these economic release series and respective price segments, we applied walk-forward linear regression (LR) models to forecast the price movement before and after the economic event release. We then tested the effect of wrapping these simple LR models with distribution-free CP [3,19,28], including Conformalised Quantile Regression (CQR) [27], to produce conformal bounds that map cleanly to trading actions such as take-profit and stop-loss. Across hundreds of economic event release series, we quantify coverage-width trade-offs and show that wider, asymmetric bands improve risk-adjusted outcomes relative to point-forecast exits produced by LR.

Keywords: Conformal prediction · Quantile regression · Price segmentation · Economic calendars · Event study · Backtesting

In financial trading, point forecasts are useful but in practice decisions often hinge on the *range* of plausible outcomes at a specific time. Traders frequently use information related to scheduled macroeconomic event announcements (such as unemployment figures or interest rate changes) to influence their trading strategy at a given point in time. The reason for this is clear - many studies have shown the ability of such announcements to move the markets [2,10,12,18,21]. What is interesting is that whilst the timing of an economic event is known in advance (in addition to approximate figures), the scale of the impact can be estimated based

K. An Nguyen and Z. Luo (Eds.): Alexander Gammerman Festschrift, LNCS 16290, pp. 345–386, 2026.
https://doi.org/10.1007/978-3-032-15120-9_16

on historical patterns, which in turn can aid trading decisions. In essence, this chapter focuses on the question: given only pre-announcement information, can we produce price-movement bounds for the subsequent impact of the economic release that a trading desk could act upon? To help us answer this question, we chose Conformal Prediction (CP) because it wraps any reasonable model and returns prediction intervals that target a chosen coverage rate without assuming a specific error distribution [3,19,28]. This is helpful because, over time, markets shift between different regimes due to factors such as geopolitical events, natural disasters, health crises and pandemics, market sentiment and so on.

We tested our approach on publicly available minute-resolution prices for a major currency pair (EUR/USD, euro versus US dollar) and a historical calendar of economic announcements [16,26]. The EUR/USD pair was chosen because it is the most liquid and widely traded currency pair, often viewed as a bellwether for gauging relative USD and EUR strength. Movements in this pair frequently reflect shifts in global risk sentiment and macroeconomic expectations, making it a natural choice for studying market reactions. The setup was intentionally simple: we treated the pre-announcement and post-announcement windows as two supervised learning tasks and translated interval endpoints into trade entry/exit rules.

To reduce sensitivity to minor price fluctuations, we applied an online "retracement" rule: as prices stream in, we open a new segment only when the path reverses by more than a small threshold. This intrinsic-time segmentation is akin to the directional-change framework and the "zig-zag" indicator used by traders [9,13,25]. It collapses millions of minutes into a few thousand price-swings while preserving localised highs and lows that are important for trading decisions and filtering noise. We could then align segments around economic event releases. For each of these economic release series and respective price segments, we applied walk-forward linear regression (LR) models to forecast the price movement before and after the economic event release. We then tested the effect of wrapping these simple LR models with distribution-free CP [3,19,28], including Conformalised Quantile Regression (CQR) [27], to produce conformal bounds that map cleanly to trading actions such as take-profit and stop-loss. Figure 1 illustrates retracement-based trend segments for EUR/USD price movements (measured in basis points, bps) from 6AM to 4PM (intraday) on 17th December 2024.

In what follows, we show how simplifying the workflow through segmentation helped manage noisy data and computational complexity. Each step remains interpretable and links clearly to execution outcomes, allowing readers to see the practical trade-offs involved.

This chapter is contributed by **David Lindsay**, a former PhD student of **Alex Gammerman**, in collaboration with his wife and research partner, **Siân Lindsay**. It draws on David's more than twenty years of applying AI/ML to real-world data in quantitative finance. Our aim is a *think-piece* rather than a claim of novel research results: we use familiar, industry-tested techniques and take the opportunity to show how conformal methods can be applied in practice, where they remain uncommon. While computational cost is often a barrier, we show that simple segmentation can make conformal approaches tractable in real workflows.

1   Markets, Releases and Price Trend Segments

Before building models, it is crucial to clearly establish the operational context by defining and connecting the key components of our analysis:

- The raw EUR/USD price time series data;
- The economic release calendar, which structures market activity around scheduled macroeconomic announcements;
- The segmentation mechanism used to convert noisy price data into meaningful trend lines, enabling clearer analysis and more effective forecasting.

1.1   Who Trades, Where and over What Horizons?

Global markets comprise several core asset classes: FX, rates/fixed income, equities, commodities, and (more recently) crypto—with a diverse ecology of participants and trading horizons. End-users (corporates, real-money asset managers, insurance/pension funds) typically trade to hedge or allocate; leveraged funds (global macro, CTA, stat-arb) express views across horizons from minutes to months; market-makers and principal trading firms provide liquidity and inventory risk transfer; and central banks set policy or occasionally intervene. These actors connect through a mix of central-limit-order-book venues (exchanges and Electronic Crossing Networks) and quote-driven Over The Counter platforms, with microstructures that vary by instrument (tick sizes, matching rules, minimum trade sizes, trading hours) [15,17,20].

This chapter focuses deliberately on **foreign exchange (FX)**—specifically EUR/USD spot—for three pragmatic reasons:

1. **Liquidity and near-continuous trading:** EUR/USD is among the most actively traded currency pairs, characterised by tight spreads and deep liquidity throughout European and US trading sessions. Practically, it trades *24/5* (Sun 22:00–Fri 22:00 UTC).
2. **Clear alignment with the macroeconomic calendar:** Scheduled economic releases (employment, inflation, GDP, monetary policy decisions, PMIs, trade data and housing statistics) are time-stamped weeks in advance, facilitating precise event-aligned learning and rigorous backtesting.

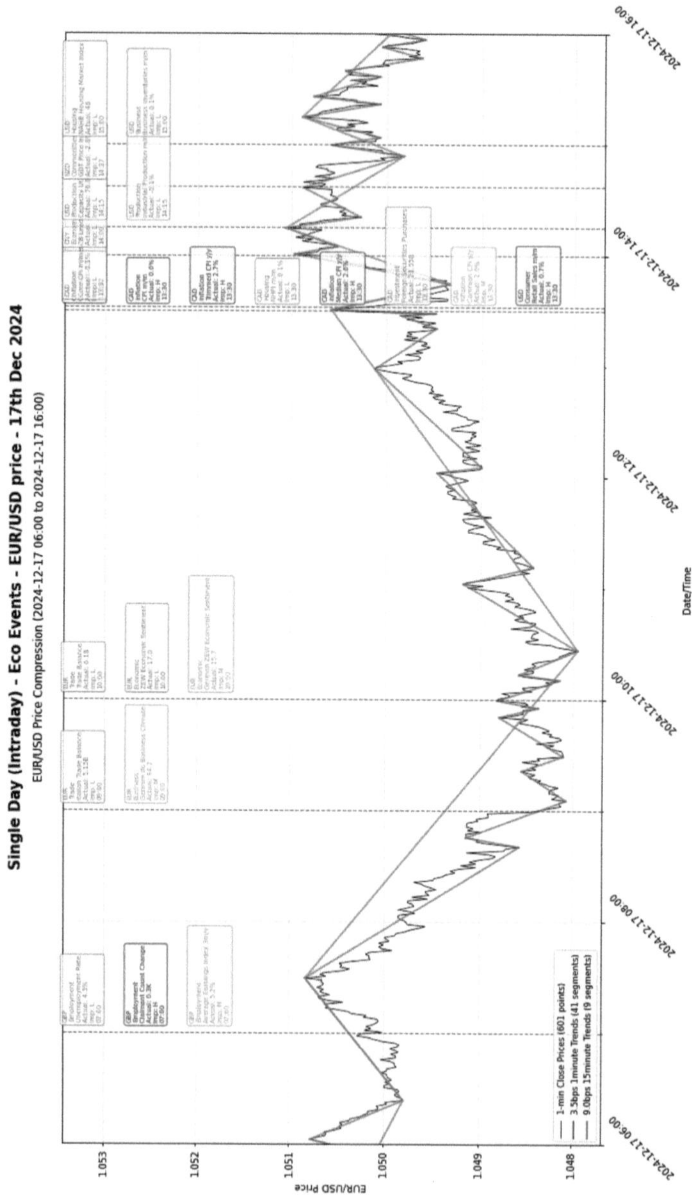

Fig. 1. Zoomed intraday view of EUR/USD on 17th Dec 2024 with retracement-based trend segments (3.5 bps in blue, 9.0 bps in orange) overlaid and all scheduled economic releases annotated. Here **bps** denotes *basis points*, a unit for tiny percentage moves (1 bps = 0.01%; 100 bps = 1%). Vertical dashed lines mark release times. Boxes are colour-coded by importance (high=red, medium=yellow, low=grey). (Color figure online)

3. **Data breadth and simplicity:** We utilise a publicly available extensive dataset that tracks minute-by-minute EUR/USD prices over 17 years [16]. We have been able to combine this with a comprehensive historical economic release calendar [26], thus ensuring that our focus remains firmly on methodological insights rather than proprietary data or market-specific nuances.

Horizons in our study are explicitly intraday operating from minutes to hours around release times. At these scales, trading decisions are sensitive to execution frictions (spread, queue position, slippage) and to the distribution of short-horizon returns (skewness, fat tails) [9]. As we will discover later, practical solutions for trading desks will ideally express uncertainty as ranges rather than point estimates, be robust to non-Gaussian residuals and time-varying volatility whilst still being computationally tractable with regular retraining (e.g. walk-forward use). These underlying principles guide the pre-processing and modelling choices in the following sections of this chapter.

Wednesday August 27 2025			Actual	Previous	Consensus	Forecast
02:00 AM	AU	Westpac Leading Index MoM JUL	0.1%	-0.03%		0.2%
02:30 AM	AU	Construction Work Done QoQ Q2	3.0%	-0.3%	0.7%	0.5%
02:30 AM	AU	Monthly CPI Indicator JUL	2.8%	1.9%	2.3%	2.0%
02:30 AM	CN	Industrial Profits (YTD) YoY JUL	-1.7%	-1.8%		-1.8%
04:35 AM	JP	BoJ JGB Purchases				
06:00 AM	SG	5-Year Bond Auction	1.53%	2.05%		
07:00 AM	DE	GfK Consumer Confidence SEP	-23.6	-21.7	-22	-21.3
10:00 AM	GB	Treasury Gilt 2028 Auction	3.991%	3.941%		
10:10 AM	IT	6-Month BOT Auction	2.012%	2.003%		
10:30 AM	DE	7-Year Bund Auction	2.46%	2.05%		
11:00 AM	FR	Unemployment Benefit Claims JUL	52.9K	-21.6K		-14.0K
11:00 AM	FR	Jobseekers Total JUL	3033.5K	2980.6K		3100.0K
11:00 AM	GB	CBI Distributive Trades AUG	-32	-34	-33	-30
12:00 PM	US	MBA 30-Year Mortgage Rate AUG/22	6.69%	6.68%		
12:00 PM	US	MBA Mortgage Applications AUG/22	-0.5%	-1.4%		
12:00 PM	US	MBA Mortgage Market Index AUG/22	275.8	277.1		
12:00 PM	US	MBA Mortgage Refinance Index AUG/22	894.1	926.1		
12:00 PM	US	MBA Purchase Index AUG/22	163.8	160.3		
12:30 PM	BR	Bank Lending MoM JUL	0.4%	0.5%		0.3%
01:00 PM	MX	Balance of Trade JUL	$-0.017B	$0.514B	$0.3B	$ 0.3B
03:30 PM	US	EIA Crude Oil Stocks Change AUG/22	-2.392M	-6.014M	-2M	
03:30 PM	US	EIA Gasoline Stocks Change AUG/22	-1.236M	-2.72M	-2.5M	
03:30 PM	US	EIA Crude Oil Imports Change AUG/22	0.299M	-1.218M		
03:30 PM	US	EIA Cushing Crude Oil Stocks Change AUG/22	-0.838M	0.419M		
03:30 PM	US	EIA Distillate Fuel Production Change AUG/22	-0.113M	0.193M		
03:30 PM	US	EIA Distillate Stocks Change AUG/22	-1.786M	2.343M	1.1M	

Fig. 2. Economic release calendar displaying scheduled macroeconomic events, their importance levels, and relevant data fields such as actual, forecast, and previous values. Taken from https://tradingeconomics.com/calendar

1.2 The Economic-Release Calendar One of the Few Things We Know in Advance

The economic calendar provides structured, forward-looking information. For each scheduled event we know the *timestamp, currency/region,* and *series description,* and in many cases the *previous, consensus forecast,* and *actual* values. The main numerical interpretation of 'surprise' is calculated by quantifying the variance between the actual and forecasted estimates for each economic release announcement:

$$\text{Surprise} = \text{Actual} - \text{Forecast}.$$

These forecasts, which estimate the value of the upcoming economic release, are derived from surveys of professional economists conducted by organisations such as Bloomberg and Reuters [6]. In the freely available version of the economic release dataset used in this study [26], these forecasts are not always fully populated. A good proxy in such cases is to compare the *Actual* value with the history of *Previous* announced values, which is what we use as a feature in our models later in Sect. 2.2.

Figures 1 and 2 illustrate how releases punctuate intraday price evolution and the economic release calendar's typical field layouts respectively. Although the economic release descriptions contain hundreds of distinct series names, they collapse into fewer than a dozen broad categories (e.g., *Employment, Inflation, GDP, Policy/Rates, Manufacturing/PMI, Trade, Housing*) as can be seen in Fig. 3. There is a natural hierarchy: for each *currency* there are many different *categories* of release which fall into specific *descriptions.* Two structural points matter for our modelling:

- **Currency-specific vs. cross-currency series.** Some releases are inherently tied to an issuing authority and therefore appear for a single specific currency (e.g., *Non-Farm Employment Change* is a US-specific payroll statistic; it does not have EUR/JPY/GBP analogues under the same name), whereas others are conceptually common across regions (e.g., *Trade Balance, CPI, Industrial Production*) and thus exist for multiple currencies such as AUD, CAD, CHF, JPY, GBP, NZD, USD.
- **Heterogeneous update frequencies.** Most headline series are *monthly* (roughly 200–230 observations over our 17+-year window), some are *quarterly* (shorter histories), and a few are *weekly.* The frequency dispersion has direct consequences for data sufficiency and model stability across Series (Sects. 3 and 3.1).

Figure 4 shows that even within a single category (e.g. "Employment"), different series have wildly different units and magnitudes: some (e.g. Unemployment Rate) fluctuate by only a few percentage points, while others (e.g. ADP Non-Farm Employment Change) involve hundreds of thousands of jobs. This makes direct comparison challenging without normalisation.

Figure 5 further highlights the heterogeneity across all categories: nominal series such as Industrial Production or Trade Balance reach into the hundreds of thousands or millions, rate or index series such as Policy Rate or PMI oscillate within single- or double-digit ranges, and percent-change measures (e.g. CPI y/y, GDP q/q) lie on yet another scale. This underscores the need for careful pre-processing (e.g. standardisation or log-scaling) when incorporating these raw numbers into a unified predictive framework.

In summary, the economic calendar provides known decision times and a taxonomy that spans both currency-specific and cross-currency phenomena, with uneven sampling across series. Our event-aligned formulation and segmentation (Sect. 2) leverages this structure: labels are defined relative to the scheduled timestamp, and the intrinsic-time segments provide a compact, noise-robust context for features and for the conformal uncertainty sets that follow [2,10,12,18,21].

1.3 The Problem of "too Much Data" and the Need for Segmentation

At 1-minute resolution, the full EUR/USD sample contains **over 6.8 million** prices. Modelling directly at this granularity is both *noisy* and *computationally heavy*, especially once we layer conformal wrappers that rely on stable residuals and repeated calibration. We therefore insert a tractable intermediate representation: a single-pass, online retracement segmentation that produces alternating up/down intrinsic-time trend lines. This is described by Algorithm 1. A segment ends and the next begins only when the price *retraces* by at least a fixed threshold θ (in basis points[1]) the most recent local high/low. This is closely related to the directional-change framework [9,13] and the technical indicator [25] used by traders.

---

[1] Here basis points also known as **bps** denotes a unit for tiny percentage moves (1 bps = 0.01%; 100 bps = 1%).

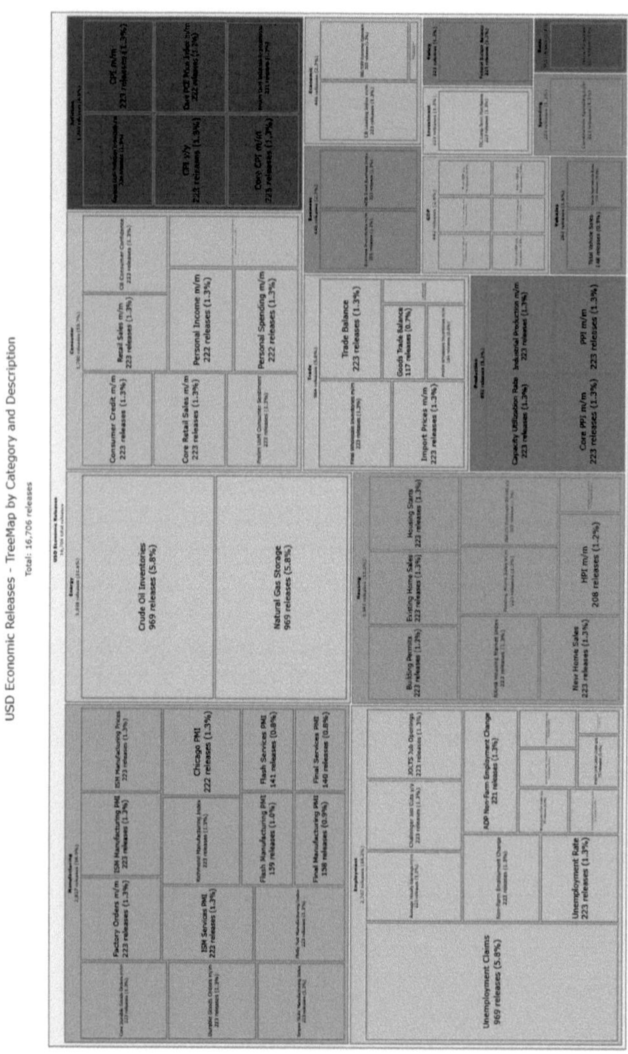

Fig. 3. Treemap of USD economic releases (2006–2025, after cleaning), showing the distribution of $\approx 16{,}700$ events by *category* (colour blocks) and *description* (sub-blocks). Area is proportional to the count of occurrences for each series. Most mainstream indicators contribute on the order of ~ 200 observations (monthly cadence), while truly high-frequency series dominate count mass: for example, weekly *Crude Oil Inventories* and *Natural Gas Storage* each contribute ~ 960 releases across the sample - nearly five times the typical monthly history. This heterogeneity explains why some Series in our ranked results enjoy rich training histories whereas others remain data-sparse. (Section 3.1)

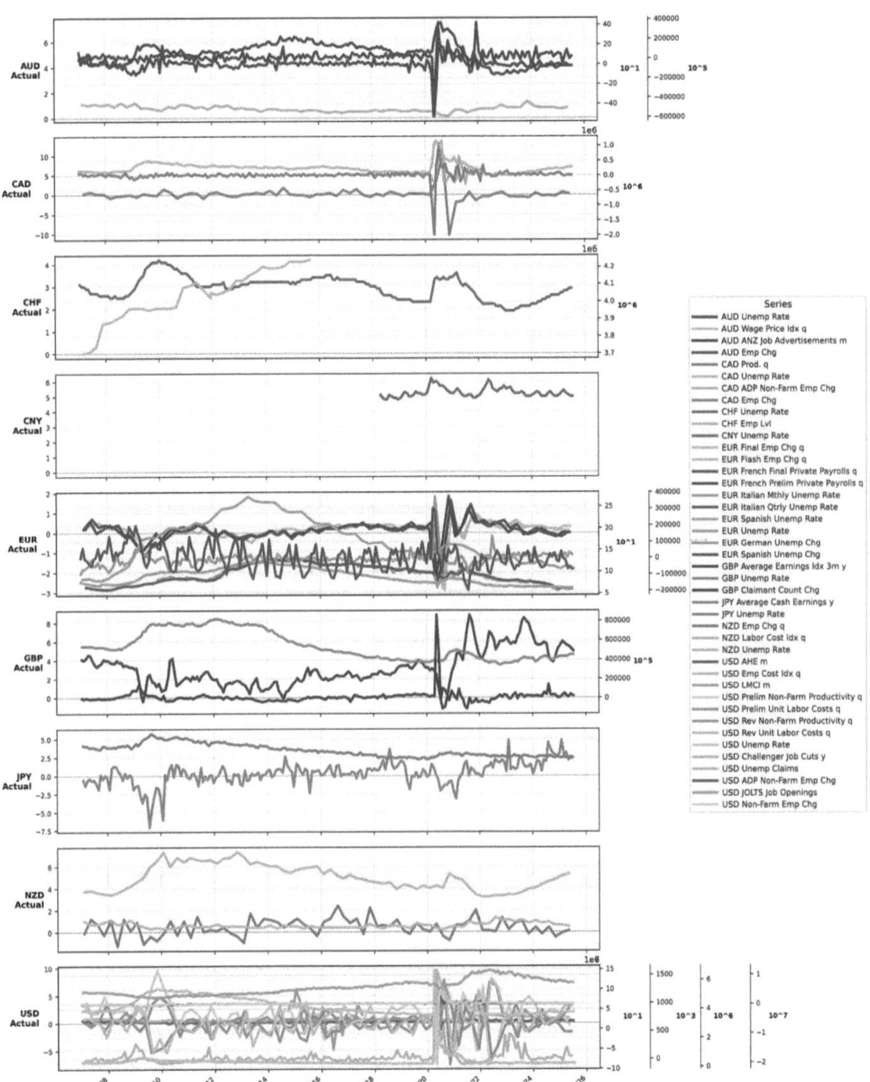

Fig. 4. Time-series of all **Employment**-category releases for each of nine major currencies (AUD, CAD, CHF, CNY, EUR, GBP, JPY, NZD, USD), spanning 2006–2025. Each line corresponds to a distinct release (e.g. Unemployment Rate, ADP Non-Farm Employment Change, Average Hourly Earnings, etc.), plotted at its actual value on the release date.

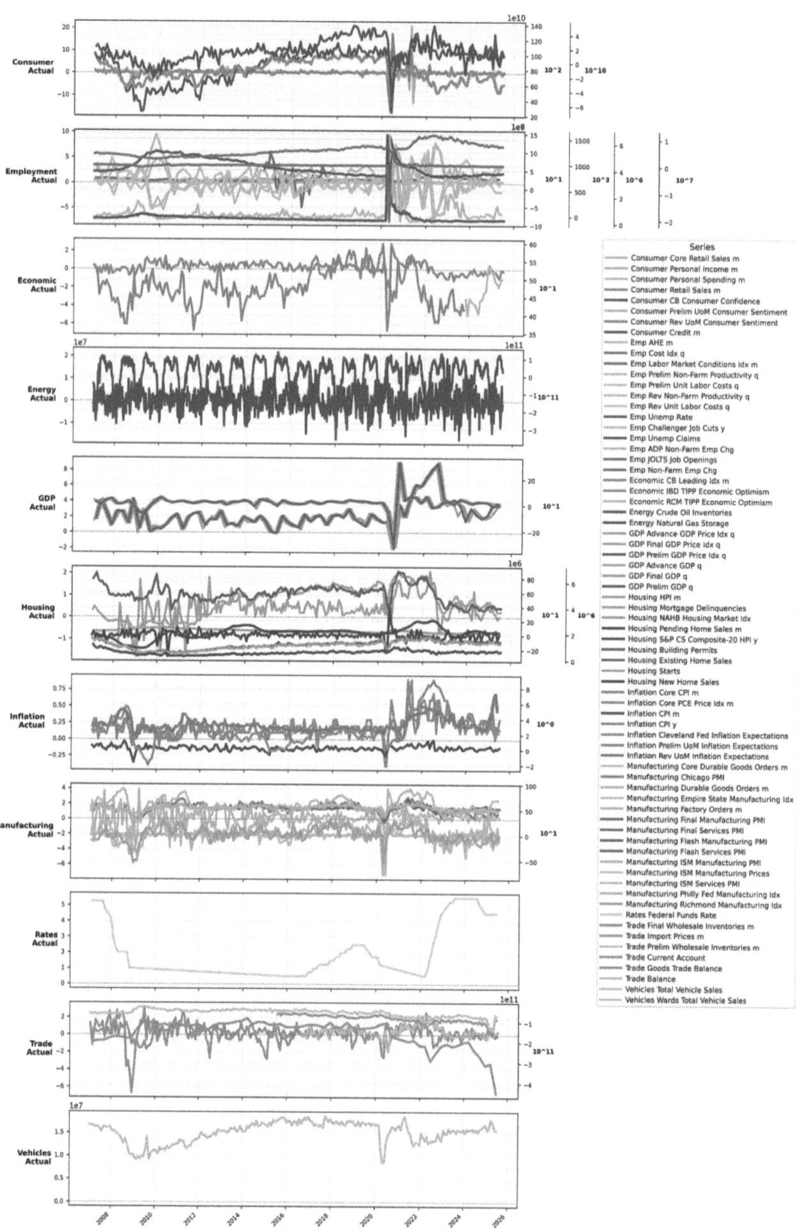

Fig. 5. Time-series of **all economic categories** for the USD calendar, 2006–2025. Each panel shows one category (Business, Consumer, Economic, Employment, Energy, GDP, Housing, Inflation, Investment, Manufacturing, Policy, Production, Rates, Spending, Trade, Vehicles), and within each panel the actual release values are plotted at their release dates.

Algorithm 1. Online Trend Line Segmentation

1: **Input:** price series $\{P_t\}_{t=1}^{T}$, threshold θ (in bps).

2: **Idea:** Build straight-line segments that connect confirmed local extrema, ignoring oscillations smaller than θ bps.

3: **Start.** Take the first price P_1. Until price has moved by at least θ bps from P_1, do nothing (no direction is confirmed).

4: **Lock the direction.** As soon as price is $\geq \theta$ bps away from the start, lock the current direction: up if price rose, down if it fell. From that moment, keep track of the running *extreme* P_{ext}: the highest high in an up-leg, the lowest low in a down-leg.

5: **Extend or flip.**

    – If price makes a new extreme in the same direction, simply extend the current segment to the new P_{ext}.

    – If price *retraces* by θ bps or more from P_{ext}, confirm (finish) the current segment at that last extreme, flip direction, and start a new segment from that extreme.

6: **Repeat.** Continue through the series. The output is an alternating sequence of straight-line segments that connect confirmed local extrema; small oscillations $< \theta$ bps are ignored.

7: *Units.* Basis-point distance from the current extreme is $10^4 \times \left| \frac{P_t}{P_{\mathrm{ext}}} - 1 \right|$.

To explain the simple Algorithm 1, below offers a walk-through, step-by-step, on a days' worth of data for the 3rd August 2012, which is also illustrated in Fig. 6:

– **Panel (a): initialisation and first leg.** A downward move establishes the first active segment (red). Although the price begins to bounce, the pullback is smaller than θ; no flip is recorded.

– **Panel (b): first confirmation.** The upward pullback reaches θ, marked by the yellow dashed line in both price and retracement panels. The algorithm closes the red segment at the last low and opens a new green upsegment from that extreme. A later pullback of size θ confirms the following red segment in the same way.

– **Panel (c): alternating structure.** As trading continues, the mechanism produces a sparse intrinsic-time skeleton: long green/red legs connected at confirmed turning points. Sub-θ noise is suppressed, which is precisely the compression we exploit downstream.

– **Panel (d): latest confirmation.** After a small down-leg, a sharp rally drives the retracement statistic beyond θ again, confirming a fresh upsegment just after 06:28. The dotted horizontals show the running extreme levels that anchor the calculation, while the retracement subplot makes the threshold-crossing rule transparent to the eye.

What Are the Practicalities of Segmenting Price Time Series? In this brief section we will demonstrate the practialities of segmenting the raw price time series in relation to the analysis in the chapter:

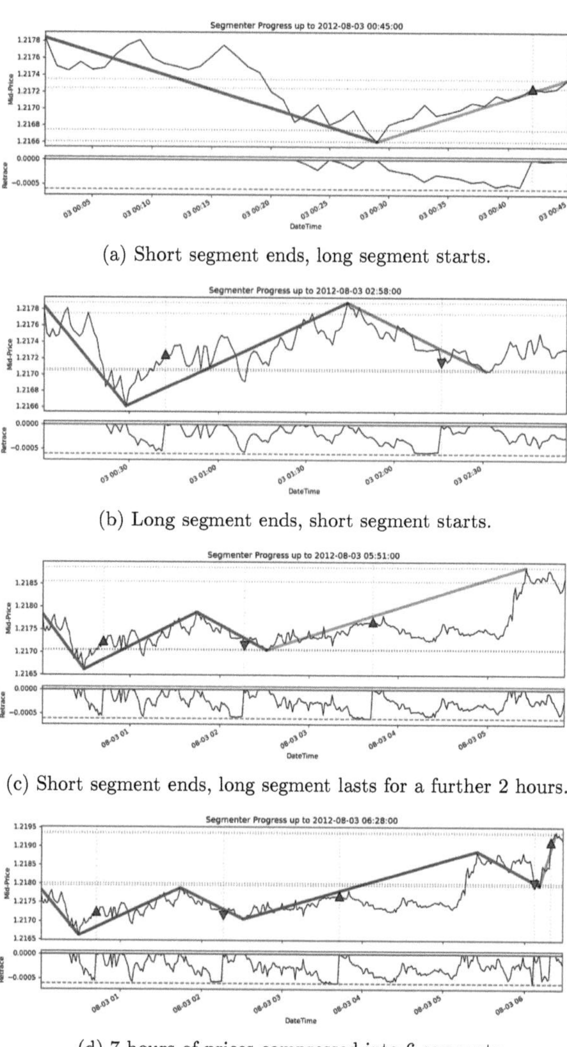

(a) Short segment ends, long segment starts.

(b) Long segment ends, short segment starts.

(c) Short segment ends, long segment lasts for a further 2 hours.

(d) 7 hours of prices compressed into 6 segments,
preserving local high and lows.

Fig. 6. Online retracement segmentation on EUR/USD 1-minute data. Price
(black) is summarised by alternating up (green) and down (red) trend-line segments. A
new segment is *confirmed* only when price retraces by at least θ from the most recent
extreme (yellow dashed verticals); dotted horizontals track the current extremes. The
lower subpanels show the retracement statistic; crossing the $\pm\theta$ bounds triggers confirmation. **Panel notes:** (a) Price retraces from local minima 1.2166 by 0.0007; short
segment ends and a long segment starts. (b) Long segment persists for nearly 2 h; price
retraces from local maxima 1.2178 and a short segment is confirmed. (c) Short segment
continues for ~30 min to local minima 1.2171; a new long segment is confirmed ~2 h
later. (d) Over ~7 h the algorithm compresses the path into 6 alternating long/short
segments. **Animated demo (GIF):** https://algolabs.com/segment-demo

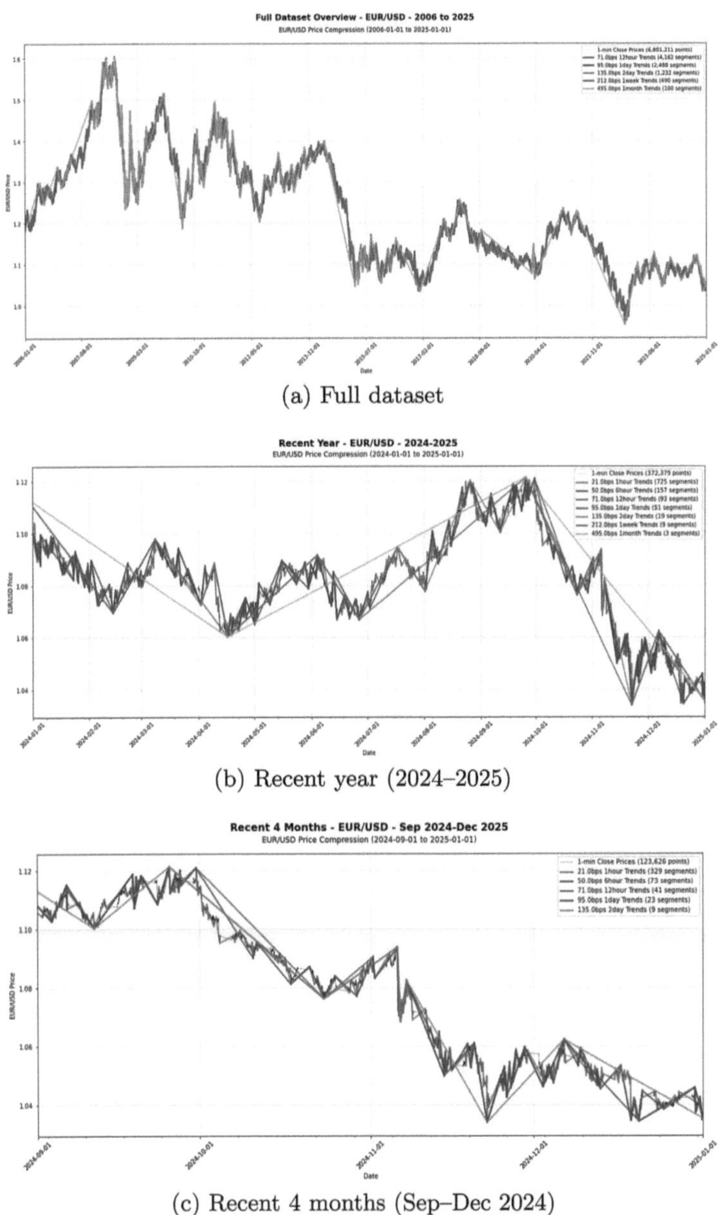

(a) Full dataset

(b) Recent year (2024–2025)

(c) Recent 4 months (Sep–Dec 2024)

Fig. 7. EUR/USD price-segment compression at broader scales: full history, recent year, and recent four months. The different line colours denote different segment resolutions.

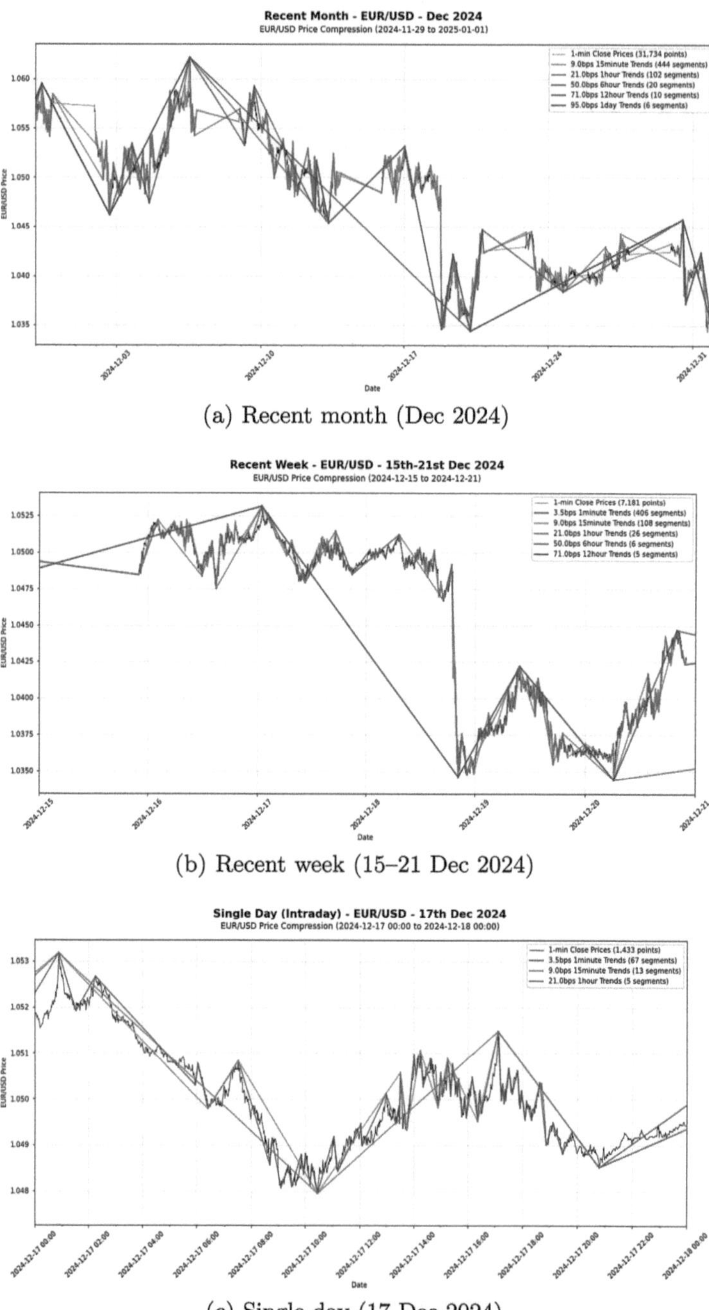

(a) Recent month (Dec 2024)

(b) Recent week (15–21 Dec 2024)

(c) Single day (17 Dec 2024)

Fig. 8. EUR/USD price-segment compression at finer scales: recent month, week, and single day. The different line colours denote different segment resolutions.

- **Compression and multi-scale views.** Running the segmenter at several thresholds (from a few bps to weekly/monthly scales) yields parallel representations. Figures 7 and 8 shows multi-scale compression from full history down to daily windows, and Table 1 quantifies how N minute bars collapse to $M \ll N$ segments while preserving the local highs and lows that matter for execution.
- **Stability and tractability.** Working at the segment level reduces the computational burden of applying machine learning methods and improves the stability of residual distributions used to form prediction intervals.
- **Scope.** A turn is only *confirmed* after a price retracement of size θ, so the segmentation is inherently lagging and is *not* a trading signal by itself. Its role in this analysis is not to provide directional forecast but to provide event-aligned bucketing of movement for the supervised learning tasks that follow.

Figures 7, 8 and Table 1 both demonstrate that we can get wildly different compression tradeoffs with different retracement parameters to the segmentation described by Algorithm 1. The choice of the retracement threshold involves a critical trade-off between data compression and signal fidelity. A small threshold, such as the 3.5 bps for the "1 min" horizon, results in over 550,000 segments and offers only a 12× compression over using the raw 1 min data. This configuration follows the price action very closely, as seen with the higher-frequency lines in Fig. 7 and 8, but retains much of the market noise and carries a high computational burden. Conversely, a large threshold like the 495.0 bps for the "1 month" horizon achieves a massive 68,012× compression, reducing the entire 17-year history to just 100 segments. While this effectively isolates the most dominant long-term trends (visible as the smoothest lines in Fig. 7 and 8), it smooths over almost all intraday and intra-week volatility, making it unsuitable for analysing short-term events.

Table 1. Segments and compression ratios across retracement scales (EUR/USD, 1-minute source data from 2007 to 2024)

Approx Time Horizon	Retrace (bps)	Segments	Compression (N/M)
1-minute bars (raw)		$Raw \approx 6.8$ million	1×
1 minute	3.5	551,738	12×
15 minutes	9.0	154,210	44×
1 hour	21.0	37,316	182×
6 hours	50.0	7,822	869×
12 hours	71.0	4,162	1,634×
1 day	95.0	2,488	2,733×
2 days	135.0	1,232	5,520×
1 week	212.0	490	13,880×
1 month	495.0	100	68,012×

For the sake of simplicity and analytical relevance, in the following sections of this chapter we will focus our analysis on a single resolution of 21 bps. This setting, which corresponds to an approximate "1-hour" horizon, provides a pragmatic balance: it achieves a significant 182× data compression while still preserving the meaningful intraday price swings that are relevant for studying the impact of economic releases. This defines the so-called *EcoTrend*—the single segment that spans each economic release time.

2   Formulating the Problem on Segments (Before/After Labels)

We now formalise how scheduled macro events interact with the segmented price path. The key construct is the *EcoTrend* i.e. the single Δ_{before} and Δ_{after} segment (at a 21 bps resolution) that *spans* a known economic release time. From this segment we derive operational labels for the signed *pre-release* and *post-release* price moves, expressed in basis points, and we specify the features and the walk-forward protocol used for all predictive experiments. Figure 9 shows an example of an EcoTrend for USD Non-Farm Employment Change on 3rd August 2012.

2.1   EcoTrend and Labels (before/after on a Single Segment)

The *EcoTrend* is defined as the unique price segment—at a chosen retracement resolution (we use the "1-hour" setting unless stated otherwise)—that spans the scheduled release time t_{eco}. Let that segment run from start (t_s, P_s) to end (t_e, P_e) with $t_s < t_{\text{eco}} < t_e$. We work on the segment's linearised scaffold (straight line between (t_s, P_s) and (t_e, P_e)), which is robust to microstructure noise yet preserves the local extrema that matter for execution.
The key quantities are:

$$\Delta_{\text{total}} \equiv \text{signed linearised move over the EcoTrend (in bps),} \tag{1}$$

$$u \equiv \frac{t_{\text{eco}} - t_s}{t_e - t_s} \in [0, 1], \qquad \text{fraction of the segment elapsed by the release,} \tag{2}$$

$$\Delta_{\text{before}} \equiv u\,\Delta_{\text{total}}, \qquad \Delta_{\text{after}} \equiv (1 - u)\,\Delta_{\text{total}}. \tag{3}$$

Equation (3) gives the *labels* for two supervised tasks: a pre-release regression with target $y = \Delta_{\text{before}}$ and a post-release regression with target $y = \Delta_{\text{after}}$. We begin with the former and return to the latter in Sect. 5 when mapping intervals to take-profit/stop-loss.

2.2   What Features Are Used by the Regression Models

We intentionally restrict ourselves to transparent features that a trading desk could defend and monitor. They fall into three themes:

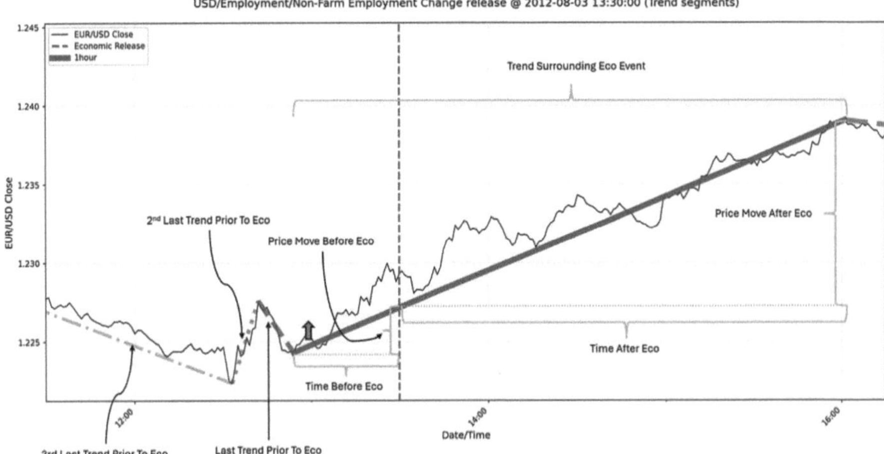

Fig. 9. Real-data example for USD *Non-Farm Employment Change* on 2012-08-03 at 13:30. At the 21 bps ("1-hour") retracement resolution, the EcoTrend is an *up* segment from $t_s \approx 12{:}54$ ($P_s \approx 1.22500$) to $t_e \approx 16{:}02$ ($P_e \approx 1.23976$). The total linearised move is $\Delta_{\text{total}} \approx +120.5\,\text{bps}$; $u \approx 36/187$; hence $\Delta_{\text{before}} \approx +23\,\text{bps}$ and $\Delta_{\text{after}} \approx +97\,\text{bps}$. Note: the straight-line scaffold slightly under/over estimate the instantaneous price at the release, as we will see later this simplification does not affect the ability to forecast effective bounds when backtesting on the original raw 1 min data.

(A) **Recent segment context.** From the three segments immediately preceding the EcoTrend we compute, for each segment:
   - duration (minutes),
   - signed move (basis points),
   - start time relative to the release t_{eco}.

   Short, fast swings versus long, slow drifts imply different near-term paths into the event.

(B) **Previous release behaviour (same series).** From the prior occurrence of the *same* economic series we reuse the EcoTrend decomposition:
   - total move,
   - pre-release portion,
   - post-release portion,
   - elapsed minutes before/after the prior t_{eco}.

   Several series (e.g., inventories, PMIs) display recurring microstructure around publication time.

(C) **Level effects from the published series.** We include:
   - a cleaned last value `Previous_Clean`,
   - a minmax oscillator of past values in $[0, 1]$,
   - `TimeMinsBeforeEco` $= t_{\text{eco}} - t_s$, the numerator of u in (2), encoding how far through the segment we are at the event.

2.3 Walk-Forward Protocol (expanding Training Window, per Economic Release Series)

Algorithm 2 outlines the protocol we use to evaluate each series online, treating each economic release series as its own time series:

Algorithm 2. Online Rolling Evaluation of Each Series

1: **Input:** For each economic series, ordered events (x_i, y_i) with publication times.
2: **Parameters:** Minimum warm-up size $n_{\min}$.

3: **Step 1. Warm-up.**
   Reserve the first $n_{\min}$ events to train the initial model; do not predict yet.

4: **for** $k = n_{\min} + 1$ **to** n **do**
5:     **Step 2. Train.**
   Fit a linear regression on past pairs $\{(x_i, y_i)\}_{i=1}^{k-1}$ using the chosen features, target $y = \Delta_{\text{before}}$.
6:     **Step 3. Predict.**
   Compute $\hat{y}_k$ using the freshly trained model; observe residual $r_k \leftarrow y_k - \hat{y}_k$.
7:     **Step 4. Expand window.**
   Include (x_k, y_k) in the training set for future steps.

8: **Note:**
   When applying any conformal wrapper (see Section 4), calibrate using residuals $\{r_i\}_{i<k}$.
   Point metrics (Pearson r, MAE, correct direction) remain identical whether or not a wrapper is used.

This protocol prevents peeking beyond t_k; it is aligned with event-time and guards against look-ahead bias. It also matches how the model would be updated and used in live operation: train on what has happened, forecast the very next scheduled economic release in that series and then roll forward [5,22].

3 Can Linear Regression Forecast Economic Release Price Moves?

We initially examined one of the simplest baselines: a linear regression (ordinary least squares, OLS) of the pre-release price move Δ_{before} on the transparent feature themes previously described in Sect. 2.2. We conducted evaluations per economic release series using the expanding walk-forward procedure described in Algorithm 2, ensuring strict causality by using only information available prior to each event.

Figure 10 visualises the top-performing series ranked by Pearson correlation r between actual and predicted values using the 21 bps segments (roughly 1-hour time horizons), highlighting where predicted and actual directions align with green ✓ marks.

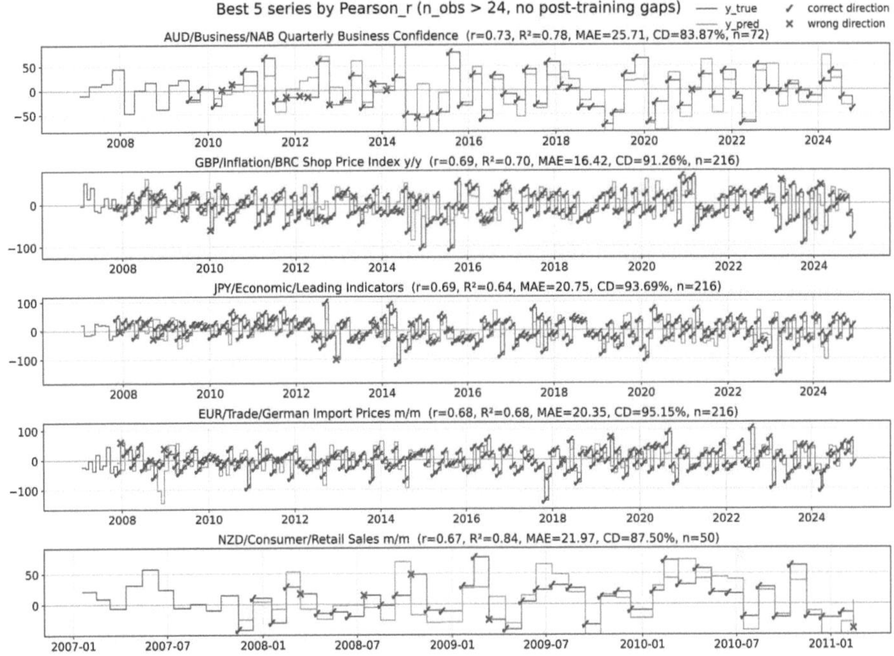

Fig. 10. Top **five** series by Pearson correlation r between y_{true} and y_{pred} for the 1-hour horizon. Blue: y_{true}; orange: y_{pred} (step lines at release times). A green ✓ marks agreement in direction (sign $\hat{y}$ = sign y); a red × marks disagreement.

Essentially, we found that this simple LR approach can achieve reasonable directional accuracy and correlation on a meaningful subset of eceonomic release series and that performance tends to improve as history accumulates (more events per series), consistent with the steadily expanding training set.

3.1   What LR Learns in Practice, Ranked Results and Intepretation

Performance for each economic release series is summarised by:

- **Pearson correlation (r):** measures linear association between predicted and actual Δ_{before}.
- **Coefficient of determination (R^2):** fraction of variance in the target explained by the model.
- **Mean absolute error (MAE):** average absolute forecast error (in basis points).
- **Correct-direction rate (CD):** percentage of events where $\mathrm{sign}(\hat{y}) = \mathrm{sign}(y)$.
- **Number of observations (n_{obs}):** count of events used per series (data sufficiency).

Across the full panel of valid economic series (see subset in Table 2, 356 in total), the distribution of key performance metrics used to evaluate the simple LR approach is summarised below:

$$\bar{r} \approx 0.36, \quad \overline{R^2} \approx 0.62, \quad \overline{\text{MAE}} \approx 36.3 \text{ bps}, \quad \overline{\text{CD}} \approx 86.7\%.$$

Additionally, 75.2% of economic release series exhibit positive correlation $r > 0$, with 78.7% surpassing $r = 0.20$ and 28.4% exceeding $r = 0.50$. This demonstrates that even this minimalistic, transparent baseline is capable of extracting meaningful signals.

Table 2. Per-series performance for the 1-hour pre-release task ($y = \Delta_{\text{before}}$): Top-5 and Bottom-5 by Pearson r.

Rank	Series	n_{obs}	r	R^2	MAE (bps)	CD (%)
1	AUD/Business/NAB Quarterly Business Confidence	72	0.73	0.78	25.7	83.8
2	GBP/Inflation/BRC Shop Price Index y/y	216	0.69	0.70	16.4	91.2
3	JPY/Economic/Leading Indicators	216	0.69	0.64	20.7	93.6
4	EUR/Trade/German Import Prices m/m	216	0.68	0.68	20.3	95.1
5	NZD/Consumer/Retail Sales m/m	50	0.67	0.84	21.9	87.5
…	…	…	…	…	…	…
356	CAD/Employment/ADP Non-Farm Employment Change	51	−0.16	0.65	158.2	72.5
355	GBP/Trade/Current Account	72	−0.12	0.67	163.9	69.3
354	USD/GDP/Prelim GDP Price Index q/q	71	−0.12	0.64	90.6	70.4
353	GBP/GDP/Final GDP q/q	72	−0.11	0.65	134.6	74.1
352	USD/Consumer/Revised UoM Consumer Sentiment	216	−0.05	0.49	77.1	88.3

Several observations emerge clearly from these initial results:

(1) **Data availability is critical.** Economic time series with longer and more regular histories consistently outperform sparse or irregular ones.
(2) **Directional accuracy outperforms magnitude precision.** Many series achieve relatively high directional hit rates (CD) despite moderate correlation, highlighting operational relevance.
(3) **Performance improves as the training dataset grows.** Expanding the historical window boosts predictive accuracy and stability over time, suggesting genuine learning rather than random performance.
(4) **Regime sensitivity.** Some bottom-ranked Series display periods of structural change or low signal-to-noise where a fixed linear mapping struggles; widening the data window helps, but ultimately we plan to transfer this baseline into conformal wrappers to expose a controllable widthcoverage trade-off (Sect. 4).

A brief scan of coefficient importance reveals commonality among impactful features, notably the timing of recent trends (`TrendStart_Prev1_TimeDiffMins_Eco`) and lightweight calendar-based fields (`dayOfQuarter`, `dayOfYear`). These findings align with microstructure intuition: short-term dynamics around scheduled economic releases are influenced by recent price momentum and cyclical timing. The OLS baseline establishes that there is learnable, stable structure in Δ_{before} for a meaningful subset of economic release series, especially as history accumulates.

Having established this basic signal, subsequent sections will extend the LR baseline by introducing conformal prediction methods to produce rigorously calibrated intervals suitable for practical trading use.

4   From Points to Bounds: Conformal Wrappers

Point forecasts convey direction and scale, but execution decisions hinge on ranges that quantify plausible outcomes over the next decision window. Conformal prediction offers a principled way to wrap any point-forecast model with distribution-free intervals that target coverage under weak assumptions [3,19,28]. We keep the base learner deliberately plain (per- economic release series LR; as shown earlier in Sect. 3.1) and study three wrappers that trade-off interval *width* for empirical *coverage*. Central to our setting is the fact that outcomes arrive in time order; hence we evaluate coverage under dependence and finite samples [30].

4.1   Evaluating Different CP Regression Approaches

We implement three lightweight constructions at each step of the expanding walk-forward protocol (detailed earlier in Algorithm 2). *Notation:* Throughout this section, $\varepsilon \in (0,1)$ denotes the *miscoverage (significance) level* of the prediction interval, so the nominal confidence/coverage is $1 - \varepsilon$. If nonconformity scores are used elsewhere, we follow the conformal-prediction convention of writing them as α_i; they are unrelated to the level ε.

- **Split-Conformal LR (symmetric).** Use absolute residuals from the OLS fitted on past rows; take the $(1 - \varepsilon)$ empirical quantile q of those residuals and report $[\hat{y} - q, \hat{y} + q]$ [3,19].
- **Split-Conformal LR (asymmetric tails).** Same as above, but allocate different tail masses with $\lambda \in (0,1)$, taking quantiles q_ℓ and q_u at levels $1 - \varepsilon\lambda$ and $1 - \varepsilon(1-\lambda)$, yielding $[\hat{y} - q_\ell, \hat{y} + q_u]$. Asymmetry improves reliability when residuals are skewed or heavy-tailed.
- **Conformalised Quantile Regression (CQR).** Fit lower/upper quantile models at $\varepsilon/2$ and $1 - \varepsilon/2$ and (optionally) conformally calibrate them with historical residuals to attain the nominal coverage $1 - \varepsilon$ [27]. In our first pass we show the *raw* quantile bands (tightest) to contrast sharpness vs. undercoverage; a calibration offset can then be added if desired.

Algorithms 3–5 formalise the above for the pre-release label Δ_{before}. As emphasised in Sect. 3.1, the wrappers *do not* alter the underlying point predictions: r,

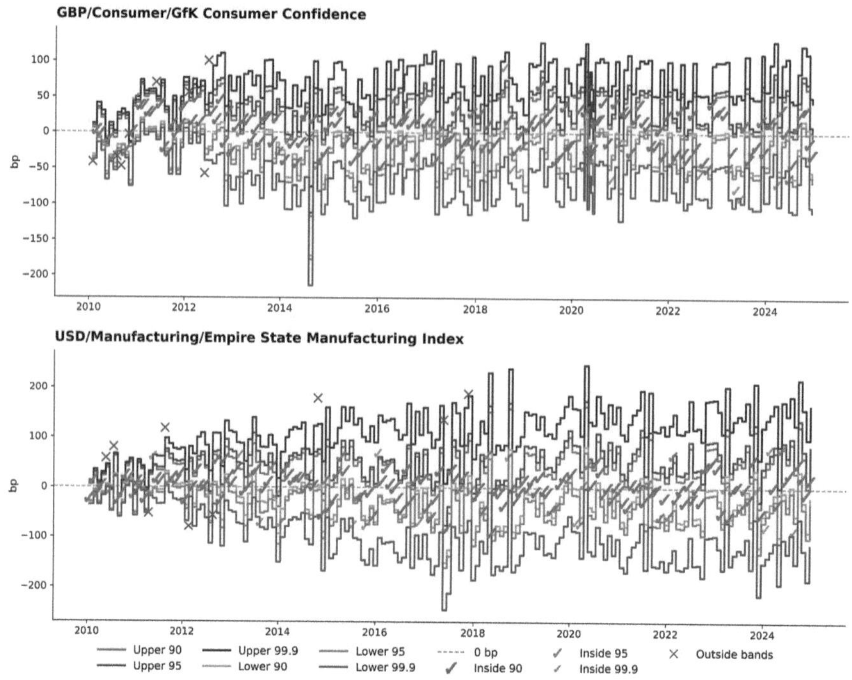

Fig. 11. Split-conformal LR with asymmetric tails (unequal q_ℓ and q_u; tail split λ). Relative to the symmetric case, coverage improves at the cost of wider intervals, especially on the upper side for positively skewed residuals.

R^2, MAE and CD remain those of the OLS baseline. Symmetric split-conformal produces compact bands with moderate coverage; asymmetric tails ($\lambda \neq 0.5$) buy coverage at the expense of width; raw CQR bands are the tightest but typically under-cover unless an additional calibration pass is applied.

Algorithm 3. Split-conformal LR: symmetric

Require: history $\mathcal{D}_{1:k-1} = \{(x_i, y_i)\}_{i=1}^{k-1}$, feature x_k, miscoverage level ε
1: Fit OLS f_{k-1} on $\mathcal{D}_{1:k-1}$; set $\hat{y}_k = f_{k-1}(x_k)$.
2: Compute absolute residuals $r_i = |y_i - \hat{y}_i|$, $i = 1, \ldots, k-1$.
3: $q \leftarrow \text{Quantile}_{1-\varepsilon}(r_1, \ldots, r_{k-1})$.
4: **Output** prediction interval (PI) $[\ell_k, u_k] = [\hat{y}_k - q, \ \hat{y}_k + q]$.

Algorithm 4. Split-conformal LR: asymmetric

Require: history $\mathcal{D}_{1:k-1}$, feature x_k, miscoverage level ε, tail split $\lambda \in (0,1)$
1: Fit f_{k-1} and compute $\hat{y}_k$ as in Alg. 3.
2: Compute absolute residuals $r_i = |y_i - \hat{y}_i|$, $i = 1, \ldots, k-1$.
3: $q_\ell \leftarrow \text{Quantile}_{1-\varepsilon\lambda}(r_{1:k-1})$; $\quad q_u \leftarrow \text{Quantile}_{1-\varepsilon(1-\lambda)}(r_{1:k-1})$.
4: **Output** PI $[\ell_k, u_k] = [\hat{y}_k - q_\ell, \; \hat{y}_k + q_u]$.

Algorithm 5. Conformalised Quantile Regression (CQR)

Require: history $\mathcal{D}_{1:k-1}$, feature x_k, miscoverage level ε
1: Fit quantile regressions f_{k-1}^ℓ at level $q_\ell = \varepsilon/2$ and f_{k-1}^u at level $q_u = 1 - \varepsilon/2$.
2: Raw bands: $\tilde{\ell}_k = f_{k-1}^\ell(x_k)$, $\tilde{u}_k = f_{k-1}^u(x_k)$ (**swap if $\tilde{\ell}_k > \tilde{u}_k$ to avoid crossing**).
3: *Optional conformal calibration (on historical data):* define residuals $e_i^\ell = \tilde{\ell}_i - y_i$
   and $e_i^u = y_i - \tilde{u}_i$; set $c_\ell = \text{Quantile}_{1-\varepsilon}(e_{1:k-1}^\ell)$, $c_u = \text{Quantile}_{1-\varepsilon}(e_{1:k-1}^u)$. Then
   $\ell_k = \tilde{\ell}_k - c_\ell$, $u_k = \tilde{u}_k + c_u$.
4: **Return** PI $[\ell_k, u_k]$. (We also report raw QR bands, which emphasise tightness but
   may under-cover.)

4.2 Coverage Vs Interval Width

Conformal wrappers add upper and lower bound intervals around a fixed point predictor, but they do not change the predictor itself. Consistent with the theory, our split-conformal constructions leave the baseline LR's point metrics essentially unchanged: R^2, Pearson r, MAE, and Correct Direction (CD) are the same as for the underlying OLS run. We focus on split-conformal methods because of their simplicity and suitability for a rolling, walk-forward, event-aligned setting. More complex resampling-based variants, such as Jackknife+ and Bootstrap+, are described in [7]. Quantile approaches can also be post-calibrated to target nominal coverage (e.g., CQR [27]), trading additional width for reliability.

Figure 11 gives us a view of what these conformal regression bounds look like in practice with the symmetric bounds on two different economic release series, namely GBP/GfK Consumer Confidence and USD/Empire State Manufacturing Index. We see the blue upper and red lower bounds are given for 90%, 95% and 99.9% confidence levels, naturally these get wider with higher confidence. This is interesting to compare with the base LR model's point forecasts shown in Fig. 10. In the conformal counterpart Fig. 11 we see that the red cross marks are where the predictions fall outside the bounds (for 90–99.9% confidence) and the intensity of the green ✓ shows at what confidence level the true regression label lay within the bounds. We can see the forecasts get generally wider with time and capture most forecasts to a 90% confidence level. We also see the red × markers where the bounds fail to capture the true price move label, these mostly occur early on in the training walk forward process, when there there the bounds are

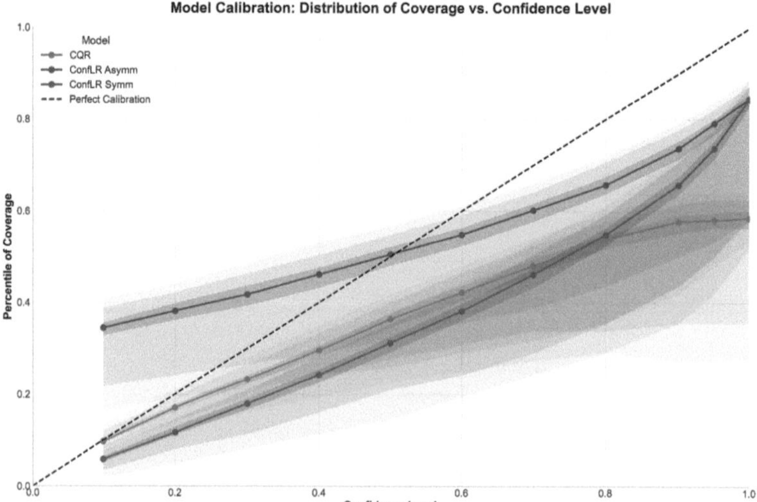

(a) Empirical coverage vs. nominal confidence. The dashed black line is perfect calibration.

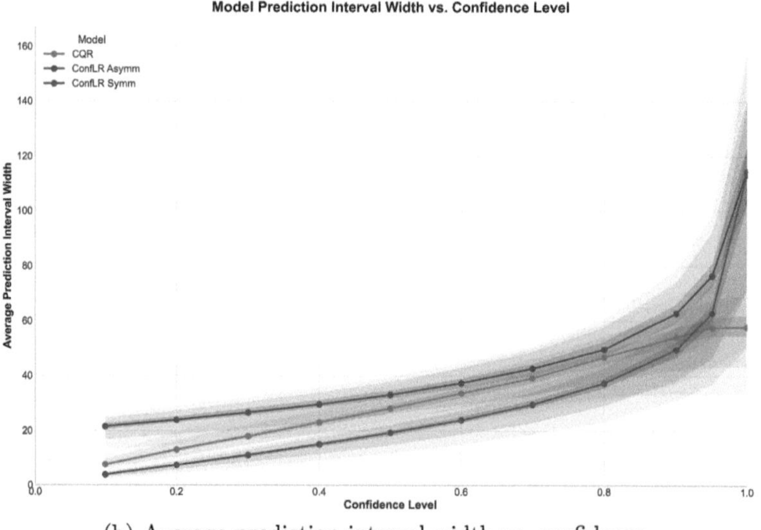

(b) Average prediction-interval width vs. confidence.

Fig. 12. Calibration and width trade-offs across all Series at the 1-hour pre-release horizon. The red curve is *split-conformal LR (symmetric)* and the **blue** curve is *split-conformal LR (asymmetric tails)*. The green curve is *CQR (conformalised quantile regression)*. Solid lines show the *median* across Series; shaded bands depict dispersion (10/90, 20/80, and 40/60 percentile envelopes). As expected, higher confidence pushes empirical coverage toward nominal (top) while requiring wider prediction intervals measured in bps (bottom). Asymmetric tails are consistently *wider* but *better calibrated* than symmetric bands—an effect that will matter when we map intervals into desk-style TP/SL rules in Sect. 5.

narrower. Even those that fall outside these bounds, near the edges, only just missing their encapsulation.

Although the bands are not strictly calibrated to their nominal levels in this simple expanding-window setup, they are meaningful, and as we will see later - practically useful. The split-conformal intervals trade coverage for width in a predictable way, while CQR provides sharp but conservative bands. In the upper plot of Fig. 12 we see the coverage rates for various confidence levels for all three CP variants tested - none are strictly calibrated (as shown by the black dashed diagonal) but are close, with asymmetric CP showing the best results in terms of calibration. Similarly the lower plot of Fig. 12 shows the mean region window at various confidence levels, CQR gives the tightest of bounds of around 50 bps but as we can see from that comes at the cost of much lower coverage at a maximum of 0.59.

5   From Forecasts to Trades: Using LR and Conformal Bounds as Take-Profit and Stop Loss

This section details the final step of our analysis: translating the statistical forecasts from our models into a simulated trading strategy. To do this, we must first define the basics of trading mechanics and performance metrics of a trading backtest.

5.1   High Level Trading Definitions and Performance Metrics

At the core of any trading decision is the direction. A trader can profit from both rising and falling markets by taking one of two positions:

- **Going Long:** A trader buys an asset with the expectation that its price will rise. The profit is realised by selling it later at a higher price. This is the conventional "buy low, sell high" approach. Our model takes a long position if its price forecast is positive.
- **Going Short:** A trader sells a borrowed asset with the expectation that its price will fall. They can then buy it back later at the new, lower price to return it, pocketing the difference. This allows one to profit from a decline in price. Our model takes a short position if its forecast is negative.

The performance of these trades is quantified using two primary metrics. The first is **Profit & Loss (PnL)**, which is the direct monetary outcome of a single trade. For a given trade size (or *notional*), it is calculated as the difference between the entry price and the exit price, adjusted for the trade's direction. We define a trade's direction, s, as $+1$ for a long position and -1 for a short position. The PnL for a single event is then:

$$\text{PnL}_{\text{event}} = (\text{Price}_{\text{exit}} - \text{Price}_{\text{entry}}) \times s$$

In our backtests, we track the *Cumulative PnL*, which is the running sum of these individual outcomes over time.

$$\text{CumPnL}_k = \sum_{i=1}^{k} \text{PnL}_i$$

This shows the strategy's overall performance, as seen in the top panel of Fig. 13 where we have shown the cumulative PnL time series for all USD/Manufacturing/ISM Services PMI eco events for all LR and conformal precitors under analysis.

The second key metric is **Drawdown**. While cumulative PnL measures overall profitability, drawdown measures the risk. It is the peak-to-trough decline in the cumulative PnL over a specific period. A strategy might be profitable in the long run, but if it experiences massive drawdowns, it may be too risky to be traded in practice. Formally, at any point in time k, the drawdown is the difference between the historical peak of the cumulative PnL and its current value:

$$\text{Drawdown}_k = \left(\max_{0 \leq j \leq k} \text{CumPnL}_j \right) - \text{CumPnL}_k$$

The *Worst Drawdown* is the largest magnitude drawdown value observed over the entire backtest and is a key indicator of a strategy's downside risk:

$$\text{WorstDrawdown}_k = \min_{1 \leq j \leq k} \text{Drawdown}_j.$$

With these concepts in mind, a common practical challenge in financial trading is deciding when to exit a trade to maximise PnL and limit drawdown. Traders frequently use predefined price levels known as *take-profit* (TP) and *stop-loss* (SL) thresholds to automate these exit decisions. A take-profit is a specified price at which a trader will close a typically profitable trade to lock in gains, whereas a stop-loss is a price at which a trade will be closed to prevent further losses. By setting TP and SL levels, traders manage their risk and reward systematically, reducing emotional influences on trading decisions.

These cumulative PnL curves (and their deepest drawdowns) provide a risk-adjusted basis for comparing the baseline linear-regression strategy with its conformal-wrapped variants (as can be seen in Fig. 13).

5.2   Formulation of Backtest Logic per Economic Event Time Series

Now that we have defined our PnL metrics which we will use to assess performance, we will formally define how we convert our conformal prediction (CP) upper- and lower-bounds as respective take-profit TP and stop-loss SL thresholds and evaluate them in a so-called *backtest* simulation; i.e. running our trading strategy rules over historic price time series data and simulating how we would have traded. For each economic event we observe minute mid prices $P(t)$ on the interval $[t_0, t_{\text{eco}}]$ where:

1. the *trade direction* $s \in \{+1, -1\}$ implied by the EcoTrend (long if $s = +1$, short if $s = -1$);
2. the trade *start time* t_0 (the segment start spanning the release);
3. the *event time* t_{eco}; and
4. the LR point forecast y_{pred} together with conformal lower/upper bps bands (lower$_\gamma$, upper$_\gamma$) at confidence $\gamma \in \{90\%, 95\%, 99.9\%\}$.

In Fig. 14, we provide examples showing how we use the original 1-minute raw prices to simulate the execution (fills) of trades placed before each economic release. Note in practice, we hold-off of $\tau_{\mathrm{pause}} = 2$ minutes after the start of the segment to determine whether a *take profit* (TP) or *stop loss* (SL) is hit first; otherwise we exit at the time of the economic release t_{eco}.

Algorithm 6. Baseline LR TP and ConformalLR TP/SL (expanding window, price scan)

Require: start time t_0, event time t_{eco}, direction s, minute prices $P(t)$; LR forecast y_{pred}; conformal bands (lower$_\gamma$, upper$_\gamma$); pause τ_{pause}.

1: Define eligible scan window $\mathcal{T} = \{t \in [t_0, t_{\mathrm{eco}}] : t \geq t_0 + \tau_{\mathrm{pause}}\}$.
2: **Baseline LR:** set $P_{\mathrm{tp}} = P(y_{\mathrm{pred}})$. If the TP condition is met at the earliest $t \in \mathcal{T}$, fill at that t with reason TAKE_PROFIT; otherwise fill at t_{eco} with reason ECO_EXIT.
3: **Conformal-LR at level γ:** map $(\mathrm{TP}_\gamma, \mathrm{SL}_\gamma)$ by direction and convert to $P_{\mathrm{tp}}, P_{\mathrm{sl}}$. Find the earliest $t \in \mathcal{T}$ that satisfies either the TP or SL condition. If neither is hit, fill at t_{eco}. Reasons are then TAKE_PROFIT, STOP_LOSS, or ECO_EXIT.
4: Compute rawBps, netBps and PnL for each method as above.

Let P_0 and P_{fill} represent the price at start time t_0 and fill times t_{fill} respectively. We can convert from basis points (which is what the forecasts from LR and CP models are based on), using this simple function:

$$BpsToPrice(x \text{ bps}) = P_0\left(1 + \frac{x}{10^4}\right)$$

Using this mapping from bps move to price, in our experiments base LR model forecasts y_{pred} can be viewed as TP prices like so:

$$P_{\mathrm{TP}} = BpsToPrice(y_{pred}),$$

For Conformal-LR at confidence level γ we map bands to TP/SL according to the trade direction (*no reversal* for longs):

$$(\mathrm{TP}_\gamma, \mathrm{SL}_\gamma) = \begin{cases} (\mathrm{upper}_\gamma, \ \mathrm{lower}_\gamma), & s = +1 \ (\mathrm{long}), \\ (\mathrm{lower}_\gamma, \ \mathrm{upper}_\gamma), & s = -1 \ (\mathrm{short}). \end{cases}$$

Therefore in our experiments the TP and SL prices for a given confidence level γ are:

$$P_{\mathrm{TP}_\gamma} = BpsToPrice(\mathrm{TP}_\gamma), P_{\mathrm{SL}_\gamma} = BpsToPrice(\mathrm{SL}_\gamma)$$

Using these prices and the raw price history we can create backtest simulations to check if these TP and SL price thresholds are breached. The exact event-level rules are given in Algorithm 6 which are:

1. For the baseline LR, there are two mutually exclusive outcomes per event: {TAKE_PROFIT, ECO_EXIT}.
2. For Conformal-LR at each confidence level γ, there are three: {TAKE_PROFIT, STOP_LOSS, ECO_EXIT}.

Figure 14a(short EcoTrend) shows price meandering before touching the *upper* (adverse) conformal band first, triggering a STOP_LOSS of $\sim$ 11bps; the LR TP would have banked a modest earlier gain of $\sim$ 48bps, and a passive ECO_EXIT would have done better gaining $\sim$ 61bps in that instance. Figure 14b(long EcoTrend) illustrates the converse: the LR TP is hit very quickly capturing just $\sim$ 4bps, while the conformal TP—farther from P_0—lets profits run and locks $\sim$ 39bps before the release which is greater than the default exit of $\sim$ 26bps. This mechanism (fewer premature exits and more TPs) explains the higher terminal CumPnL and shallower drawdowns for the best conformal predictors in Fig. 13.

5.3   What Trade Executions Actually Happen?

Table 3 reports counts and shares of event outcomes by model and confidence. Two robust patterns emerge:

- **Higher $\gamma \Rightarrow$ wider bands $\Rightarrow$ more ECO_EXITs.** As nominal confidence increases from 90% to 99.9%, both symmetric and asymmetric bands widen. The share of ECO_EXIT rises markedly (e.g., ConfLR_Sym: 56% $\rightarrow$ 82%), with corresponding declines in early TP/SL hits. This is expected: larger TP/SL targets require larger pre-release moves to trigger.
- **Asymmetry reduces STOP_LOSS frequency.** Allocating more weight to the heavier tail ($\lambda \neq 0.5$) systematically lowers STOP_LOSS counts at a given γ relative to symmetric bands, at the price of slightly wider intervals. This is consistent with the residual skew observed in several Series and the coverage gains documented in Sect. 4.2.

5.4   PnL Case Studies and Drawdowns

To understand how practically useful the conformal regression bounds were in this context, we examine economic release series-level cumulative PnL and worst drawdowns under identical notional and cost assumptions across the base LR model and all conformal models. Table 4 presents the cumulative PnL for the top 40 economic release series, comparing the baseline LR model against its various conformal-wrapped counterparts. The results reveal several clear and insightful patterns.

Table 3. Counts and shares of event fill reasons by model.

Model & Conf	LRpred	ECO_EXIT	STOP_LOSS	TAKE_PROFIT
BaseLR @ N/A	28,779 (45%)	34,717 (55%)		
ConfLR_Sym @ 90		35,258 (56%)	19,039 (30%)	9,199 (14%)
ConfLR_Sym @ 95		43,030 (68%)	12,876 (20%)	7,590 (12%)
ConfLR_Sym @ 99.9		52,348 (82%)	6,645 (10%)	4,503 (7%)
ConfLR_ASym @ 90		43,030 (68%)	12,876 (20%)	7,590 (12%)
ConfLR_ASym @ 95		47,638 (75%)	9,547 (15%)	6,311 (10%)
ConfLR_ASym @ 99.9		52,405 (83%)	6,621 (10%)	4,470 (7%)
CQR @ 90		30,930 (49%)	23,432 (37%)	9,134 (14%)
CQR @ 95		31,812 (50%)	22,756 (36%)	8,928 (14%)
CQR @ 99.9		32,054 (50%)	22,543 (36%)	8,899 (14%)

Most notably, conformal wrappers significantly enhance performance for the majority of high-frequency, high-impact series. The baseline LR model, while profitable for some series, incurs substantial losses on others, such as USD/Energy/Natural Gas Storage or USD/Manufacturing/ISM Services PMI. In these cases, applying conformal bands transforms a losing strategy into a highly profitable one. The asymmetric version at 99.9% confidence (ConfLR_Asym @ 99.9%) often delivers the best risk-adjusted outcomes, effectively managing the skewed, heavy-tailed residuals typical of financial data by preventing premature stop-losses and allowing profits to run.

However, the benefit of conformal prediction is closely tied to data availability. The method performs well for series with a rich history of over 100 releases (i.e., at least bi-monthly over the 17-year dataset). For these data-rich series, the historical residuals provide a stable basis for calibrating the prediction intervals. Conversely, for data-sparse series, the baseline LR model often remains the best performer. For instance, for releases with very few historical data points, such as NZD/Business/Prelim ANZ Business Confidence (16 events) or CNY/Rates/1-y Loan Prime Rate (17 events), the conformal wrappers fail to add value and can even degrade performance. In these scenarios, the calibration set of residuals is too small to be reliable, and the resulting intervals may be unstable or inappropriately wide, leading to suboptimal exits.

Finally, Table 4 shows that the Conformalised Quantile Regression (CQR) models consistently underperform. As implemented here with raw, uncalibrated bands, CQR tends to produce intervals that are too tight to achieve their nominal coverage. This results in an excessive number of stop-losses, eroding profitability, as seen in the results for series like EUR/Inflation/Italian Prelim CPI m/m, where CQR turns a profitable strategy into a losing one. This highlights the critical importance of the calibration step for CQR, which was omitted here to contrast it with the simpler split-conformal approaches.

Figure 13 presents a compelling case study for the USD/Manufacturing/ISM Services PMI series, where the baseline LR model struggles but is dramatically improved by the application of conformal wrappers. The top panel, showing

Table 4. Top 40 Series by SUM(EventPnL) for ConfLR_Sym @ 99. The **best** PnL in each row is shaded; negatives are in red.

Rank	Series	BaseLR @ N/A	ConfLR Sym @ 95%	ConfLR Sym @ 99.9%	ConfLR ASym @ 95%	ConfLR ASym @ 99.9%	CQR @ 95%	CQR @ 99.9%	Num Eco Events
1	USD/Employment/Unemployment Claims	2,620,482	3,077,388	3,151,597	3,142,281	3,151,597	2,458,814	2,423,433	939
2	CAD/Manufacturing/Manufacturing PMI	984,982	984,982	984,982	984,982	984,982	984,982	984,982	163
3	NZD/Commodities/GDT Price Index	659,916	659,916	659,916	659,916	659,916	659,916	659,916	371
4	USD/Trade/Goods Trade Balance	640,868	640,868	640,868	640,868	640,868	640,868	640,868	111
5	USD/Rates/Federal Funds Rate	536,316	504,000	504,000	504,000	504,000	504,000	504,000	147
6	JPY/Meetings/BOJ Policy Rate	501,714	501,714	501,714	501,714	501,714	501,714	501,714	73
7	USD/Manufacturing/Flash Manufacturing PMI	501,382	501,382	501,382	501,382	501,382	501,382	501,382	152
8	USD/Employment/Average Hourly Earnings m/m	210,675	770,138	1,046,908	895,830	1,032,927	218,711	232,920	216
9	GBP/Monetary Policy/Asset Purchase Facility	471,665	471,665	471,665	471,665	471,665	471,665	471,665	136
10	EUR/Inflation/Italian Prelim CPI m/m	291,083	560,711	663,503	623,694	663,503	-79,371	-79,371	216
11	USD/Employment/Unemployment Rate	253,867	784,829	1,039,671	894,692	1,029,718	498,619	507,930	216
12	USD/Energy/Natural Gas Storage	-2,725,771	1,588,633	2,428,075	1,908,768	2,442,591	70,238	14,151	939
13	CAD/Inflation/Trimmed CPI y/y	348,128	348,128	348,128	348,128	348,128	348,128	348,128	97
14	CAD/Inflation/Median CPI y/y	348,128	348,128	348,128	348,128	348,128	348,128	348,128	97
15	CAD/Inflation/Common CPI y/y	337,830	337,830	337,830	337,830	337,830	337,830	337,830	97
16	USD/Employment/Non-Farm Employment Change	341,216	742,222	1,047,473	933,261	1,033,876	234,092	302,924	216
17	CNY/Manufacturing/Non-Manufacturing PMI	319,656	319,656	319,656	319,656	319,656	319,656	319,656	213
18	GBP/GDP/GDP m/m	318,301	318,301	318,301	318,301	318,301	318,301	318,301	78
19	CHF/Meetings/SNB Policy Rate	270,235	270,235	270,235	270,235	270,235	270,235	270,235	23
20	EUR/Rates/Main Refinancing Rate	45,683	478,824	649,606	529,281	649,606	94,375	94,375	177
21	USD/Manufacturing/ISM Services PMI	-389,465	605,583	927,544	765,390	928,555	-219,786	-219,786	216
22	AUD/Trade/Goods Trade Balance	170,982	170,982	170,982	170,982	170,982	170,982	170,982	31
23	USD/Employment/Labor Market Conditions Index m/m	115,701	115,701	115,701	115,701	115,701	115,701	115,701	34
24	EUR/Economic/ZEW Economic Sentiment	155,021	151,411	330,454	236,392	340,839	48,197	48,197	217
25	CAD/Housing/Housing Starts	-275,652	194,658	377,177	273,341	377,177	-239,953	-280,970	216
26	EUR/Rates/Long Term Refinancing Operation	65,517	65,517	65,517	65,517	65,517	65,517	65,517	14
27	USD/Inflation/Cleveland Fed Inflation Expectations	38,124	38,124	38,124	38,124	38,124	38,124	38,124	2
28	CAD/Employment/ADP Non-Farm Employment Change	37,686	37,686	37,686	37,686	37,686	37,686	37,686	51
29	CHF/Economic/UBS Economic Expectations	34,096	34,096	34,096	34,096	34,096	34,096	34,096	8
30	USD/Spending/Construction Spending m/m	-257,206	243,162	609,498	482,455	609,498	-379,914	-318,361	216
31	AUD/Consumer/Quarterly Retail Sales q/q	-530	-530	-530	-530	-530	-530	-530	2
32	EUR/Employment/German Unemployment Change	-18,565	425,246	810,358	654,368	810,358	466,324	500,746	217
33	CNY/Housing/New Home Prices m/m	-25,703	-37,166	-37,166	-37,166	-37,166	-37,166	-37,166	15
34	CNY/Rates/1-y Loan Prime Rate	637	-47,953	-47,953	-47,953	-47,953	-47,953	-47,953	17
35	CNY/Rates/5-y Loan Prime Rate	-1,441	-47,953	-47,953	-47,953	-47,953	-47,953	-47,953	17
36	GBP/Housing/Prelim Mortgage Approvals	-53,469	-85,634	-85,634	-85,634	-85,634	-85,634	-85,634	18
37	USD/Energy/Crude Oil Inventories	-2,961,873	830,124	1,634,862	1,214,490	1,638,313	186,950	443,286	938
38	AUD/Housing/Housing Starts q/q	-17,009	-111,267	-111,267	-111,267	-111,267	-111,267	-111,267	23
39	NZD/Business/Prelim ANZ Business Confidence	-105,224	-112,730	-112,730	-112,730	-112,730	-112,730	-112,730	16
40	USD/Manufacturing/Chicago PMI	-328,157	270,212	633,313	537,815	633,494	-823,480	-813,826	216

cumulative PnL, reveals that the baseline LR model incurs steady losses (-\$250k) over the entire sample period. In stark contrast, the conformal methods transform the strategy into a profitable one making over \$850k, with the asymmetric variant at high confidence, *ConfLR_Asym @ 99.9*, showing a particularly strong and steady positive trend from 2014 onwards.

The mechanics behind this improvement are detailed in the other panels. The middle panel shows that for the best-performing models, there is a healthy frequency of TAKE_PROFIT events ($\triangle$) relative to the number of STOP_LOSS hits ($\times$). This indicates that the wider, direction-aware bounds successfully allow profitable trades to run while filtering out premature exits on noise. The profound impact on risk is quantified in the bottom panel, which plots the drawdown. While the baseline LR and CQR models suffer severe maximum drawdowns exceeding \$600k, the best conformal models contain this risk to less than half that amount. This case study effectively demonstrates that for a challenging series such as this, conformal prediction can provide a risk management framework that turns an unprofitable signal into a viable and robust trading strategy.

Table 5. SUM(EventPnL) by *EcoEventCategory* and model. The **best** PnL in each row is shaded; negatives are in red.

Rank	EcoEventCategory	Base LR (N/A)	ConfLR_Sym @99.9%	ConfLR_ASym @99.9%	CQR @99.9%	Num Eco Events
1	Manufacturing	-22,655,448	6,862,294	6,985,162	-39,766,535	10,970
2	Energy	-5,687,643	4,062,937	4,080,904	457,437	1,877
3	Inflation	-13,992,649	2,956,542	3,019,687	-29,476,551	6,960
4	Employment	-10,092,959	2,274,831	2,182,901	-18,266,117	6,731
5	Commodities	-927,946	797,341	811,509	-2,511,654	1,020
6	Meetings	771,949	771,949	771,949	771,949	96
7	Spending	-257,206	609,498	609,498	-318,361	216
8	Monetary Policy	471,665	471,665	471,665	471,665	136
9	Money Supply	-4,596,008	304,989	306,402	-10,646,204	2,157
10	Rates	-1,879,313	250,012	245,239	-4,183,868	1,183
11	Services	-1,503,172	-148,682	-138,867	-4,458,083	799
12	Policy	-2,624,026	-630,741	-622,074	-2,809,892	648
13	Trade	-11,405,065	-800,053	-766,323	-23,855,981	4,937
14	Production	-14,422,888	-1,025,831	-1,003,115	-22,019,995	5,259
15	Economic	-9,008,882	-1,389,984	-1,334,996	-13,462,101	3,214
16	Business	-4,189,794	-1,391,027	-1,374,983	-7,779,239	1,585
17	Consumer	-16,355,241	-1,595,594	-1,446,093	-28,881,809	6,739
18	Vehicles	-2,009,250	-1,936,057	-1,935,787	-3,785,401	453
19	Investment	-3,595,682	-1,945,042	-1,951,840	-5,679,243	916
20	Housing	-14,919,324	-2,354,789	-2,319,317	-23,525,507	5,288
21	GDP	-5,046,615	-7,443,849	-7,377,322	-14,112,945	2,312

In the next section we aggregate these backtest results from an economic release series level into per-economic release category, looking at CumPnL and WorstDD statistics and comparing the results from the Conformal-LR strategies with the baseline LR point-forecast rule.

5.5 Category-Level Aggregation

Table 5 aggregates economic release PnL across broad categories (Manufacturing, Energy, Inflation, Employment, Rates, Trade, Production, Consumer, etc.). The pattern mirrors that seen in USM Manufacturing case study seen in Fig. 13:

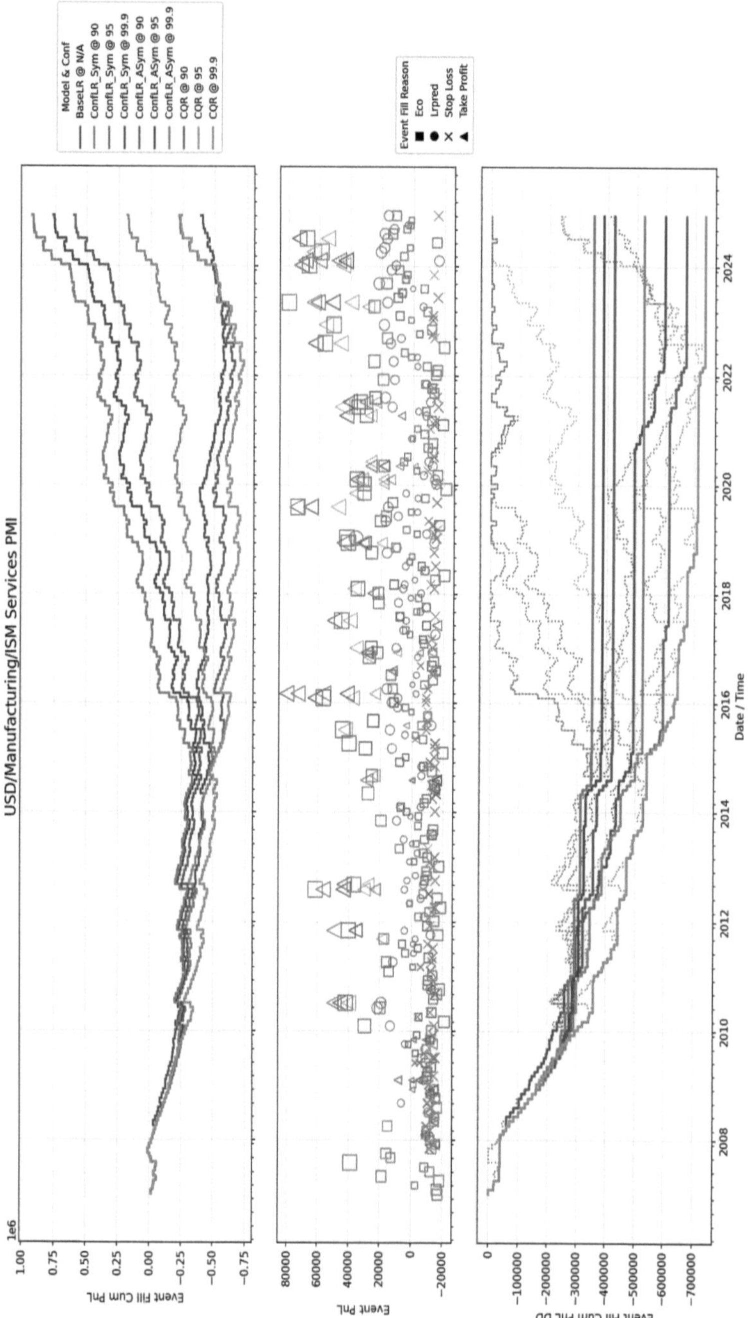

Fig. 13. USD/Manufacturing/ISM Services PMI. The baseline remains negative. Conformal LR improves outcomes materially; *ConfLR_Asym @ 99.9* again dominates with a steady ascent and contained drawdowns, while Sym 90/95 and CQR 90/95 are more fragile. Event PnL markers show a healthy mix of TPs with relatively fewer SLs for the best lines. The combination of the same LR point forecast and wider, direction-aware bounds yields more profitable exits for this PMI series.

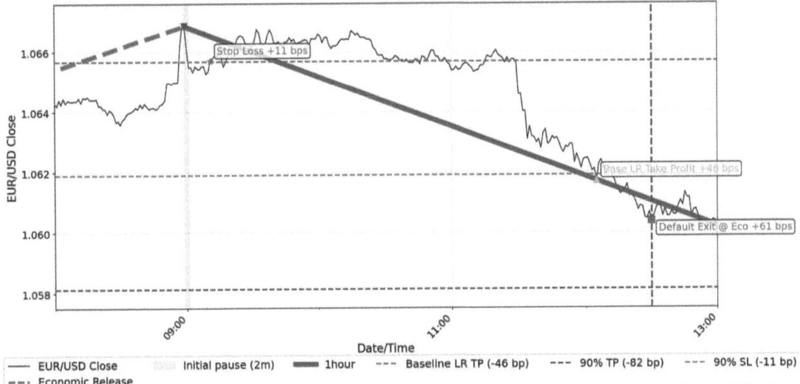

(a) Stop-loss example (USD Employment Cost Index q/q, 2023-10-31 12:30).

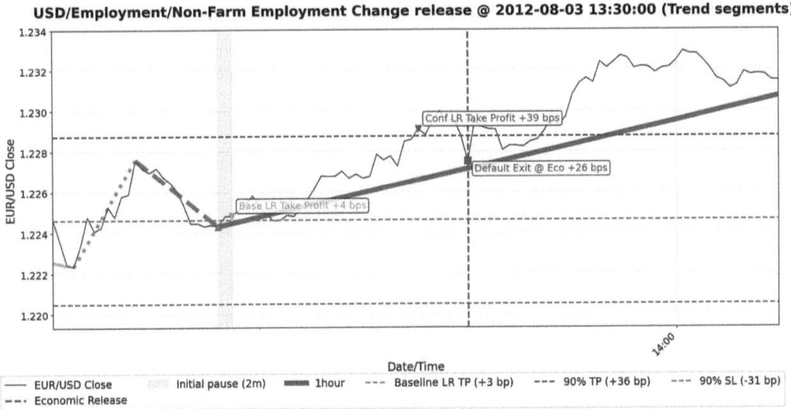

(b) Take-profit example (USD Non-Farm Employment Change, 2012-08-03 13:30).

Fig. 14. Illustrative fills using conformal LR bands on 1-minute EUR/USD prices. The thick blue line is the 1-hour (21 bps) EcoTrend segment; the vertical red dashed line is the scheduled release. **(a) Stop-loss:** The surrounding segment is short. Price stalls around 1.066 for more than two hours; the conformal upper band (orange dashed) is touched first, triggering STOP_LOSS, whereas the baseline LR TP (green dashed) exits earlier at TAKE_PROFIT with $\approx$ +48 bps. A passive ECO_EXIT would have captured $\approx$ +61 bps. **(b) Take-profit:** The segment is long. The baseline LR TP is reached very quickly, banking only $\approx$ +4 bps, whereas the conformal TP (purple dashed) lets the trade run and locks in $\approx$ +39 bps, beating the default ECO_EXIT ($\approx$ +26 bps). (Color figure online)

– **Where bands help most.** Categories prone to pre-release oscillations and larger shocks (e.g., *Manufacturing, Energy, Employment*) tend to improve materially when moving from the LR point-TP to conformal bands, with asymmetric high-γ settings often topping the category.

- **Where benefits are muted.** *Rates/Policy* categories show limited dispersion across models because pre-release prices cross TP/SL thresholds less frequently; exits cluster at ECO_EXIT.
- **Inflation/Trade/Production.** Mixed behaviour: some series are learnable and reward wider bands; others are regime-sensitive, where tight symmetric bands can be breached frequently (excess STOP_LOSS).

Conformal wrappers are not detrimental to the base LR model's point estimate performance, R^2, MAE, and CD which are essentially unchanged, but they allow us to tune widthcoverage control (via the confidence level) that maps directly into execution outcomes: TAKE_PROFIT, STOP_LOSS or ECO_EXIT, increasing cumulative PnL trajectories, and minimising drawdown profiles. For economic release series with skewed or heavy-tailed residuals, asymmetric high-confidence bands offer a robust default. For slow, calendar-dominated economic release series, simple LR point-TP rules suffice, as most exits occur at the release time.

6 Conclusions: Summary, Limitations and Future Directions

This chapter has shown that a straightforward combination of event-aligned price segmentation, linear regression and conformal prediction can generate statistically sound and practically actionable price-movement bounds around scheduled macroeconomic announcements. Using more than seventeen years of high-frequency EUR/USD data and a historical economic release calendar, we demonstrated that an online retracement-based segmentation compresses millions of minute-by-minute prices into compact, interpretable structures without sacrificing the essential directional dynamics that drive market behaviour.

The study's central finding is that simple, transparent models—when paired with conformal prediction—can provide empirically valid uncertainty bounds that translate naturally into trading actions such as take-profit and stop-loss levels. Conformal wrappers, particularly asymmetric variants, improved both calibration and risk-adjusted profitability across many economic release series by widening prediction bands to accommodate market asymmetries and heavy-tailed residuals. The analysis further highlighted that performance depends strongly on data richness: series with longer, regular histories deliver more reliable calibration and superior trading outcomes.

6.1 Addressing Limitations: Early Detection of Economic Release Price Segments and Trading Post Economic Announcements

Whilst this study demonstrated that retrospective, event-aligned segmentation provides a tractable way to structure noisy price data around macroeconomic announcements, it also exposed several operational limitations when considering live implementation. Chief among these is the reliance on *ex-post* identification of the segment spanning an event, defined as the EcoTrend in Sect. 2, equations (1)–(3). In practice, traders must operate under real-time constraints, where the precise start and end of a price leg are unknown until after the fact. Consequently, there is an inevitable delay between a true market turning point and its formal confirmation under the existing retracement rule.

To mitigate this, Algorithm 7 proposes an *early-turn detector* that integrates multiple retracement resolutions to infer provisional EcoTrends prior to full confirmation. The algorithm runs two concurrent online segmenters—a finer "micro" scale and a coarser "macro" scale—and uses their combined state to promote a macro leg to a provisional EcoTrend, when short-term reversals coincide with high retracement oscillation on the longer horizon. In essence, this acts as an anticipatory signal that a genuine directional change is forming before it is formally locked in. Such early detection could significantly reduce latency between market reversals and model recognition, improving pre-release positioning and enabling more responsive dynamic calibration of conformal prediction intervals.

Algorithm 7. Early-turn detector for provisional EcoTrend (two-resolution tracker)

Require: thresholds $\theta_{\text{micro}} < \theta_{\text{macro}}$; trigger $\tau \in [0.7, 0.9]$; stream of 1-minute mid-prices $P(t)$

1: Initialize two online segmenters using Alg. 1: $\mathsf{S}_{\text{micro}} \leftarrow \text{ZIGZAG}(\theta_{\text{micro}})$, $\mathsf{S}_{\text{macro}} \leftarrow \text{ZIGZAG}(\theta_{\text{macro}})$.

2: Set *provisional* $\leftarrow$ **false**; unset (t_s, P_s).

3: **for** each new minute t with price $P(t)$ **do**

4:     Update both segmenters with $(t, P(t))$.

5:     Compute the macro oscillator

$$o_{\text{macro}}(t) \leftarrow \frac{|P_{\text{macro}}^{\text{ext}} - P(t)|}{\theta_{\text{macro}}}$$

where $P_{\text{macro}}^{\text{ext}}$ is the current extreme tracked by $\mathsf{S}_{\text{macro}}$.

6:     **if** $\mathsf{S}_{\text{micro}}$ **confirms** a flip ($\uparrow \leftrightarrow \downarrow$) **and** $o_{\text{macro}}(t) \geq \tau$ **then**

7:         Promote the currently active macro leg to a *provisional EcoTrend*; freeze its start (t_s, P_s).

8:         Set *provisional* $\leftarrow$ **true**.

9:     **if** *provisional* **and** $\mathsf{S}_{\text{macro}}$ **confirms** a new leg before t_{eco} **then**

10:         Roll forward (t_s, P_s) to the new macro leg; refresh the pre-release features.

11:     **if** $t = t_{\text{eco}}$ **then**

12:         Define the *EcoTrend* as the macro leg active at t_{eco}.

13:         Record Δ_{before} using Eqs. (1)(3); **break.**

Further refinements could employ probabilistic decision thresholds or reinforcement learning techniques to adapt θ_{micro}, θ_{macro}, and τ dynamically based on volatility regimes or liquidity conditions. The multi-resolution logic aligns naturally with event-time learning, as it allows for "soft" detection of regime shifts in the minutes leading up to a macroeconomic announcement. In doing so, it bridges the gap between the statistical robustness of ex-post segmentation and the practical exigencies of real-time execution.

Beyond early detection, an equally important extension concerns the trading period *after* an economic release. The present analysis focused primarily on forecasting and trading *before* scheduled events, modelling pre-release price behaviour as a function of prior EcoTrend dynamics. However, numerous studies have shown that post-announcement price trajectories often exhibit complex, asymmetric adjustments and partial mean reversion once the initial informational shock has been absorbed by the market [11,23,24,29].

For instance, [11] identify strong state-dependent asymmetries in how macroeconomic surprises affect intraday exchange rates, while [23,24] show that market activity and volatility after news releases can be modelled as self-exciting "intensity bursts" using Hawkes processes. These bursts correspond to cascades of reactions by heterogeneous traders, where initial shocks are often followed by periods of adjustment and stabilisation. More recently, [29] simulated order flow dynamics under heterogeneous agent responses, finding that institutional and retail traders react at different timescales, producing layered waves of liquidity withdrawal and return. Together, this literature suggests that post-release dynamics are not merely noise, but structured processes that can be modelled and exploited.

A promising extension of the current framework would therefore mirror the pre-release formulation but apply it to the post-release EcoTrend (the Δ_{after} component of Eq. 3). This would allow the model to quantify uncertainty around the subsequent normalisation or overshoot of prices after major announcements. From a trading perspective, this corresponds to identifying opportunities for short-term mean reversion once the initial burst of volatility subsides, an area where conformal prediction's distribution-free guarantees could provide calibrated confidence bands for both breakout and reversion strategies.

Finally, future work should incorporate more realistic market frictions into these analyses. Our current backtests assume mid-market execution without accounting for transaction costs, bidask spreads, or slippage. Integrating these into simulations would improve the operational realism of profitability estimates. Similarly, adaptive recalibration of conformal intervals under regime changes—perhaps using rolling or weighted residual buffers—would help maintain reliable coverage rates despite evolving market conditions.

We next consider how the segmentation methodology introduced here could be applied more broadly to other financial time series beyond exchange rates.

6.2   Application of Segmentation to Other Financial Time Series

While the previous sections focused primarily on the segmentation of mid-price data to study event-aligned dynamics, the same methodology can be extended to a much wider range of financial time series that drive or reflect trading behaviour. The online retracement segmentation introduced earlier is a general-purpose, model-free transformation that identifies intrinsic turning points and compresses high-frequency data into interpretable regimes. Crucially, it makes no assumptions about the semantic meaning of the underlying variable—whether it represents price, position, volume, or cumulative PnL—making it particularly suitable for multi-stream analysis in market microstructure research [4,14].

Figure 15 illustrates this cross-domain generality. Each panel depicts a different anonymised series from a trading system, segmented using the same retracement thresholds as those applied to the EcoTrend framework. The resulting piecewise-linear representation compresses thousands of observations into a few dozen regimes without losing essential structure. This provides a unified view across disparate metrics—inventory changes (*NetPosUsd*), realised profitability (*CumNetPnLUsd*), execution intensity (*AbsVolumeUsd*), and market prices (*Price*)—all expressed in a common intrinsic-time basis.

Such alignment opens new analytical possibilities. It enables questions that cut across execution, risk, and behavioural dimensions, for example:

- Do inventory adjustments systematically precede directional price shifts, implying anticipatory trading or information advantage?
- Is the curvature or convexity of execution volume within a segment predictive of volatility clustering around macroeconomic events?
- How do drawdowns or risk exposure (*CumNetPnLUsd*) evolve within identified structural regimes relative to price cycles?

By unifying these data under the same segmentation framework, one can derive multi-signal event studies in which price, volume, and PnL are synchronised in both time and state. This supports richer inference on market microstructure phenomena and the behaviour of heterogeneous agents, linking observable trading patterns to the underlying event cycle [4,8].

More broadly, this generalisation demonstrates that segmentation is not simply a data compression technique but a bridge between time-series abstraction and interpretability. It transforms continuous-valued, high-frequency signals into structured symbolic sequences suitable for both quantitative and qualitative analysis—including visualisation, clustering, and even auditory rendering.

The next subsection explores this final theme by applying segmented representations to *data sonification*, where changes in financial signals can be expressed audibly. This approach offers an intuitive means of detecting structural transitions, potentially enabling new forms of interactive analytics and trader decision support.

6.3   Segmentation for Data Sonification

An unexpected but enjoyable side project that emerged from developing the visualisations in Fig. 6 involved exploring how segmented time series could be expressed not only visually but also sonically. As we animated the evolving price segments, it became apparent that the discretised structure—comprised of turning points, direction changes, and relative magnitudes—possessed an intrinsic

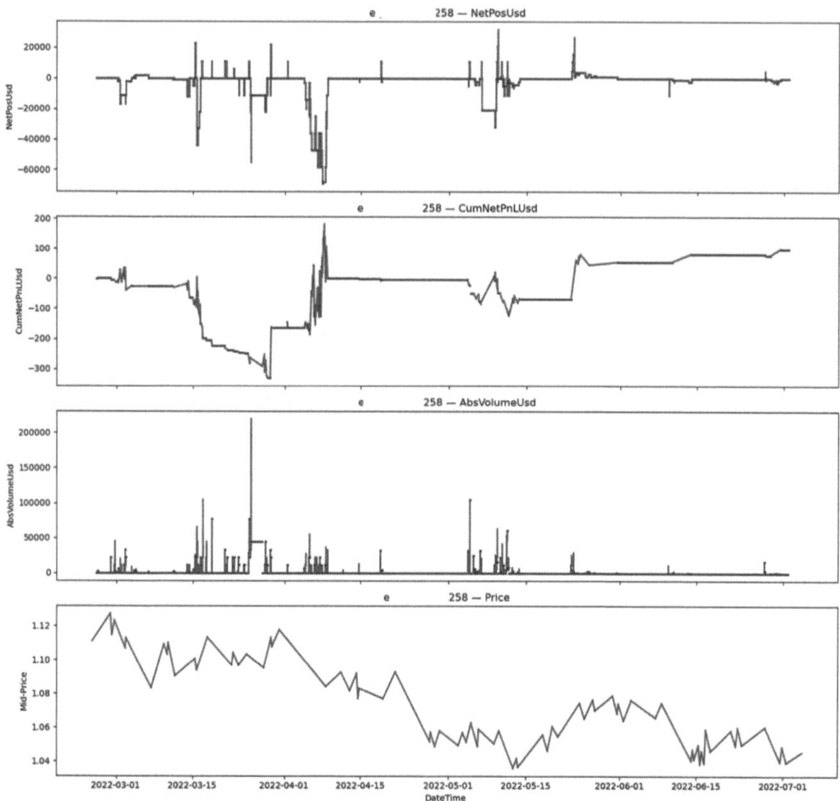

Fig. 15. Segmenting non-price financial time series (anonymised example). Each panel shows the raw series (green/red bars) with fitted piecewise-linear segments overlaid in blue. From top to bottom: *NetPosUsd* (net trading inventory), *CumNetPnLUsd* (cumulative profit and loss), *AbsVolumeUsd* (execution intensity), and *Price* (mid-quote). The segmentation algorithm captures local structural changes consistently across these heterogeneous data streams, enabling aligned, interpretable representations for downstream analysis. (Color figure online)

rhythm that could naturally lend itself to musical interpretation. This observation inspired an experiment in *data sonification*, in which financial data were transformed into sound and, ultimately, into fully composed musical pieces.

The process, described in greater detail at the URL below, followed three main stages:

https://algolabs.com/data-sonification

First, segmented price trajectories and event markers were algorithmically converted into structured MIDI sequences. Each economic releases currency was mapped to a specific instrument family, and different macroeconomic event categories—such as GDP, inflation, or employment—triggered unique rhythmic or percussive elements. The magnitude and direction of price changes determined melodic intervals and harmonic intensity, effectively encoding market volatility as musical tension and release.

Second, these MIDI structures served as inputs to generative AI composition models, guided through natural-language prompting via the Suno AI music platform[2]. Human supervision played a central role in specifying the desired composition styles, refining harmonic structure, and iteratively reviewing outputs. Through this interactive process, the AI expanded the data-driven motifs into rich, layered musical arrangements, with each piece given a distinct identity and title. In total, over seventy musical segments were produced, each reflecting the dynamics and emotional character of the underlying financial data.

Finally, the resulting musical tracks were synchronised with the original price-segmentation animations to create an immersive audio-visual narrative—an example of which depicts the unfolding of the 2008 financial crisis as a three-hour symphonic journey through two years of market data. The experience revealed a striking parallel between the cyclical patterns of market movements and the emotional arcs inherent in musical form: both are structured sequences of tension, release, and transformation.

While this sonification project lies outside the conventional domain of econometric modelling, it underscores the expressive power of segmentation as a unifying abstraction. By translating market structure into sound, it bridges quantitative analysis and artistic interpretation, demonstrating how structured data representations can engage not only analytical reasoning but also sensory and creative perception. What began as an exploratory animation evolved into an interdisciplinary experiment—a reminder that clarity in data often transcends medium, and that even the most technical models can inspire new ways of listening to the patterns that drive financial systems.

In closing, this chapter has shown that segmentation and conformal prediction provide both a rigorous statistical framework and a flexible foundation for creative exploration. Whether applied to forecasting price movements, detecting early structural shifts, or transforming markets into music, the underlying principle remains the same: by imposing structure on complexity, we make uncertainty interpretable—and in doing so, we open new pathways for understanding, decision-making, and even artistic expression (Fig. 16).

---

[2] https://suno.com.

A personal note. The first author, **David Lindsay**, is deeply grateful to **Alex Gammerman** for his teaching, support, and enduring belief over the past 25 years. Alex's encouragement helped shape David's career at the intersection of academia and industry, without which this chapter could not have been written. As a "then and now" coda, the left image shows Alex presenting David with the Best Computer Science Finalist Medal in 2002 at Royal Holloway University of London; the right image is from a reunion at COPA, Brighton, UK (July 2022), where together with David's PhD student Najim al-Baghdadi, they presented a paper entitled *Online Portfolio Hedging with the Weak Aggregating Algorithm* [1].

Fig. 16. "Then and now" reunion photos, twenty years apart.

References

1. Al-Baghdadi, N., Kalnishkan, Y., Lindsay, D., Lindsay, S.: Online portfolio hedging with the weak aggregating algorithm. In: Johansson, U., Boström, H., Nguyen, K.A., Luo, Z., Carlsson, L. (eds.) Proceedings of the Eleventh Symposium on Conformal and Probabilistic Prediction with Applications (COPA 2022). Proc. Mach. Learn. Res., vol. 179, pp. 149–168. PMLR (2022)
2. Andersen, T.G., Bollerslev, T., Diebold, F.X., Vega, C.: Micro effects of macro announcements: real-time price discovery in foreign exchange. Am. Econ. Rev. **93**(1), 38–62 (2003). https://doi.org/10.1257/000282803321455151
3. Angelopoulos, A.N., Bates, S.: A gentle introduction to conformal prediction and distribution-free uncertainty quantification. arxiv preprint arXiv:2107.07511 (2021)
4. Bacry, E., Mastromatteo, I., Muzy, J.-F.: Hawkes processes in finance. Market Microstruct. Liquidity **1**(1), 1550005 (2015). https://doi.org/10.1142/S2382626615500057
5. Bailey, D.H., Borwein, J.M., López de Prado, M., Zhu, Q.J.: The deflated Sharpe ratio: correcting for selection bias, backtest overfitting, and non-normality. J. Portfolio Manag. **40**(5), 94–107 (2014). https://doi.org/10.3905/jpm.2014.40.5.094
6. Balduzzi, P., Elton, E.J., Green, T.C.: Economic news and bond prices: evidence from the U.S. Treasury market. J. Financ. Quant. Anal. **36**(4), 523–543 (2001). https://doi.org/10.2307/2676223

7. Barber, R.F., Candès, E.J., Ramdas, A., Tibshirani, R.J.: Predictive inference with the jackknife+. Ann. Stat. **49**(1), 486–507 (2021). https://doi.org/10.1214/20-AOS1965

8. Bouchaud, J.-P., Bonart, J., Donier, J., Gould, M.: Trades, Quotes and Prices: Financial Markets Under the Microscope. Cambridge University Press, Cambridge (2018)

9. Dacorogna, M.M., Gençay, R., Müller, U.A., Pictet, O.V., Olsen, R.B.: An Introduction to High-Frequency Finance. Academic Press, San Diego (2001)

10. Ederington, L.H., Lee, J.H.: How markets process information: news releases and volatility. J. Finance **48**(4), 1161–1191 (1993). https://doi.org/10.1111/j.1540-6261.1993.tb04750.x

11. Fatum, R., Hutchison, M.M., Wu, T.: Asymmetries and state dependence: the impact of macro surprises on intraday exchange rates. J. Jpn. Int. Econ. **26**(4), 542–560 (2012). https://doi.org/10.1016/j.jjie.2012.08.004

12. Faust, J., Rogers, J.H., Wang, S.Y.B., Wright, J.H.: The high-frequency response of exchange rates and interest rates to macroeconomic announcements. J. Monet. Econ. **54**(4), 1051–1068 (2007). https://doi.org/10.1016/j.jmoneco.2006.05.015

13. Glattfelder, J.B., Dupuis, A., Olsen, R.B.: Patterns in high-frequency FX data: discovery of 12 empirical scaling laws. Quant. Finance **11**(4), 599–614 (2011). https://doi.org/10.1080/14697688.2010.481632

14. Hardiman, S.J., Bercot, N., Bouchaud, J.-P.: Critical reflexivity in financial markets: a Hawkes process analysis. Eur. Phys. J. B **86**(10), 1–9 (2013). https://doi.org/10.1140/epjb/e2013-40107-3

15. Hasbrouck, J.: Empirical Market Microstructure: The Institutions, Economics, and Econometrics of Securities Trading. Oxford University Press, Oxford (2007)

16. HistData.com: EUR/USD 1-minute quotes. https://www.histdata.com/ (2025)

17. King, M.R., Osler, C.L., Rime, D.: The foreign exchange market. In: James, J., Marsh, I.W., Sarno, L. (eds.) Handbook of Exchange Rates, pp. 3–44. Wiley, Hoboken (2012)

18. Kuttner, K.N.: Monetary policy surprises and interest rates: evidence from the Fed funds futures market. J. Monet. Econ. **47**(3), 523–544 (2001). https://doi.org/10.1016/S0304-3932(01)00055-1

19. Lei, J., G'Sell, M., Rinaldo, A., Tibshirani, R.J., Wasserman, L.: Distribution-free predictive inference for regression. J. Am. Stat. Assoc. **113**(523), 1094–1111 (2018). https://doi.org/10.1080/01621459.2017.1307116

20. Lyons, R.K.: The Microstructure Approach to Exchange Rates. MIT Press, Cambridge (2001)

21. MacKinlay, A.C.: Event studies in economics and finance. J. Econ. Lit. **35**(1), 13–39 (1997). https://www.jstor.org/stable/2729691

22. López de Prado, M.: Advances in Financial Machine Learning. Wiley, Hoboken (2018)

23. Rambaldi, M., Filimonov, V., Lillo, F.: Detection of intensity bursts using Hawkes processes: an application to high-frequency financial data. Phys. Rev. E **97**, 032318 (2018). https://doi.org/10.1103/PhysRevE.97.032318

24. Rambaldi, M., Pennesi, P., Lillo, F.: Modeling FX market activity around macroeconomic news: a Hawkes process approach. Phys. Rev. E **91**, 012819 (2015). https://doi.org/10.1103/PhysRevE.91.012819

25. RealTrading.com: Zigzagindicator. https://realtrading.com/trading-blog/zig-zag-indicator/ (2025)

26. Robots4Forex: Historical forex economic calendar (2007–present). https://robots4forex.com/historical-forex-economic-calendar-2007-present-csv-format/. Accessed 20 Jan 2025
27. Romano, Y., Patterson, E., Candès, E.J.: Conformalized quantile regression. In: Advances in Neural Information Processing Systems (NeurIPS), vol. 32 (2019)
28. Vovk, V., Gammerman, A., Shafer, G.: Algorithmic Learning in a Random World. Springer, New York (2005)
29. Wang, H.K.: Heterogeneous trader responses to macroeconomic surprises: simulating order flow dynamics. arxiv preprint arXiv:2505.01962 (2025)
30. Xu, C., Xie, Y.: Conformal prediction intervals for time series. arxiv preprint arXiv:2104.07795 (2021)

Reliable Train Delay Forecasting
with Conformal Prediction

Xu Feng$^{(\boxtimes)}$, Khuong An Nguyen , and Zhiyuan Luo

Department of Computer Science, Royal Holloway University of London, Egham,
Surrey TW20 0EX, UK
Xu.Feng@rhul.ac.uk

Abstract. Train delays cause significant economic and operational
impacts. Accurate prediction of these delays is therefore essential for
improving railway service reliability and supporting effective decision-
making. Thus, this chapter introduces a novel approach combining tree-
based machine learning (ML) models with Conformal Prediction (CP)
to estimate train delays by predicting train travel times between consecutive stations. Using real-world data from over 12.8 million production
service records collected over a period of 3 years and 8 months, the proposed approach achieved a prediction accuracy of 20 s, 90% of the time.
Furthermore, CP delivers statistically valid prediction intervals with an
average width of 19.3 s at a 90% confidence level, successfully meeting
the target coverage as demonstrated by empirical results.

Keywords: Train delay prediction · Timetabling optimisation ·
Conformal Prediction

1 Introduction

The railway system is a key component of modern transportation infrastructure
that plays a pivotal role in connecting people and goods over long distances [1, 44,
58, 59]. One of the most critical aspects of the railway operations is punctuality,
as it represents the reliability and integrity of the entire system [48, 58–60, 68].
Punctuality, in the context of railway systems, refers to the ability of trains
to operate according to the timetable with minimal deviation [36]. However,
high variability in real-world railway operations and unexpected disruptions such
as weather conditions, train malfunctions and infrastructure issues like signal
failures often result in train delays [75]. Train delays may result in economic
losses, passenger inconvenience and unsatisfactory, and a cascade of congestion
across the entire railway network [71].

Given the far-reaching impacts of train delays, predicting and mitigating
them is of paramount importance. Over the past decades, there has been a concerted effort to employ various Machine Learning (ML) models and methods to
train delay predictions. In [57], XGBoost combined with Bayesian Optimisation

K. An Nguyen and Z. Luo (Eds.): Alexander Gammerman Festschrift, LNCS 16290, pp. 387–410, 2026.
https://doi.org/10.1007/978-3-032-15120-9_17

were utilised to predict train arrival delays based on spatio-temporal, operational, and infrastructure data. A long short-term memory (LSTM) model was proposed in [69], leveraging train operational features to predict train arrival delays. In [61], the researchers used artificial neural networks, random forest regression, gradient boosting regression with features from planned and actual train operation data to predict arrival and departure times at each main station.

However, train journeys do not happen in isolation. Thus, modelling the train delay prediction problem requires capturing the sequential relationships between stations along the route, which complicates dataset construction and model training. In addition, current train delay prediction models lack the ability to provide meaningful confidence measures for their predictions, making it difficult for railway operators to assess the reliability of the forecasts and potentially leading to suboptimal operational decisions.

To this end, this chapter proposes a novel train delay prediction model that leverages Machine Learning (ML) methods and Conformal Prediction (CP) to estimate train travel times between stations, thereby generating reliable delay forecasts at the following stations. In this way, the proposed approach preserves the dependencies between train arrival and departure events, as well as the sequential relationship between stations, while ensuring efficient dataset construction and model training. Furthermore, conformal prediction provides prediction intervals with user-specified confidence level for each delay estimate, enabling uncertainty quantification and more transparent and trustworthy decision-making. The contributions of this study are summarised as follows:

- We propose a novel machine learning-based approach for train delay prediction that leverages operational data to estimate the travel time between consecutive stations. This model preserves the dependencies between train events and the sequential relationships between stations, while also ensuring efficient dataset construction and model training.
- We leverage conformal prediction to generate rigorous, statistically valid prediction intervals for each train delay forecast, thereby enhancing the confidence, reliability, transparency, and trustworthiness of the decision-making process in railway systems.
- We validate the performance of the proposed machine learning model using real-world, large-scale train operational data, comprising 12,840,590 train service records collected over a period of 3 years and 8 months. Furthermore, we compare the performance of the most widely used machine learning models in the literature, offering an in-depth analysis of their reliability for train delay prediction.

The remainder of this paper is structured as follows: Sect. 2 overviews the most popular train delay prediction approaches. Section 3 provides a detailed description of the problem formulation and the system architecture. Section 4 offers in-depth introduction to the machine learning model used for train delay prediction and the conformal prediction framework. The empirical experiments and performances analysis are presented in Sect. 5. Finally, Sect. 6 concludes our work and outlines future work.

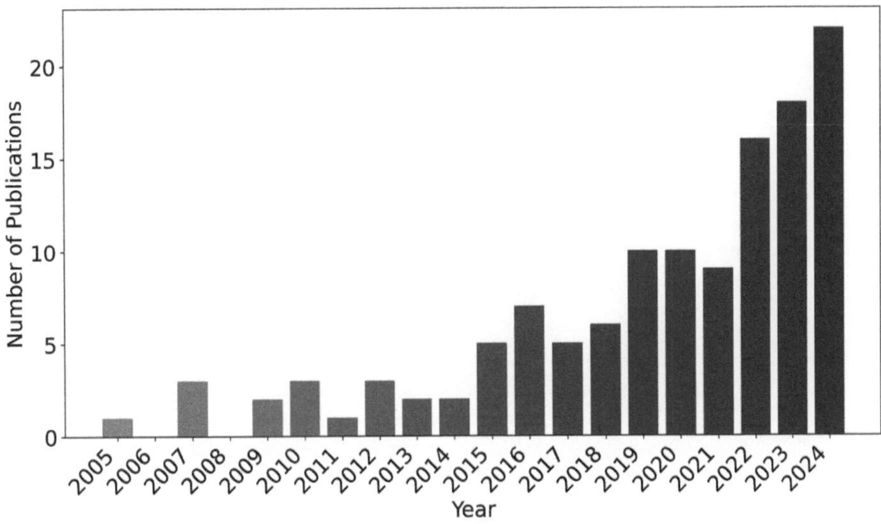

Fig. 1. Number of published papers on train delay prediction by year.

2   Literature Review

Over the past two decades, there has been a continuously growing trend in the number of train delay prediction contributions in the academic world, as discussed in Sect. 1 and illustrated in Fig. 1. The most popular train delay prediction approaches in the literature can be categorised into two groups, namely the event-driven models and data-driven models [58]. Event-driven approaches explicitly model the dependencies between the different train events (e.g., arrivals, departures, dwells, passing-by, etc.) and create a chain or network of events, capturing how delays propagate through the railway system. In contrast, data-driven approaches use historical data and machine learning or statistical methods to directly predict the delay at a given future point or station, based on available features (e.g., current delay, infrastructure status, weather information, etc.) in a single step. Event-driven approaches are generally more interpretable, as the operator can trace how delays propagate through the railway system, while data-driven methods typically rely more on large datasets and can capture complex hidden relationships in the data.

Consequently, typical event-driven train delay prediction approaches employ models such as Graph Models, Markov Chain (MC) models, Bayesian Networks (BN), and Equation Systems (EQS), among others, to highlight the dependencies between various train events during prediction. To explore the underlying patterns in the large-scale datasets, machine learning models like Neural Networks (NN), Random Forests (RF), Decision Trees (DT), Support Vector Machines (SVM), Linear Regression (LNR) were employed for data-driven methods. The proportions of different prediction models among the included references in the literature are shown in Fig. 2. To provide a comprehensive comparison of the approaches in the literature, we categorise and analyse them using Mean Absolute Error (MAE) and Root Mean Squared Error (RMSE), as shown in Table 1.

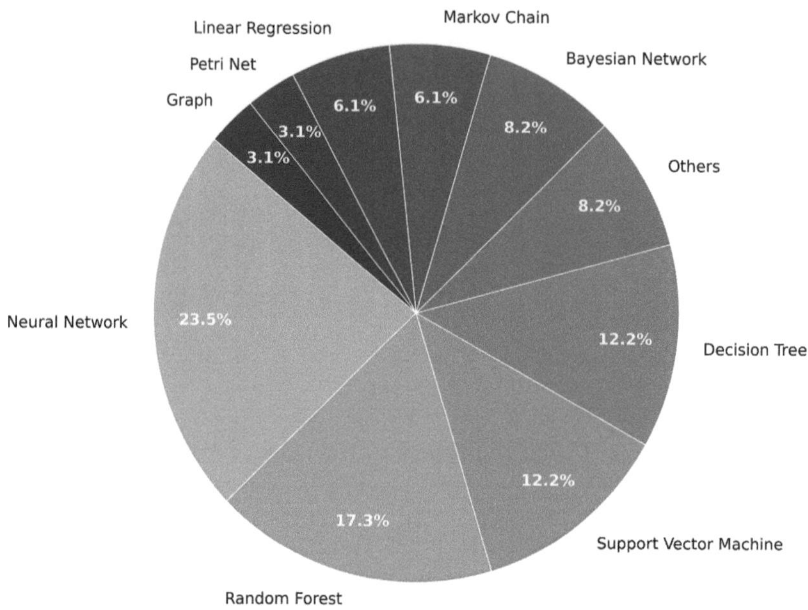

Fig. 2. Proportions of different prediction models in the literature.

Event-driven approaches heavily rely on the logic chains of the train events along the route, thus building and calibrating the dependency structure is time-consuming and challenging [13,42,58]. Upon using the highly specified data structure, event-driven methods may not fully utilise the richness of historical data beyond parameter fitting and may produce a less flexible prediction model that struggles with capturing complex, hidden relationships in the datasets. Data-driven methods can generate the train delay prediction directly with promising accuracies. However, the black-box training makes the final estimation less interpretable for real-world operators.

To address this issue, this chapter proposes a novel machine learning-based approach incorporating the most widely used tree-based models and conformal prediction [38,40,41]. The proposed approach leverages historical train operational data to predict the travel time between the current station and the next, rather than focusing solely on train delays. By estimating the train travel time between consecutive stations, the approach captures the dependencies between train departure and arrival events, which can then be used to more accurately estimate train delays at following stations along the route. At the same time, conformal prediction provides prediction intervals for estimated train travel time that helps train operators take proactive measures to stay on schedule with confidence. Leveraging the underlying patterns in large-scale, real-world datasets

Table 1. Comparisons of different train delay prediction approaches in the literature.

Paper	Model	Country	Data size	Metrics	Prediction Error	Prediction horizon
[17]	BN	CHN	98,982	MAE, RMSE	MAE 3.79 min, RMSE 9.03 min	Multiple
[28]	BN	CHN	378,510	MAE, MSE, RMSE	MAE 30 sec	Multiple
[9]	BN	SWE	486,000	MAE	1.40 min	Multiple
[33]	BN	N/A	N/A	N/A	N/A	One
[75]	BN	NLD	1,307	RMSE	100.15 min	One
[72]	EQS	N/A	N/A	MAE	33.16 sec	One
[15]	MC	IND	194,400	RMSE	11.75 min	One
[52]	MC	TUR	421	N/A	N/A	Multiple
[23]	MC	NLD	6,803	MAE	2.40 min	Multiple
[5]	MC	CHE	17,000	N/A	N/A	Multiple
[74]	PN	CHN	N/A	MAE	2 min	Multiple
[27]	Graph	DEU	2M	Acc	98.90%	One
[25]	Graph	DEU	223,873	N/A	N/A	One
[32]	Max-plus	CHN	N/A	N/A	N/A	Multiple
[24]	TEG	NLD	49,200	MAE	40 sec	Multiple
[3]	NN	CHN	400,000	MAE, RMSE	MAE 0.35 min, RMSE 1.04 min	One
[18]	NN	CHN	975,592	MAE	3.24 min	One
[73]	NN	CHN	2M	RMSE	0.50 min	One
[35]	NN	NLD	66 days	MAE, RMSE	30 sec, 87%	One
[69]	NN	NLD	66,178	MAE, RMSE	30 sec, 86%	One
[46]	NN	ITA	6 months	Avg Acc	1.70 min	Multiple
[31]	NN	CHN	1 day	RMSE	4.91 sec	One
[70]	NN	IRN	179,982	Acc	93.34%	One
[29]	NN	CHN	400,000	MAE, RMSE	MAE 0.89 min, RMSE 2.13 min	One
[20]	NN	CHN	229,320	MAE, RMSE	<0.6 min	One
[57]	DT	CHN	183,207	MAE, RMSE	MAE 0.85 min, RMSE 2.29 min	One

(continued)

Table 1. (*continued*)

Paper	Model	Country	Data size	Metrics	Prediction Error	Prediction horizon
[56]	DT	CHN	330,787	MAE, MSE, RMSE	MAE 0.46 min, RMSE 1.37 min	One
[65]	DT	CHN	2.7M	MAE	25 min	Multiple
[67]	DT	JPN	N/A	N/A	N/A	Multiple
[26]	DT	TUN	12,350	MAE, RMSE	MAE 9.88 min, RMSE 18.29 min	One
[8]	SVM	CHN	2,025	MAE, RMSE	MAE 1.68 min, RMSE 2.24 min	One
[66]	SVM	CHN	17,371	MAE, MSE	MAE 0.32 min, MSE 0.44 min	One
[19]	SVM	CHN	57,796	MAE	0.55 min	One
[53]	SVM	FRA	5 years	Acc	85.38%	One
[4]	SVM	USA	150,000	MAE	N/A	One
[2]	RF	IND	1,170	MAE, RMSE	MAE 49.28 min, RMSE 80.07 min	One
[14]	RF	CHN	1 year	MAE	<3 min	One
[30]	RF	NLD	5.6M	MAE, RMSE	MAE 0.69 min, RMSE 1.68 min	One
[37]	RF	DEU	3.3 years	RMSE	369.50 sec	One
[36]	RF	NLD	10M	Acc	93%	One
[21]	RF	CHN	29,662	MAE	1 min, 80.4%	One
[22]	RF	SWE	2.2M	RMSE	3.33 min	One
[50]	LNR	IND	1 year	N/A	N/A	One
[16]	LNR	DEU	N/A	MSE	<75 min	Multiple
[12]	Clustering	NLD	525,600	MSE	90 min	Multiple
[49]	TSA	THA	182 days	MAE	5.86 min	One, multiple
[11]	Simulation	RUS	N/A	Mean, STD	N/A	Multiple

Model abbreviations: BN=Bayesian Network; DT=Decision Tree; EQS=Equation System; LNR=Linear Regression; MC=Markov Chain; NN=Neural Network; PN=Petri Network; RF=Random Forest; SVM=Support Vector Machine; TEG=Timed Event Graph; TSA=Time Series Analysis.
Country codes: Standard ISO ALPHA-3 country codes (e.g. CHN=China, DEU=Germany).
Metrics: MAE=Mean Absolute Error; RMSE=Root Mean Square Error; MSE=Mean Square Error; Acc=Accuracy; Avg Acc=Average Accuracy; STD=Standard Deviation.
Units: min=minutes; sec=seconds; M=millions.
Prediction Horizon: One=one station ahead; Multiple=multiple stations ahead.

while providing well-calibrated confidence measures for each prediction, the proposed machine learning-based approach delivers accurate and reliable train delay predictions.

3   Problem Statement and System Architecture

This section begins by presenting the formulation of the train delay prediction problem that the proposed method aims to solve. Subsequently, a general overview of the proposed method's system architecture is provided.

3.1   Problem Formulation

To understand the train delay prediction problem in a systematic way, assume a targeted train service i is scheduled to arrive at station j at time t_s. Then at a current time $t \leq t_s$, the predicted train delay $\hat{\Phi}$ is defined as:

$$\hat{\Phi}_{i,j,t,t_s} = f(\Omega_{i,j}) \tag{1}$$

where $\Omega_{i,j}$ is historical train operational data of train service i at station j, namely the input to the model, and f maps the input information $\Omega_{i,j}$ to the predicted train delay $\hat{\Phi}_{i,j,t,t_s}$. This formulation naturally lends itself to a regression task.

In the proposed method, the predicted train delay $\hat{\Phi}$ is calculated from the scheduled train arrival time t_s and the difference in the train travel time $\Delta T_{i,j-1,j}$ from the previous station $j-1$ to the current station j. The estimated train travel time difference $\Delta \hat{T}_{i,j-1,j}$ is generated by the machine learning model M based on train operational data $\Omega_{i,j}$, defined as

$$\Delta \hat{T}_{i,j-1,j} = M(\Omega_{i,j}) - T_{j-1,j} \tag{2}$$

where $T_{i,j-1,j}$ is the scheduled train travel time between these consecutive stations $j-1$ and j of the train service i.

Therefore, the predicted train delay $\hat{\Phi}$ in the proposed method is defined as:

$$\hat{\Phi}_{i,j,t,t_s} = M(\Omega_{i,j}) - T_{i,j-1,j}. \tag{3}$$

3.2   System Architecture

To deliver accurate train delay estimates along with prediction intervals that provide meaningful confidence measures of each prediction $\hat{\Phi}_{i,j,t,t_s}$, the proposed system is structured into three main steps: data preparation, model training, and test sample prediction, as shown in Fig. 3. The brief description of each step is as follows:

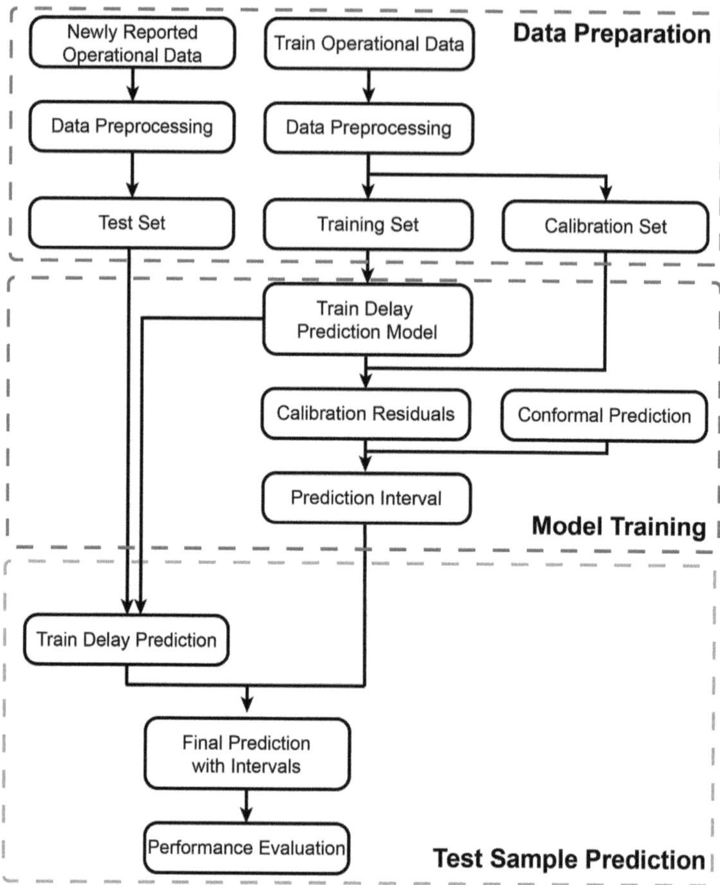

Fig. 3. The overview of the proposed train delay prediction approach.

Data Preparation: We begin by preprocessing the raw historical train operational records. Specifically, we first identify and remove any complete train services that contain missing values or outliers in their arrival time, departure time, dwell time, or train travel time at any station.

To facilitate numerical processing, ISO timestamps in the aforementioned operational data are converted to Unix timestamps. Next, categorical attributes, those identifying train services, engine models, and specific stations, are label-encoded to transform them into numeric form.

Finally, the dataset is divided into training, calibration, and test sets for model training, prediction interval generation, and final evaluation, respectively.

Model Training: In the model training step, a machine learning-based model for train delay prediction is developed, and conformal prediction is employed to generate statistically valid prediction intervals. In the proposed approach, the output is the estimated travel time to the next station, which is subsequently used to generate the final train delay predictions (see Eq. 3).

First, tree-based models, identified in Sect. 2 as the most widely adopted algorithms in the literature, are trained on the training dataset. Next, the trained model is used to predict outcomes for the calibration set. The residuals from these predictions, calculated as the differences between the predicted and actual values in the calibration set, are then used within the conformal prediction framework to construct prediction intervals with user specified confidence level. Finally, the trained machine learning model is applied to the test set to generate train delay estimations.

Test Sample Prediction: In the final step, the accuracy of the machine learning train delay prediction model and the quality of the prediction intervals are evaluated.

Performance metrics such as Root Mean Squared Error (RMSE) and Mean Absolute Error (MAE) are used to assess the predictive accuracy of the machine learning model. Subsequently, the model's performance is compared with other popular machine learning approaches for train delay prediction. Finally the quality of the prediction intervals is analysed.

4 Machine Learning Models and Conformal Prediction

This section offers a comprehensive description to the most widely used machine learning models in the literature for train delay prediction tasks and an in-depth introduction to conformal prediction that produces statistical valid prediction intervals with a predefined confidence level.

4.1 Machine Learning Models in Train Delay Prediction

To provide accurate and reliable train delay predictions, a large-scale real-world train operational dataset spanning 3 years and 8 months (see Sect. 5.1 for details) is leveraged in this study. To effectively uncover underlying patterns in this dataset, the data-driven approach is adopted in the proposed system.

As discussed in Sect. 2 and illustrated in Fig. 2, the most commonly used machine learning algorithms for data-driven approaches in the literature are Neural Networks (NN), Random Forests (RF), and Support Vector Regression (the regression extension of SVM). Given the dataset's size of 12,840,590 records, training NNs is extremely time-consuming and computationally expensive. Therefore, the more efficient RF model is selected as the primary machine learning algorithm in the proposed approach.

Random Forest (RF). Random Forest is a robust ensemble learning method that consists of multiple decision trees and combines their outputs to improve prediction accuracy and control overfitting in train delay prediction [6]. A standard RF constructs numerous decision trees during training, each trained on a random subset of the data and a random subset of features at each split, and outputs the average of individual tree predictions.

A standard decision tree works by repeatedly splitting the feature space into smaller regions. Each internal node applies a rule based on one of the input features, and each leaf node gives a final prediction [7, 34, 45, 51]. Thus, the ensemble of trees is defined as:

$$\{Tree_d(\mathbf{\Omega}_{i,j})\}_{d=1}^{D} \tag{4}$$

where $Tree_d$ denotes the individual decision tree, D is the total number of trees in the ensemble, and $\mathbf{\Omega}_{i,j}$ is a bootstrap sample of the input train operational data $\Omega_{i,j}$. A bootstrap sample is a sub-dataset created by randomly sampling with replacement from the original data, helping the model mitigate overfitting during training. For a tree node n_{tree} with $N_{n_{tree}}$ samples, the training objective is to minimise the MSE, defined as:

$$\text{MSE}(n_{tree}) = \frac{1}{N_{n_{tree}}} \sum_{k \in n_{tree}} \left(T_{i,j-1,j} - \bar{T}_n\right)^2 \tag{5}$$

where $T_{i,j-1,j}$ is the ground truth train travel time of the i-th train service at station j in the node, and $\bar{T}_n$ is the average of target train travel time values in the node. Subsequently, the final prediction $\hat{T}_{i,j-1,j}$ of the train travel time from the previous station $j - 1$ to the current station j is calculated as:

$$\hat{T}_{i,j-1,j} = \frac{1}{D} \sum_{d=1}^{D} Tree_d(\mathbf{\Omega}_{i,j}). \tag{6}$$

In the proposed train delay prediction approach, the input train operational data $\Omega_{i,j}$ to the RF model contains service identifiers that uniquely distinguish the train service, as well as operational details that describe the train's arrival and departure at each station stop. After the estimation of the train travel time, the final train delay prediction $\hat{\Phi}_{i,j}$ at station j is derived.

To provide a comprehensive comparison of the performance of popular machine learning models for train delay prediction, we also implement distinguishing machine learning models such as Support Vector Regression (SVR) and Linear Regression (LNR).

Support Vector Regression (SVR). SVR, a regression extension of SVM, is a robust supervised learning models widely used in the literature to enhance the accuracy of train delay prediction [10, 39, 58, 62]. A standard SVR aims to find an optimal hyperplane that best fits the data within a predefined margin. The hyperplane is determined by support vectors, which are the training samples nearest to the decision boundary and play a crucial role in its definition.

For the regression model is defined by the function:

$$f(\Omega_{i,j}) = \langle \mathbf{w}, \Omega_{i,j} \rangle + b. \tag{7}$$

The primal optimisation objective for SVR model for train delay prediction is given by:

$$\min_{\mathbf{w},b,\xi_k,\xi_k^*} \frac{1}{2}\|\mathbf{w}\|^2 + C \sum_{i=1}^{N_{train}} \sum_{j=1}^{N_{station}} (\xi_{i,j} + \xi_{i,j}^*) \tag{8}$$

subject to:

$$\hat{T}_{i,j-1,j} - \langle \mathbf{w}, \Omega_{i,j} \rangle - b \leq \epsilon + \xi_{i,j}, \tag{9}$$

$$\langle \mathbf{w}, \Omega_{i,j} \rangle + b - \hat{T}_{i,j-1,j} \leq \epsilon + \xi_{i,j}^*, \tag{10}$$

$$\xi_{i,j} \geq 0, \quad \xi_{i,j}^* \geq 0 \tag{11}$$

where $\mathbf{w}$ is the weight vector defining the orientation of the regression hyperplane, b is the bias term shifting the hyperplane from the origin, $\xi_{i,j}$ and $\xi_{i,j}^*$ are slack variables measuring prediction errors above and below the margin tolerance ϵ respectively, C is the regularisation parameter controlling the trade-off between model simplicity and error tolerance, $\|\mathbf{w}\|^2$ is the squared L2-norm regularisation term promoting model flatness, N_{train} is the total number of train services, and $N_{station}$ is the total number train stations in the railway network.

Linear Regression (LNR). Linear Regression is a popular supervised learning algorithm for modelling the relationship between a the historical train operational data and the train delays [43,54,59]. The goal is to find a linear function that best predicts train delay times based on input features.

In the proposed approach, to predict the train travel time to the next station, the model assumes:

$$T_{i,j-1,j} = \mathbf{w}^T \Omega_{i,j} + b + \epsilon \tag{12}$$

where $\mathbf{w}$ is the weight vector, b is the bias term, and ϵ is the residual error. The LNR model is optimised by minimising the loss, defined as:

$$\min_{\mathbf{w},b} \mathcal{L}(\mathbf{w},b) = \min_{\mathbf{w},b} \frac{1}{N_{train} \times N_{station}} \sum_{i=1}^{N_{train}} \sum_{j=1}^{N_{station}} \left(T_{i,j-1,j} - (\mathbf{w}^T \Omega_{i,j} + b)\right)^2 \tag{13}$$

where N_{train} is the total number of train services, and $N_{station}$ is the total number train stations in the railway network.

Additionally, a range of tree-based models, including Decision Trees (DT), Gradient Boosting (GB), Histogram-based Gradient Boosting (histGB), and eXtreme Gradient Boosting (XGB), as well as regularised LNR-based models such as Ridge Regression (Ridge) and Lasso Regression (Lasso), are included in the empirical experiments.

4.2 Conformal Prediction

While machine learning models provide promising predictions of train delays, they do not guarantee the uncertainty associated with individual predictions. In high-stakes systems like railway networks, the ability to provide not only accurate but also interpretable and uncertainty-guaranteed predictions is crucial. Thus, conformal prediction is leveraged to quantify prediction confidence, allowing train operators to assess how often true train delays fall within a specified bounds, thus enabling more informed and reliable decision-making.

Conformal Prediction (CP) is a statistically rigorous framework for uncertainty quantification that produces prediction intervals with guaranteed coverage, thereby improving the reliability of machine learning model predictions [47,55,63,64].

Recall that a machine-learning model M is trained on train operational data $\Omega_{i,j}$ for train travel time prediction of train service i at station j, defined as:

$$\hat{T}_{i,j-1,j} = M(\Omega_{i,j}). \tag{14}$$

Given a user-specified confidence level α, CP constructs prediction intervals that are guaranteed to contain the true target value with probability at least $1 - \alpha$, under the assumption of exchangeable data. In our experiments, we use a computationally efficient variant of CP, namely Inductive Conformal Prediction (ICP). To quantify uncertainty non-parametrically, ICP first uses a calibration dataset to compute nonconformity scores, which measures of how unusual or nonconforming a new observation is relative to the known data distribution. The nonconformity score is calculated by the calibration residuals defined as:

$$R_k^{\mathrm{cal}} = \left| T_k - M(\Omega_k) \right|, \quad k = 1, \ldots, m \tag{15}$$

where R_k^{cal} is the residuals of the calibration set, m represents the number of samples in the calibration dataset, T_k and Ω_k is the ground truth label (i.e., train travel time) and input features of the k-th sample in the calibration set, respectively, M is the applied machine learning model. Subsequently, the critical quantile $\hat{q}$ which determines how wide the prediction interval should be to meet the specified confidence level α is calculated as:

$$\hat{q} = \mathrm{quantile}\left(\left\{ R_k^{\mathrm{cal}} \right\}_{k=1}^{m} \cup \{\infty\}, \frac{\lceil (m+1)(1-\alpha) \rceil}{m} \right). \tag{16}$$

Once the critical quantile $\hat{q}$ is computed, a prediction interval for a new train operational data input Ω_{test} is defined as:

$$[M(\Omega_{test}) - \hat{q}, M(\Omega_{test}) + \hat{q}]. \tag{17}$$

Under the assumption that the training and test data are exchangeable, this prediction interval comes with a rigorous coverage guarantee [47,55]. Specifically, the interval will contain the true outcome with probability at least $1 - \alpha$, formalised as:

$$\mathbb{P}\left(T_{\text{test}} \in [M(\Omega_{test}) - \hat{q}, M(\Omega_{test}) + \hat{q}]\right) \geq 1 - \alpha. \tag{18}$$

This implies that, across many predictions, the fraction of the prediction intervals that capture the true train travel time values will be at least $1 - \alpha$, therefore providing a statistically sound and transparent measure of prediction uncertainty.

5   Empirical Experiments

This section begins with a detailed introduction to the train operational dataset used in this study, followed by a brief overview of the evaluation metrics employed to assess the performance of the proposed train delay prediction approach. Finally, it presents the empirical results and a performance comparison with other popular machine learning models in the literature.

5.1   Dataset Description

To evaluate the accuracy of the proposed train delay prediction method and the quality of the prediction intervals, a large-scale real-world train operational dataset is utilised. This dataset encompasses approximately 3 years and 8 months of data, spanning from February 12, 2020, to October 8, 2023. It includes records for 12,621 unique train services operating between specific stations along a designated railway line in the southeast of the United Kingdom. Each of these services may have run multiple times during the recorded period. An example of a raw train service record is shown in Table 2. A brief description of the train operational record attributes is as follows:

`Headcode` Used nationally to determine a train service between specific stations and on a prespecified line.

`UnitNumber` Identifies which engine is working on this train service; this is the engine that "drives" the train.

`TrainModel` Represents the service family inside the railway system of the operator company.

`Stops` Consists of the train operational records at each station where it stops, including:

    `Name` The name of the station.

    `CRS` The Computer Reservation System code, used to identify railway stations for ticketing and passenger information systems.

    `Tiploc` The Timing Point Location code, used to identify specific timing points on the railway network, including stations, junctions, sidings, and other timing points (e.g., for arrival, departure, dwell).

    `BookedDeparture`, `ActualDeparture`, `DepartureDiff` Information about departure at this station.

    `BookedArrival`, `ActualArrival`, `ArrivalDiff` Information about train arrival at this station.

`DwellBooked, DwellActual, DwellDiff` Information about train dwell time at this station.

`UntilNextLocationBookedTime` Booked train travel time until the next stop.

`UntilNextLocationActualTime` Actual train travel time until the next stop.

`UntilNextLocationTimeDiff` Difference in train travel time until the next stop.

`Delayed` The delayed time identified for this stop.

Note that the time '0001-01-01T00:00:00.000Z' in the booked and actual arrival times indicates that this is the first station/stop in this train service, while the same time appears in the booked and actual arrival times represents the very last stop in the route.

To construct a training dataset suitable for machine learning models, each train service record is segmented into multiple data samples. Each sample includes key identifiers (e.g., 'Headcode', 'UnitNumber') to uniquely distinguish the train service, as well as operational details (e.g., 'Tiploc', 'BookedDeparture', 'ActualDeparture', 'BookedArrival', 'ActualArrival', 'DepartureDiff', 'ArrivalDiff', 'DwellBooked', 'DwellActual', 'DwellDiff', 'UntilNextLocationBookedTime', 'UntilNextLocationActualTime', 'UntilNextLocationTimeDiff') that describe the train's movements at each station stop. After preprocessing, the final dataset consists of 12,840,590 individual train service records.

5.2   Evaluation Metrics

To assess the prediction performance of the train delay prediction approach, popular evaluation metrics including Mean Absolute Error (MAE) and Root Mean Squared Error (RMSE) are utilised. To assess the quality of prediction intervals, Residual Coverage Score (RCS) is leveraged.

Mean Absolute Error (MAE)

The Mean Absolute Error (MAE) is a straightforward metric for evaluating the accuracy of a train delay prediction model. It calculates the mean of the absolute differences between predicted and actual train delays, as shown by:

$$MAE = \frac{1}{n} \sum_{k=1}^{n} |y_k - \hat{y}_k|. \tag{19}$$

Here, n is the number of observations, y_k represents the actual delay, and $\hat{y}_k$ denotes the predicted delay for each observation. A lower MAE suggests better model accuracy, offering a direct measure of the average prediction error.

Root Mean Squared Error (RMSE)

The Root Mean Squared Error (RMSE) is a widely used metric for evaluating the accuracy of train delay predictions. It is derived from the Mean Squared Error (MSE), which measures the average of the squared differences between the actual and predicted values:

$$MSE = \frac{1}{n} \sum_{k=1}^{n} (y_k - \hat{y}_k)^2. \tag{20}$$

Table 2. A snapshot of the raw train operational records.

Attribute	Value
Headcode	1V84BA
UnitNumber	376016
TrainModel	465
Stops	
Name	Charing Cross
Crs	CHX
Tiploc	CHRX
BookedDeparture	2020-03-17T23:37:00.000Z
BookedArrival	0001-01-01T00:00:00.000Z
ActualDeparture	2020-03-17T23:37:04.000Z
ActualArrival	0001-01-01T00:00:00.000Z
DwellBooked	0
DwellActual	0.0
DepartureDiff	4.0 s
ArrivalDiff	0.0 s
DwellDiff	0.0 s
UntilNextLocationActualTime	143.0 s
UntilNextLocationBookedTime	120.0 s
UntilNextLocationTimeDiff	23.0 s
Delayed	0
Stops	
Name	London Bridge
. . .	. . .

By taking the square root of MSE, RMSE expresses this error in the same units as the original data:

$$RMSE = \sqrt{\frac{1}{n}\sum_{k=1}^{n}(y_k - \hat{y}_k)^2}. \tag{21}$$

RMSE penalises larger errors more heavily due to the squaring step, making it particularly sensitive to significant deviations between predicted and actual delays. Its interpretability—being in the same scale as the target variable—makes it especially valuable for assessing the magnitude of prediction errors in a practical and intuitive manner. A lower RMSE indicates better predictive performance and is therefore a central metric in evaluating the effectiveness of delay prediction models.

Residual Coverage Score (RCS)

To quantify how frequently the true train delay values fall within the predicted intervals, Residual Coverage Score (RCS) is introduced, defined as:

$$\text{RCS} = \frac{1}{n} \sum_{k=1}^{n} \mathbb{I}\left(\hat{y}_k^{\text{low}} \leq y_k \leq \hat{y}_k^{\text{up}}\right) \tag{22}$$

where y_k is the true value for the k-th sample, $\hat{y}_k^{\text{low}}$ and $\hat{y}_k^{\text{up}}$ are the lower and upper bounds of the prediction interval, respectively. For each of the test samples, the indicator function $\mathbb{I}$ returns 1 if the interval contains the true value and 0 otherwise. The RCS is then the average of these outcomes across the entire dataset. In repeated sampling, the proportion of times that the interval contains the true train delay values should match the predefined confidence level $1 - \alpha$. Thus, RCS assesses whether the prediction intervals are statistically valid.

5.3   Empirical Results

To evaluate the performance of the proposed train delay prediction approach, Random Forest (RF) and several widely used machine learning models are implemented on the aforementioned train operational dataset. These include Support Vector Regression (SVR), Linear Regression (LNR), Decision Trees (DT), Gradient Boosting (GB), Histogram-based Gradient Boosting (HistGB), and eXtreme Gradient Boosting (XGB), along with regularised linear models such as Ridge Regression (Ridge) and Lasso Regression (Lasso).

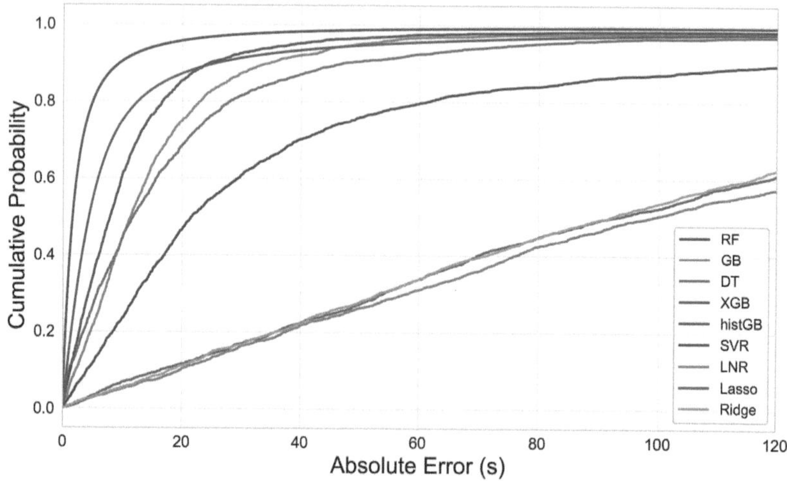

Fig. 4. The CDF curves of the absolute errors generated by each machine learning model. The RF model produces the most accurate and robust prediction, achieving an accuracy of 20 s, 95% of the time. However, SVR and LNR-based models fail to effectively map the correlation between the train operational data to the train travel time between stations.

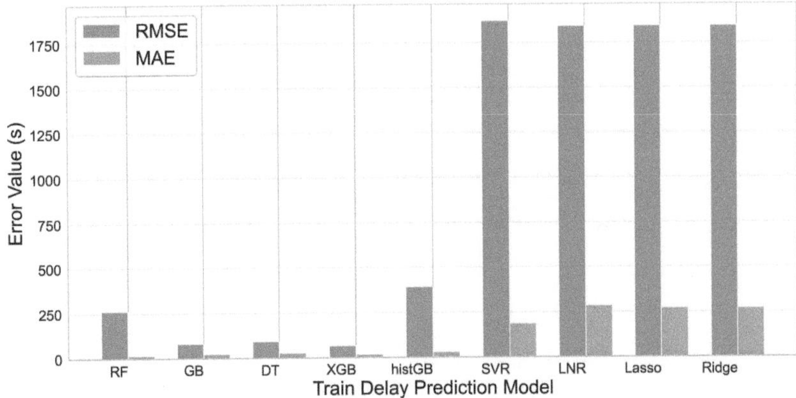

Fig. 5. The performance comparison of the popular machine learning models. The results show that tree-based models outperform SVR and LNR models in predicting train delays on large-scale operational data.

In order to predict the train travel time to the next station (i.e., 'UntilNext-LocationActualTime'), 'Headcode', 'UnitNumber', 'Tiploc', 'BookedDeparture', 'ActualDeparture', 'BookedArrival', 'ActualArrival', 'DepartureDiff', 'ArrivalD-iff', 'DwellBooked', 'DwellActual', 'DwellDiff', and 'UntilNextLocationBooked-Time' are leveraged as the input features to the machine learning model. A desktop PC equipped with an Intel i9-12900k @ 4.90 GHz CPU and 32GB DDR4 4000MHz memory was used to analyse the results.

The cumulative distribution function (CDF) curves, along with the RMSE and MAE of the employed models, are presented in Table 3 and Figs. 4 and 5. It is observed that the RF model produces the most accurate and robust predictions, achieving an accuracy of 20 s, 95% of the time. SVR, another popular machine learning model in train delay prediction, however, struggles at a mean error of 185.83 s. Given the extremely time-consuming training process on real-world train operational data, SVR models are not considered as a suitable method for train delay prediction on large-scale datasets. Additionally, LNR-based models fail to effectively map the correlation between the train operational data to the train travel time between stations. This is because, in real-world railway systems, operational features often lack a clear linear relationship with the train travel time between stations. In contrast, most tree-based models offer accurate and robust estimations of train delays, although their performance can vary across models. Tree-based models excel in train delay prediction because they effectively capture complex, non-linear relationships between diverse factors such as train services identifier, train operational records and station information.

However, while machine learning models predict train delays effectively, they lack guarantees on individual uncertainty. Therefore, conformal prediction is leveraged to quantify uncertainty, helping train operators assess how often true delays fall within predicted bounds for more reliable decision-makings. By apply-

Table 3. Train delay prediction model performance comparison.

Model	RMSE (s)	MAE (s)
RF	**261.59**	**14.71**
GB	**261.38**	22.99
DT	265.69	27.14
XGB	289.50	17.09
histGB	389.96	29.78
SVR	1869.18	185.83
LNR	1839.20	284.96
Lasso	1840.17	271.07
Ridge	1841.20	268.53

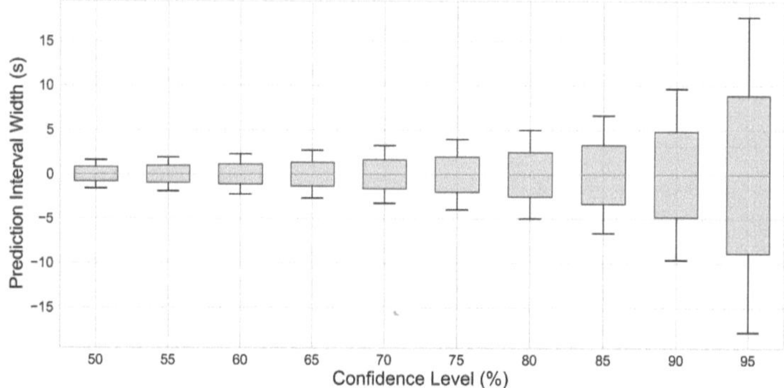

Fig. 6. The prediction interval widths (blue boxplot) at different confidence levels were evaluated within the conformal prediction framework. The orange line indicates the prediction generated by the ML model. At a 90% confidence level, the resulting interval width was 19.3 s, while the interval width increases to 35.5 s at the confidence level of 95%. This demonstrates the expected trade-off in conformal prediction, where higher confidence levels yield wider prediction intervals.(Color figure online)

ing conformal prediction with confidence levels of 90% and 95%, the resulting prediction intervals are 19.3 s and 35.5 s, respectively, as illustrated in Fig. 6. This demonstrates the expected trade-off in conformal prediction, where higher confidence levels yield wider prediction intervals. This means that for any train delay prediction $\hat{\Phi}_{test}$ made by the proposed approach, the probability that the true delay time falls within $[\hat{\Phi}_{test} - 9.65\,\text{s},\ \hat{\Phi}_{test} + 9.65\,\text{s}]$ and $[\hat{\Phi}_{test} - 17.75\,\text{s},\ \hat{\Phi}_{test} + 17.75\,\text{s}]$ is 90% and 95%, respectively.

To investigate the validity of the prediction intervals, the Residual Coverage Score is also analysed, as shown in Fig. 7. It illustrates the actual empirical coverage scores within each bin, compared to the target coverage rates of 90% (red dashed line) and 95% (yellow dashed line). It is observed that the conformal prediction intervals achieve the expected coverage for lower predicted values below 263.4 s. However, they exhibit under-coverage in the highest prediction bin, where the predicted values exceed 425.4 s. This suggests that for larger and longer train delays, the model is unable to provide estimations with enough confidence.

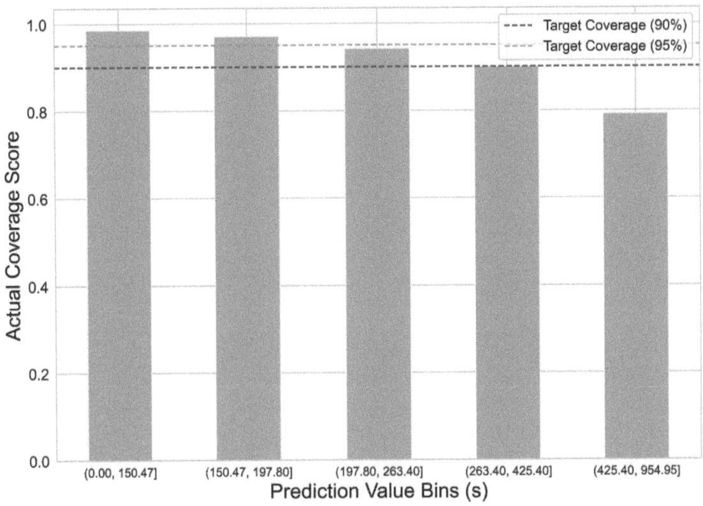

Fig. 7. The coverage scores of conformal prediction intervals across different predicted value bins. It is observed that the prediction interval achieve the desired coverage for lower predicted train delay values below 263.4 s, but under-cover for the predicted values above 425.4 s.

6   Conclusion

This study presents a robust machine learning approach for train delay prediction with quantified uncertainty. By modelling train travel times between consecutive stations using tree-based machine learning models, the proposed approach effectively captures sequential dependencies while ensuring efficient processing of large-scale datasets and facilitating the learning of underlying patterns within the dataset. The integration of Conformal Prediction provides statistically rigorous uncertainty quantification, delivering prediction intervals (e.g., 19.3 s at the 90% confidence level) that enhance decision-making transparency for railway operators. Validated on over 12.8 million real-world train production service records, the proposed method demonstrates an accuracy below 20 s, 90% of the time, outperforming widely used train delay prediction models such as LNR and SVR alternatives. While the framework achieves target coverage for typical delays below 263 s, extreme delays longer than 425 s exhibit under-coverage.

Further research could investigate methods for dynamically adjusting prediction intervals for individual test samples, particularly in relation to varying levels of delay severity.

Acknowledgments. The authors would like to thank Distributed Analytics Solutions Ltd for the financial support of this research.

References

1. Abdi, A., Amrit, C.: A review of travel and arrival-time prediction methods on road networks: classification, challenges and opportunities. PeerJ Comput. Sci. **7**, e689 (2021)
2. Arshad, M., Ahmed, M.: Train delay estimation in indian railways by including weather factors through machine learning techniques. Recent Adv. Comput. Sci. Commun. (Formerly: Recent Patents on Computer Science) **14**(4), 1300–1307 (2021)
3. Bao, X., Li, Y., Li, J., Shi, R., Ding, X.: Prediction of train arrival delay using hybrid elm-pso approach. J. Adv. Transp. **2021**, 1–15 (2021)
4. Barbour, W., Mori, J.C.M., Kuppa, S., Work, D.B.: Prediction of arrival times of freight traffic on us railroads using support vector regression. Transp. Res. Part C: Emerg. Technol. **93**, 211–227 (2018)
5. Barta, J., Rizzoli, A.E., Salani, M., Gambardella, L.M.: Statistical modelling of delays in a rail freight transportation network. In: Proceedings of the 2012 Winter Simulation Conference (WSC), pp. 1–12. IEEE (2012)
6. Breiman, L.: Random forests. Mach. Learn. **45**, 5–32 (2001)
7. Breiman, L., Friedman, J., Olshen, R.A., Stone, C.J.: Classification and Regression trees. Routledge (2017)
8. Chen, Z., Wang, Y., Zhou, L.: Predicting weather-induced delays of high-speed rail and aviation in china. Transp. Policy **101**, 1–13 (2021)
9. Corman, F., Kecman, P.: Stochastic prediction of train delays in real-time using bayesian networks. Transp. Res. Part C: Emerg. Technol. **95**, 599–615 (2018)
10. Cortes, C., Vapnik, V.: Support-vector networks. Mach. Learn. **20**, 273–297 (1995)
11. Davydov, B., Chebotarev, V., Kablukova, K.: Online train traffic adjustments: probabilistic modeling and estimating. In: Advanced Solutions of Transport Systems for Growing Mobility: 14th Scientific and Technical Conference Transport Systems. Theory & Practice 2017 Selected Papers, pp. 50–60. Springer (2018)
12. Dekker, M.M., Panja, D., Dijkstra, H.A., Dekker, S.C.: Predicting transitions across macroscopic states for railway systems. PLoS ONE **14**(6), e0217710 (2019)
13. Feng, X., Nguyen, K.A., Luo, Z.: A survey of deep learning approaches for wifi-based indoor positioning. J. Inf. Telecommun. **6**(2), 163–216 (2022)
14. Gao, B., Ou, D., Dong, D., Wu, Y.: A data-driven two-stage prediction model for train primary-delay recovery time. Int. J. Software Eng. Knowl. Eng. **30**(07), 921–940 (2020)
15. Gaurav, R., Srivastava, B.: Estimating train delays in a large rail network using a zero shot markov model. In: 2018 21st International Conference on Intelligent Transportation Systems (ITSC), pp. 1221–1226. IEEE (2018)

16. Hauck, F., Kliewer, N.: Data analytics in railway operations: using machine learning to predict train delays. In: Operations Research Proceedings 2019: Selected Papers of the Annual International Conference of the German Operations Research Society (GOR), Dresden, Germany, 4–6 September 2019, pp. 741–747. Springer (2020)

17. Huang, P., et al.: A bayesian network model to predict the effects of interruptions on train operations. Transp. Res. Part C: Emerg. Technol. **114**, 338–358 (2020)

18. Huang, P., et al.: Modeling train operation as sequences: a study of delay prediction with operation and weather data. Transp. Res. Part E: Logist. Transp. Rev. **141**, 102022 (2020)

19. Huang, P., Wen, C., Fu, L., Peng, Q., Li, Z.: A hybrid model to improve the train running time prediction ability during high-speed railway disruptions. Saf. Sci. **122**, 104510 (2020)

20. Huang, P., Wen, C., Fu, L., Peng, Q., Tang, Y.: A deep learning approach for multi-attribute data: a study of train delay prediction in railway systems. Inf. Sci. **516**, 234–253 (2020)

21. Jiang, C., Huang, P., Lessan, J., Fu, L., Wen, C.: Forecasting primary delay recovery of high-speed railway using multiple linear regression, supporting vector machine, artificial neural network, and random forest regression. Can. J. Civ. Eng. **46**(5), 353–363 (2019)

22. Jiang, S., Persson, C., Akesson, J.: Punctuality prediction: combined probability approach and random forest modelling with railway delay statistics in sweden. In: 2019 IEEE Intelligent Transportation Systems Conference (ITSC), pp. 2797–2802. IEEE (2019)

23. Kecman, P., Corman, F., Meng, L.: Train delay evolution as a stochastic process. In: 6th International Conference on Railway Operations Modelling and Analysis (RailTokyo2015). IVT, ETH Zurich, Orange Labs (2015)

24. Kecman, P., Goverde, R.M.: Online data-driven adaptive prediction of train event times. IEEE Trans. Intell. Transp. Syst. **16**(1), 465–474 (2014)

25. Keyhani, M.H., Schnee, M., Weihe, K., Zorn, H.P.: Reliability and delay distributions of train connections. In: 12th Workshop on Algorithmic Approaches for Transportation Modelling, Optimization, and Systems. Schloss Dagstuhl-Leibniz-Zentrum fuer Informatik (2012)

26. Laifa, H., Ghezalaa, H.H.B., et al.: Train delay prediction in tunisian railway through lightgbm model. Procedia Comput. Sci. **192**, 981–990 (2021)

27. Lemnian, M., Rückert, R., Rechner, S., Blendinger, C., Müller-Hannemann, M.: Timing of train disposition: towards early passenger rerouting in case of delays. In: 14th Workshop on Algorithmic Approaches for Transportation Modelling, Optimization, and Systems. Schloss Dagstuhl-Leibniz-Zentrum fuer Informatik (2014)

28. Lessan, J., Fu, L., Wen, C.: A hybrid bayesian network model for predicting delays in train operations. Comput. Ind. Eng. **127**, 1214–1222 (2019)

29. Li, Y., Xu, X., Li, J., Shi, R.: A delay prediction model for high-speed railway: an extreme learning machine tuned via particle swarm optimization. In: 2020 IEEE 23rd International Conference on Intelligent Transportation Systems (ITSC), pp. 1–5. IEEE (2020)

30. Li, Z., Wen, C., Hu, R., Xu, C., Huang, P., Jiang, X.: Near-term train delay prediction in the dutch railways network. Int. J. Rail Transp. **9**(6), 520–539 (2021)

31. Liu, Y., Tang, T., Xun, J.: Prediction algorithms for train arrival time in urban rail transit. In: 2017 IEEE 20th International Conference on Intelligent Transportation Systems (ITSC), pp. 1–6. IEEE (2017)

32. Ma, H., Qin, Y., Han, G., Jia, L., zhu, T.: Forecast of train delay propagation based on max-plus algebra theory. In: Proceedings of the 2015 Chinese Intelligent Systems Conference, vol. 1, pp. 661–672. Springer (2016)

33. Martin, L.J.W.: Predictive reasoning and machine learning for the enhancement of reliability in railway systems. In: Lecomte, T., Pinger, R., Romanovsky, A. (eds.) RSSRail 2016. LNCS, vol. 9707, pp. 178–188. Springer, Cham (2016). https://doi.org/10.1007/978-3-319-33951-1_13

34. Meister, J.A., Nguyen, K.A.: Conformalised data synthesis. Mach. Learn. **114**(3), 1–37 (2025)

35. Mou, W., Cheng, Z., Wen, C.: Predictive model of train delays in a railway system. In: Proceedings of the 8th International Conference on Railway Operations Modelling Analysis (RailNorrköping), pp. 913–929 (2019)

36. Nabian, M.A., Alemazkoor, N., Meidani, H.: Predicting near-term train schedule performance and delay using bi-level random forests. Transp. Res. Rec. **2673**(5), 564–573 (2019)

37. Nair, R., et al.: An ensemble prediction model for train delays. Transp. Res. Part C: Emerg. Technol. **104**, 196–209 (2019)

38. Nguyen, K., Luo, Z.: Conformal prediction for indoor localisation with fingerprinting method. In: IFIP International Conference on Artificial Intelligence Applications and Innovations, pp. 214–223. Springer (2012)

39. Nguyen, K.A.: A performance guaranteed indoor positioning system using conformal prediction and the wifi signal strength. J. Inf. Telecommun. **1**(1), 41–65 (2017)

40. Nguyen, K.A., Luo, Z.: Enhanced conformal predictors for indoor localisation based on fingerprinting method. In: IFIP International Conference on Artificial Intelligence Applications and Innovations, pp. 411–420. Springer (2013)

41. Nguyen, K.A., Luo, Z.: Reliable indoor location prediction using conformal prediction. Ann. Math. Artif. Intell. **74**(1), 133–153 (2015)

42. Nguyen, K.A., Luo, Z., Li, G., Watkins, C.: A review of smartphones-based indoor positioning: challenges and applications. IET Cyber-Syst. Robot. **3**(1), 1–30 (2021)

43. Nguyen, K.A., Luo, Z., Watkins, C.: Epidemic contact tracing with smartphone sensors. J. Location Based Serv. **14**(2), 92–128 (2020)

44. Nguyen, K.A., Watkins, C., Luo, Z.: Co-location epidemic tracking on london public transports using low power mobile magnetometer. In: 2017 International Conference on Indoor Positioning and Indoor Navigation (IPIN), pp. 1–8. IEEE (2017)

45. Obayemi, A., Nguyen, K.A.: Uncertainty quantification of multimodal models. In: International Conference on Industrial, Engineering and Other Applications of Applied Intelligent Systems, pp. 272–280. Springer (2025)

46. Oneto, L., et al.: Train delay prediction systems: a big data analytics perspective. Big Data Res. **11**, 54–64 (2018)

47. Papadopoulos, H., Vovk, V., Gammerman, A.: Regression conformal prediction with nearest neighbours. J. Artif. Intell. Res. **40**, 815–840 (2011)

48. Parbo, J., Nielsen, O.A., Prato, C.G.: Passenger perspectives in railway timetabling: a literature review. Transp. Rev. **36**(4), 500–526 (2016)

49. Pongnumkul, S., Pechprasarn, T., Kunaseth, N., Chaipah, K.: Improving arrival time prediction of thailand's passenger trains using historical travel times. In: 2014 11th International Joint Conference on Computer Science and Software Engineering (JCSSE), pp. 307–312. IEEE (2014)

50. Pradhan, R., Kumar, A., Kumar, M., Sharma, B.: Simulating and analysing delay in indian railways. In: IOP Conference Series: Materials Science and Engineering, vol. 1116, p. 012127. IOP Publishing (2021)

51. Quinlan, J.R.: Induction of decision trees. Mach. Learn. **1**, 81–106 (1986)
52. Şahin, İ: Markov chain model for delay distribution in train schedules: assessing the effectiveness of time allowances. J. Rail Transp. Plann. Manage. **7**(3), 101–113 (2017)
53. Sara, L., Houda, J., Mohamed, A., et al.: Predict france trains delays using visualization and machine learning techniques. Procedia Comput. Sci. **175**, 700–705 (2020)
54. Seber, G.A., Lee, A.J.: Linear Regression Analysis. Wiley (2003)
55. Shafer, G., Vovk, V.: A tutorial on conformal prediction. J. Mach. Learn. Res. **9**(3) (2008)
56. Shi, R., Wang, J., Xu, X., Wang, M., Li, J.: Arrival train delays prediction based on gradient boosting regression tress. In: Proceedings of the 4th International Conference on Electrical and Information Technologies for Rail Transportation (EITRT) 2019: Rail Transportation Information Processing and Operational Management Technologies, pp. 307–315. Springer (2020)
57. Shi, R., Xu, X., Li, J., Li, Y.: Prediction and analysis of train arrival delay based on xgboost and bayesian optimization. Appl. Soft Comput. **109**, 107538 (2021)
58. Spanninger, T., Trivella, A., Büchel, B., Corman, F.: A review of train delay prediction approaches. J. Rail Transp. Plann. Manage. **22**, 100312 (2022)
59. Tiong, K.Y., Ma, Z., Palmqvist, C.W.: A review of data-driven approaches to predict train delays. Transp. Res. Part C: Emerg. Technol. **148**, 104027 (2023)
60. Tsolaki, K., Vafeiadis, T., Nizamis, A., Ioannidis, D., Tzovaras, D.: Utilizing machine learning on freight transportation and logistics applications: a review. ICT Express (2022)
61. Vafaei, S., Yaghini, M.: Online prediction of arrival and departure times in each station for passenger trains using machine learning methods. Transp. Eng. **16**, 100250 (2024)
62. Vapnik, V.: The Nature of Statistical Learning Theory. Springer (2013)
63. Vovk, V.: Conditional validity of inductive conformal predictors. In: Asian Conference on Machine Learning, pp. 475–490. PMLR (2012)
64. Vovk, V., Gammerman, A., Shafer, G.: Algorithmic Learning in a Random World. Springer Nature (2022)
65. Wang, P., Zhang, Q.p.: Train delay analysis and prediction based on big data fusion. Transp. Saf. Environ. **1**(1), 79–88 (2019)
66. Wang, Y., Wen, C., Huang, P.: Predicting the effectiveness of supplement time on delay recoveries: a support vector regression approach. Int. J. Rail Transp. **10**(3), 375–392 (2022)
67. Watanabe, S., Mori, Y., Takatori, Y., Yonemoto, K., Tomii, N.: Train traffic simulation algorithm based on historical train traffic records. Comput. Railways XVI pp. 285–32 (2018)
68. Wen, C., Huang, P., Li, Z., Lessan, J., Fu, L., Jiang, C., Xu, X.: Train dispatching management with data-driven approaches: a comprehensive review and appraisal. IEEE Access **7**, 114547–114571 (2019)
69. Wen, C., Mou, W., Huang, P., Li, Z.: A predictive model of train delays on a railway line. J. Forecast. **39**(3), 470–488 (2020)
70. Yaghini, M., Khoshraftar, M.M., Seyedabadi, M.: Railway passenger train delay prediction via neural network model. J. Adv. Transp. **47**(3), 355–368 (2013)
71. Zhang, D., Du, C., Peng, Y., Liu, J., Mohammed, S., Calvi, A.: A multi-source dynamic temporal point process model for train delay prediction. IEEE Trans. Intell. Transp. Syst. (2024)

72. Zhang, L., Feng, X., Ding, C., Liu, Y.: Mitigating errors of predicted delays of a train at neighbouring stops. IET Intel. Transport Syst. **14**(8), 873–879 (2020)
73. Zhou, P., Chen, L., Dai, X., Li, B., Chai, T.: Intelligent prediction of train delay changes and propagation using rvflns with improved transfer learning and ensemble learning. IEEE Trans. Intell. Transp. Syst. **22**(12), 7432–7444 (2020)
74. Zhuang, H., Feng, L., Wen, C., Peng, Q., Tang, Q.: High-speed railway train timetable conflict prediction based on fuzzy temporal knowledge reasoning. Engineering **2**(3), 366–373 (2016)
75. Zilko, A.A., Kurowicka, D., Goverde, R.M.: Modeling railway disruption lengths with copula bayesian networks. Transp. Res. Part C: Emerg. Technol. **68**, 350–368 (2016)

Conformal Mining for Multiple Error Correction in the Data

Ilia Nouretdinov$^{(\boxtimes)}$

Centre for Reliable Machine Learning, Royal Holloway University of London,
Egham Hill, Egham, Surrey TW20 0EX, UK
`i.r.nouretdinov@rhul.ac.uk`

Abstract. In this work, we discuss a possible extension of the recently developed Conformal Association Rule Mining (CARM) technique to address multiple errors in the data. We utilize the calibration properties of the Conformal Prediction framework for reliable machine learning to bound the false alarm rate and introduce a flexible stopping criterion for the iterations. We developed new non-conformity measures for this task and provided a case study with images of handwritten digits.

Keywords: Conformal prediction · Data correction · Non-conformity measures

1 Introduction

In this work, we apply a combination of the conformal prediction framework and the association rule technique to the data correction task. This is done in a more general form than earlier in [5].

Conformal prediction (CP) [6] is a recently developed framework for reliable machine learning. In the area of classification, it complements a prediction with an individual measure of confidence. The core detail of CP is the function called the non-conformity (strangeness) measure (NCM), which inputs a sequence of data examples and quantifies how different each of them is from the others.

In [5], we measure the strangeness by a method that can be called Rule-Exception. Assume that we have a set of examples presented as feature vectors. Additionally, a set of possible association rules exists, i.e., regular IF-AND-THEN dependencies between the features. Analysing the set of examples, we assess the applicability of each rule and assign a weight accordingly. A rule is accepted if it is true for the majority of examples, although it may have exceptions. The fewer exceptions it has, the greater the weight assigned to the rule. Then, for a concrete example, its relative strangeness is calculated as the sum of the weights of the rules for which it is an exception.

Another source of inspiration for this work is the notion of regularity. It is an extension of the consistency developed in [4] and the Gestalt profile presented in [3]. The principle of regularity means that the non-conformity measure is derived from a measure of the regularity of the whole set of examples, as a kind

K. An Nguyen and Z. Luo (Eds.): Alexander Gammerman Festschrift, LNCS 16290, pp. 411–431, 2026.
https://doi.org/10.1007/978-3-032-15120-9_18

of its discrete gradient. Like NCM, regularity is a general concept that can be explored in various ways. In the context of data correction, regularity means a measure of the quantity of rules that the sequence of data examples follows. This approach is more general and can be extended beyond using the concrete basis of association rules. Having a regular measure of the data sequence that increases at each step, we initially prevent the iterative data correction procedure from falling into a loop.

In [5], a data set with binary labels was used, and a number of mistakes were introduced by changing the label to its opposite. Then, by reviewing the data in online mode, the strangest examples were identified using CP techniques and recommended for correction. The principal advantage of the CP framework was to limit the number of false alarms in error detection.

Now, in the case of multiple error correction, its role is extended to the selection of the stopping point of the correction procedure. A feature vector is adjusted until it no longer appears unusual compared to the others.

A challenge in this area is the evaluation. What is the real goal of the data correction? This is conventional and depends on the concrete details of the problem. Unlike [5], we do not artificially impute errors into the data, but instead concentrate on denoising the data. Therefore, correcting a data image means improving it as a training example (which can be measured by the accuracy after training); an error is an image (or pixel) that impairs the prediction quality. The correction is successful to the extent that it improves the accuracy of the classification.

In Sect. 2, we recall the key notions and earlier achievements that are necessary to understand this work. In Sect. 3, we present new contributions and algorithms, including the multiple error correction scheme, and necessary amendments to non-conformity. In Sect. 4, we show the experimental output of a case study on images of handwritten digits and discuss it with respect to the evaluation goal. In this application, we also use more complex composite association rules than those used earlier. Section 5 concludes the chapter.

2   Background

2.1   Data Correction Task

Error detection and correction are known to be tools for data editing. In a narrow sense, error detection/correction refers to the discovery/elimination of a real mistake that may occur during the data collection process. This is how the task was treated in [5]. There, it was assumed that some of the labels were occasionally replaced with the wrong opposite ones. Therefore, they had to be corrected or inverted to the right state.

More generally, data correction involves smoothing the data and removing noise that is not erroneous but is practically useless for further data processing. This may include tools such as removing outliers from the data, but we prefer the approach of improving them, because an outlier may still be informative in some aspects.

In particular, if the data set is a collection of images (as in our example), this means an improvement in its visual quality. The task is to examine the images and identify any pixels that appear strange and should have their value inverted from 0 to 1 or from 1 to 0 (if the images are on the binary scale).

To be practical and have measurable performance, this task should be linked to the usual goal of data editing, which can be formulated as the ability of the data sequence to serve as a training set for machine learning models. So, by *data correction*, we mean its editing with the aim of improving the accuracy of the prediction after training on the data. Correspondingly, by an *error* we will mean a record (a feature of one of the data examples) such that its replacement makes a positive contribution to the quality of the training set as far as it can be assessed by the prediction accuracy.

Returning to the sample task, when the data examples are binary images, we can discuss the separation of two classes of images, such as handwritten digits 5 and 8. This is typically done after processing a training set that includes examples of both classes. Therefore, the final accuracy of the separation depends on the quality of the training set and may be used to validate whether an error correction attempt was truly beneficial for its improvement.

2.2 Conformal Prediction

Conformal prediction (CP) [6] is a framework for making reliable predictions without the use of complex probabilistic models, with the only assumption that the data are *i.i.d.* (randomly generated by a power distribution) or *exchangeable* (the examples follow in random order).

It is applied within a machine learning setting. There is a sequence of objects $z_1, \ldots, z_l$ and the task is to provide a prediction for a new example z_{l+1}. All examples belong to an object space Z. In supervised learning, each $z_i \in Z$ is a pair (x_i, y_i) consisting of a feature vector x_i and a label y_i, so Z is the product of a vector space X and a set Y of possible labels.

Conformal predictors can operate on top of another ('underlying') ML algorithm. This can be any classification or regression algorithm; the only requirement is that we can extract a 'score' from it, from which we can develop a *non-conformity measure (NCM)*. The NCM numerically evaluates how different a new example is from a set of other examples.

Typically, the non-conformity score is defined as some kind of redundancy, or a difference between the real label and the prediction made by a standard machine learning method such as Nearest Neighbours, Neural Nets, Decision Trees, Random Forest, or Support Vector Machine. For example, if the underlying method is *k-nearest neighbors*, the strangeness of an example with respect to a bag may be the proportion of its k nearest neighbors that have a different label. Intuitively, if an example with one label is surrounded by examples with a different label, then we can consider it to be strange (anomalous).

Formally, a NCM is a function $A : Z^{(*)} \times Z \to \mathbb{R}$ where Z is the set of all possible examples and $Z^{(*)}$ is the set of all bags of examples of Z. A tells us how different an example $z_i \in Z$ is from a bag (multi-set) $Z^{(*)}$ by assigning it a score.

Consider now a new example z_{l+1}. Our null hypothesis is that z_{l+1} was drawn *i.i.d.* from the same distribution as the observed examples $z_1, \ldots, z_l$. Adding this new example z_{l+1} to our data set gives us an extended bag $\{z_1, \ldots, z_l, z_{l+1}\}$. Now we can use an NCM to compute

$$\alpha_i = A(\{z_1, \ldots, z_{i-1}, z_{i+1}, \ldots, z_{l+1}\}, z_i) \tag{1}$$

for each example $i = 1, \ldots, l+1$. Note that it is applied to each of the examples, not just to the new one.

The scores produced by the NCM for each example may not be particularly informative. But if we compare α, for example, z_{l+1} with that of all other examples, then we can quantify how unusual it is by using the notion of a p-value:

$$p(z_{l+1}) = \frac{\#\{i = 1, \ldots, l+1 : \alpha_i \geq \alpha_{l+1}\}}{l+1} \tag{2}$$

This p-value tests the null hypothesis stated above. Usually, a threshold or *significance level* ε is selected, and if the p-value of the example z_{l+1} is higher than this threshold, then it is included in the prediction set.

2.3 Conformal Predictors in Data Correction

How can the CP framework be used in error detection? According to its validity (calibration) in [6], if z_{l+1} is actually generated by the same mechanism as $z_1, \ldots, z_l$, then the probability that p-value falls below ε is at most ε. This can be used as a tool for anomaly detection for both labeled and unlabeled examples [2].

In the context of data correction, this means that a possible mistake in z_{l+1} can cause a high value of NCM α_{l+1} and a low value of $p(z_{l+1})$. On the other hand, the validity property makes a natural limitation on the false alarm rate. If there is no real error in z_{l+1}, and an alarm is raised if $p < \varepsilon$, the probability of the false alarm is below the same threshold ε.

However, in this way, only the whole example is marked as abnormal, and there is no answer to what error caused its abnormality. In [5], one of the tools was to develop an NCM that is more intuitive to explain. Therefore, NCM engineering is important for this task.

2.4 Association Rules

Another fundamental approach is the *Association Rule* learning [1]. It expands the idea that errors in the data can be caught as exceptions to the rules that are true for most of the data. More references to related work can be found in [5].

Although these rules are less popular than Decision Trees, in the context of CP, they appear to be interesting from a new perspective, when the goal of making an immediate conclusion is replaced with an assessment of strangeness. The advantage of association rules is their transparency, which is needed to move from the stage of error detection to the stage of error correction.

The most basic form of association rules is the following.

- **IF** A-**th feature has a value of** a, **THEN** B-**th feature has a value of** b

This corresponds to the initial form of the association rules. The next step is to include more complex logical clauses.

- **IF** A-**th feature** $= a$ **AND** B-**th feature** $= b$ **THEN** C-**th feature** $= c$**.**

It is important to fix some *pool* of association rules that are considered. If the data size is limited, it should not be too large or too complex. For example, in the C-th case, the feature may be limited to only the label if the task is classification or correction of the label. However, such a restriction does not make sense if the correction task is more general.

These rules are basic because the involved features are fixed. In the context of image processing, we may be interested in *composite* rules that are invariant on geometric shifts.

2.5   Rule-Exception NCM

The solution suggested in [5] is summarised in Algorithm 1.1.

A *Rule-Exception* algorithm was used to define an NCM based on association rules. For each rule of the pool, the hypothesis of its irrelevance was checked. If it was discarded with a small p-value, the rule was assigned a weight back proportional to the corresponding p-value on the logarithmic scale. This statistical p-value is different from that used in the CP framework. Here, the p-value is related to checking the distribution of labels within the area of a rule's applicability against the Bernoulli distribution of the labels by the relevant statistical test. The weight assigned to a rule is added to the NCM of each *exception* that is an example where the rule is discarded.

This was the way to revise the understanding of the role of association rules. However, we will not need to use this method directly in this study. Therefore, we refer to [5] for further details.

2.6   Consistency NCM

Another useful idea we use is *consistency* from [4]. It was an alternative to the definition of NCM as a redundancy function. In that study, the strangeness of an example with respect to the set was estimated by examining the consistency of the remaining data after excluding that example. The consistency was understood in a statistical sense as the number of features that show a significant difference between two classes in the distribution. This was defined in a very specific case where the number of features (voxels) is extremely large. In that case, consistency was directly related to the separability of the classes, although this is not the only possible understanding.

Algorithm 1.1. Association Rule NCM (Rule-Exception version)

INPUT: example $z = (x_{l+1}, y_{l+1})$ where $x \in \{0,1\}^m$, $y \in \{0,1\}$

INPUT: bag of (same kind) examples $Z = \{(x_1, y_1), \ldots, (x_l, y_l)\}$

INPUT: pool (set) P of association rules $r =$'IF $\ldots$ THEN $y = Y$"

$y_a = \frac{\sum_{i=1}^{m+1} y_i}{m+1}$

$\alpha := 0$

for $r \in P$ **do**

    support base $S := \{i = 1, \ldots, l+1 : $ (conditions of r) are true $\}$

    support base size $s := |S|$

    $p_r :=$ p-value of the empirical distribution of y_i within the support base, tested against the Bernoulli distribution with probability y_a;

    calculate average of y_i over the support base:

$$y'_a := \frac{\sum_{i \in S} y_i}{s}$$

    if $y'_a > y_a$ **then**

      $C := 1$

    else

      $C := 0$

    end if

    if $y_{l+1} \neq C$ **then**

      $\alpha := \alpha - \log p_r$

    end if

end for

OUTPUT: non-conformity score $\mathcal{A}(z, Z) = \alpha$

Exploring this idea in a more general way, we prefer to call it *regularity*. In general, regularity is a function of a data set only, not dependent on the order of the examples.

$$r = R(\{z_1, ..., z_n\}) \tag{3}$$

Like NCM, it is a general concept without formal restriction, but with expectation to be defined in accordance with some intuitive understanding.

The strangeness (non-conformity) of an example with respect to a set is defined as the difference between the regularity of the set without and with this example included. In other words, an example is strange with respect to a set if its addition to this set makes it less regular.

So, the core parameter of CP shifted from NCM, which is the distance between an example and a set, to regularity as a function of a set only. The connection between these functions can be expressed as follows:

$$A(\{z_1, ..., z_{i-1}, z_{i+1}, ..., z_{l+1}\}, z_i)$$

$$= R(\{z_1, ..., z_{i-1}, z_{i+1}, ..., z_{l+1}\}) - R(\{z_1, ..., z_{i-1}, z_i, z_{i+1}, ..., z_{l+1}\})$$

where A is the NCM, and R is the regularity function.

3   Methodology

In this section, we present the principal contributions of this work: regularity and its connection to CP; regularity related to association rules; and a multiple error prediction scheme based on the CP framework.

These topics are closely related to one another. Although they develop the ideas of [5], they do not explore them in a straightforward way and can be studied independently of that work.

3.1   Regularity as a Source of NCM

The Rule-Exception approach for NCM has the disadvantage that it does not provide any guarantees against falling into a loop if used sequentially to detect and correct multiple errors. A way to avoid this is to define a value that is a function of the data sample as a whole, which strictly increases (or decreases) after each correction.

Therefore, we adopt the idea of *regularity* mentioned in Sect. 2.6. In line with our goal of data correction, we aim to shift our focus to a more constructive objective. An example is strange if it can be modified into an example that is less strange, and the whole sequence of examples would become more regular. Therefore, we prefer to speak of the correction (modification) of the examples rather than their exclusion from the data.

Therefore, we compare the regularity of the data sequence not to a shorter data sequence, but to a data sequence of the same length, differing only in one or another feature. To clarify this idea, we assume that all features are presented in binary form. So, a correction means that one of the features of one of the examples is changed to the opposite.

Schematically, this can be expressed as

$$A(\{z_1, ..., z_{i-1}, z_{i+1}, ..., z_{l+1}\}, z_i)$$

$$= |\{j : R(\{z_1, ..., z_{i-1}, C_j(z_i), z_{i+1}, ..., z_{l+1}\}) > R(\{z_1, ..., z_{i-1}, z_i, z_{i+1}, ..., z_{l+1}\})\}|$$

where A is the NCM, and R is the regularity function, and $C_j(z_i)$ means that z_i is changed in its j-th feature. In other words, an example is strange with respect to the set if it has a better influence on the regularity in a modified (corrected) form than in the original form.

Algorithm 1.2 gives a scheme of how the NCM function can be linked to a regularity function. As far as the examples are presented as binary feature vectors, we implement here a possible discrete analogue of gradient descent. This is done as follows. Each pair (example, feature) can have the property that the overall regularity of the data sequence increases if the feature of this example is changed (from 0 to 1, or from 1 to 0). We interpret it as a feature where correction is desirable. For each example, we can find the number of such features. The meaning of such strangeness is that the example has a relatively large number of features that require correction.

Algorithm 1.2. Regularity NCM scheme

INPUT: example $z_{l+1} \in \{0,1\}^m$
INPUT: bag of (same kind) examples $Z = \{z_1, \ldots, z_l\}$.
INPUT: regularity function $\mathcal{E}$ on bags of examples
$E := \mathcal{E}\left(Z \cup \{z_{l+1}\}\right)$
$\alpha := 0$
for $j := 1, \ldots, m$ **do**
  $z_{l+1}^j := 1 - z_{l+1}^j$
  $E' := \mathcal{E}\left(Z \cup \{z_{l+1}\}\right)$
  $z_{l+1}^j := 1 - z_{l+1}^j$
  if $E' > E$ **then**
    $\alpha := \alpha + 1$
  end if
end for
OUTPUT: non-conformity score $\mathcal{A}(z_{l+1}, Z) = \alpha$

3.2 Regularity with Association Rules

In general, the regularity function is a kind of black-box, and it is not bound strictly to a concrete approach. However, we present here the scheme of a regularity function based on a pool of association rules. We understand the regularity of a data sequence in the following sense. The more rules that are satisfied (over the whole picture set, with a few exceptions), the larger the value of this function.

Algorithm 1.3. Regularity function based on Association Rules

INPUT: bag of (same kind) examples $Z = \{z_1, \ldots, z_{l+1}\} \subset \{0,1\}^m$
INPUT: pool P of association rules r
$E := 0$
for $r \in P$ **do**
  $N :=$ number of cases were the conditions of r are true
  $K :=$ number of cases were the conditions of r are true and the clause is true
  $p_r :=$ p-value of binomial test applied to (N, K) with probability $\frac{1}{2}$
  $E := E - \log_2 p_r$
end for
OUTPUT: regularity $\mathcal{E}(Z) = E$

We present it for the case of data with binary features (valued 0 or 1), as most other cases can be converted to this form. Thus, we assume that a data sample includes $l+1$ binary m-dimensional vectors. Next, assume that there is a pool P of association rules. The regularity of the data sample is understood as the total number of rules that are in force for this data set. Here, we do not mean that they are strict and allow exceptions, although these exceptions decrease their weight. Formally, the binomial test is applied to measure how strongly a rule is held within the data. We calculate the number N of examples where the rule is defined (its conditions are true), and K is the number of them that really support the rule. The binomial test on the log-scale measures the significance of this event. This is summarised in Algorithm 1.3.

3.3   Multiple Error Correction

The method from [5] was concentrated on the correction of *one* error in the data. One of the challenges for further usage of conformal prediction is the following. Once an example is corrected, the configuration loses its symmetry property: the new example becomes 'marked' and is no longer generated by the same mechanism as the training examples.

Therefore, we need to impose more symmetry on the correction procedure. Although the primary aim is correction of the *new* example, we have to note that all the examples are processed in an equal way. This is necessary to preserve the validity properties of the conformal prediction: the corrected examples remain exchangeable, and the probability of a limited false alarm remains bounded in each step. This means that the second correction of the new examples should take more than one step: in the intermediate steps, the corrections are applied to other examples.

Algorithm 1.4 gives the new scheme of the whole loop with the stopping criterion related to a significance level. The essence of the iterative process is the growth of regularity by a discrete analogue of gradient descent. The scheme does not distinguish between the new example and the training example. Therefore, the exchangeability holds, and the calibration property remains true at the stage of the second and further corrections, as well as at the first one.

This process may end when it reaches a local maximum, but we may prefer to stop it earlier when the new example is no longer unusual compared to the others.

4   A Case Study

4.1   Data

In this work, we utilize the well-known United States Postal Service (USPS) benchmark dataset [7]. It contains 9,298 grayscale images of handwritten digits of size 16×16 pixels, belonging to 10 classes $\{0, 1, \ldots, 9\}$.

For our purposes, it will be sufficient to select a small subset of the randomly chosen data examples from $n = 100$. We take examples only from two classes: 5 and 8, because this pair of digits is known for its relatively hard separability. We do not divide it into training and testing sets, but we use it in the leave-one-out cross-validation mode. We convert the features (the pixels) into a binary format. Each value is converted to 1 or 0, according to its sign.

Algorithm 1.4. Sequential error correction scheme

INPUT: example $z_{l+1} \in \{0,1\}^m$

INPUT: bag of (same kind) examples $Z = \{z_1, \ldots, z_l\}$.

INPUT: regularity function $\mathcal{E}$

INPUT: NCM function $\mathcal{A}$

INPUT: significance threshold ε

$E := \mathcal{E}\left(Z \cup \{z_{l+1}\}\right)$

$p := 0$

while $p < \varepsilon$ **do**

    for $i := 1, \ldots, l+1$ **do**

        for $j := 1, \ldots, m$ **do**

            $z_i^j := 1 - z_i^j$

            $E_{ij} := \mathcal{E}\left(Z \cup \{z_{l+1}\}\right)$

            $z_i^j := 1 - z_i^j$

        end for

    end for

    find $(i,j) : E_{ij} \to \max$

    $z_i^j := 1 - z_i^j$

    for $i := 1, \ldots, l+1$ **do**

        $\alpha_i := A(z_i, Z \cup \{z_{l+1}\} \setminus \{z_i\})$

    end for

    $p := \dfrac{|\{i=1,\ldots,l+1 : \alpha_i \geq \alpha_{l+1})\}|}{l+1}$

end while

4.2 Basic Settings

In our setting, we model the multiple error correction. This is rather different from [5], as it detects a single mistake within an example. Now, an example takes the form of an image, and correcting one pixel does not change it significantly.

Unlike the example in [5], we do not use artificial error imputation for this data set. We assume that the images already include some noise. The purpose of the correction is understood practically. The goal is to make the classes of images more separable after correcting the individual images. If such an improvement is observable, this confirms that the correction is really 'correct' independently of the cause of the mistake.

Another difference from [5] is that we are looking for errors in all the features of the objects, not in their labels. More exactly, everywhere with the exception of labels. We will not involve the labels in the image correction stage at all. The correction will be performed in a 'blind' mode, solely on a sequence of unlabeled images, while the labels will be reserved for the evaluation stage of the correction quality.

4.3 Association Rules

In the pool of association rules, we consider the dependencies between neighbouring pixels in two-dimensional images. The rules have the following form, similar to that used in cellular automata. In the following description, we mean that the number of features $m = k^2$ and (i,j)-th pixel is the same as the $((i-1)k + j)$-th feature.

- **IF** $(i-1, j)$**th pixel** $= Y_U$ **AND** $(i+1, j)$**th pixel** $= Y_D$
  AND $(i, j-1)$**th pixel** $= Y_L$ **AND** $(i, j+1)$**th pixel** $= Y_R$
  THEN (i, j)**th pixel** $= Y$
  where $Y, Y_U, Y_D, Y_L, Y_R \in \{0, 1\}$.

Unlike the example in [5], these rules have a sliding composite form, not bound to concrete features, because both i and j can take values in the range of $\{2, \ldots, k-1\}$.

4.4   Details of Running

For our experiment, we randomly selected examples from two classes of digits: 5 and 8. The total size of the selection is 100. As mentioned, at this stage we do not use labels, so the input at this stage is a sequence of $n = 100$ binary data images (black and white) of size 16×16.

The work begins by calculating the overall regularity (Algorithm 1.3) of the actual data sequence, using the pool of rules described in Sect. 4.3. Recall that the labels are not used as part of the input at this stage. For the core algorithm, they are all considered as various images from the same collection.

During each iteration, the plan is to identify the most unusual pixel (within any of the images) and correct it. This is done by checking each possible correction (alteration of a pixel, either from 0 to 1, or from 1 to 0).

The aim of this analysis is twofold. The first aim is to assign each pixel a corresponding increment of the overall regularity value. This will show the direction of the best correction. The change is applied to all of the pixels where the increment is the highest. Typically, there are more than one.

The second aim is to calculate a non-conformity score for each individual image (not just a pixel) using Algorithm 1.2.

In this way, two processes are observed in parallel: first, all images are corrected step by step, and second, the p-value assigned to each image is either increasing or decreasing.

4.5   Evaluation

We observe the first 50 iteration steps of the correction process. Due to our way of resolving the ties, many corrections are made in each of these steps.

We assume that the aim of the correction is to improve the separability of the data. For evaluation, we compare the output with the accuracy of leave-one-out 1-nearest-neighbour two-class separation. This is the only stage at which the labels play a role.

By evaluation, we mean to determine whether the goal of increasing regularity shares something in common with the goal of increasing separability. We observe it on concrete examples. If a correction is necessary to change how the example is classified, it may be either useful or harmful to accuracy.

4.6   Results

Not surprisingly, most of the images (91 out of 100) are classified in a constant manner, without any changes during the image correction process. Therefore, we pay special attention to the remaining 9 of them. These cases, along with their corresponding numbers, are shown in Figs. 1, 2, 3, 4, 5, 6, 7, 8 and 9.

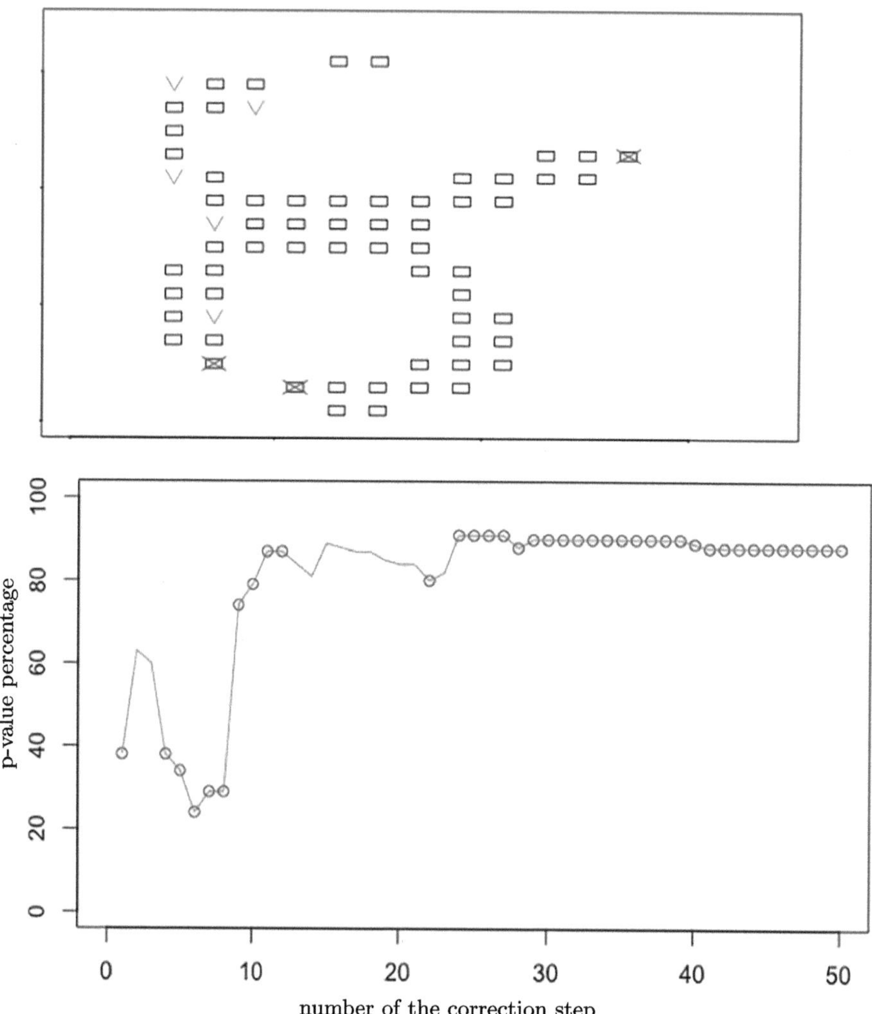

Fig. 1. Image **17**/100: on this and following graphs, the ticked pixels are included, the crossed are excluded during the correction; the plot shows how p-value changes during 50 iterations of the correction, and how this corresponds to an error of the leave-one-out classification (a circle for an error, nothing otherwise). The box above the graph shows the image under correction. The squares represent the pixels, which are initially set to 1. The crosses indicate the changes from 1 to 0, while the ticks denote the changes from 0 to 1.

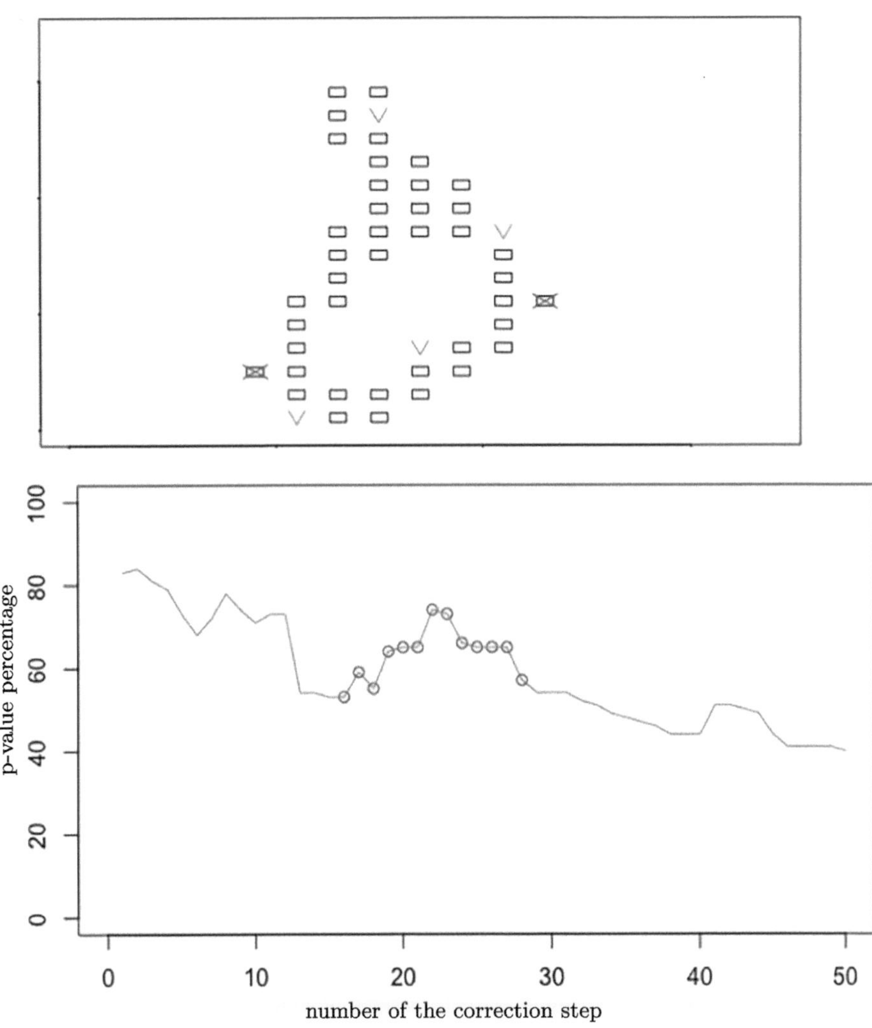

Fig. 2. Image **20**/100

Each plot consists of two parts. The upper part shows the difference between the image at the starting and ending points of the correction. The crosses represent deleted pixels (changed from 1 to 0), and the ticks represent added pixels (changed from 0 to 1). The general tendency is that increasing the regularity makes images smoother and of a more uniform width.

The lower part shows how the p-value assigned for this concrete example is changing, meanwhile, and how this corresponds to the accuracy of the prediction. A circle means that the leave-one-out 1-nearest-neighbour classifier would make a mistake on this concrete example after training on all the others (which are also being changed during the correction process).

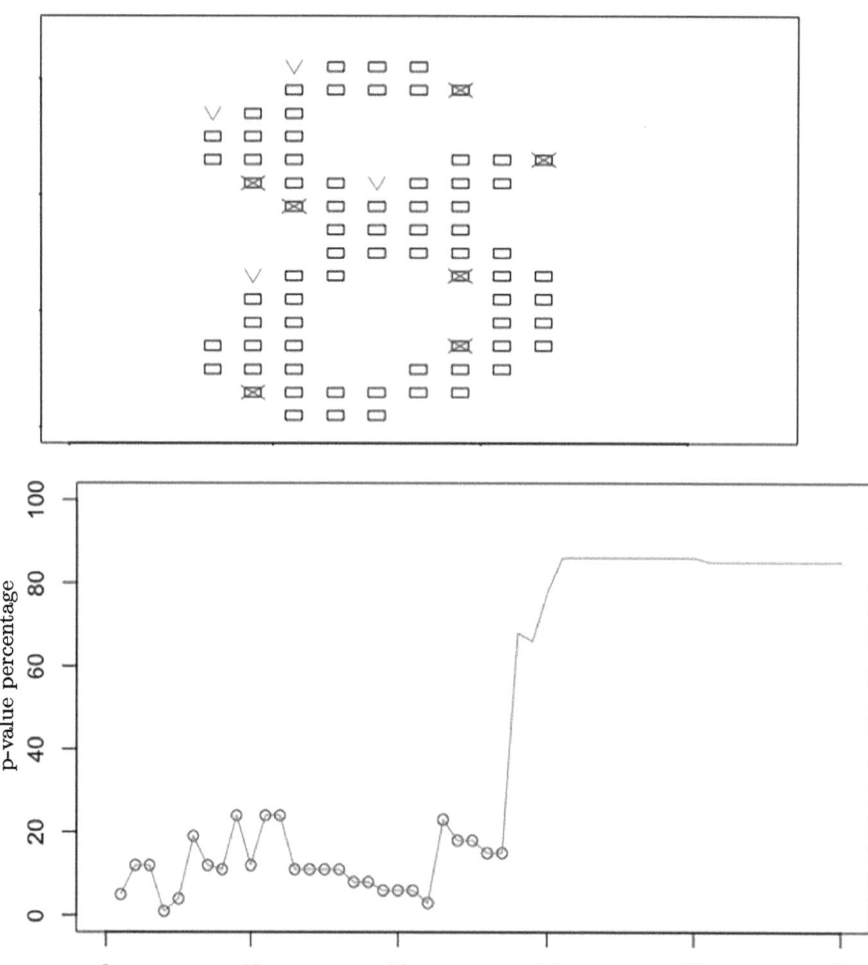

Fig. 3. Image **30**/100

4.7   Observations

Let us start with Fig. 3, which is for the image number **30**. This is an encouraging observation with the effect that we are interested in seeing: correction of an incorrectly classified image causes both an increase in the assigned p-value and a repair of its classification. Furthermore, it is observed that once the p-value reaches a high value of 50% (at step 28), confirming that the example no longer appears stranger than the average one, there is no need for further correction in this manner. However, with such small data, we can not expect that all observations are equally nice from this point of view. Let us look at the rest.

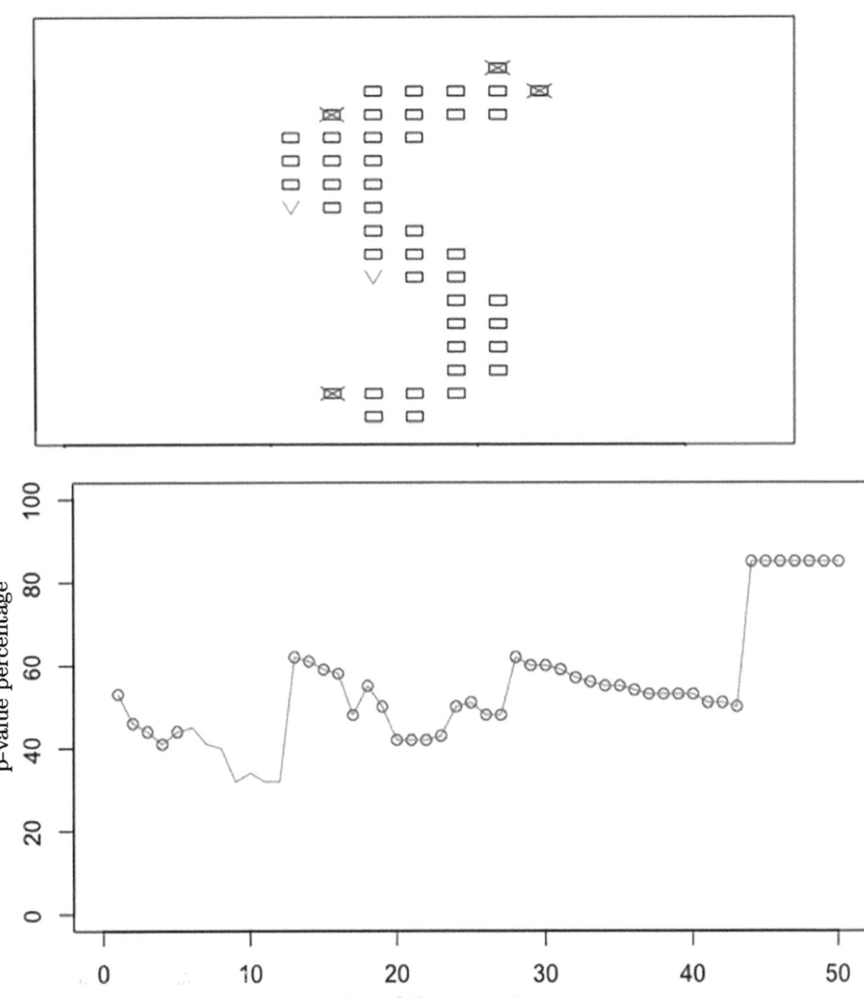

Fig. 4. Image **35**/100.

In most observed cases, we can see the following trend: the threshold of 50% may be insufficient, but a good stopping point for the correction process is when the p-value reaches 80%. The images in Figs. 2, 5, 7 and 9 follow this law. For image **20**, the p-value is above 80% and the prediction is correct initially. For the rest, it should be stopped immediately once this threshold is reached, as the later correction is useless or harmful for correct classification.

The image in Fig. 8 roughly follows the same trend. If stopped at 80% (step 2), the error will not be corrected, but it is corrected very soon after (step 4), and becomes stable for quite a long time.

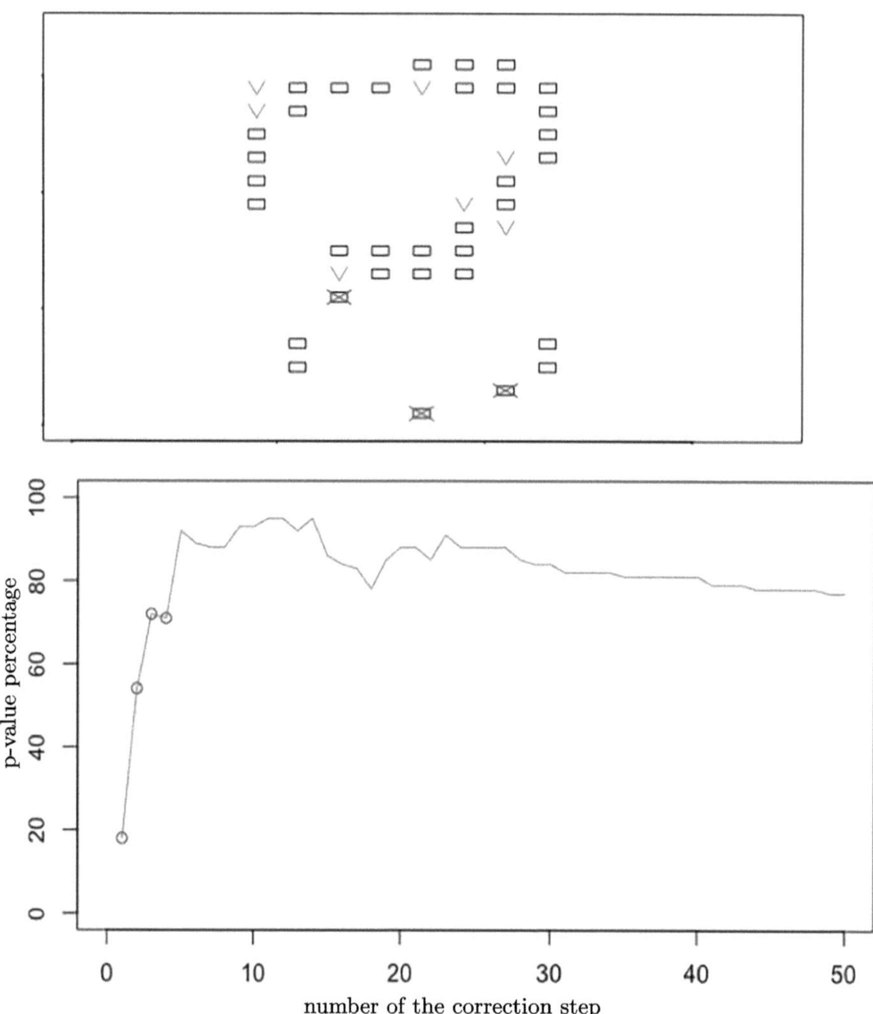

Fig. 5. Image **37**/100

The three remaining cases are more abnormal. The images in Figs. 1 and 4 are quite strange and hardly recoverable by small corrections. Therefore, dynamic shows only some occasional and unstable improvements.

The image in Fig. 6 also has an unusual configuration, and in the initial steps, the trend is opposite to the desirable one: by increasing the p-value, we make the prediction incorrect. However, in the later stage it becomes correct and stable, but only after the p-value reaches 90% (not 80%).

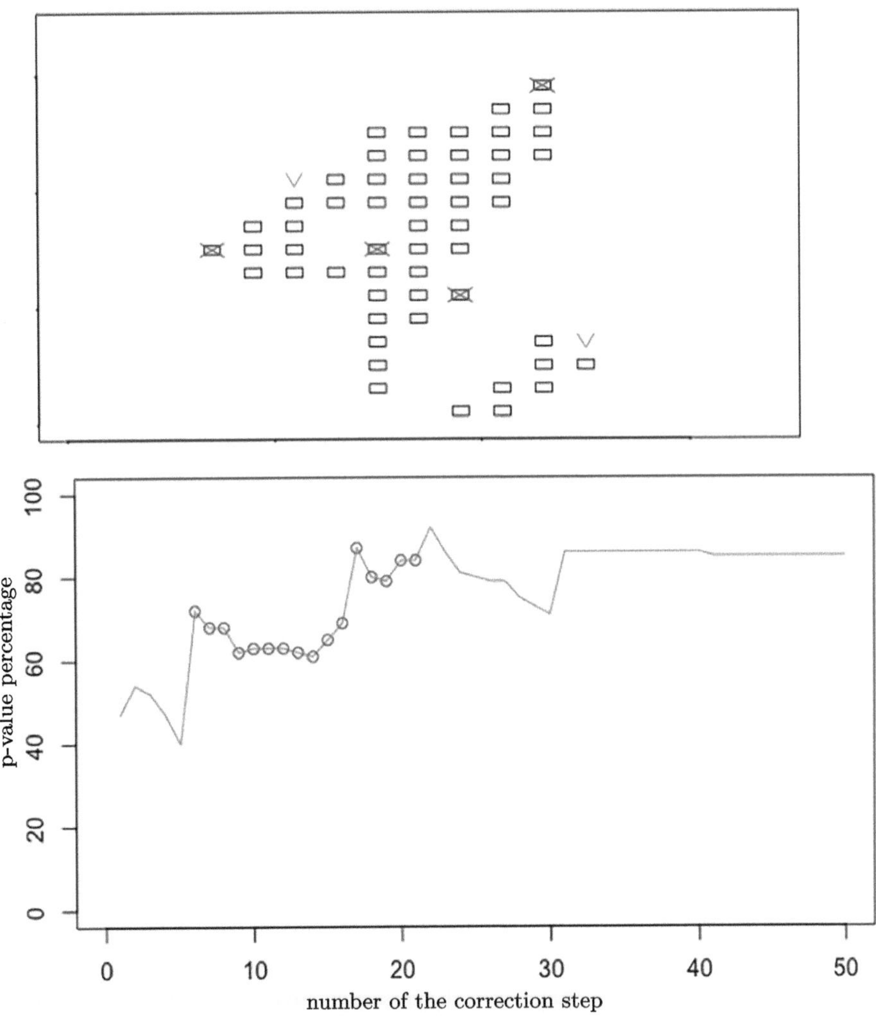

Fig. 6. Image **44**/100.

Together, these cases demonstrate a surprising finding: the optimal value of the stopping threshold is not very low, but rather approximately 0.8. When corrections are made below this level, they are useful for improving the accuracy of the classification. However, attempts to make the p value even higher than that lead to unnatural changes in the picture and are likely to prevent correct classification.

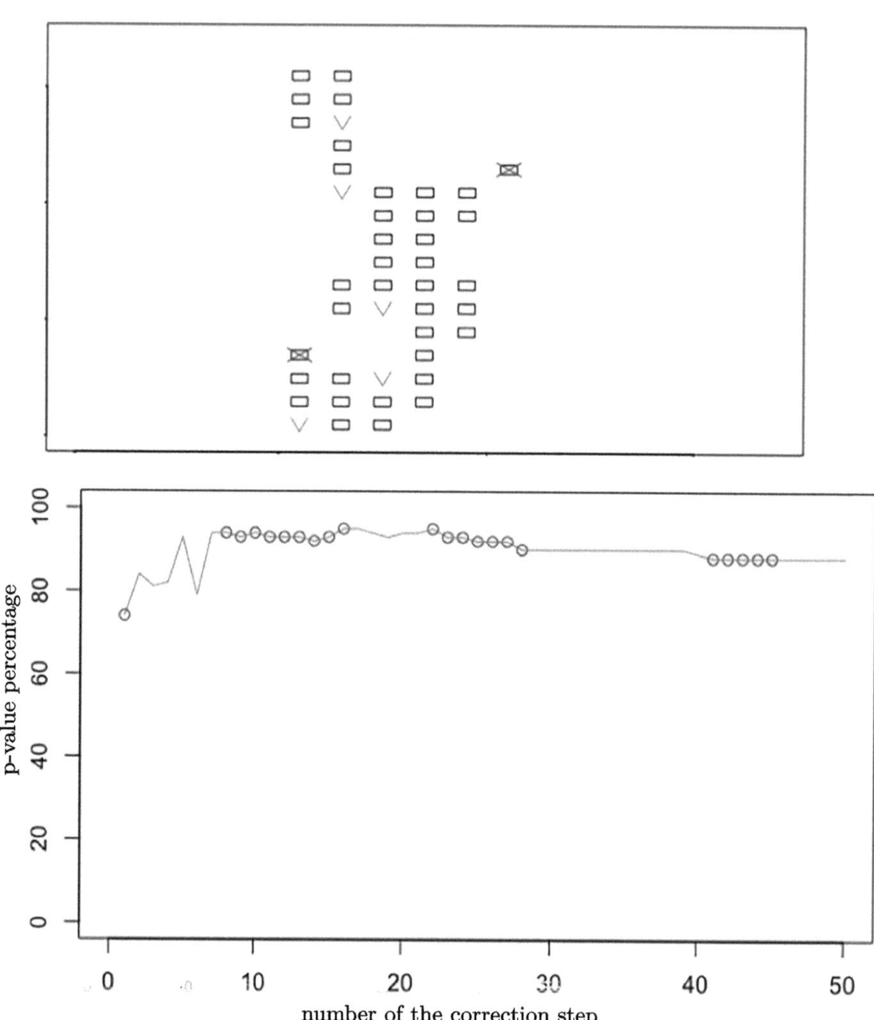

Fig. 7. Image **61**/100.

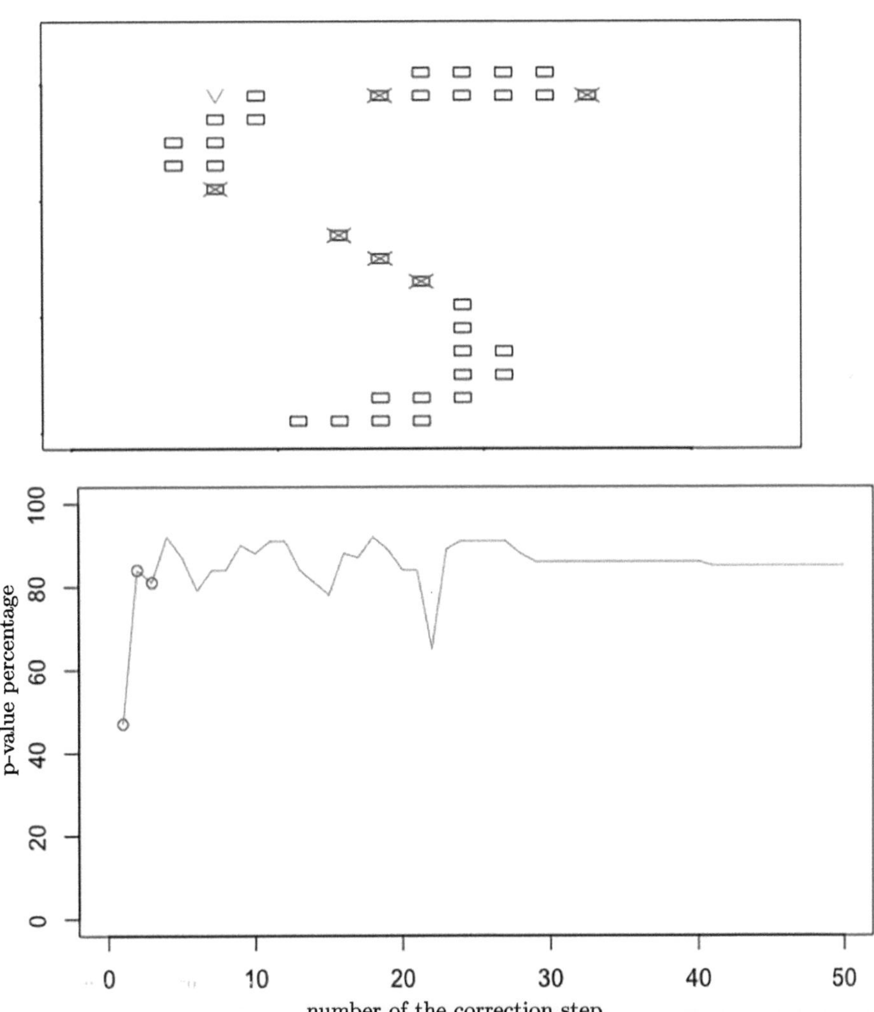

Fig. 8. Image **91**/100.

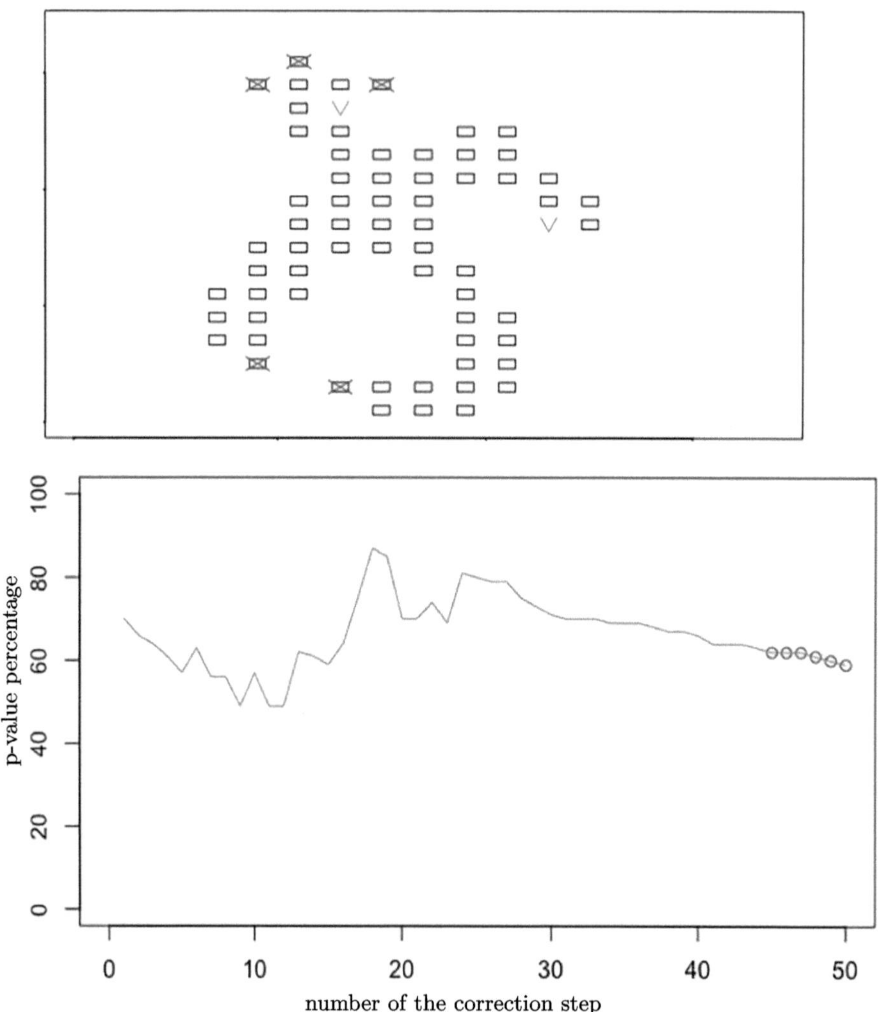

Fig. 9. Image **93**/100.

5   Conclusion and Further Development

We have developed a technique that combines conformal prediction and association rule mining to detect anomalies and applied it to find possible errors in the database. Like conformal prediction, they are valid under the assumption that the data are exchangeable, and we extended their applicability from correcting one error, as shown in [5], to multiple error correction, while maintaining the same assumption of exchangeability.

In addition to illustrating the methods, we made some initial observations on the effect of data correction on the accuracy of image classification. This was done for hand-written digits, and the next step would be to try this on other kinds of images.

Multiple error correction of the image always carries a risk of oversimplification and loss of essential content. We have observed that setting a threshold on the growth of the conformal p-value is a reasonable stopping criterion in most cases.

However, we have also observed some limitations on its optimal values: it should neither be too small nor too large. Specifically, its best value is around $80 - 90\%$. This was somewhat surprising, given that it initially met expectations based on the understanding of strangeness as having a very small p-value. These hypotheses will require further verification in larger datasets.

Another interesting topic for further development is adaptive feature extraction. It means marking the most informative features within a concrete data example. The difference from the feature selection task is that the set of selected features is not fixed and may depend on the example.

References

1. Agrawal, R., Imielinski, T. and Swami, A.: Mining association rules between sets of items in large databases. In: Proceedings of the ACM SIGMOD International Conference on Management of Data (1993)
2. Laxhammar, R.: Conformal Anomaly Detection: Detecting Abnormal Trajectories in Surveillance Applications. PhD Thesis. University of Skövde (2014)
3. Nouretdinov, I., Balinsky, A. and Gammerman, A.: Conformal anomaly detection for visual reconstruction using gestalt principles. In: Proceedings of the Ninth Symposium on Conformal and Probabilistic Prediction and Applications, vol. 128, pp. 151–170 (2020)
4. Nouretdinov, I., et al.: Machine learning classification with confidence: application of transductive conformal predictors to MRI-based diagnostic and prognostic markers in depression. Neuroimage **56**(2), 809–813 (2011)
5. Nouretdinov, I., Gammerman, J.: Conformal Association Rule Mining (CARM): a novel technique for data error detection and probabilistic correction. In: Proceedings of the Twelfth Symposium on Conformal and Probabilistic Prediction with Applications, vol. 204, pp. 267–286 (2023)
6. Vovk, V., Gammerman, A. and Shafer, G.: Algorithmic Learning in a Random World. 2nd edition. Springer (2022)
7. Handwritten digits USPS dataset. https://www.kaggle.com/datasets/bistaumanga/usps-dataset

Protected Probabilistic Classification Library

Ivan Petej[(⊠)]

Centre for Reliable Machine Learning, Royal Holloway, University of London,
Egham, Surrey TW20 0EX, UK
`i.petej@rhul.ac.uk`

Abstract. This chapter introduces a new **Python** package specifically designed to address calibration of probabilistic classifiers under dataset shift. The method, originally co-authored by Professor Gammerman in 2021, is here demonstrated in binary and multi-class settings and its effectiveness is measured against a number of existing post-hoc calibration methods. The empirical results are promising and suggest that our technique can be helpful in a variety of settings for batch and online learning classification problems where the underlying data distribution changes between the training and test sets.

Keywords: Machine learning classification · Probabilistic calibration · Online learning

1 Introduction

In machine learning, predictive models often output probability estimates that guide decision-making in high-stakes applications such as healthcare, finance, and autonomous systems. Such probability estimates must be well-calibrated, meaning that the predicted probabilities accurately reflect the true likelihood of events [3]. Poor calibration can lead to overconfident or under-confident predictions potentially causing erroneous decisions. Unfortunately, many standard probabilistic machine learning algorithms lack inherent calibration properties [4].

Probabilistic calibration seeks to improve the reliability of model confidence scores by aligning predicted probabilities with actual observed frequencies. A number of post-hoc calibration approaches (i.e., methods applied after initial algorithm training), such as *Platt scaling* [14], *isotonic regression* [22], *binning* [21], *Venn-ABERS* calibration [18] and *beta* calibration [7] have been used to convert underlying classifier scores or probabilities into well-calibrated probabilities. Such methods have shown to perform well, especially when the underlying data in the test set follows the same distribution as the data in training and calibration sets [17].

In many real world applications however, as soon as the algorithm is trained, the data distribution changes and the prediction algorithm may need to be retrained or recalibrated. In this setting, it may be advantageous to utilize an

K. An Nguyen and Z. Luo (Eds.): Alexander Gammerman Festschrift, LNCS 16290, pp. 432–448, 2026.
https://doi.org/10.1007/978-3-032-15120-9_19

algorithm which prevents a drop in the quality of calibration without the need to retrain the underlying classifier. In this chapter we specifically address calibration of probabilistic classifiers under such dataset shift. Our work is an extension of the study originally reported in [20] and co-authored by Professor Gammerman, extending it further to artificial and real-life multi-class problems. We further compare the performance of our method with other recently reported post-hoc calibration techniques, including those specifically aimed at streaming data, with some promising results.

2   Background and Related Work

Good probabilistic calibration ensures that a model's predicted probabilities align with the true likelihood of outcomes, which is vital for applications involving risk-sensitive decisions. Over the years, research in this domain has addressed various aspects, including over and under-confidence of traditional machine learning algorithms, the loss of calibration under data distributional shift and the resulting design of a number of post-hoc calibration methods.

In addition to traditional post-hoc calibration methods described in Sect. 1, Guo et al. [4] were among the first to rigorously analyse the probabilistic calibration of deep neural networks. They observed that despite improvements in accuracy, deep neural network models tend to produce poorly calibrated probabilities. They proposed *temperature scaling*, a simple yet effective post-processing technique to improve calibration without altering accuracy. In other work, Ovadia et al. [13] evaluated the reliability of predictive uncertainty under dataset shift through large-scale empirical analysis. They demonstrated that while some models perform well under i.i.d. settings, they often fail to maintain calibrated uncertainty estimates under distributional shift, thereby highlighting the need for shift-resilient calibration methods. Li et al. [9] proposed a Bayesian online learning approach for detecting and adapting to irregular distribution shifts. Their model combines Bayesian inference with change-point detection, enabling dynamic adaptation to evolving data distributions while maintaining probabilistic integrity.

In the context of classification under label shift, Podkopaev and Ramdas [15] developed a distribution-free uncertainty quantification framework using conformal prediction. This method provides prediction sets with finite-sample guarantees, without requiring strong distributional assumptions. In other work Álvarez et al. [1] introduced a probabilistic load forecasting approach using adaptive online learning. Their framework updates forecasts in real time, achieving calibrated predictions in non-stationary environments. Gupta and Ramdas [5] extended calibration techniques to online settings by proposing *Online Platt Scaling with Calibeating*. This method enables real-time updating of calibration parameters, outperforming static methods by adapting to streaming data. Finally, a recent survey by Minderer et al. [10] reviews the state of classifier calibration, covering both classical and modern techniques, metrics, and practical considerations. This study serves as a valuable resource for both theoreticians and practitioners looking to assess and improve calibration performance.

Together, these contributions represent an evolving body of work aimed at improving the reliability, adaptivity, and robustness of probabilistic predictions in machine learning systems to date. The aim of this work is to add and expand on a complementary method in the growing body of literature tackling machine learning algorithm adaptation under dataset shift.

2.1 Main Contributions

The main contributions of this chapter are:

- We introduce a new **Python** package for protected classification[1] and use it to extend a set of experimental results across a range of real and artificial datasets, in binary and multi-class settings.
- We compare the method of protected classification in terms of calibration with a recently published alternative method [5].
- We demonstrate the effectiveness of protected classification in an online setting for streaming data problems - to our knowledge this is the first such study to date.

3 Theoretical Background

Theoretical exposition of protected probabilistic classification has been extensively covered in [20] and here we briefly summarize the key results for context.

Our starting assumption is that we have access to a prediction model which is obtained by training an algorithm. The algorithm maps a new object from the object space $\mathbf{X}$ to a label from the label space $\mathbf{Y}$ using past data of $x_1, \ldots, x_n \in \mathbf{X}$ and labels $y_1, \ldots, y_n \in \mathbf{Y}$. In the simple case of binary classification for example, such a system maps a given object $\mathbf{x}$ with an unknown label $y \in [0, 1]$ to a number $p \in [0, 1]$, interpreted as the predicted probability that $y = 1$. We call such a system the *base predictive system*.

We have no guarantees that the base predictive system is calibrated - implying that for any given $P(y|x) = p$ the average percentage of the true label of Y is p. To calibrate the base predictive system, we define a family Θ of calibrating functions $f_\theta : [0, 1] \to [0, 1], \theta \in \Theta$. The intuition behind any given calibration function f_θ is that we are trying to improve the base predictions p, by using a new prediction $f(p)$ instead of p.

In our library, and in line with [20] we make use of the Cox [2] calibrating functions, which for a general multi-class case are defined in Eq. 1 as:

$$f_{\alpha,\beta}(\mathbf{p})_y := \frac{p_y^\beta \exp(\alpha(y))}{\sum_{y' \in \mathcal{Y}} p_{y'}^\beta \exp(\alpha(y'))},$$
(1)

where the parameters $\alpha \in \mathbb{R}^{\mathbb{Y}}, \beta \in \mathbb{R}$. We are mainly interested in values of $\beta \in \{1, 0.5, 2\}$ and $\alpha(y) \in \{0, 1, -1\}$ for each class label y.

---

[1] available at https://pypi.org/project/protected-classification.

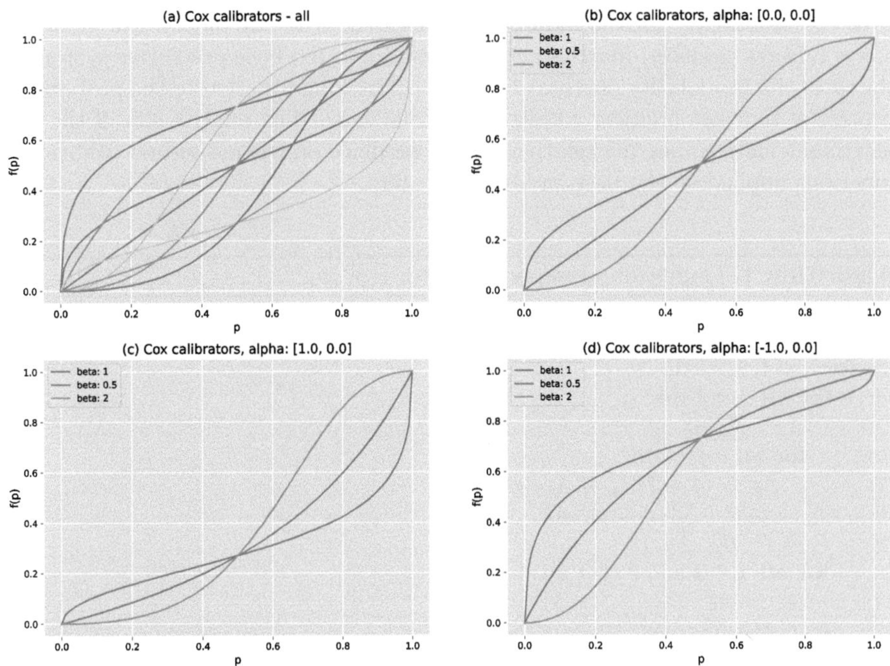

Fig. 1. Cox calibrator functions for binary class datasets

Figure 1 illustrates Cox calibrating functions for the binary class problem with subplot (b) showing the case of $\alpha(y) = [0.0, 0.0]$, subplot (c) showing the case of $\alpha(y) = [1.0, 0.0]$ and subplot (d) showing the case of $\alpha(y) = [-1.0, 0.0]$ separately for clarity. By setting α to zero for all y (subplot (b)), we obtain the one parameter subfamily where $\beta = 1$ is the neutral value indicating no calibration, $\beta = 0.5$ corrects the overconfidence of underlying base predictive system and $\beta = 2.0$ the under-confidence. The equivalent plots in Fig. 1 (c) and (d) show the Cox calibrator plots aimed for correcting any possible asymmetric over and under-confidence of the base predictive system.

In contrast to a number of post-hoc calibration algorithms, which are applied to the base predictive system only once post initial training, our method of calibration is applied online. Since we do not know in advance which of the calibrating functions will be best for a given base predictive system, we use a betting framework to make this decision in an online fashion. To do so, we define a *probability process* as a function mapping any finite sequence of observations to the product $B_{p_1}(y_1), \ldots, B_{p_n}(y_n)$ (i.e., the probability attached to this sequence by the predictive system), where n is the number of observations in the sequence, $y_1, \ldots, y_n$ are their labels, $p_1, \ldots, p_n$ are the predictions for those observations and B_p, is the Bernoulli distribution on $0, 1$ with $B_p(1) = p$. We regard this probability process as the capital process of a player playing a betting game where their capital cannot go up, and for it not to go down they have to predict with the probability measure concentrated on the true outcome. Next we define a test (always positive) martingale with respect to the base predictive system as the ratio of a probability process of a given calibrating function to the probability process of the base predictive system and vice versa. The relative size of the test martingale determines the relative weight we place on the individual calibrating functions applied to the base predictive system.

Algorithm 1. Composite Jumper predictor $((p_1, p_2, \ldots) \mapsto (p'_1, p'_2, \ldots))$

1: $P := \pi$
2: $A_\theta^J := \frac{1-\pi}{|\mathbf{J}|} \mathbb{1}_{\theta=0}$ {for all $J \in \mathbf{J}$ and $\theta \in \Theta$}
3: **for** $n = 1, 2, \ldots$ **do**
4:     **for all** $J \in \mathbf{J}$ **do**
5:        $A := \sum_\theta A_\theta^J$
6:        **for all** $\theta \in \Theta$ **do**
7:           $A_\theta^J := (1 - J)A_\theta^J + AJ/|\Theta|$
8:     $p'_n := p_n P + \sum_{J,\theta} f_\theta(p_n)A_\theta^J$
9:     $P := P \cdot B_{p_n}(\{y_n\})$
10:     **for all** $J \in \mathbf{J}$ and $\theta \in \Theta$ **do**
11:        $A_\theta^J := A_\theta^J \cdot B_{f_\theta(p_n)}(\{y_n\})$
12:     $C := P + \sum_{J,\theta} A_\theta^J$
13:     $P := P/C$
14:     **for all** $J \in \mathbf{J}$ and $\theta \in \Theta$ **do**
15:        $A_\theta^J := A_\theta^J/C$

The intuition behind method above is that we start from a base predictive system, design a way of gambling against it (a test martingale) and then use the martingale (applying the idea of "tracking the best expert" [6]) as protection against the kind of changes that the test martingale benefits from. If and when those changes happen, the product of base probability process and the test martingale (defined as the protected probability process) outperforms the base probability process in terms of the log loss function. The method further makes use of the concept of *Composite Jumper* martingale with a two-parameter set $\pi \in (0, 1)$ and a finite set $\mathbf{J}$ of non-zero jumping rates J. Each jumping rate characterises the transition function between maintaining the same state with probability $1 - J$ and, with probability J, choosing a new state from the uniform probability measure on Θ. The full method, summarised in Algorithm 1 and described fully in [20] is refereed to as *protected probabilistic classification*. By continually correcting the base predictive system using a calibration function which obtains the highest capital under a martingale betting process described in Algorithm 1 we calibrate (protect) in an adaptive fashion. The cost of such a calibration is given by $\log \frac{1}{\pi}$ (Vovk et al. [20, Section 4]).

4 Python `protected-Classification` package

In order to enable the method of protected probabilistic classification more widely available we introduce an open source package `protected-classification` shared at `PyPi`[2] under the MIT License. The package is flexible and caters for binary and multi-class settings as well as settings compatible with streaming data problems analogous to those used by a popular machine learning package `river` [11].

Further details and instruction on use of the package are available in the `github`[3] repository. Here we illustrate the key features by applying it to the `Bank marketing` dataset [12] from the UCI Machine Learning Repository, thereby ensuring consistency with previous reported work in [20]. The dataset represents results of a marketing campaign of a Portuguese banking institution and the goal is to predict whether a client will subscribe to a term deposit, making it useful for classification tasks. We take the first 10,000 observations as the training set and use the remaining 35,211 as the test set, without shuffling the data. We train the `scikit-learn` random forest (RF) algorithm with a random seed of 2021 and all hyper-parameters set to default except for `no_estimators` which was set to a value of 1000. This is in slight contrast to the equivalent experiment in [20], but it does avoid the need for truncation of underlying random forest probabilities as described in Sect. 5, Equation (14). We use the same set of Cox calibrating function parameters as described in Sect. 3, namely $\beta \in \{1, 0.5, 2\}$ and $\alpha(y) \in \{0, 1, -1\}$ for each class label y and set the Composite Jumper jumping rates to $J = \{10^{-2}, 10^{-3}, 10^{-4}\}$. These parameters are set as default in the `protected-classification` library. As this is a binary classification problem the full set of Cox calibrating functions are the same as those depicted in Fig. 1.

Figure 2 shows the results of application of the `protected-classification` algorithm (*protected base*) to the test set results generated by the underlying random forest algorithm (*base*). Figure 2(a) shows comparison of the ROC/AUC curves between the two with a clear outperformance of *protected base* over *base*. Figure 2(b) shows the relative increase of the positive class label in the later part of the test set and the resulting increase of the test martingale indicating the outperformance of one or more of the Cox calibrator functions vs. the base algorithm in terms of log loss. A closer inspection of the individual Cox calibrator function with the highest final martingale value is shown in Fig. 2(d) - its shape is similar to the shape of the underlying random forest algorithm calibration plot shown in Fig. 2(c), indicating that the algorithm correctly tries to protect against the miscalibration caused in part by the distribution shift of the labels shown in Fig. 2(b).

---

[2] https://pypi.org/project/protected-classification/.
[3] https://github.com/ip200/protected-classification.

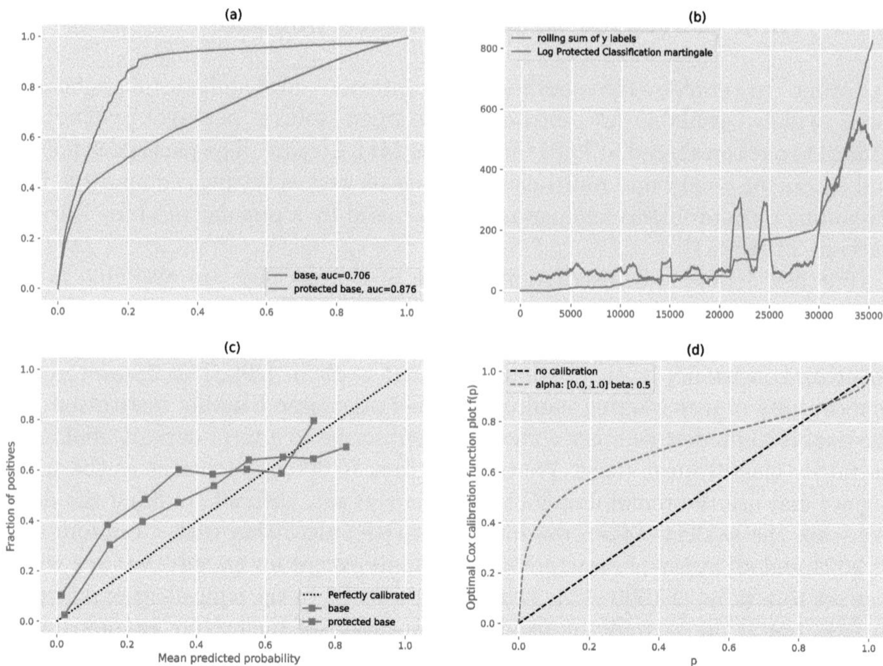

Fig. 2. (a) ROC/AUC comparison between the base algorithm and its protected equivalent, (b) cumulative sum of the positive class label in the test set, (c) calibration comparison between the base algorithm and its protected equivalent, (d) Cox calibrator function with the highest final martingale value

Table 1 shows the relative outperformance of the protected calibration over the base algorithm across accuracy, Brier loss and log loss (with **bold** indicating a better outcome, the convention used in the remainder of this chapter). The code illustrating results in this section can be found in the `github` repository under `notebooks/protected-classification`. The same repository shows examples of protected classification applied in three other settings - *batch* mode (`notebooks/protected-batch`), where the calibration is applied to a validation set only, analogous to standard post-hoc calibration algorithms such as Platt scaling ([14]) or isotonic regression ([21]), *multiclass* (`notebooks/protected-multiclass`), where we illustrate the application of the algorithm to multiclass classification problems and *streaming* (`notebooks/protected-streaming`), where we illustrate the application of the algorithm to streaming data problems compatible with `river` library [11]. The latter two applications will be further illustrated in the next section (Table 1).

Table 1. Performance comparison between for the underlying RF algorithm and its protected version on the `Bank Marketing` dataset

	accuracy	Brier loss	log loss
base	0.861	0.118	0.589
protected base	**0.871**	**0.091**	**0.363**

5 Experimental Results

The experimental results section consists of three sets of experiments. In the first section we compare the performance of `protected-classification` algorithm vs. traditional machine learning algorithms and existing post-hoc calibration methods in binary and multi-class settings under dataset shift, i.e., where the data in the test set is induced to have a different distribution to that in the training/calibration sets. In the second section we compare our method with a recently reported equivalent technique of *Online Platt Scaling with Calibeating* [5]. In the final section, we apply our algorithm within streaming data problems, comparing its effectiveness relative to other state-of-the art techniques in this setting.

5.1 Experiments Under Dataset Shift

Artificial Datasets. In the set of experiments using synthetically generated data, we used the `scikit-learn` functionality to create a set of datasets. Each dataset consists of 20 input features, and contains 2000 training/calibration examples and 1000 test examples. We consider 2, 3, 5 and 10-class problems, in each case generating five independent datasets using a different random seed.

In order to carry our protected-classification experiments under dataset shift, we consider four scenarios: the *unperturbed* dataset, where we assume that the training and test have the same probability distribution, the *concept shift*, where we swap the binary class labels for the second half of the test dataset only (last 500 examples), the *X-imbalance*, where we simulate the change in the underlying feature distribution by including examples only where one of the relevant numerical features has a value of less than 0 for the last 500 examples, and finally the *y-imbalance*, where we simulate the change in the frequency of labels and exclude the examples whose labels are equal to 0 for the last 500 test points.

We use 5 different underlying classifiers: Gaussian naïve Bayes, logistic regression, random forests, support vector machines and gradient boosted trees, all sourced from `scikit-learn` using the default set of hyperparameters. We consider each underlying algorithm in its base form (b) and also calibrated using *Platt scaling* (s), *isotonic Regression* (i) and *Venn-ABERS* calibration [19] (v) using two-fold cross-validation for all. The former two calibration methods were carried out using the `scikit-learn calibration` functionality and the latter using the `venn-abers` package from PyPi[4]. Our goal is to assess whether the pro-

---

[4] `pypi.org/project/venn-abers`.

tected classification algorithm can outperform the base and calibrated underlying algorithms on the test set under the log loss, Brier loss and expected calibration error (ECE, defined as in [8]) when the test examples are exchangeable (where the modification to the test set is applied only to the last 500 examples and the order of the test set examples is then randomly permuted). The code to reproduce all experiments can be found in the Online Appendix[5]. Tables 2-5 show the average ECE for each classifier, number of classes and base or calibration algorithm with and without protected classification applied on top of the underlying method (Tables 3 and 4).

Table 2. Expected calibration error (ECE) - *unperturbed* dataset experiments

classes	classifier	standard				protected			
		b	i	s	v	b	i	s	v
2	Logistic	0.043	**0.000**	0.047	0.041	0.042	0.000	0.046	0.000
	Naive Bayes	0.108	0.076	0.083	**0.000**	0.046	0.000	0.075	0.000
	Random forest	0.056	0.033	**0.000**	0.038	0.036	0.008	0.000	0.000
	SVM	0.050	0.053	0.050	**0.000**	0.050	0.000	0.050	0.000
	XGBoost	0.017	0.031	0.029	**0.000**	0.023	0.000	0.030	0.000
3	Logistic	0.027	0.034	0.055	**0.021**	0.026	0.035	0.056	0.025
	Naive Bayes	0.118	**0.026**	0.092	0.027	0.050	0.030	0.090	0.029
	Random forest	0.055	0.007	0.034	**0.000**	0.035	0.006	0.032	0.000
	SVM	0.034	**0.022**	0.034	0.023	0.035	0.024	0.035	0.023
	XGBoost	0.023	0.024	0.028	0.004	0.019	0.025	0.029	**0.000**
5	Logistic	0.042	0.063	0.068	0.040	0.040	0.047	0.047	**0.039**
	Naive Bayes	0.114	0.048	0.053	0.044	0.056	0.048	0.051	**0.044**
	Random forest	0.080	0.041	0.041	0.033	**0.017**	0.040	0.041	0.031
	SVM	0.066	0.039	0.066	0.038	0.043	0.039	0.043	**0.038**
	XGBoost	0.030	0.080	0.085	0.024	0.028	0.025	0.028	**0.024**
10	Logistic	0.037	**0.029**	0.037	0.031	0.036	0.029	0.036	0.031
	Naive Bayes	0.098	**0.036**	0.036	0.037	0.042	0.036	0.037	0.038
	Random forest	0.078	0.043	0.039	0.038	**0.018**	0.043	0.036	0.037
	SVM	0.038	0.021	0.038	0.031	0.038	**0.021**	0.038	0.031
	XGBoost	0.019	0.054	0.065	0.034	0.019	**0.018**	0.020	0.035
	Average	0.057	0.038	0.049	0.025	0.035	0.024	0.041	**0.021**

The protected classification results for the *unperturbed* dataset are mostly in line with the base algorithm results, with the mild improvement largely a rebuilt of the better calibration of the underlying algorithms themselves rather than protection against any dataset shift. In other experiment sets with dataset shift, the protected calibration algorithm generally results in a lower calibration error across most perturbations with protected classification applied on top of one of the calibration methods yielding the best results on average. The results are more significant for lower number of classes. It is important to note that the ECE measure used in our experiments sourced from the `uncertainty-calibration` package [8] measures the calibration error of all classes (per-class calibration error), rather than just the equivalent for the top-prediction. For the multi-class experiments the former measure is a stronger notion of calibration.

---

[5] https://github.com/ip200/chapter-protected-classification.

Table 3. Expected calibration error (ECE) - *concept shift* dataset experiments

classes	classifier	standard				protected			
		b	i	s	v	b	i	s	v
2	Logistic	0.373	0.384	0.367	0.371	0.278	0.278	0.270	**0.267**
	Naive Bayes	0.462	0.388	0.393	0.392	0.383	0.301	**0.255**	0.288
	Random forest	0.429	0.456	0.446	0.443	0.333	0.357	**0.323**	0.337
	SVM	0.366	0.382	0.366	0.381	**0.266**	0.275	0.266	0.266
	XGBoost	0.456	0.451	0.454	0.447	0.338	0.375	**0.330**	0.344
3	Logistic	0.210	0.187	0.168	0.198	0.138	0.121	**0.115**	0.126
	Naive Bayes	0.258	0.188	0.202	0.171	0.195	0.126	0.114	**0.113**
	Random forest	0.201	0.228	0.217	0.225	0.143	0.144	**0.134**	0.144
	SVM	0.176	0.214	0.176	0.203	**0.116**	0.129	0.116	0.127
	XGBoost	0.232	0.217	0.212	0.220	0.145	0.143	**0.138**	0.139
5	Logistic	0.068	0.058	**0.055**	0.074	0.066	0.055	0.056	0.067
	Naive Bayes	0.156	0.086	0.084	0.096	0.086	**0.074**	0.080	0.088
	Random forest	0.113	0.114	0.103	0.115	0.105	0.104	**0.090**	0.099
	SVM	**0.054**	0.059	0.054	0.073	0.056	0.061	0.056	0.068
	XGBoost	0.121	0.093	0.093	0.110	0.101	0.094	**0.089**	0.093
10	Logistic	0.048	**0.034**	0.038	0.045	0.042	0.034	0.036	0.044
	Naive Bayes	0.111	0.052	**0.042**	0.048	0.056	0.049	0.042	0.047
	Random forest	0.066	0.048	0.046	0.043	**0.038**	0.045	0.043	0.043
	SVM	0.041	**0.032**	0.041	0.045	0.041	0.032	0.041	0.044
	XGBoost	0.041	0.058	0.060	0.043	0.041	0.042	**0.037**	0.040
	Average	0.199	0.187	0.181	0.187	0.148	0.142	**0.132**	0.139

Table 4. Expected calibration error (ECE) - *X-imbalance* dataset experiments

classes	classifier	standard				protected			
		b	i	s	v	b	i	s	v
2	Logistic	0.045	**0.000**	0.044	0.036	0.044	0.000	0.044	0.000
	Naive Bayes	0.110	0.014	0.084	**0.000**	0.026	0.000	0.075	0.000
	Random forest	0.057	**0.000**	0.022	0.000	0.041	0.000	0.016	0.015
	SVM	0.049	0.047	0.049	0.047	0.051	**0.000**	0.051	0.000
	XGBoost	0.023	**0.000**	0.011	0.000	0.027	0.000	0.008	0.000
3	Logistic	0.167	0.154	0.134	0.160	0.116	**0.104**	0.107	0.112
	Naive Bayes	0.213	0.123	0.179	0.115	0.121	**0.081**	0.099	0.108
	Random forest	0.183	0.179	0.213	0.140	0.149	**0.132**	0.144	0.133
	SVM	0.130	0.171	0.130	0.164	**0.108**	0.116	0.108	0.115
	XGBoost	0.208	0.195	0.197	0.215	**0.117**	0.124	0.139	0.143
5	Logistic	0.036	0.064	0.067	0.036	0.035	0.051	0.040	**0.034**
	Naive Bayes	0.113	0.053	0.052	0.042	0.049	0.053	0.051	**0.041**
	Random forest	0.079	0.041	0.045	0.031	**0.016**	0.039	0.042	0.026
	SVM	0.064	0.042	0.064	0.032	0.056	0.040	0.056	**0.032**
	XGBoost	0.030	0.081	0.088	0.027	0.029	**0.018**	0.020	0.021
10	Logistic	0.031	0.022	0.030	0.028	0.030	**0.021**	0.029	0.028
	Naive Bayes	0.098	**0.028**	0.034	0.032	0.041	0.029	0.034	0.032
	Random forest	0.071	0.037	0.040	0.030	**0.014**	0.033	0.036	0.030
	SVM	0.035	**0.022**	0.035	0.033	0.035	0.022	0.035	0.032
	XGBoost	0.020	0.051	0.059	0.031	0.020	0.022	**0.019**	0.031
	Average	0.088	0.066	0.079	0.060	0.056	**0.044**	0.058	0.047

Table 5. Expected calibration error (ECE) - *y-imbalance* dataset experiments

classes	classifier	standard				protected			
		b	i	s	v	b	i	s	v
2	Logistic	0.088	0.068	0.092	0.074	0.026	**0.010**	0.028	0.011
	Naive Bayes	0.105	0.110	0.114	0.076	0.059	0.035	0.040	**0.032**
	Random forest	0.019	**0.000**	0.034	0.027	0.029	0.022	0.025	0.024
	SVM	0.090	0.085	0.090	0.069	0.034	**0.009**	0.034	0.031
	XGBoost	0.037	0.040	0.048	0.053	0.030	**0.018**	0.040	0.040
3	Logistic	0.089	0.094	0.106	0.085	0.058	0.069	0.058	**0.047**
	Naive Bayes	0.149	0.092	0.124	0.092	0.060	0.047	0.065	**0.040**
	Random forest	0.054	0.074	0.072	0.088	0.051	**0.043**	0.052	0.057
	SVM	0.089	0.074	0.089	0.081	0.074	0.053	0.074	**0.044**
	XGBoost	0.065	0.053	0.054	0.077	0.040	**0.038**	0.053	0.050
5	Logistic	0.077	0.095	0.101	0.069	0.055	0.048	**0.048**	0.060
	Naive Bayes	0.112	0.076	0.082	0.068	**0.059**	0.067	0.061	0.060
	Random forest	0.067	0.041	0.049	0.038	**0.033**	0.037	0.045	0.037
	SVM	0.095	0.072	0.095	0.067	**0.049**	0.053	0.049	0.056
	XGBoost	0.036	0.083	0.097	0.039	0.036	0.026	**0.021**	0.033
10	Logistic	0.046	**0.043**	0.051	0.045	0.045	0.043	0.051	0.045
	Naive Bayes	0.104	0.047	0.050	0.046	0.050	0.046	0.050	**0.045**
	Random forest	0.074	0.049	0.049	0.038	**0.024**	0.039	0.043	0.040
	SVM	0.052	0.045	0.052	0.042	0.050	0.043	0.050	**0.042**
	XGBoost	0.028	0.064	0.074	0.047	**0.028**	0.030	0.028	0.046
	Average	0.074	0.065	0.076	0.061	0.045	**0.039**	0.046	0.042

Distorted Image Classification. In this set of experiments we use the standard MNIST digit classification dataset which we analyse using a standard neural network (3-layer CNN with pooling), commonly used as an architecture for MNIST digit classification. The standard MNIST dataset consists of 50,000 training points and 10,000 test points. The idea of these protected calibration experiments is to apply incremental distortion transformations to the objects in the test set in order to shift the distribution of data used by the CNN classifier to make a prediction. The transformations made are image rotation (from 10 degrees to a maximum of 90 degrees), after the paper *Failing Loudly: An Empirical Study of Methods for Detecting Dataset Shift* [16]. Only the last 5000 examples are distorted, each of the experiments for each transformation was repeated 5 times and we report the average results. We show the performance of the underlying neural network under ECE, Brier and log loss without (labelled *b*) and with a commonly used calibration method for neural networks, *temperature scaling* after Guo et al. [4] (labelled *t*) (Tables 6, 7 and 8).

Table 6. MNIST distorted image dataset - Brier loss

	standard		protected	
	b	t	b	t
10 rotation	0.0712	0.0739	**0.0677**	0.0697
30 rotation	0.1261	0.1335	**0.1225**	0.1227
50 rotation	0.2656	0.2841	0.2535	**0.2458**
90 rotation	0.4486	0.4796	**0.3916**	0.4048
Average	0.2279	0.2428	**0.2088**	0.2108

Table 7. MNIST distorted image dataset - log loss

	standard		protected	
	b	t	b	t
10 rotation	0.1517	0.1711	**0.1491**	0.1573
30 rotation	0.2773	0.3310	0.2777	**0.2744**
50 rotation	0.6912	0.8327	0.6552	**0.5791**
90 rotation	1.5246	1.8207	**1.0706**	1.0974
Average	0.6612	0.7889	0.5382	**0.5271**

Table 8. MNIST distorted image dataset - ECE

	standard		protected	
	b	t	b	t
10 rotation	0.0198	0.0193	0.0144	**0.0138**
30 rotation	0.0240	0.0260	**0.0175**	0.0176
50 rotation	0.0499	0.0581	0.0377	**0.0332**
90 rotation	0.0806	0.0919	**0.0452**	0.0570
Average	0.0435	0.0488	**0.0287**	0.0304

The results suggest that protected classification significantly improves the calibration and log and Brier losses with and without temperature scaling with higher improvement for larger distortions. This could be a helpful technique therefore in real-life imaging classifications where some dataset shift is expected to occur post initial training and calibration.

5.2 Comparison with an Alternative Online Post-Hoc Calibration Method

In the work so far above we compared the implementation of the protected classification algorithm solely against post-hoc calibration techniques which are applied once post initial training. In this section we directly compare the protected classification algorithm to an equivalent online post-hoc calibration algorithm recently published in *Online Platt Scaling with Calibeating* [5]. This method is theoretically guaranteed to be calibrated for adversarial outcome sequences. In particular we follow a set of experiments described in the paper where the authors introduce synthetic drifts in the data based on covariate values, i.e. an instance of covariate drift. We show results here comparing our methods - applying protected classification to the underlying base algorithm (*BM - protected*), the equivalent but after first calibrating using the Venn-abers method (*Venn-abers protected*), as well as applying protected classification to the authors' underlying method of Online Platt Scaling with hedging (*HOPS - protected*). The experiments were carried out on four real life data sets reported in the paper (`bank`, `credit`, `churn` and `fetal`) with synthetically induced covariate shift using the authors publicly available code[6]. The underlying algorithms used were random forests *(rf)*, gradient boosted tress *(xb)* and logistic regression *(lr)* to which the authors applied five different post-hoc calibration methods described in Gupta and Ramdas [5, Section 4].

The results in Tables 9 and 10 compare the calibration error (CE, as defined by the authors in the study) between the protected classification method and other equivalent algorithms. The values in each cell represent the average CE for a given algorithm and dataset across the full length of the sample containing the induced covariate shift. The protected classification methods are comparable to the methods reported in [5], with protected classification applied on top of OPS + hedging (*HOPS-protected*) yielding the lowest average CE across all datasets and classifiers and the *Venn-abers protected* method the second lowest. The results suggest therefore that applying protected classification in combination with other similar online calibrating algorithms may help to improve calibration further in problems when the underlying data distribution changes in the test set.

Table 9. CE for the bank and credit datasets, with competing calibration algorithms applied to the underlying classifiers

dataset	bank			credit			average
classifier calibrator	lr	rf	xb	lr	rf	xb	
Base model (BM)	0.065	0.108	0.091	0.071	0.040	0.034	0.068
Fixed-batch Platt scaling (FPS)	0.025	0.051	0.043	0.037	0.039	0.024	0.036
Online Platt scaling (OPS)	0.027	0.023	0.018	0.018	0.031	0.013	0.022
OPS + tracking (TOPS)	0.021	0.016	0.019	0.005	0.006	0.006	0.012
OPS + hedging (HOPS)	0.022	0.017	0.015	0.012	0.021	0.021	0.018
Windowed Platt scaling (WPS)	0.026	0.024	0.020	0.037	0.034	0.016	0.026
BM -protected	0.031	0.037	0.031	0.033	0.037	0.031	0.033
HOPS - protected	0.016	0.012	**0.009**	**0.005**	**0.006**	**0.006**	**0.009**
Venn-abers protected	**0.011**	**0.010**	0.011	0.011	0.010	0.011	0.010

---

[6] https://github.com/aigen/df-posthoc-calibration.

Table 10. CE for the churn and fetal datasets, with competing calibration algorithms applied to the underlying classifiers

dataset	churn			fetal			average
classifier calibrator	lr	rf	xb	lr	rf	xb	
Base model (BM)	0.050	0.039	0.055	0.139	0.160	0.129	0.095
Fixed-batch Platt scaling (FPS)	0.051	0.040	0.042	0.126	0.125	0.129	0.085
Online Platt scaling (OPS)	0.014	0.026	0.012	0.039	0.036	0.039	0.028
OPS + tracking (TOPS)	0.013	0.015	0.010	0.038	0.022	0.029	0.021
OPS + hedging (HOPS)	0.019	0.018	0.017	0.048	0.054	0.055	0.035
Windowed Platt scaling (WPS)	0.042	0.033	0.034	0.084	0.071	0.081	0.057
BM -protected	0.019	0.026	0.029	0.085	0.087	0.092	0.056
HOPS - protected	**0.012**	0.015	0.010	0.032	**0.019**	**0.023**	**0.018**
Venn-abers protected	0.013	**0.012**	**0.009**	**0.029**	0.027	0.028	0.019

5.3   Comparison with Online Algorithms for Streaming Data

In this section we directly compare protected classification applied to algorithms which are designed to learn online. For this we make use of the **river** library [11], which provides multiple state-of-the-art learning methods, data generators/transformers, performance metrics and evaluators for different streaming data learning problems. In order to perform equivalent online classification we created a modified version of the protected classification algorithm which is compatible with the **river** package so that our results are directly comparable. We illustrate an example of the `learn_one` and `predict_one` in the `protected-classification` **github** repository.

We apply our modified protected classification method to 8 classification algorithms within the **river** library for one representative binary dataset (`Phishing`, a spam message classification problem) and one multi-class dataset (`ImageSegments`, an artificial dataset for classifying types of images). Each underlying algorithm is designed to learn online with two in particular (ARF and SRP) designed to handle datasets with shift. More details on the underlying algorithms and datasets can be found in the online documentation for the **river** package at https://riverml.xyz/latest/.

The results in Table 11 show the average log loss over five different runs with different random seeds for each classifier and dataset combination.

Table 11. Log loss for streaming data experiments

model	Phishing		ImageSegments	
	base	protected	base	protected
ADWIN	0.2337	**0.1895**	0.4316	**0.4298**
ARF	0.2336	**0.2038**	0.4598	**0.4315**
Bagging	0.2286	**0.1837**	0.4371	**0.4310**
DriftARF	0.2242	**0.2033**	0.4553	**0.4310**
DriftSRP	0.1778	**0.1768**	0.4394	**0.4314**
EFTree	0.3621	**0.3518**	2.6494	**2.2573**
Hoeffding	0.4333	**0.4120**	0.4302	**0.4298**
SRP	0.1871	**0.1858**	0.4392	**0.4312**
Average	0.2601	**0.2383**	0.7178	**0.6591**

The results suggest that as in the case of comparison in the previous section, applying protected classification to underlying online learning algorithms designed for streaming data can help to produce better calibrated outputs. Finally, Fig. 3 shows an example of the `protected-classification` applied on top of the `ARF` classification algorithm applied to the `Phishing` dataset. We

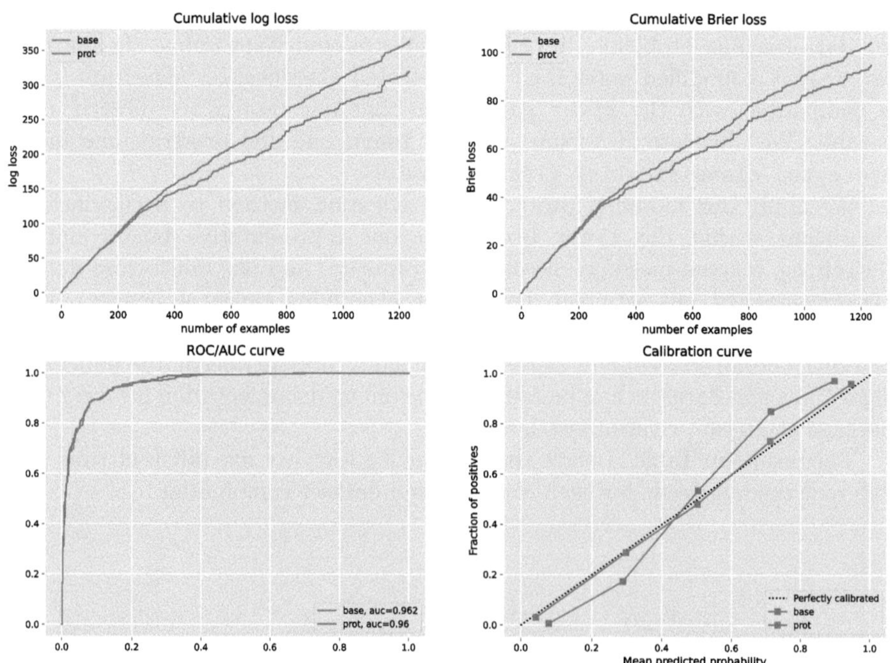

Fig. 3. Cumulative log loss, Brier loss, total ROC/AUC and calibration curve for the `ARF` algorithm (base) and the protected classification equivalent (protected) for the `Phishing` dataset

can see the relatively constant cumulative log loss improvement of the protected classification algorithm, which also results in better calibration overall.

6   Conclusions and Further Work

This chapter introduced the extension to original work on protected classification reported in [20] through the use of a novel library which can be applied to a range of binary and multi-class classification experiments with and without dataset shift. The empirical results suggest that the method is applicable to a broad range of problems and can help to produce more calibrated outputs when applied on top of the underlying classifiers or other equivalent post-hoc calibration methods. Further work could extend this technique to some of the more recent developments generative AI, for example, calibrating probabilistic token outputs of Large Language Models (LLM) in an online setting. We hope that the research community can benefit from the use of this library as an additional tool in handing dataset shifts in other machine learning problems.

References

1. Álvarez, V., Mazuelas, S., Lozano, J.A.: Probabilistic load forecasting based on adaptive online learning. IEEE Trans. Power Syst. **36**(4), 3668–3680 (2021)
2. Cox, D.R.: Two further applications of a model for binary regression. Biometrika **45**(3/4), 562–565 (1958)
3. Dawid, A.P.: The well-calibrated Bayesian. J. Am. Stat. Assoc. **77**(379), 605–610 (1982)
4. Guo, C., Pleiss, G., Sun, Y., Weinberger, K.Q.: On calibration of modern neural networks. In: International Conference on Machine Learning, pp. 1321–1330, PMLR (2017)
5. Gupta, C., Ramdas, A.: Online Platt scaling with calibeating. In: International Conference on Machine Learning, pp. 12182–12204, PMLR (2023)
6. Herbster, M., Warmuth, M.K.: Tracking the best expert. Mach. Learn. **32**(2), 151–178 (1998)
7. Kull, M., Silva Filho, T., Flach, P.: Beta calibration: a well-founded and easily implemented improvement on logistic calibration for binary classifiers. In: Artificial Intelligence and Statistics, pp. 623–631, PMLR (2017)
8. Kumar, A., Liang, P., Ma, T.: Verified uncertainty calibration. In: Advances in Neural Information Processing Systems (NeurIPS) (2019)
9. Li, Y., Ma, T., Blei, D.M.: Detecting and adapting to irregular distribution shifts in Bayesian online learning. In: International Conference on Learning Representations (2020)
10. Minderer, M., Sriram, A., Djolonga, J., et al.: Classifier calibration: a survey on how to assess and improve predicted class probabilities. arXiv preprint arXiv:2307.10802 (2024)
11. Montiel, J., Halford, M., Mastelini, S.M., Bolmier, G., Sourty, R., Vaysse, R., Zouitine, A., Gomes, H.M., Read, J., Abdessalem, T., et al.: River: machine learning for streaming data in python. J. Mach. Learn. Res. **22**(110), 1–8 (2021)

12. Moro, S., Cortez, P., Rita, P.: A data-driven approach to predict the success of bank telemarketing. Decis. Support Syst. **62**, 22–31 (2014)
13. Ovadia, Y., et al.: Can you trust your model's uncertainty? evaluating predictive uncertainty under dataset shift. In: Advances in Neural Information Processing Systems, vol. 32 (2019)
14. Platt, J.: Probabilistic outputs for support vector machines and comparisons to regularized likelihood methods. Adv. Large Margin Classifiers **10**(3), 61–74 (1999)
15. Podkopaev, A., Ramdas, A.: Distribution-free uncertainty quantification for classification under label shift. In: Uncertainty in Artificial Intelligence, pp. 844–853, PMLR (2021)
16. Rabanser, S., Günnemann, S., Lipton, Z.: Failing loudly: an empirical study of methods for detecting dataset shift. In: Advances in Neural Information Processing Systems, vol. 32 (2019)
17. Silva Filho, T., Song, H., Perello-Nieto, M., Santos-Rodriguez, R., Kull, M., Flach, P.: Classifier calibration: a survey on how to assess and improve predicted class probabilities. Mach. Learn. **112**(9), 3211–3260 (2023)
18. Vovk, V., Petej, I.: Venn-abers predictors. In: Proceedings of the Thirtieth Conference on Uncertainty in Artificial Intelligence, pp. 829–838 (2014)
19. Vovk, V., Petej, I., Fedorova, V.: Large-scale probabilistic predictors with and without guarantees of validity. In: Advances in Neural Information Processing Systems, vol. 28 (2015)
20. Vovk, V., Petej, I., Gammerman, A.: Protected probabilistic classification (2021). https://arxiv.org/abs/2107.01726
21. Zadrozny, B., Elkan, C.: Learning and making decisions when costs and probabilities are both unknown. In: Proceedings of the seventh ACM SIGKDD International Conference on Knowledge Discovery and Data Mining, pp. 204–213 (2001)
22. Zadrozny, B., Elkan, C.: Transforming classifier scores into accurate multiclass probability estimates. In: Proceedings of the eighth ACM SIGKDD International Conference on Knowledge Discovery and Data Mining, pp. 694–699 (2002)

GPyConform: Conformal Prediction with Gaussian Process Regression in Python

Harris Papadopoulos[✉][ID]

Computational Intelligence Research Lab (COIN), Department of Electrical
Engineering, Computer Engineering and Informatics, Frederick University, Nicosia,
Cyprus
h.papadopoulos@frederick.ac.cy

Abstract. This chapter presents GPyConform, a Python package that
extends the GPyTorch framework with Conformal Prediction (CP) for
Gaussian Process Regression (GPR). CP is a Machine Learning framework for constructing prediction regions, or Prediction Intervals (PIs) in
the case of regression, with distribution-free, finite-sample coverage guarantees under the sole assumption of data exchangeability. GPyConform
implements both the Transductive and Inductive versions of CP with a
GPR-specific normalized nonconformity measure that exploits the predictive variance of Gaussian Processes to produce adaptive uncertainty
quantification for each instance. It also supports symmetric and asymmetric PIs in both framework versions through a unified and intuitive
interface. We provide detailed examples showing how to set up models,
construct Transductive and Inductive CP regressors, retrieve and evaluate PIs, and assess their empirical performance. An empirical study
on the Abalone dataset investigates the coverage and widths of the provided PIs, as well as the practical computational efficiency of the package
through GPU acceleration. Together, these features make GPyConform a
practical tool for practitioners seeking reliable uncertainty quantification
with GPR and a foundation for future research in CP.

Keywords: Conformal Prediction · Gaussian Process Regression ·
Uncertainty Quantification · Prediction Intervals · GPyConform ·
GPyTorch

1 Introduction

Uncertainty quantification is a central component of modern machine learning,
especially in safety-critical tasks where point predictions alone are insufficient.
Gaussian Process Regression (GPR) is a widely used Bayesian approach that
naturally provides predictive distributions for each test instance. In addition,
it has many other desirable properties, including its nonparametric nature, the
flexibility of designing covariance functions for specific data and problems, and

© The Author(s), under exclusive license to Springer Nature Switzerland AG 2026
K. An Nguyen and Z. Luo (Eds.): Alexander Gammerman Festschrift, LNCS 16290, pp. 449–466, 2026.
https://doi.org/10.1007/978-3-032-15120-9_20

the ability to use standard Bayesian methods for model selection. The predictive distributions provided by GPR however, rely on the assumption that the underlying model is correctly specified. In practice, this assumption is almost always violated, since full prior knowledge is rarely available. As a result, the predictive distributions provided by GPR can be misleading, for example the derived Prediction Intervals (PIs) at the 95% confidence level may cover far less than 95% of the true labels.

Conformal Prediction (CP) [11] is a Machine Learning framework for producing prediction regions with finite-sample coverage guarantees without assuming anything more than data exchangeability. The framework was first introduced in its transductive form in [9], now known as Transductive CP (TCP) or Full CP. Shortly after, a much more computationally efficient inductive version was proposed in [7], now referred to as Inductive CP (ICP) or Split CP. TCP leverages all available data for both training and calibration, achieving higher statistical efficiency but with a substantially greater computational overhead. ICP in contrast, partitions the data into a *proper training* set and a *calibration* set, maintaining the computational efficiency of its base method, but sacrificing some statistical efficiency. Additionally, there is the Cross-Conformal Prediction (CCP) version of the framework [10] (for regression see [4]), which combines the ideas of ICP and cross validation. CCP falls somewhere in between TCP and ICP in terms of statistical and computational efficiency. However, it is important to note that CCP does not enjoy the same theoretical guarantees of the other two versions.

A central ingredient of CP is a real-valued function, called *nonconformity measure*, that evaluates how strange or nonconforming a candidate instance is from a set of known instances according to the underlying technique. In regression the standard nonconformity measure is the model residual (difference between actual and predicted value). However, plain residuals do not reflect instance-specific uncertainty. Normalized nonconformity measures, introduced in [6,8], address this by dividing residuals by a measure of instance difficulty, consequently providing more adaptive and informative PIs. For GPR in particular, a normalized nonconformity measure utilizing the predictive variance of GPR was recently proposed in [5].

This chapter introduces the `GPyConform` Python package which extends the popular and flexible `GPyTorch` framework with both transductive and inductive CP for GPR. Specifically it implements CP with the normalized nonconformity measure of [5], and supports two modes for each version of the framework: one using the absolute residuals for symmetric PIs, and one using raw residuals for asymmetric PIs. The package provides a unified interface for both TCP and ICP, as well as utilities for data splitting and evaluating the resulting PIs.

The general CP software landscape includes a number of noteworthy Python libraries such as `crepes` [1], `MAPIE` [2], and `PUNCC` [3], that provide a range of CP approaches, mostly based on scikit-learn models. In contrast, `GPyConform` is developed specifically for GPR, and to our knowledge, it is the sole Python package that implements a Transductive CP for regression. Being built directly on `GPyTorch`, it inherits a modular structure that accommodates a wide range of kernels, mean functions, likelihoods, approximate inference schemes, and deep GPs. Importantly, as it is based entirely on `torch` tensors, it can leverage GPU acceleration, enabling scalability to larger datasets and faster experimentation.

In summary, `GPyConform` offers a comprehensive and efficient implementation of Conformal Prediction for Gaussian Process Regression, featuring:

- A unified interface for both Transductive and Inductive CP;
- Support for symmetric and asymmetric PIs;
- A normalized nonconformity measure for GPR leveraging predictive variance for adaptive uncertainty quantification;
- Seamless integration with the `GPyTorch` ecosystem, including kernels, mean functions, likelihoods, and models;
- GPU-accelerated computation for efficient scaling to larger datasets;
- To our knowledge, the only available implementation of Transductive CP for regression in a Python package;
- Utilities for dataset splitting, calibration, and quantitative evaluation of prediction intervals.

The remainder of this chapter presents the `GPyConform` functionality through detailed examples. Section 2 explains how to install the package and import the required dependencies. Section 3 covers loading the dataset and splitting it into training and test sets. Section 4 introduces the TCP implementation, showing how to set up and apply transductive conformal predictors. Section 5 then describes the ICP implementation, covering data partitioning and setting up, calibrating and applying inductive conformal predictors using both a wrapper class and a lower-level interface. In Sect. 6 we turn to the `PredictionIntervals` class, which provides tools for retrieving and evaluating the intervals. Section 7 presents an empirical illustration on the Abalone dataset, highlighting both coverage guarantees and computational efficiency. Finally, Sect. 8 concludes the chapter with a summary of contributions and directions for future work.

2   Installing and Importing GPyConform

The source code of `GPyConform`, licensed under the permissive *BSD-3-Clause* license, can be found on GitHub: github.com/harrisp/GPyConform, while its documentation is available on Read the Docs.

The package can be installed from the Python Package Index (PyPI)[1] or from conda-forge[2], with the commands shown below.

From PyPI:

```
pip install gpyconform
```

---

[1] https://pypi.org/.

[2] https://anaconda.org/conda-forge.

From conda-forge:

```
conda install -c conda-forge gpyconform
```

These commands will install the most recent version of GPyConform, which you can then import into your code. In practice, you will typically import it together with other required packages, as shown in the example below. Specifically, we import the packages needed for the examples in the next sections of the chapter: (i) torch, since both GPyTorch and GPyConform operate on torch.Tensors; (ii) GPyTorch, which GPyConform extends and whose classes we will use to define our GPR models; (iii) NumPy, for data loading; and (iv) GPyConform plus two utility functions it provides for splitting the data in tensor form.

```
import torch
import gpytorch
import numpy as np
import gpyconform
from gpyconform.utils import tensor_train_test_split,
    tensor_train_cal_split
```

3   Data Loading and Splitting

All examples in the following sections use the Airfoil Self-Noise dataset from the UCI repository. We first load a preprocessed copy of this dataset, which has been scaled and normalized appropriately for convenience.

```
# Load preprocessed Airfoil CSV and convert to torch tensors
url = (
    "https://raw.githubusercontent.com/harrisp/"
    "GPyConform/master/examples/data/airfoil.csv"
)
data = np.loadtxt(url, delimiter=",", skiprows=1)

all_data_x = torch.tensor(data[:, :-1], dtype=torch.float32)
all_data_y = torch.tensor(data[:, -1], dtype=torch.float32)
```

Now we can split the dataset into training and test sets using the utility function tensor_train_test_split of GPyConform. Here we assign 30% of the data to the test set and the other 70% to the training set:

```
train_x, test_x, train_y, test_y = tensor_train_test_split(
    all_data_x, all_data_y, test_size=0.3, shuffle=True, seed=None
)
```

The tensor_train_test_split function divides the paired input and target tensors into two disjoint training and test sets. The test_size parameter determines the size of the test set, either as a proportion (if given as a float in the range $(0,1)$) or as an absolute number of instances (if given as an integer ≥ 2). The function also provides two optional keyword-only parameters: shuffle (default: True), which controls whether the data are shuffled before splitting, and seed (default: None), which sets the random seed for reproducibility when shuffling is enabled.

4   Transductive GPR Conformal Prediction

This section presents the implementation of Gaussian Process Regression (GPR) Transductive (or Full) Conformal Prediction in GPyConform. This functionality is provided by the gpyconform.ExactGPCP class, an extension of GPyTorch's gpytorch.models.ExactGP, since the TCP version requires modifications to the model's internal prediction process.

In particular, ExactGPCP implements the approach proposed in [5] for symmetric PIs, as well as its corresponding asymmetric version following the approach in Sect. 2.3 of [11]. As is the nature of these approaches, ExactGPCP assumes a gpytorch.likelihoods.GaussianLikelihood; other likelihoods are not supported.

In practical use, ExactGPCP behaves identically to ExactGP: models can be defined and trained in exactly the same way as in GPyTorch. The key distinction is that, in addition to standard GP predictions, ExactGPCP can also produce conformal PIs.

4.1   Setting Up the Model

The underlying GPR model can be constructed in almost the same way as with GPyTorch, but by subclassing the gpyconform.ExactGPCP class in place of gpytorch.models.ExactGP. Additionally, the constructor includes an extra CP-specific argument, cpmode with three possible values: 'symmetric', 'asymmetric', or None, with the default being set to 'symmetric' in the example below; the role of cpmode and other CP-specific parameters will be detailed in the next subsection. Apart from this, the model definition process is identical to that of ExactGP. Any mean function from gpytorch.means and any covariance module from gpytorch.kernels that is compatible with the ExactPredictionStrategy (e.g., RBF, Matern, SpectralMixture, and standard Scale/Add/Product compositions) can be used with GPyConform. Note however that approximation kernels (e.g., GridInterpolationKernel, InducingPointKernel) are not supported. For details on defining Exact GPR models, see the GPyTorch documentation: gpytorch.readthedocs.io/en/latest/.

In the example below we construct a rather simple GPR model with a constant mean function and a scaled RBF covariance function. We then initialize the likelihood and model. As mentioned before, for ExactGPCP the likelihood should always be set to gpytorch.likelihoods.GaussianLikelihood().

```python
# Define ExactGPCP-based model
class ExactGPCPModel(gpyconform.ExactGPCP):
    def __init__(self, train_x, train_y, likelihood, cpmode='symmetric'):
        super(ExactGPCPModel, self).__init__(
            train_x, train_y, likelihood, cpmode)
        self.mean_module = gpytorch.means.ConstantMean()
        self.covar_module = gpytorch.kernels.ScaleKernel(
            gpytorch.kernels.RBFKernel()
        )

    def forward(self, x):
        mean_x = self.mean_module(x)
        covar_x = self.covar_module(x)
        return gpytorch.distributions.MultivariateNormal(mean_x, covar_x)

# Initialize likelihood and model
likelihood = gpytorch.likelihoods.GaussianLikelihood()
model = ExactGPCPModel(train_x, train_y, likelihood)
```

After constructing the GPR model we proceed to optimize its hyperparameters. This is also performed in exactly the same way as in GPyTorch. In the example below we perform optimization for 50 iterations, with the torch.optim.Adam optimizer, minimizing the negative marginal log likelihood on the training data. We print the negative loss together with the RBF lengthscale and noise hyperparameters at each step. Note that, like in GPyTorch, for performing the hyperparameter optimization we first have to put the model and likelihood in .train() mode. Additionally, for this step, we can choose any of the available torch optimizers for minimizing any of the available GPyTorch measures.

```python
training_iter = 50

model.train()
likelihood.train()

optimizer = torch.optim.Adam(model.parameters(), lr=0.1)
mll = gpytorch.mlls.ExactMarginalLogLikelihood(likelihood, model)

for i in range(training_iter):
    optimizer.zero_grad()
    output = model(train_x)
    loss = -mll(output, train_y)
    loss.backward()
    print('Iter %d/%d - Loss: %.3f   lengthscale: %.3f   noise: %.3f' % (
        i + 1, training_iter, loss.item(),
        model.covar_module.base_kernel.lengthscale.item(),
        model.likelihood.noise.item()
    ))
    optimizer.step()
```

4.2 Making Predictions: Obtaining Prediction Intervals

Prediction is the stage where GPyConform extends the standard GPyTorch functionality by producing conformal PIs. Like in GPyTorch, the model must first be put in .eval() mode, and gradient calculation should be disabled for efficiency with torch.no_grad() during prediction.

As mentioned in the previous subsection, ExactGPCP objects have an additional property cpmode, which determines the prediction approach:

– cpmode='symmetric': (default) employs the absolute residual nonconformity measure approach described in [5];
– cpmode='asymmetric': employs the asymmetric version of the nonconformity measure defined in [5], following Chap. 2.3 of [11];
– cpmode=None: reverts to the standard ExactGP behavior (see GPyTorch documentation).

The cpmode property can be changed at any time with model.cpmode=... without the need of retraining the model. Note that for cpmode=None we should also put the likelihood in .eval() mode and call both modules on the test data. This is required because in standard ExactGP likelihood evaluation is explicit.

After putting the model in .eval() mode, and if needed changing cpmode, we can then call the model on the test objects to obtain the corresponding PIs. In addition to the test objects, the model (in .eval() mode) has the following optional parameters:

– gamma: (float; default: 2), the γ parameter of the nonconformity measure. Controls the sensitivity of the nonconformity measure to predictive variance differences, see [5].
– confs: (torch.Tensor, numpy.array, or list; default: [0.95]), a list of the confidence levels for which to return PIs.

Calling a trained GPyConform model in .eval() mode (with cpmode set to either 'symmetric' or 'asymmetric') returns a PredictionIntervals object containing the PIs for all requested confidence levels. This object provides direct access to the PIs as well as their evaluation in terms of empirical calibration and informational efficiency.

In the following example we put the model in .eval() mode and obtain the PredictionIntervals object containing the (symmetric) PIs with gamma set to 2 at the 99%, 95% and 90% confidence levels.

```
model.eval()  # Get into .eval() mode

with torch.no_grad():  # Disable gradient calculation
    symmetric_pis = model(test_x, gamma=2, confs=[0.99, 0.95, 0.9])
```

To obtain the corresponding asymmetric PredictionIntervals object we only need to change cpmode before calling the model (below we assume that the model is already in .eval() mode):

```
model.cpmode = 'asymmetric'    # Change cpmode to 'asymmetric'

with torch.no_grad():
    asymmetric_pis = model(test_x, gamma=2, confs=[0.99, 0.95, 0.9])
```

We can now extract a torch tensor with the PIs for any of the three confidence levels by calling the returned `PredictionIntervals` object with that confidence level as argument (here 95%):

```
print(symmetric_pis(0.95))
```

```
tensor([[0.6039, 0.9468],
        [0.4827, 0.8260],
        [0.0886, 0.4435],

        ...,

        [0.2601, 0.6033],
        [0.1361, 0.4918],
        [0.4258, 0.7764]], dtype=torch.float64)
```

The output is a tensor with a row for each test instance, where the first column is the lower bound and the second is the upper bound of the corresponding PI.

Additionally, we can evaluate the PIs at any of the three confidence levels using the `evaluate` method. Below we evaluate on all available metrics (the default) at the 99% confidence level. We round all values to three decimals for readability.

```
res = symmetric_pis.evaluate(0.99, y=test_y)

# Round to 3 decimals and print
pretty_res = {k: f"{v:.3f}" for k, v in res.items()}
print(pretty_res)
```

```
{'mean_width': '0.564', 'median_width': '0.560', 'error': '0.007'}
```

The output is a dictionary with the metrics as keys. The complete functionality of `PredictionIntervals` is described in Sect. 6.

5 Inductive GPR Conformal Prediction

This section presents the GPR Inductive (or Split) Conformal Prediction implementation in `GPyConform`. Unlike the Transductive version of the framework, Inductive Conformal Prediction (ICP) does not need to modify the underlying model's internal prediction process. Instead, it divides the available training data into two parts: the *proper training set*, which is used for training the underlying model, and the *calibration set*, which is used for constructing the conformal PIs based on the trained model's predictions. Therefore, the ICP implementation of `GPyConform` can be used with any GPR model. It is not restricted to `ExactGP` or `GaussianLikelihood`.

The package provides two classes offering the ICP functionality at different levels: (i) the `gpyconform.GPRICPWrapper`, which provides a simple interface for building and calibrating a GPR ICP on top of a trained `GPyTorch` regression model; and (ii) the `gpyconform.InductiveConformalRegressor`, which works at a lower level by taking calibration and test predictions (means and variances) together with the corresponding ground truth targets directly, rather than through a wrapped model. The latter makes it possible to apply ICP not only to `GPyTorch` models, but also to any other regression framework that exposes predictive means and variances.

5.1 Dividing the Training Data

In ICP, the first step is to split the available training data into two disjoint subsets: a *proper training set* and a *calibration set*. This can be performed with `GPyConform`'s `tensor_train_cal_split` utility function, which behaves in the same way as `tensor_train_test_split` (Sect. 3). Here we assign 25% of the training data to the calibration set and the other 75% to the proper training set:

```
prop_train_x, cal_x, prop_train_y, cal_y = tensor_train_cal_split(
    train_x, train_y, cal_size=0.25, shuffle=True, seed=None
)
```

5.2 Setting Up the Model

Any `GPyTorch` regression model can serve as the basis for `GPyConform`'s ICP, without requiring special treatment in its definition or hyperparameter optimization. This includes `ExactGP`, `ApproximateGP`, and `DeepGP` models, with any mean function from `gpytorch.means` and any covariance module from `gpytorch.kernels`, including both exact (e.g., RBF, Matern) and approximate strategies (e.g., GridInterpolationKernel, InducingPointKernel). In contrast to TCP, ICP base models can also use any regression likelihood from `gpytorch.likelihoods`. For consistency, in the example below we define the same `ExactGP` model as in the transductive version (Sect. 4). Note, however, that in this case we subclass the original `GPyTorch` model class and train it only on the proper training set, rather than the full training data.

```python
# Define standard ExactGP base model for ICP
class ExactGPModel(gpytorch.models.ExactGP):
    def __init__(self, train_x, train_y, likelihood):
        super(ExactGPModel, self).__init__(train_x, train_y, likelihood)
        self.mean_module = gpytorch.means.ConstantMean()
        self.covar_module = gpytorch.kernels.ScaleKernel(
            gpytorch.kernels.RBFKernel()
        )

    def forward(self, x):
        mean_x = self.mean_module(x)
        covar_x = self.covar_module(x)
        return gpytorch.distributions.MultivariateNormal(mean_x, covar_x)

# Initialize likelihood and model (on proper training data)
likelihood = gpytorch.likelihoods.GaussianLikelihood()
base_model = ExactGPModel(prop_train_x, prop_train_y, likelihood)
```

We can now optimize the model hyperparameters, following the standard procedure of GPyTorch, with the only difference that the optimization must be performed on the proper training data. Again for consistency, we perform the same optimization process as in the transductive version. Remember that we first have to put the model and likelihood in `.train()` mode.

```python
# Optimize hyperparameters on proper training data
training_iter = 50

base_model.train()
likelihood.train()

optimizer = torch.optim.Adam(base_model.parameters(), lr=0.1)
mll = gpytorch.mlls.ExactMarginalLogLikelihood(likelihood, base_model)

for i in range(training_iter):
    optimizer.zero_grad()
    output = base_model(prop_train_x)
    loss = -mll(output, prop_train_y)
    loss.backward()
    print('Iter %d/%d - Loss: %.3f   lengthscale: %.3f   noise: %.3f' % (
        i + 1, training_iter, loss.item(),
        base_model.covar_module.base_kernel.lengthscale.item(),
        base_model.likelihood.noise.item()
    ))
    optimizer.step()
```

5.3 Constructing the ICP

After setting up the base model, we can construct an Inductive CP using the
gpyconform.GPRICPWrapper class, by providing the base model and calibration
data. As required by GPyTorch, both the base model and likelihood must first
be put in .eval() mode.

Like the TCP ExactGPCP objects, GPRICPWrapper objects also have the cpmode
property. This property takes the same values ('symmetric', 'asymmetric' or
None) as in ExactGPCP, with two differences in behavior: (1) when cpmode=None,
GPRICPWrapper objects return a pair of tensors with the predictive means and
variances of test objects, as opposed to the MultivariateNormal predictive distribution returned by ExactGPCP (and in general GPyTorch models); (2) if cpmode
is changed after calibration, the ICP must be recalibrated.

In the example below we construct a GPRICPWrapper with our base model
and calibration data.

```
base_model.eval()    # Get into .eval() mode
likelihood.eval()

gpricp_model = gpyconform.GPRICPWrapper(base_model, cal_x, cal_y)
```

The GPRICPWrapper constructor also has three optional arguments:

- cpmode: (default='symmetric'), see Sect. 4.2;
- device: (torch.device or str), the device on which to place the model and
  perform tensor computations. If omitted, the model device is used.
- batch_size: (int), the batch size to be used when evaluating the calibration
  predictions. If omitted, all calibration instances will be processed in a single
  batch.

After constructing the GPRICPWrapper, we should calibrate it by calling the
calibrate method with the desired nonconformity measure parameter γ as argument. Calibration is also dependent on the current cpmode (not passed as an argument). Here we calibrate with cpmode='symmetric' (default) and $\gamma = 2$ (default):

```
gpricp_model.calibrate(gamma=2)
```

Now the gpricp_model is ready to produce PIs for any given test set. Note
that if the base model is modified (e.g. trained further) at any point, the wrapper
must be refreshed using refresh_model so that the modifications are reflected
in the GPRICPWrapper object. This method takes the same arguments as the
constructor with the exception of cpmode.

Alternatively, we can construct a calibrated InductiveConformalRegressor,
which does not need to have direct interaction with the model. The constructor of
InductiveConformalRegressor takes two required and four optional arguments:

- cal_targets: (torch.Tensor), the calibration targets;
- cal_preds: (torch.Tensor), the predictive means at the calibration inputs;

- cal_vars: (torch.Tensor; optional), the predictive variances at the calibration inputs. If omitted the non-normalized residuals are used as nonconformity measure (equivalent to setting $\gamma = \infty$);
- gamma: (float; optional; default: 2), the nonconformity measure parameter γ;
- cpmode: ('symmetric' or 'asymmetric'; default: 'symmetric'; optional), see Sect. 4.2. Note that None is not accepted here;
- device: (torch.device or str; optional), the device to place all tensors. If omitted, the device is inferred from the inputs.

In the example below, we manually compute the calibration predictions of our base model and construct an InductiveConformalRegressor equivalent to the (symmetric) gpricp_model above.

```
# Compute calibration predictive means and variances
with torch.no_grad():
    cal_obs_pred = likelihood(base_model(cal_x))
    cal_preds = cal_obs_pred.mean
    cal_vars = cal_obs_pred.variance

# Construct a (calibrated) InductiveConformalRegressor
calibrated_icr = gpyconform.InductiveConformalRegressor(
    cal_y, cal_preds, cal_vars, gamma=2.0, cpmode='symmetric'
)
```

The resulting calibrated_icr can now produce PIs for any test set, when provided with the predictive means and variances from the base model at the test inputs.

5.4   Making Predictions: Obtaining Prediction Intervals

Once calibrated, the GPRICPWrapper can be used to obtain conformal PIs at any set of confidence levels by calling its predict method with the test inputs. This returns a PredictionIntervals object, as in the transductive case. In addition to the test inputs, the predict method accepts the following optional parameters:

- confs: (torch.Tensor, numpy.array, or list; default: [0.95]), a list of the confidence levels for which to return PIs.
- batch_size: (int), the batch size to be used when evaluating the test predictions. If omitted, all test instances will be processed in a single batch.

If no test inputs are provided, cached predictions from the previous call are reused to avoid recomputation.

In the following example we call predict to obtain the PredictionIntervals object containing the PIs (with the cpmode and gamma used for calibration) at the 99%, 95% and 90% confidence levels. We disable gradient calculation for computational efficiency.

```
with torch.no_grad():
    symmetric_pis = gpricp_model.predict(test_x, confs=[0.99, 0.95, 0.9])
```

To obtain the corresponding asymmetric `PredictionIntervals` object we can change `cpmode` and recalibrate before calling `predict`. We can also recalibrate with a different `gamma` either after changing `cpmode` or without changing it. Here we obtain the asymmetric `PredictionIntervals` with the same γ:

```python
gpricp_model.cpmode = 'asymmetric'    # Change cpmode to 'asymmetric'
gpricp_model.calibrate()              # Recalibrate

with torch.no_grad():
    asymmetric_pis = gpricp_model.predict(confs=[0.99, 0.95, 0.9])
```

Note that here we omit the test objects, reusing the cached predictions from the previous call. Also since $\gamma = 2$ is the default we do not need to provide it when calling `calibrate`. Finally, calling the object directly provides the same functionality of `predict`, so calling `gpricp_model(confs=[0.99, 0.95, 0.9])` would have the same effect.

In the case we have constructed a calibrated `InductiveConformalRegressor`, we can obtain the PIs for our test set by calling the corresponding object with the test predictions of our base model (predictive means and variances) and the required confidence levels. In particular, the arguments for calling an `InductiveConformalRegressor` are:

- `test_preds`: (torch.Tensor), the predictive means at the test inputs;
- `test_vars`: (torch.Tensor; optional), the predictive variances at the test inputs. Can only be omitted if `cal_vars` were not provided for construction;
- `confs`: (torch.Tensor, numpy.array, or list; default: [0.95]), a list of the confidence levels for which to return PIs.

In the following example, we manually compute the test predictions of our base model and use them to obtain the (symmetric) `PredictionIntervals` for our test objects by calling the `calibrated_icr` we constructed in the previous subsection. Again at the 99%, 95% and 90% confidence levels.

```python
# Compute test predictive means and variances
with torch.no_grad():
    test_obs_pred = likelihood(base_model(test_x))
    test_preds = test_obs_pred.mean
    test_vars = test_obs_pred.variance


# Obtain PIs from the calibrated ICR
symmetric_pis = calibrated_icr(test_preds, test_vars, [0.99, 0.95, 0.9])
```

This will return the same PIs we obtained with the `gpricp_model` above in the symmetric case. Note that for a different `cpmode` or `gamma`, we need to construct a new `InductiveConformalRegressor`.

We can now extract a torch tensor with the PIs or evaluate them at any of the three confidence levels in the same way as in Sect. 4. The complete functionality of `PredictionIntervals` is described in Sect. 6.

6 Retrieving and Evaluating Prediction Intervals

This section describes the `PredictionIntervals` class, explaining how the PIs returned by the Transductive or Inductive GPR CP can be retrieved and evaluated. All examples will use the `symmetric_pis` produced by the `gpricp_model` of Sect. 5 (with $\gamma = 2$).

We can retrieve the PIs from a `PredictionIntervals` object by calling it directly. Specifically, such a call accepts four optional parameters, of which the last three are keyword-only:

- `conf_level`: (`float` in $(0,1)$), the confidence level for which to return the corresponding PIs. If omitted, the PIs for all confidence levels will be returned as a dictionary;
- `y_min`: (`float`), if provided the returned PIs are cut to exclude values below `y_min`;
- `y_max`: (`float`), if provided the returned PIs are cut to exclude values above `y_max`;
- `dp`: (`int` in $[1,6]$; default: 6), number of decimals to show in the string keys when returning all confidence levels.

If a `conf_level` is specified, the output is a tensor with one row for each test instance, where the first column is the lower bound and the second is the upper bound of the corresponding PI. Otherwise, the output is a dictionary with confidence levels as string keys and the corresponding PI tensors as values. Note that if a confidence level is specified, it should match one of those originally requested when constructing the object.

In the example below we retrieve and print the PIs at the 99% confidence level.

```
print(symmetric_pis(0.99))
```

```
tensor([[ 0.4853,   1.0575],
        [ 0.3596,   0.9327],
        [-0.0289,   0.5637],
        ...,
        [ 0.1602,   0.7345],
        [ 0.0186,   0.6106],
        [ 0.3151,   0.8956]], dtype=torch.float64)
```

By not specifying a `conf_level` we retrieve a dictionary with the PIs for all confidence levels. Here we also cut the PIs in the range $[0,1]$ and request only 2 decimals in the dictionary keys for readability:

```
print(symmetric_pis(y_min=0, y_max=1, dp=2))
```

```
{'0.99': tensor([[0.4853, 1.0000],
                 [0.3596, 0.9327],
```

```
                    [0.0000, 0.5637],

                    ...,
                    [0.1602, 0.7345],
                    [0.0186, 0.6106],
                    [0.3151, 0.8956]], dtype=torch.float64),
  '0.95': tensor([[0.5901, 0.9527],
                    [0.4646, 0.8278],
                    [0.0797, 0.4552],

                    ...,
                    [0.2654, 0.6294],
                    [0.1270, 0.5022],
                    [0.4214, 0.7893]], dtype=torch.float64),
  '0.90': tensor([[0.6239, 0.9189],
                    [0.4984, 0.7940],
                    [0.1146, 0.4202],

                    ...,
                    [0.2993, 0.5955],
                    [0.1619, 0.4672],
                    [0.4557, 0.7550]], dtype=torch.float64)}
```

The `evaluate` method provides the functionality for evaluating the PIs against three metrics:

- `'mean_width'`: the average width of the PIs;
- `'median_width'`: the median width of the PIs;
- `'error'`: the percentage of PIs that do not contain the true target value (i.e. $1 - $ coverage).

The method accepts five arguments, of which the last two are keyword-only:

- `conf_level`: (`float` in $(0, 1)$), the confidence level of the PIs to evaluate;
- `metrics`: (list of `str` or `str`; optional), the metrics to calculate. If omitted all metrics are calculated;
- `y`: (`tensor`; optional), the true target values of the test set, required for the `'error'` metric. If omitted, `'error'` is not calculated;
- `y_min`: (`float`; optional), if provided PIs are evaluated after cutting values below `y_min`;
- `y_max`: (`float`; optional), if provided PIs are evaluated after cutting values above `y_max`.

The output is a dictionary with a key for each metric and the calculated result as its value.

In the example below we evaluate on all metrics (the default) at the 99% confidence level. All values are rounded to three decimals for readability.

```
res = symmetric_pis.evaluate(0.99, y=test_y)


pretty_res = {k: f"{v:.3f}" for k, v in res.items()}
print(pretty_res)
```

```
{'mean_width': '0.579', 'median_width': '0.575', 'error': '0.007'}
```

We can also perform the same evaluation with the PIs cut to a minimum of 0 and a maximum of 1:

```
res = symmetric_pis.evaluate(0.99, y=test_y, y_min=0, y_max=1)

pretty_res = {k: f"{v:.3f}" for k, v in res.items()}
print(pretty_res)
```

```
{'mean_width': '0.565', 'median_width': '0.572', 'error': '0.007'}
```

This does not affect the error metric since all target values in our data lie within $[0, 1]$.

7 Investigating Prediction Intervals and Computational Efficiency

In this section we empirically examine two aspects of GPyConform: the coverage of the PIs it produces and its computational efficiency when run on modern hardware. For consistency, we use the simple GPR models defined in our examples in the previous sections, but apply these to the larger Abalone dataset (4177 instances). We repeat the experiments 100 times with different random divisions of the data to ensure robustness of the results. Specifically, for each experiment we randomly divide the data into training and test sets, consisting of 80% and 20% of the whole dataset respectively, and evaluate the Transductive CP. We then further divide the training data into proper training and calibration sets with the calibration set consisting of 999 instances, resulting in 1000 instances with the addition of the test instance. These are used to construct and evaluate the corresponding Inductive CP. The code for these experiments is available in the examples folder of the GPyConform GitHub repo.

To examine the practical usability of the package, the experiments were run on Kaggle Notebook with the GPU P100 accelerator and we report the mean runtime measurements for each version and mode of the framework. Despite TCP being computationally inefficient, our runtime results show that GPyConform with GPU acceleration enables its usage with moderately sized datasets in seconds.

Table 1 presents the mean error percentage and width of the PIs produced by each approach, as well as the corresponding mean runtimes in seconds. The error percentages show that, as expected, all approaches achieve a coverage $(1 - \text{error})$ almost identical to the nominal. In terms of the interval widths, an interesting observation is that the widths of TCP and ICP are very close, while typically TCP produces narrower PIs due to its more effective utilization of the data. In this case however, the proper training and calibration data used by ICP appear sufficient to provide equally good GPR model performance and calibration. This is a property of the particular dataset and task, rather than a general conclusion. In comparing symmetric and asymmetric PIs, we observe that the asymmetric intervals are narrower at the 99% confidence level, while the symmetric ones are slightly narrower at the lower levels. This is again a data and task dependent outcome, which nevertheless demonstrates that different modes can provide better performance in different cases.

With regards to runtimes, these demonstrate the important computational advantage provided by GPyConform when combined with GPU acceleration. In particular, the computationally inefficient TCP can process more than 3k training and 800 test instances in under 25 s. This makes exploratory experiments with multiple runs feasible within minutes. On the other hand, the computationally efficient ICP runs in split seconds (about 200 times faster than TCP) and can readily scale to much larger datasets.

Table 1. Mean error percentage, interval width, and runtime over 100 random splits of the Abalone dataset.

Approach	Error (%)			Mean Width			Runtime
	99%	95%	90%	99%	95%	90%	(secs/run)
TCP (symmetric)	1.01	5.10	10.07	0.5064	0.3148	0.2322	24.5495
TCP (asymmetric)	0.95	5.08	10.04	0.4581	0.3170	0.2441	21.6728
ICP (symmetric)	1.00	5.14	10.14	0.5039	0.3146	0.2334	0.1229
ICP (asymmetric)	1.04	5.09	10.10	0.4584	0.3188	0.2451	0.1040

8   Conclusions

This chapter introduced GPyConform, a Python package that extends GPyTorch with Conformal Prediction for Gaussian Process Regression. The package implements both Transductive and Inductive CP with a GPR specific normalized nonconformity measure, supporting symmetric and asymmetric PIs through a unified and intuitive interface. By being built entirely on torch tensors, GPyConform benefits from GPU acceleration, making it efficient for large-scale experimentation.

Through worked examples and an empirical study on the Abalone dataset, we showed how GPyConform can be seamlessly integrated into Gaussian Process workflows. The results confirmed that both TCP and ICP achieve coverage close to the nominal levels, while runtimes demonstrated the practical scalability of the package: TCP, although computationally intensive, remains feasible with GPU acceleration, and ICP runs in split seconds even on thousands of instances. Taken together, these features make GPyConform a valuable tool for practitioners seeking reliable uncertainty quantification and for researchers exploring new developments in conformal prediction for GPR.

Future extensions could include support for Conformal Predictive Systems, which generalize conformal regressors by producing predictive distributions.

Another direction is the inclusion of alternative frameworks such as Mondrian CP, which provides conditional guarantees, and Cross-Conformal Prediction. Finally, the addition of further normalized nonconformity measures remains a natural direction for future development.

References

1. Boström, H.: Crepes: a python package for generating conformal regressors and predictive systems. In: Johansson, U., Bostrm, H., An Nguyen, K., Luo, Z., Carlsson, L. (eds.) Proceedings of the Eleventh Symposium on Conformal and Probabilistic Prediction and Applications. Proceedings of Machine Learning Research, vol. 179. PMLR (2022)
2. Cordier, T., Blot, V., Lacombe, L., Morzadec, T., Capitaine, A., Brunel, N.: Flexible and systematic uncertainty estimation with conformal prediction via the MAPIE library. In: Conformal and Probabilistic Prediction with Applications (2023)
3. Mendil, M., Mossina, L., Vigouroux, D.: Puncc: a python library for predictive uncertainty calibration and conformalization. In: Conformal and Probabilistic Prediction with Applications, pp. 582–601. PMLR (2023)
4. Papadopoulos, H.: Cross-conformal prediction with ridge regression. In: Gammerman, A., Vovk, V., Papadopoulos, H. (eds.) SLDS 2015. LNCS (LNAI), vol. 9047, pp. 260–270. Springer, Cham (2015). https://doi.org/10.1007/978-3-319-17091-6_21
5. Papadopoulos, H.: Guaranteed coverage prediction intervals with gaussian process regression. IEEE Trans. Pattern Anal. Mach. Intell. **46**(12), 9072–9083 (2024). https://doi.org/10.1109/TPAMI.2024.3418214
6. Papadopoulos, H., Gammerman, A., Vovk, V.: Normalized nonconformity measures for regression conformal prediction. In: Proceedings of the IASTED International Conference on Artificial Intelligence and Applications (AIA 2008), pp. 64–69. ACTA Press (2008)
7. Papadopoulos, H., Proedrou, K., Vovk, V., Gammerman, A.: Inductive confidence machines for regression. In: Elomaa, T., Mannila, H., Toivonen, H. (eds.) ECML 2002. LNCS (LNAI), vol. 2430, pp. 345–356. Springer, Heidelberg (2002). https://doi.org/10.1007/3-540-36755-1_29
8. Papadopoulos, H., Vovk, V., Gammerman, A.: Regression conformal prediction with nearest neighbours. J. Artif. Intell. Res. **40**, 815–840 (2011). https://doi.org/10.1613/jair.3198
9. Saunders, C., Gammerman, A., Vovk, V.: Transduction with confidence and credibility. In: Proceedings of the 16th International Joint Conference on Artificial Intelligence, vol. 2, pp. 722–726. Morgan Kaufmann, Los Altos (1999)
10. Vovk, V.: Cross-conformal predictors. Ann. Math. Artif. Intell. **74**, 9–28 (2015)
11. Vovk, V., Gammerman, A., Shafer, G.: Algorithmic Learning in a Random World, 2nd edn. Springer, Heidelberg (2022)

Author Index

A
Ahlberg, Ernst 227
Aitken, Colin G. G. 75

B
Balasubramanian, Vineeth N. 269
Bellotti, Anthony 177
Boström, Henrik 118

C
Carlsson, Lars 227
Cherkassky, Vladimir 157
Cherubin, Giovanni 306

F
Feng, Xu 387

G
Gammerman, Alexander 28
Gammerman, James 227

H
Hallberg Szabadváry, Johan 227

K
Kalnishkan, Yuri 325
Kancheti, Sai Srinivas 269
Krishnan, Sai Mathura 269

L
Lee, Eng Hock 157
Li, Guang 198
Lindsay, David 325, 345
Lindsay, Siân 325, 345

Liu, Xiaohui 56
Liu, Yifan 198
Luo, Zhiyuan 1, 198, 387

M
Michaelson, Greg 46

N
Nguyen, Khuong An 198, 387
Nouretdinov, Ilia 411

P
Papadopoulos, Harris 449
Petej, Ivan 432

S
Schlembach, Filip 144
Smirnov, Evgueni 144
Stavan, Christian 269

T
Tilaganji, Kunal 269

V
Vovk, Vladimir 87

W
Wang, You 198
Wisniewski, Wojciech 325

Z
Zhang, Zhirui 198
Zhao, Shuo 198
Zhao, Xindi 177